VOLUME 13

Membrane Science and Technology Series

Inorganic Membranes: Synthesis, Characterization and Applications

Membrane Science and Technology Series

Volume 1: *Pervaporation Membrane Separation Processes*
Edited by R.Y.M. Huang (1991)

Volume 2: *Membrane Separations Technology, Principles and Applications*
Edited by R.D. Noble and S.A. Stern (1995)

Volume 3: *Inorganic Membranes for Separation and Reaction*
By H.P. Hsieh (1996)

Volume 4: *Fundamentals of Inorganic Membrane Science and Technology*
Edited by A.J. Burggraaf and L. Cot (1996)

Volume 5: *Membrane Biophysics*
Edited by H. Ti Tien and A. Ottova-Leitmannova (2000)

Volume 6: *Recent Advances in Gas Separation by Microporous Ceramic Membranes*
Edited by N.K. Kanellopoulos (2000)

Volume 7: *Planar Lipid Bilayers (BLMs) and their Applications*
Edited by H.T. Tien and A. Ottova-Leitmannova (2003)

Volume 8: *New Insights into Membrane Science and Technology: Polymeric and Biofunctional Membranes*
Edited by D. Bhattacharyya and D.A. Butterfield (2003)

Volume 9: *Ion-Exchange Membrane Separation Processes*
By H. Strathmann (2004)

Volume 10: *Nano and Micro Engineered Membrane Technology*
By C.J.M van Rijn (2004)

Volume 11: *Membrane Contactors: Fundamentals, Applications and Potentialities*
By E. Drioli, A. Criscuoli and E. Curcio (2006)

Volume 12: *Ion Exchange Membranes: Fundamentals and Applications*
By Y. Tanaka

VOLUME 13

MEMBRANE SCIENCE AND TECHNOLOGY SERIES

Inorganic Membranes: Synthesis, Characterization and Applications

Edited by

REYES MALLADA
MIGUEL MENÉNDEZ
Chemical and Environmental Engineering Department
Faculty of Sciences
University of Zaragoza, Spain

ELSEVIER

Amsterdam • Boston • Heidelberg • London • New York • Oxford
Paris • San Diego • San Francisco • Singapore • Sydney • Tokyo

Elsevier
Radarweg 29, PO Box 211, 1000 AE Amsterdam, The Netherlands
Linacre House, Jordan Hill, Oxford OX2 8DP, UK

First edition 2008

Library of Congress Cataloging-in-Publication Data
A catalog record for this book is available from the Library of Congress

British Library Cataloguing in Publication Data
A catalogue record for this book is available from the British Library

ISBN: 978-0-444-53070-7
ISSN: 0927-5193

Printed and bound by CPI Group (UK) Ltd, Croydon, CR0 4YY

Transferred to Digital Print 2011

Contents

Contributors

José R. Álvarez
Department of Chemical and Environmental Engineering, University of Oviedo, 33071 Oviedo, Spain.

André Ayral
Institut Européen des Membranes, UMR no 5635 CNRS-ENSCM-UMII, CC047, Université Montpellier II, Place Eugène Bataillon, 34095 Montpellier cedex 5, France.

Giuseppe Barbieri
National Research Council - Institute for Membrane Technology (ITM-CNR), The University of Calabria, cubo 17/C, 87030 Rende CS, Italy.

Angelo Basile
Institute on Membrane Technology, ITM-CNR, c/o University of Calabria, 87030 Rende (CS), Italy.

Rune Bredesen
SINTEF Materials and Chemistry, NO-0314 Oslo, Norway.

Adele Brunetti
National Research Council - Institute for Membrane Technology (ITM-CNR), The University of Calabria, cubo 17/C, 87030 Rende CS, Italy
Department of Chemical Engineering and Materials, The University of Calabria, cubo 44/A, 87030 Rende CS, Italy.

Anita Buekenhoudt
VITO (Flemish Institute for Technological Research), Process Technology, Boeretang 200, B-2400 Belgium.

Clara Casado Coterillo
European Membrane Institute Twente, University of Twente, 7500AE Enschede, The Netherlands.

Joaquín Coronas
Department of Chemical and Environmental Engineering, University of Zaragoza, 50018 Zaragoza, Spain
Nanoscience Institute of Aragon, University of Zaragoza, 50009 Zaragoza, Spain.

Óscar de la Iglesia
Department of Chemical and Environmental Engineering, University of Zaragoza, 50018 Zaragoza, Spain
Nanoscience Institute of Aragon, University of Zaragoza, 50009 Zaragoza, Spain.

Jean Durand
Institut Européen des Membranes, UMR no 5635 CNRS-ENSCM-UMII, CC047, Université Montpellier II, Place Eugène Bataillon, 34095 Montpellier cedex 5, France.

Marie-Laure Fontaine
SINTEF Materials and Chemistry, NO-0314 Oslo, Norway.

Daniel Gómez
Department of Chemical and Environmental Engineering, University of Oviedo, 33071 Oviedo, Spain.

Fausto Gallucci
Institute on Membrane Technology, ITM-CNR, c/o University of Calabria, 87030 Rende (CS), Italy.

Patricia Gorgojo
Department of Chemical and Environmental Engineering, University of Zaragoza, 50018 Zaragoza, Spain
Nanoscience Institute of Aragon, University of Zaragoza, 50009 Zaragoza, Spain.

Tor Grande
Department of Material Science and Engineering. Norwegian University of Science and Technology (NTNU), Trondheim, Norway.

Ahmad Fauzi Ismail
Membrane Research Unit, Faculty of Chemical and Natural Resources Engineering, Universiti Teknologi Malaysia, 81310 Skudai, Johor, Malaysia.

Anne Julbe
Institut Européen des Membranes, UMR no 5635 CNRS-ENSCM-UMII, CC047, Université Montpellier II, Place Eugène Bataillon, 34095 Montpellier cedex 5, France.

Yngve Larring
SINTEF Materials and Chemistry, NO-0314 Oslo, Norway.

Kang Li
Department of Chemical Engineering, Imperial College London, South Kensington Campus, London SW7 2AZ, United Kingdom.

Susana Luque
Department of Chemical and Environmental Engineering, University of Oviedo, 33071 Oviedo, Spain.

Ana María Urtiaga Mendia
Department of Chemical Engineering, Universidad de Cantabria, 39005 Santander (Cantabria), Spain.

Truls Norby
Department of Chemistry, University of Oslo, Centre for Materials Science and Nanotechnology, Gaustadalleen 21, NO-0349 Oslo, Norway.

Inmaculada Ortiz Uribe
Department of Chemical Engineering, Universidad de Cantabria, 39005 Santander (Cantabria), Spain.

Suzana Pereira Nunes
Institute of Polymer Research, GKSS Research Centre, Geesthacht, Germany.

Stéphanie Roualdes
Institut Européen des Membranes, UMR no 5635 CNRS-ENSCM-UMII, CC047, Université Montpellier II, Place Eugène Bataillon, 34095 Montpellier cedex 5, France.

Vincent Rouessac
Institut Européen des Membranes, UMR no 5635 CNRS-ENSCM-UMII, CC047, Université Montpellier II, Place Eugène Bataillon, 34095 Montpellier cedex 5, France.

Francesco Scura
National Research Council - Institute for Membrane Technology (ITM-CNR), The University of Calabria, cubo 17/C, 87030 Rende CS, Italy.

Silvano Tosti
ENEA, Dipartimento Fusione Tecnologie e Presidio Nucleare, C.R. ENEA Frascati, Frascati, Roma 00044, Italy.

PREFACE

This book has its origin in the XXII Summer School of the European Membrane Society that took place in Jaca (Spain) in 2005, devoted to "Inorganic Membranes: Preparation, Characterization and Applications." Experts on inorganic membranes from both industry and the academic world attending the course all agreed that there was a need for a new general book updating the literature in the field.

BOOKS ON INORGANIC MEMBRANES

A review of books devoted to inorganic membranes starts with those published by Burggraaf in 1992 and by Burggraaf and Cot in 1996, "Inorganic Membranes" and "Fundamentals of Inorganic Membrane Science and Technology," respectively. These two books described the basis of ceramic membranes. Also, in 1996, Hsieh's book "Inorganic Membranes for Separation and Reaction" discussed the possibilities of inorganic membranes in what was then the emerging field of inorganic membrane reactors. Subsequent major advances were in the microporous membrane field, and "Recent Advances in Gas Separation by Microporous Ceramic Membranes" edited by Kanellopoulus appeared in 2000. There is no recent general book devoted to inorganic membranes with an overview including all such membranes and their main applications, although some aspects have been discussed in recent books such as "Catalytic Membranes and Catalytic Membrane Reactors" by Sanchez and Tsotsis (2002), "Nonporous Inorganic Membranes: For Chemical Processing" by Sammells and Mundschau (2006), "Membrane Technology: In the Chemical Industry" by Nunes and Peinemann (2001 and 2006), and "Ceramic Membranes for Separation and Reaction" by Li (2007).

AIM OF THE BOOK

The main purpose of this book is to gather together in one work the main advances in the field of inorganic membranes, both porous and dense, provided by leading experts in the field. In the case of porous membranes, these advances are related to the synthesis of microporous materials such as silica, carbon, and zeolites in the form of nondefective thin layers and the production of hollow fiber membranes. In addition, the book discusses the increasingly relevant hybrid membranes, which combine the stability of inorganic membranes and the processability of polymers. Attention is also devoted to dense inorganic membranes divided into the hydrogen-selective palladium membranes and the dense ceramic mixed conducting membranes, all experiencing constantly improved fluxes and new synthesis methods for obtaining thinner layers and higher temperature stability. The applications of all these membranes are considered throughout the book.

Book Organization

The book is divided into two parts, the first dealing with porous inorganic membranes and the second with dense inorganic membranes. This division is very important given the very different permeation mechanisms of these two types of membrane and the consequent differences in selectivity and flux. In general, fluxes are higher in the case of porous membranes while higher selectivity can be achieved using dense membranes.

Porous membranes

The part of the book devoted to porous inorganic membranes consists of seven chapters that evaluate the "state of the art" and recent advances in this field. The first chapter serves as an introduction to these membranes in terms of their structure and provides a general overview of the stability of ceramic membranes, the principal advantage of inorganic membranes. Porous ceramic membranes are mainly applied in harsh circumstances in terms of temperature or when aggressive chemicals such as solvents or highly acidic or caustic solutions are used. In these circumstances, cheaper polymer membranes have limited stability. They can generally only stand temperatures below ~100 °C, and they swell or partly dissolve in many solvents and in strong acids or bases. Ceramics, on the other hand, are generally known for their very high chemical and thermal stability, although this stability can be limited, due mainly to sintering processes, corrosion, and hydrothermal stability.

Chapters 2–5 describe the inorganic materials processed as planar, tubular, or even hollow fiber membranes. Chapter 2 deals with microporous silica membranes. The two main routes for the synthesis of these membranes, sol–gel techniques and chemical vapor deposition (CVD), are explained in detail. The main applications and goals achieved with these membranes as well as the transport mechanisms are revised. The preparation of inorganic membranes as hollow fibers provides a high surface to volume ratio and a good opportunity for these materials to be implemented in the industry. Chapter 3 describes the preparation of carbon and alumina membranes as hollow fibers.

The possibility of combining polymers and inorganic components in one material has emerged in recent decades with the birth of soft inorganic chemistry and the introduction of the concept of "hybrid organic–inorganic" materials, which allows tailoring innovative materials with organic and inorganic phases coexisting on a nano-scale and with multifunctionalization. These hybrid organic–inorganic membranes could be applied, for example, in the field of fuel cells. Chapter 4 describes the preparation of these composites either with nonpermeable or permeable fillers.

Zeolite membranes are presented in Chapter 5. This chapter tackles the peculiarities of these solids, with more than a hundred different structures, widely used as ion-exchangers and catalysts. There have been many significant advances from the first zeolite polycrystalline layer prepared in 1987 to achieving complete control of the zeolite layer growth and orientation, all of which are reviewed in this chapter. The different zeolite synthesis methods and other important aspects such as support selection, calcinations, and posttreatments are also described in this chapter. The last part of the chapter is dedicated to the characterization and some of the main applications of zeolite membranes not only in the field of separation but also as membrane reactors, microreactors, and sensors.

Porous inorganic membrane applications are described in Chapters 6 and 7. Chapter 6 considers applications in pressure-driven membrane processes, that is, microfiltration, ultrafiltration, and nanofiltration, using mainly macro- and mesoporous membranes. These processes are applicable in various industries such as the chemical industry, food and beverage processing, and wastewater treatment. Chapter 7 deals with applications of microporous inorganic membranes in pervaporation and gas separation. Moreover, there is a section in this chapter devoted to transport modeling in microporous membranes.

Dense membranes

The last three chapters are devoted to dense membranes and their applications. Palladium is a material that for many years has been proposed for use in hydrogen separation, including application in membrane reactors. Recent advances leading to thinner palladium films with higher flux and lower cost, as described in Chapter 8, are paving the way toward many applications. Since membrane reactors are a key application of this material, the design of the reactor must be considered. Chapter 9 presents a wide-ranging review of the different possible approaches to the mathematical modeling of such membrane reactors.

Finally, the last chapter (i.e., Chapter 10) reviews current knowledge of dense ceramic membranes, which hold the promise of solving one of the major problems of humankind: sustainable energy. This kind of membrane offers high selectivity to hydrogen or oxygen which is transported as charged species in the membrane matrix. The main application is in solid oxide fuel cells, but there may also be potential uses in the generation of gases or the separation of high purity oxygen or hydrogen from mixed streams.

In conclusion, this book provides researchers and users in the field of inorganic membranes with an updated review of the state of the art and points to future needs and prospective applications. As such, we hope that it will be useful both for researchers starting out in the field and for those who, having contributed to recent developments in inorganic membranes, wish to have a reference book where current knowledge is conveniently assembled.

Reyes Mallada
Miguel Menéndez

CHAPTER ONE

Stability of Porous Ceramic Membranes

Anita Buekenhoudt

Contents

1. Introduction

The development of porous ceramic membranes started in the beginning of the 1980s with membranes suitable for the enrichment of U^{235} [1]. In the period between 1980 and 1985, the development of industrial liquid filtration membranes, namely, ultrafiltration (pore diameter between 2 and 50 nm) and microfiltration (pore diameter >50 nm) membranes, started. Nowadays, microporous ceramic membranes are also available in the market. These membranes with pores of ~1 nm, and below, are suitable for nanofiltration, pervaporation, and/or gas separation. Besides the porous ceramic membranes, dense ion-conducting membranes have also been developed during the last decades. These materials are used as electrolytes in fuel cells, or as hydrogen- and oxygen-selective membranes. The use and stability of these dense membranes will be described in Chapter 10.

1.1. General considerations on porous ceramic membranes

Nowadays, a whole range of materials is used to produce porous ceramic membranes. The majority of the porous membranes, including almost all commercially available membranes, are made up of metal oxides. The oxides preferably used are

VITO (Flemish Institute for Technological Research), Process Technology, Boeretang 200, B-2400 Belgium

Membrane Science and Technology, Volume 13
ISSN 0927-5193, DOI: 10.1016/S0927-5193(07)13001-1

aluminum oxide or alumina (Al_2O_3), zirconium oxide or zirconia (ZrO_2), titanium oxide or titania (TiO_2), and silicium oxide or silica (SiO_2). Also, mixtures of these metal oxides are frequently used. In academic laboratories, membranes are also made up of more exotic metal oxides as tin (Sn) and hafnium (Hf) oxide. For the microporous membranes suitable for pervaporation and gas separation, besides SiO_2, other materials are used, namely, zeolites and microporous carbon. These membranes are dealt with in two separate chapters (3 and 4). In this chapter, we will concentrate on the majority of the porous ceramic membranes, namely, the metal oxide membranes. SiO_2 is also treated in more detail in Chapter 2, but where relevant we will compare SiO_2 with the other metal oxide materials in this chapter.

In order to combine a high permeability with the wanted selectivity, all commercially available porous ceramic membranes, and most of the laboratory-made membranes, are asymmetric. This means that they consist of a support with a very open pore structure, and on top of the support one or more thin layers with smaller pores. The top layer contains the smallest pores and defines the real selectivity of the porous membrane. The support is used to enhance the mechanical strength of the membrane. The intermediate layers are used to gradually decrease the pore size and the surface roughness of the membrane. A typical picture of the multilayer structure of a ceramic membrane is shown in Fig. 1.1.

Membrane supports are mainly produced by slip casting or extrusion. The main method for the preparation of metal oxide intermediate and top layers is the sol–gel technique [2]. Mesoporous intermediate layers are made by colloidal sols and microporous top layers by polymeric sols. The sols are prepared from metal salts or metal organic precursors. The layers are deposited on the supports or on the previous multilayer structure by dip coating. The porosity of the supports or of the previous multilayer structure leads to the gelling of the sols. These gel layers are further dried and thermally treated to form the final membrane layers. This final

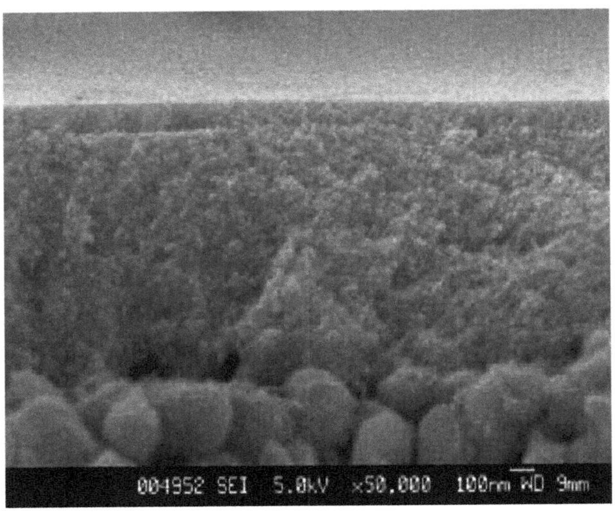

Figure 1.1 SEM picture of the multilayer structure of a ceramic nanofiltration membrane.

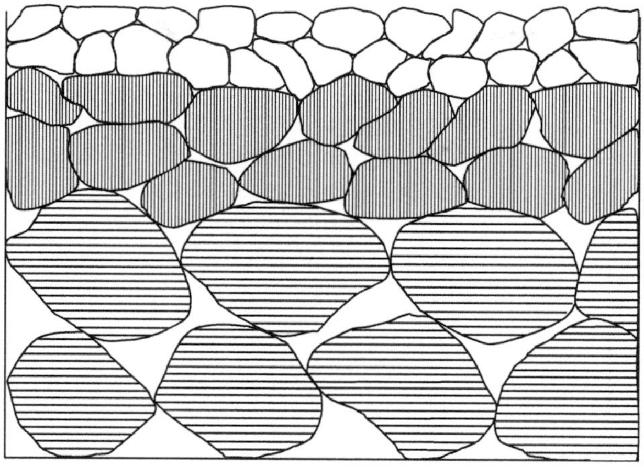

Figure 1.2 Schematic representation of a multilayer membrane consisting of grains, grain boundaries, and pores.

thermal treatment is called calcination or sintering [2]. Its aim is to stabilize the crystallographic and morphological structure of the dried layer.

Because of the preparation described, all layers are polycrystalline and consist of a packing of particles or grains, with pores between them. The size of the grains from intermediate and top layers depends on the sol–gel procedure and the calcination temperature, and is correlated to the pore size. Because of the calcination/sintering, the contact points between the grains form necks and more extended grain boundaries. Figure 1.2 shows a schematic representation of a multilayer membrane consisting of grains, grain boundaries, and pores.

1.2. Stability of porous ceramic membranes

Porous ceramic membranes are mainly used for applications in harsh circumstances of temperature, or aggressive chemicals as solvents and high acid or caustic solutions. In these circumstances, cheaper polymer membranes have limited stability: they can generally stand temperatures only below ~100 °C, and they swell or partly dissolve in many solvents, and in strong acids or bases. Ceramics, on the contrary, are generally known for their very high chemical and thermal stability. Specifically, ceramic membranes are insensitive to swelling in any solvent; they are highly resistant to a whole range of cleaning agents in order to get rid of fouling; and they can be sterilized at high temperature. However, the high stability of ceramic membranes is definitely not ideal. They do show corrosion and thermal instabilities. It is the purpose of this chapter to give a good idea of the real stability of porous ceramic membranes.

As is clear from the above discussion, the chemical and thermal stability of ceramic membranes is a very important parameter for their application. Unfortunately, these terms are not very well defined in literature. This is mainly because the processes involved are very complex. In order to discuss the stability of porous ceramic membranes properly, the different general properties of these membranes, as described in Section 1.1, need to be taken into account, namely, the membrane material(s), the multilayer structure, and the membrane morphology.

The first important aspect determining the corrosion resistance of a membrane is the material or materials used to build up the membrane. For the different parts of the asymmetric membranes, different materials can be used. The support, very often, consists of Al_2O_3 (in the α-Al_2O_3 phase). However, nowadays more and more supports are made of TiO_2 because of its higher chemical stability (see Section 2.2.1). The intermediate layers are mainly made of Al_2O_3 (in the γ-Al_2O_3 phase), TiO_2 or ZrO_2, or their mixtures. For the top layer, Al_2O_3 (in the boehmite phase), TiO_2, ZrO_2, or SiO_2 all are used.

For the stability of ceramic membranes, the chemical and thermal resistance of each of the layers of the asymmetric structure is important. For thermal resistance, we will see that the interplay between support and subsequent layers also plays a role.

Moreover, the membrane morphology plays an important role in the stability of the membrane. As described in Section 1.1, all membrane layers are polycrystalline and consist of a packing of particles or grains, with pores and grain boundaries between them. In most cases, the grains consist of a crystalline material (exceptions are SiO_2 and TiO_2 calcinated at low temperature; these materials have amorphous grains), while grain boundaries are usually noncrystalline and contain a relatively large amount of impurities or glassy material, and defects. We will see that the existence of these "messier" grain boundaries has an effect on the thermal and chemical stability of the material.

2. Chemical Stability

2.1. Background

The chemical stability of a material or a membrane can be defined as its inertness against corrosive attack by aggressive liquids. As every reaction, corrosion is determined as well by thermodynamics, as by kinetics. From a thermodynamic point of view, the free energy and the total energy of formation of a material play an important role for its stability. Materials with a strongly negative free energy and total energy of formation, for example, metal oxides as yttria and thoria, are expected to be chemically very stable in different environments [3]. Al_2O_3, TiO_2, and ZrO_2 are also recognized for their potential chemical stability [3]. From a kinetic point of view, aspects such as contact time, porosity, temperature, and working pressure or flow velocity are of major importance. These parameters determine, for example, the supply of the corrosive reactants to the membrane, and the discharge of corrosion products from the membrane. This influences the kinetics of the corrosion reaction.

The corrosion of or chemical attack on a membrane is further complicated, by the membrane morphology. Indeed, the chemical attack starts on the surface, that is, the complete specific pore surface of the membrane. As pointed out in Section 1.1 and Fig. 1.2, this surface consists of the surface of the membrane grains, and the grain boundaries, with usually a slightly different composition than that of the bulk grain crystalline material, including some impurities. Moreover, grain surfaces and grain boundaries contain more defects, as kinks (steps) and dislocations. Both the nonideal crystal composition and the defects make sure that a grain surface and a grain boundary usually contain more energetically favorable positions for dissolution than does the bulk material. As a consequence, the corrosion that occurs there will be faster than in the rest of the material. This preferential corrosion of grain surfaces and grain boundaries has also been shown experimentally [4, 5]. This effect can lead to the removal of whole grains from the membrane under the action of a shear stress exerted by the streaming fluid, without dissolution of the grains themselves. In this case, the membrane top layers will disintegrate very rapidly (see also Section 2.2.2).

Moreover, the roughness and curvature of the specific surface of the membrane also plays a role in the corrosion reaction. Indeed, material from strongly curved convex parts of the membrane surface is preferentially dissolved [6]. As a consequence, membranes with smaller grains, and correlated, smaller pore sizes, will show more corrosion. This leads to another reason why grain surfaces corrode faster than the bulk of the grains: grain surfaces are, in general, rougher, and these rough parts will dissolve fast, until they are smoothed.

The corrosion of the crystalline grain material of a membrane is a complex process by itself. This is illustrated by the example of dissolution of α-Al_2O_3 in aqueous H_2SO_4 solutions [7]. In principle, three different reaction steps can be found in this corrosion reaction. The first one is hydration of Al_2O_3, with the possible formation of two different products:

$$\begin{aligned} Al_2O_3 + 3H_2O &\rightarrow 2Al(OH)_3 \\ Al_2O_3 + H_2O &\rightarrow 2AlOOH \end{aligned}$$

After the hydration, hydrogen ion adsorption takes place:

$$\begin{aligned} AlOOH + H^+ &\rightarrow Al(OH)_2^+ \\ Al(OH)_3 + H^+ &\rightarrow Al(OH)_2^+ + H_2O \end{aligned}$$

And, finally, the reaction products are further attacked or react back:

$$\begin{aligned} Al(OH)_2^+ + 2H^+ &\rightarrow Al^{3+} + 2H_2O \\ 2Al(OH)_2^+ + S_4^{2-} &\rightarrow [Al(OH)_2]_2SO_4 \\ Al(OH)_2^+ &\rightarrow AlOOH + H^+ \end{aligned}$$

It is clear from this example that the possible reactions are very complex and strongly depend on the reaction conditions. Important parameters are the temperature, leading to different reaction rates, and the pH, leading to different reactants and reactant concentrations. It can be expected that this holds for all the reactions of Al_2O_3 and other metal oxides, making the prediction of the reaction rate and therefore of the chemical stability very complicated. In the above-described dissolution of crystalline material, the thermodynamical stability of the crystalline material is also important. This means that a thermodynamically unstable phase as γ-Al_2O_3 will show faster corrosion reactions than would the stable α-Al_2O_3. And in the extreme, noncrystalline amorphous material, like amorphous SiO_2 or TiO_2, will always corrode more rapidly than crystalline phases.

From everything that is mentioned above, it is clear that corrosion of a porous ceramic membrane is definitely a very complex process that can hardly be theoretically predicted in a quantitative way. However, it is possible to design experiments to get a better idea of the chemical stability of porous ceramic membranes. This is described in the next section.

2.2. Experiments

In the literature, only a few papers deal with corrosion experiments to determine the chemical stability of porous ceramics or porous ceramic membranes [3–6, 9–17]. Most of the described tests are not performed on complete membranes, but are done on unsupported membrane material. This material is made in the same way as a specific membrane layer, but is dip coated on, for instance, a glass support, removed from the glass, and used as flakes in the test. This material is very similar to the membrane layer, but not completely the same, for example, the pores are normally somewhat bigger, while on a support the grains, and therefore the pores, grow less at the same sintering temperature [8]. In the corrosion tests, the unsupported membrane layer flakes are normally immersed, for a certain time, in the corrosion liquid in a stirred or unstirred cell. The amount of corrosion is then judged from the weight loss of the membrane material, or from the change of the corrosion liquid (analyzed, e.g., from its ion content).

To determine the chemical stability of complete porous membranes among the corrosion experiments, there are two main test methods. The tests most frequently used are static tests where the membrane is immersed only in the corrosion liquid [3,9–14,16]. The severity of the corrosion, in this case, is again measured from the weight loss of the membrane, or from the change in the corrosion liquid. The second test method involves dynamic corrosion tests where the corrosion liquid is filtrated through the membrane. The filtration can be done in dead-end mode or in cross flow. It is clear that in dynamic tests with cross-flow filtration, the corrosion test approaches closest to the conditions of real applications. Dynamic filtration tests were first performed by J.M. Hofman-Züter as discussed in her Ph.D. thesis [6]. These first dynamic tests were done in dead-end mode, and the exposure of the membrane to the corrosion liquid was rather short and limited to a maximum of 16 h. The first cross-flow dynamic corrosion tests were done by Van Gestel [17]. He used long-term exposure of several weeks to the corrosion liquid, and checked the membrane stability by regular cut-off measurements.

In corrosion experiments, the experimental conditions play a major role in the correct interpretation of the results. The corrosion circumstances in both static and dynamic tests are markedly different: the supply of reactants and the discharge of reaction products differ widely. When the membranes are tested by just immersing them in a liquid, the concentration of corrosion reactants at the membrane–liquid interface decreases with time. Hence, at a certain moment, the corrosion reaction continues only by diffusion through the pores from and to the bulk of the liquid, and this is a slow process. In dynamic tests, on the contrary, the reactants are continuously refreshed and the reaction products are continuously transported out of the membrane, so that the reaction can proceed as long as the reaction conditions are such that the membrane material can be dissolved.

As we have seen in Section 2.1, the membrane corrosion depends on the size of the membrane pores. Therefore, we will separately describe the experimental results of

1. membrane supports or macroporous membranes and
2. membrane intermediate and top layers or mesoporous and microporous membranes.

2.2.1. Membrane supports or macroporous membranes

Hsieh [3] was one of the first to perform static corrosion tests on macroporous membranes. He did a very interesting test, comparing the corrosion behavior of an α-Al_2O_3 macroporous membrane with that of similar membranes made of two totally different membrane materials, namely, stainless steel 316 and PTFE. Table 1.1 summarizes the weight loss of the membranes in 35% HCl solutions. From this table, it is clear that the ceramic membrane is the most stable of the three.

Corrosion tests on α-Al_2O_3 macroporous membranes have also been done by Hofman-Züter and Van Gestel [6,16]. The former used dynamic tests on the complete membrane; the latter used static tests on α-Al_2O_3 powder. Table 1.2 shows the corrosion solutions used, and the results of the aluminum (Al) content found in the corrosion liquids. From the results, it is clear that the dissolution of α-Al_2O_3 in strongly acidic HNO_3 is small, but not zero.

It is obvious that membrane producers are specifically interested in knowing the corrosion behavior of their membranes. Consequently, they have a lot of information on the stability of their macroporous membranes, also used as supports for denser membranes. Table 1.3 gives an overview of the stability of different macroporous membranes of Inocermic GmbH in a series of corrosion liquids [18]. This table shows that TiO_2 membranes are the most stable. None of the membranes, though, is stable in HF or in concentrated hot H_2SO_4. Their results further show that cordierite or mullite supports are less stable than α-Al_2O_3, and that membranes with smaller pore size have a lower stability [18].

Table 1.1 Weight loss of three different macroporous membranes due to corrosion with a 35% HCl solution [3]

Membrane material	Weight loss (%/cm^2/day)
Stainless steel 316	0.4–0.04
PTFE	0.0006
α-Al_2O_3	0.00002

Table 1.2 Corrosion results on α-Al_2O_3 membrane [6] and α-Al_2O_3 powder [16] after dynamic and static corrosion tests, respectively

Corroded material	Alumina content in permeate after filtration of HNO_3 at pH 2 (μg/liter)	Alumina content in dip solution after filtration of HNO_3 at pH 1 (μg/liter)
α-Al_2O_3 support [6]	40	
α-Al_2O_3 powder [16]		570

Table 1.3 Stability of the macroporous membranes available at Inocermic GmbH in a series of corrosion liquids [18]

Membrane material	pH stability	Conc. HCl Conc. HNO_3	HF	H_3PO_4	Conc. hot H_2SO_4	Conc. hot NaOH
α-Al_2O_3	0–14	O.K.	–	–	–	–
TiO_2	0–14	O.K.	–	O.K.	–	O.K.
ZrO_2	0–14	O.K.	–	–	–	O.K.
SiO_2	0–10	O.K.	–	–	–	–

The dash "–" means not O.K.

Because of the high chemical resistance of TiO_2 macroporous membranes, Inocermic GmbH has also fabricated complete TiO_2 membranes [19]. Extra incentives to change to TiO_2 supports instead of the usual α-Al_2O_3 supports are the lower sintering costs due to the lower sintering temperature of TiO_2, and the lower abrasion of the preparation tools (mixer, kneader, extruder) due to the lower hardness of TiO_2. For complete TiO_2 membranes also, a better thermal shock resistance is expected because of an almost perfect adaptation of the thermal expansion coefficients of all layers in the membrane. In Table 1.4, one can find more detailed corrosion data for the Inocermic TiO_2 supports, underlining again the excellent chemical stability of this material in the whole of the pH range. In concentrated acid and alkaline solutions and at high temperatures, a small weight loss was observed. However, this corrosion is smaller than in the case of α-Al_2O_3 [19].

The extra high chemical stability of TiO_2 macroporous membranes has also inspired membrane producer Tami Industries to develop complete TiO_2 membranes in a whole range of pore sizes [20]. It mentions an extra advantage of TiO_2 as membrane material, namely, the nontoxicity of titanium (Ti) ions. Al ions, on the contrary, show a clear neurological toxicity effect. This makes TiO_2 membranes also applicable in the environmental and pharmaceutical sectors.

The German membrane producer Atech Innovations GmbH tests the stability of its α-Al_2O_3 macroporous membranes by measuring the burst pressure before and after long-term filtration of a corrosive liquid at high temperature (80 °C) [21]. Table 1.5 gives an overview of their results. The corrosive filtration at high temperature is used to fasten the corrosion process and hence to have a much faster

Table 1.4 Stability of the macroporous TiO_2 membranes of Inocermic GmbH in NaOH and H_2SO_4 in different circumstances of concentration and temperature [19]

Corrosion liquid	Concentration (%)	Time (h)	Temperature (°C)	Weight loss (%)
NaOH	4	6	100	0.00
NaOH	45	6	25	0.20
NaOH	45	6	75	0.30
NaOH	45	6	90	0.35
H_2SO_4	1	6	100	0.00
H_2SO_4	70	6	175	2.40
H_2SO_4	70	12	175	2.45

Table 1.5 Burst pressure of the macroporous membranes of Atech Innovations GmbH before and after filtration of corrosive liquids [21]

	Burst pressure (bar)		
Support	New	After 14 × 24 h KOH 5% at 80 °C	After 23 × 9 h HNO_3 5% at 80 °C
α-Al_2O_3	100–120	40–50	40

impression of the stability of the tested membranes. The results show a clear decrease in the burst pressure after the dynamic corrosion tests. This is mainly due to the attack on the grain boundaries, weakening the contact points between the grains, and hence weakening the mechanical strength of the membranes.

A very recent publication [22] studies the corrosion of macroporous TiO_2 membranes as a function of their preparation. Besides the conclusions already mentioned in this section, they show that defect-free membranes have a superior stability than poor quality membranes. In the last case, the corrosion, preferentially around defects, leads to pore blockage and fouling-layer formation, seriously deteriorating the performance of these membranes.

Because of its decade-long experience with the use of porous ceramic membranes, membrane manufacturer Atech Innovations GmbH also has a good view on the stability of its ceramic membranes toward a nonchemical corrosion process, namely, abrasion. One of the applications vulnerable to abrasion is the filtration of a sugar solution in the presence of active carbon (in order to absorb some unwanted constituents of the sugar solution as e.g. color). The membrane used in this case is an open ultrafiltration membrane. The active carbon particles, measuring from 20 to 100 nm in size (see SEM picture Fig. 1.3), are really "polishing" the surface of the membrane.

This process is visualized in the SEM pictures of Fig. 1.4, showing the new membrane, after 8 weeks, and after 6 months of filtration. After 8 weeks of filtration,

Figure 1.3 SEM picture of the active carbon particles used in the filtration of sugar solutions [21].

only the 50-nm intermediate layers remain. After 6 months, the active carbon particles have polished the membrane up to the 0.1 micrometer support.

Moreover, the material polished away from the membrane causes fouling of the membrane. This can be seen in Fig. 1.5, which shows the flux of the filtration installation at the start of the filtration, when the membranes are intact, and after several months of filtration, when the membranes are damaged.

2.2.2. Mesoporous and microporous membranes

2.2.2.1. Static corrosion tests First of all, we will consider the results of some static corrosion tests on mesoporous and microporous membrane material, made in the same way as the intermediate or top layers of fine-porous membranes. Van Gestel [16] tested γ-Al_2O_3 and TiO_2, both mesoporous and microporous, in the same way. He dipped unsupported membrane flakes in HNO_3 and in NaOH at pH 1 and 12, respectively, for 4 days. The tests were done at room temperature, and with stirring. Each time 100 mg of membrane material was dipped into 100 ml of corrosion liquid. The severity of the corrosion was evaluated by measuring the metal dissolution, that is, in this case the Ti or Al ion content, in the corrosion liquid with inductively coupled plasma–mass spectroscopy (ICP-MS). Table 1.6 summarizes the results. The metal content in the corrosion liquids is mentioned in micrograms per liter of corrosion liquid. In the table is also stated the formula of the material tested; the calcination temperature used to produce the material; and for TiO_2, the crystallographic phase the material is in. The results show the superior stability of the mesoporous TiO_2 material in the anatase phase. However, when the calcination temperature is too low (200 °C) and the TiO_2 remains in an amorphous phase, the corrosion is definitely more severe. This is due to the thermodynamic unstability of amorphous phases, as mentioned in Section 2.1. Compared to TiO_2, γ-Al_2O_3 is a material with an intrinsically low thermodynamic stability. Therefore, it shows clear dissolution in the static corrosion tests shown here. This dissolution is clearly varying with the calcination temperature, and correlated to the crystallographic phase.

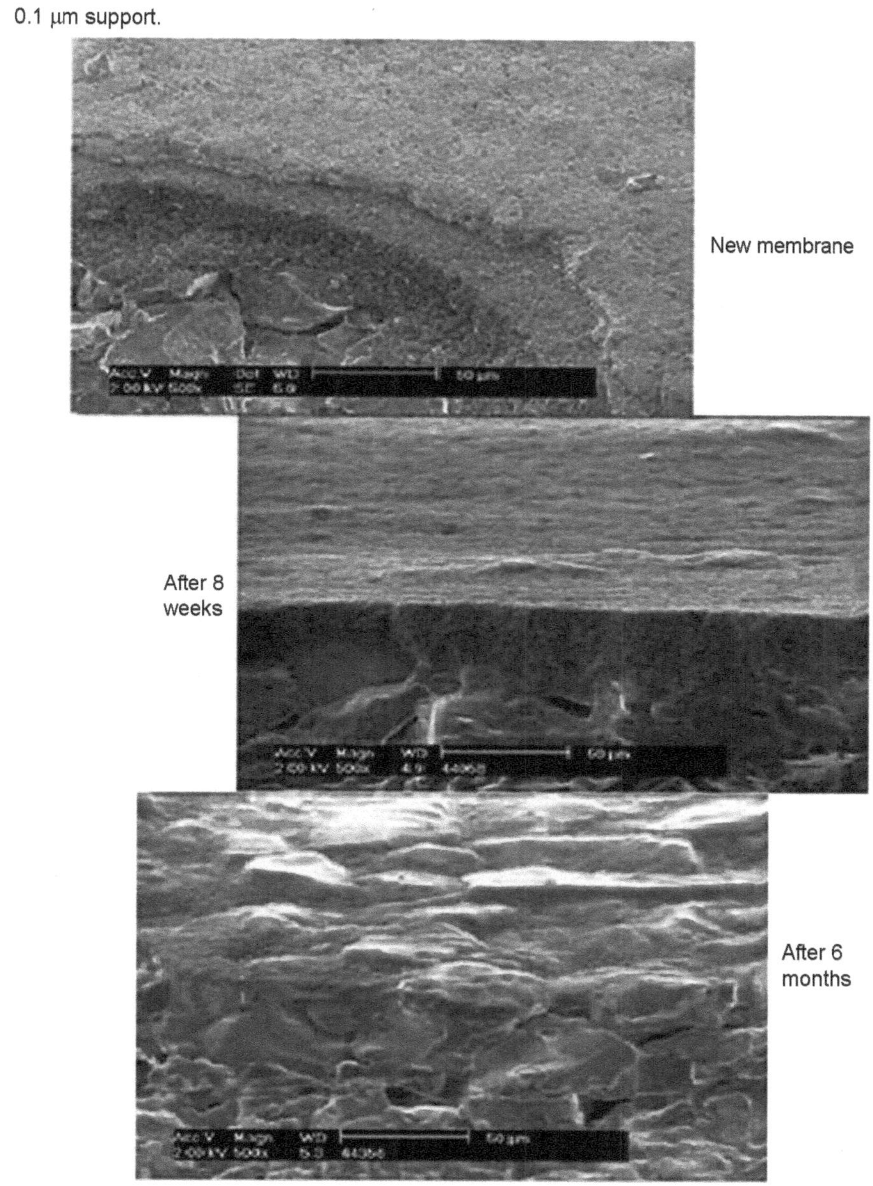

Figure 1.4 SEM picture of a multilayer ceramic membrane before and after filtration of a sugar solution containing the abrasive active carbon particles of Fig. 1.3 [21].

When the calcination temperature is low (400 °C), the material is produced in the boehmite phase (γ-AlOOH). When the calcination temperature is increased to 600 °C, the material turns into the γ-Al_2O_3 phase, which has a lower crystallinity than the boehmite phase. This leads to an increase in the corrosion. At a calcination temperature of 800 °C, the γ-Al_2O_3 phase remains, but the porosity decreases,

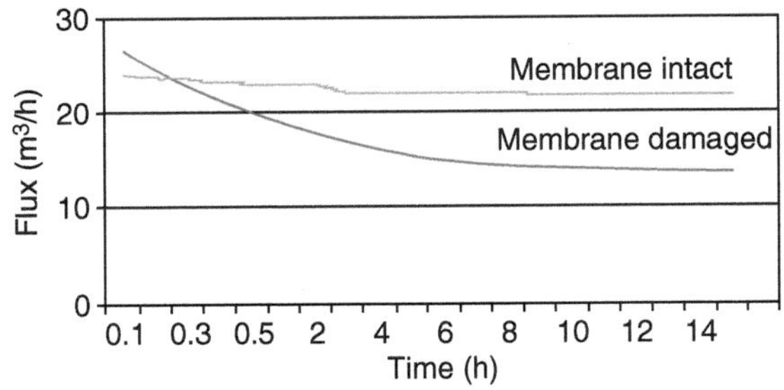

Figure 1.5 Flux of the installation filtrating a sugar solution containing the abrasive active carbon particles of Fig. 1.3 with intact membranes and when the membranes are damaged [21].

Table 1.6 Results of static corrosion tests on different Al_2O_3 and TiO_2 materials [16]

	Metal content		
Material tested (calcination temperature)	In H_2SO_4 at pH 1	In NaOH at pH 12	Pore size (nm)
α-Al_2O_3	570		100
γ-AlOOH (400 °C)	45,000		3.7
γ-Al_2O_3 (600 °C)	80,000	48,000	3.8
γ-Al_2O_3 (800 °C)	42,000		4.6
TiO_2 (400 °C) anatase	80	35	3.4
TiO_2 (300 °C) anatase	610	<10	2.4
TiO_2 (200 °C) amorphous	22,000		1.6

The metal contents mentioned are in μg/liter.

leading to a decrease in the corrosion. All this is consistent with the general principles mentioned in Section 2.1.

In static corrosion tests, SiO_2 shows a clear dissolution in diluted alkaline solutions. At higher temperatures, the effect is even bigger, and solubility is even seen at neutral pH. This explains the limited use of microporous SiO_2 membranes in aqueous solutions or wet atmospheres. However, SiO_2 can be used in more organic media. Indeed, the solubility of SiO_2 strongly decreases for methanol-rich water–methanol mixtures, and for pure alcohols, the solubility also decreases with the length of the alkyl chain. These last data contribute to explain the preservation of the SiO_2 microporosity in organic media, and explains the use of SiO_2 as pervaporation membrane. Details on the magnitude of the dissolution in these different media and references can be found in Chapter 2 attributed to SiO_2.

2.2.2.2. Dynamic corrosion tests in dead-end mode J.-M. Hofman-Züter performed the first dynamic corrosion tests on mesoporous and microporous membranes [6]. The tests were done in dead-end filtration mode. The membranes were subjected to two times filtration of HNO_3 at pH 3, followed by two times filtration of HNO_3 at pH 2, followed by three times filtration of HNO_3 at pH 1. Each filtration cycle was performed at room temperature and lasted for 16 h. Hofman-Züter evaluated the corrosion by measuring the pure water permeability and the pore size of the membranes after each corrosive filtration cycle, and by following the metal content in the feed and permeates of each filtration cycle (using ICP-MS). The results of her tests on γ-Al_2O_3 and on α-Al_2O_3 membranes are shown in Fig. 1.6 and Table 1.7. In this figure and table, the subsequent filtration cycles are labeled a up to g. Fig. 1.6 shows a clear water permeability increase: after the complete experiment of six filtration cycles, the water permeability has increased up to 46%. At the same time, Al can be found in all feeds and permeates, not only for the acid filtrations but also for the water filtrations, as can be seen in Table 1.7. Both results are a sign of the severe corrosion of mesoporous γ-Al_2O_3 membranes. This is further confirmed by the change in the pore size of the membranes during the complete test: from 1.6 to 1.9 nm, as measured with permporometry. These results of dynamic corrosion tests are consistent with the static corrosion tests of Van Gestel [16].

Hofman-Züter has also modified her γ-Al_2O_3 membranes, in order to enhance their chemical corrosion stability: she used heterogeneous precipitation to prepare mixed γ-Al_2O_3/ZrO_2 membranes with the γ-Al_2O_3 particles coated by ZrO_2 [6]. She performed the same dynamic corrosion tests on these membranes as on the unmodified γ-Al_2O_3 membranes. The results showed that the corrosion was clearly decreased and the pore size of the membrane was now unchanged after the whole series of filtration cycles. The signs of corrosion that could still be found (increase in the water permeability, Al, and very small amounts of zirconium (Zr) in the filtration feeds and permeates) were interpreted as corrosion of the support and local corrosion of the top layer at strongly curved places (see Section 2.1). This local corrosion clearly decreases as a function of time. It is clear from these results that the intrinsically more stable metal oxide ZrO_2 acts here as a protecting layer for the less stable

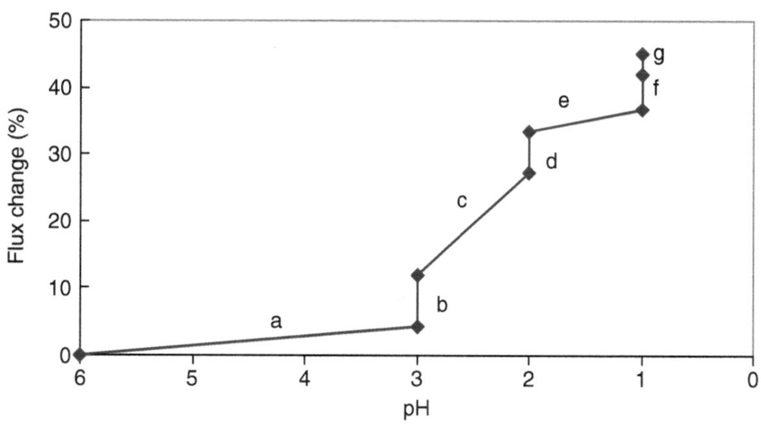

Figure 1.6 Variation in the water permeability of a γ-Al_2O_3 membrane after subsequent corrosion filtration tests at pH 3, 2, and 1. The letters a up to g label the subsequent filtration periods of 16 h at a certain pH (see text). Reprinted from [6] with permission from Hofman-Züter.

Table 1.7 Aluminum content of the feeds and permeates of the subsequent water permeability and corrosion filtration experiments performed on mesoporous γ-Al_2O_3 membranes

		Corrosion experiment, Al^{3+} (μg/liter)	Water experiment, Al^{3+} (μg/liter)	ΔF (%)
Before experiment	Feed		n.d.	
	Permeate		n.d.	
After pH 3 (1)	Feed	n.d.	119.3	4.3 (a)
	Permeate	n.d.	402.4	
After pH 3 (2)	Feed	n.m.	n.m.	7.6 (b)
	Permeate	1596.3	242.9	
After pH 2 (1)	Feed	195.9	150.7	15.3 (c)
	Permeate	358.7	388.0	
After pH 2 (2)	Feed	n.m.	n.m.	6.2 (d)
	Permeate	1815.14	195.8	
After pH 1 (1)	Feed	166.2	n.m.	3.4 (e)
	Permeate	2277.7	203.6	
After pH 1 (2)	Feed	120.3	n.m.	5.3 (f)
	Permeate	1101.1	116.5	
After pH 1 (3)	Feed	109.8	n.m.	2.9 (g)
	Permeate	861.5	84.7	

Reprinted from [6] with permission from Hofman-Züter. n.d., below detection limit; n.m., not measured; the letters a up to g label the subsequent filtration periods (see text).

γ-Al_2O_3. On the contrary, when the mixed γ-Al_2O_3/ZrO_2 membranes are made as such that the ZrO_2 does not form a coating around the Al_2O_3 particles, but is homogeneously distributed in the Al_2O_3, the dynamic corrosion tests show that the chemical stability is, in this case, not increased.

Hofman-Züter also performed the same series of dynamic corrosion tests on α-Al_2O_3–supported ZrO_2 membranes with mesoporous TiO_2 as intermediate layers. Figure 1.7 and Table 1.8 show again the change in the water permeabilities after each corrosion filtration, and the metal content in the feeds and permeates of every corrosion filtration and water filtration, measured by ICP-MS (the subsequent filtration cycles are again labeled a up to f). The results show that the water permeability is decreasing, while the pore size of the membrane remains 4.0 nm. This is not fully understood but is probably due to fouling by a small amount of corrosion products. The ICP-MS results show that Al ions can be found in all feeds and permeates of the corrosion filtrations, while Ti and Zr ions are found only from pH 1 onwards. Al ions can come only from the supports, and are due to its relative instability (see Section 2.2.1). TiO_2 and ZrO_2 layers, on the contrary, are chemically stable up to pH 1. Of the two materials, TiO_2 is the most stable, as was also found for the macroporous membranes in Section 2.2.1. The stability of mesoporous TiO_2 is again in agreement with the results found in the static corrosion tests, as described above.

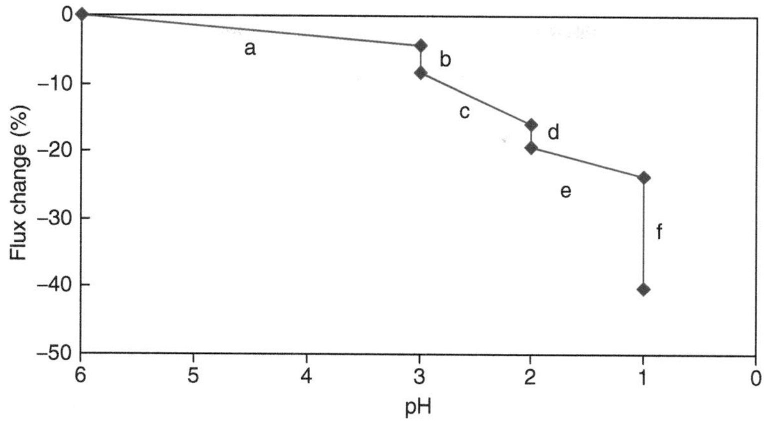

Figure 1.7 Variation in the water permeability of an α-$Al_2O_3/TiO_2/ZrO_2$ membrane after subsequent corrosion filtration tests at pH 3, 2, and 1. The letters a up to f label the subsequent filtration periods of 16 h at a certain pH (see text). Reprinted from [6] with permission from Hofman-Züter.

2.2.2.3. Dynamic corrosion tests using cross-flow filtration Dynamic corrosion tests using cross-flow filtration have been performed by Van Gestel [17]. He used filtration with HNO_3 and NaOH solutions with pH values from 1 to 13. The filtrations were performed at room temperature, for several days and weeks. A recycle mode was used, recycling the permeate back to the retentate. The severity of the corrosion was evaluated using regular molecular weight cut-off (MWCO) measurements. In this way, different microporous nanofiltration membranes were tested. Figure 1.8 summarizes the results of a boehmite (γ-AlOOH) membrane. The figure shows the different MWCO curves after 1 week of filtration with corrosive solutions of different pHs. In mild acids (pH $\geq$ 3), the MWCO decreases (from 800 Da at start to 200 Da after filtration at pH 3). This behavior in mild acid is unexpected, and not clarified. After filtration with a pH 2 solution, the MWCO of the membrane clearly increases (to 1700 Da). Once more it is clear that γ-Al_2O_3 membranes are chemically unstable.

Van Gestel also performed tests on microporous TiO_2 membranes [17]. The first tested membrane consisted of a TiO_2 top layer calcined at 300 °C. This leads to a top layer that is not completely crystalline, as was measured with X-ray diffraction (XRD). This membrane was filtrated for several days with a pH 2 solution. Figure 1.9 shows the rapid degradation of the MWCO of this membrane, indicating severe membrane corrosion. After 1 day of acid filtration, the MWCO already increased from 250 to 450 Da. This corrosion is again consistent with the results of the static corrosion tests [16] mentioned before.

The clear corrosion of this membrane was visualized by Field Emission Scanning Electron Microscopy (FESEM). Figure 1.10 shows the surface of the untreated membrane, and of the corroded membrane. The corroded membrane reveals a surface with nonuniform coverage of the NF top layer: parts of the top layer have been pealed off. As explained in Section 2.1, this is due to attack on the grain

Table 1.8 Metal content of the feeds and permeates of the subsequent water permeability and corrosion filtration experiments performed on α-$Al_2O_3/TiO_2/ZrO_2$ membranes

		Corrosion experiment			Water			
		Al^{3+} (μg/liter)	Zr^{4+} (μg/liter)	Ti^{4+} (μg/liter)	Al^{3+} (μg/liter)	Zr^{4+} (μg/liter)	Ti^{4+} (μg/liter)	ΔF (%)
Before treatment	Feed				n.d.	n.d.	n.d.	
	Permeate				35.4	n.d.	n.d.	
After pH 3 (1)	Feed	46.0	n.d.	n.d.	n.d.	n.d.	n.d.	–4.29 (a)
	Permeate	1613.0	n.d.	n.d.	93.3	n.d.	n.d.	
After pH 3 (2)	Feed	29.0	n.d.	n.d.	30.7	n.d.	n.d.	–4.05 (b)
	Permeate	449.5	n.d.	n.d.	50.4	n.d.	n.d.	
After pH 2 (1)	Feed	n.d.	n.d.	n.d.	n.d.	n.d.	n.d.	–7.57 (c)
	Permeate	310.1	n.d.	n.d.	72.9	n.d.	n.d.	
After pH 2 (2)	Feed	n.d.	n.d.	n.d.	n.d.	n.d.	n.d.	–3.37 (d)
	Permeate	193.4	n.d.	n.d.	60.0	n.d.	n.d.	
After pH 1 (1)	Feed	n.d.	n.d.	20.4	n.d.	n.d.	n.d.	–4.24 (e)
	Permeate	360.4	645.4	56.8	101.5	n.d.	12.9	
After pH 1 (2)	Feed	26.4	599.1	36.9	n.d.	n.d.	n.d.	–16.6 (f)
	Permeate	232.5	n.d.	n.d.	68.0	n.d.	n.d.	

Reprinted from [6] with permission from Hofman-Züter. n.d., below detection limit; n.m., not measured; the letters a up to f label the subsequent filtration periods (see text).

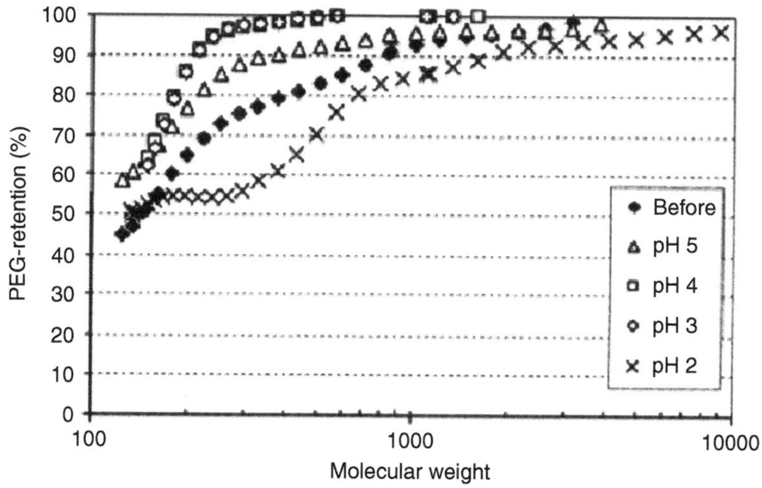

Figure 1.8 Molecular weight cut-off (MWCO) curves for corrosion tests on a γ-AlOOH microporous membrane with HNO_3 solutions at pH 5 to 2. Reprinted from [17] with permission from Elsevier.

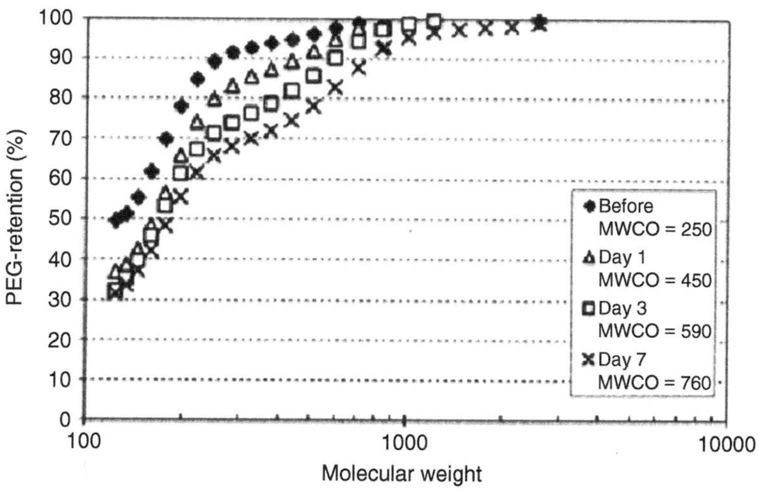

Figure 1.9 Molecular weight cut-off (MWCO) curves for corrosion tests on a TiO_2 microporous membrane calcined at 300 °C with a HNO_3 solution at pH 2. Reprinted from [17] with permission from Elsevier.

boundaries, and other noncrystalline places in the top layer, leading to a removal of whole grains in the top layer, and a fast disintegration of this layer.

Van Gestel also tested a second microporous TiO_2 membrane, with a top layer calcined at 450 °C. In this case, the top layer is completely crystalline anatase. This membrane was tested for several weeks with HNO_3 solutions of pH 3 and 2. Figure 1.11 summarizes the results. Filtration at pH 3 showed a small decrease in

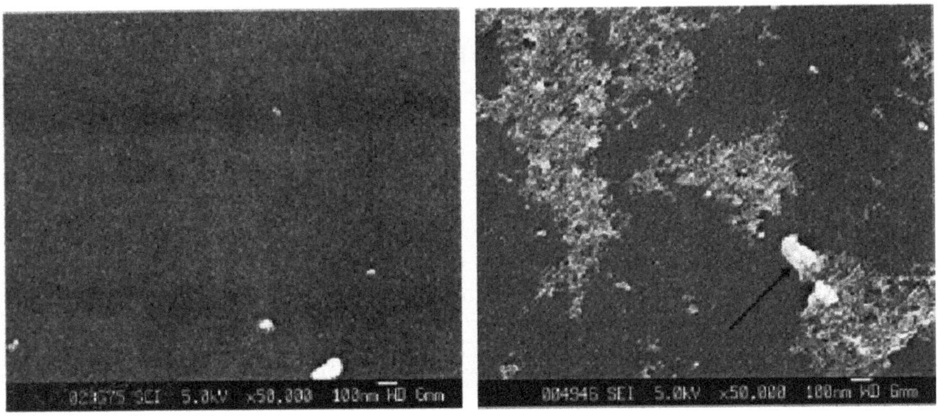

Figure 1.10 FESEM surface micrographs of a TiO_2 microporous membrane calcined at 300 °C. The picture on the left represents an untreated membrane, and the picture on the right represents a membrane corroded with a HNO_3 solution at pH 2. Reprinted from [17] with permission from Elsevier.

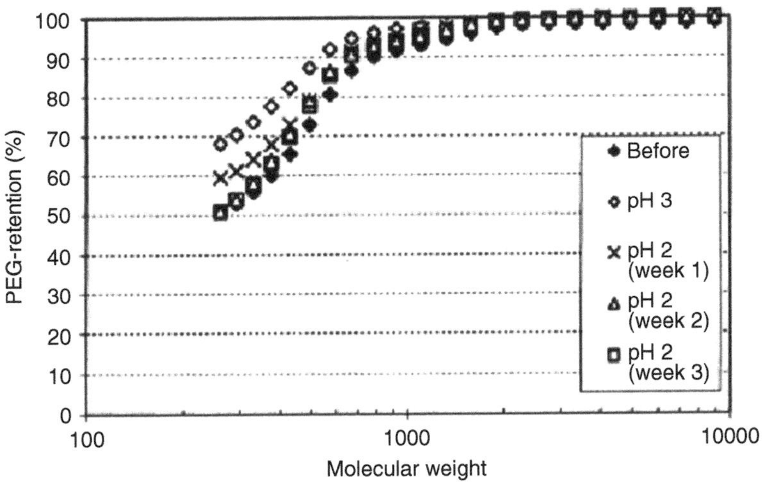

Figure 1.11 Molecular weight cut-off (MWCO) curves for corrosion tests on a TiO_2 microporous membrane calcined at 450 °C with HNO_3 solutions at pH 3 and 2. Reprinted from [17] with permission from Elsevier.

the MWCO (from 800 to 540 Da, similar to the behavior of the boehmite membrane). The first week of filtration at pH 2 showed a small increase up to 650 Da in the MWCO, but this value remained constant after 3 weeks of corrosive filtration. Van Gestel has corroded a similar membrane even further, up to 6 weeks of pH 2 filtration, followed by 2 weeks extra of pH 1.5 filtration. Even during these further corrosive tests, the MWCO of the membrane remained unchanged, proving the superb chemical stability of this microporous crystalline anatase membrane.

A microporous crystalline anatase membrane was also subjected to caustic corrosive filtration with NaOH solution of pH from 11 to 13 [17]. The results in Fig. 1.12 show that no significant membrane degradation takes place in caustic solution. All these tests on this type of membrane are again consistent with the results of the static corrosion tests mentioned before.

2.2.2.4. *Corrosive chemicals in combination with thermoshock* The highly stable microporous crystalline TiO_2 membrane can, however, be corroded when corrosive chemicals are combined with a temperature shock. This has been shown in unpublished tests done at VITO. FESEM pictures of membrane surfaces reveal that a thermoshock from room temperature to 100 °C (boiling water) does not deteriorate the membrane, neither does an acid shock from pH 7 to 1. But when the membrane is transferred from boiling water (100 °C, pH 7) to room temperature acid (pH 1), the top layer of the membrane is cracked. The same happens when the membrane is put in an acid ultrasound bath. FESEM micrographs of the membranes subjected to these different situations can be seen in Fig. 1.13.

2.2.2.5. *Solvent stability* To end this section, here are some words about the solvent stability of mesoporous and microporous membranes. Unpublished static corrosion tests at VITO have shown that no dissolution of material can be found in a whole range of different solvents, confirming that porous ceramic membranes have a broad solvent resistance. However, to use ceramic membranes in solvents, correct sealing material has to be used. In order to be useful in many solvents, teflon (PTFE) sealings or glass sealings are recommended. Despite the solvent stability of ceramic membranes, it is seen more frequently in solvent filtration that solvent fluxes take a

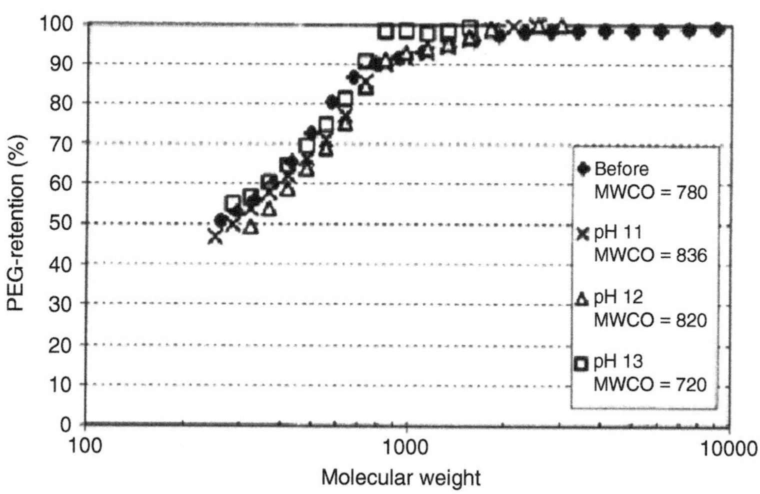

Figure 1.12 Molecular weight cut-off (MWCO) curves for corrosion tests on a TiO_2 microporous membrane calcined at 450 °C with NaOH solutions at pH 11–13. Reprinted from [17] with permission from Elsevier.

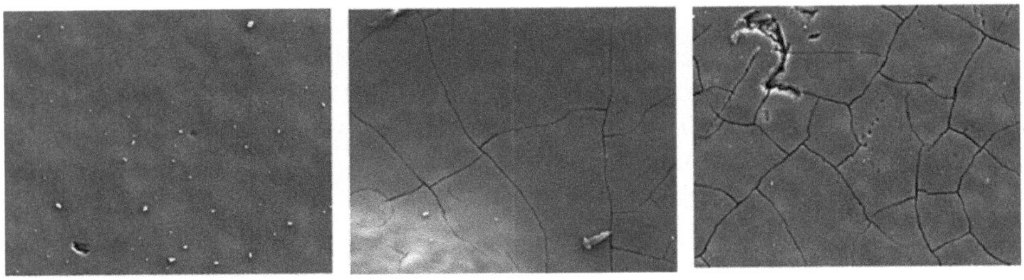

Figure 1.13 FESEM micrographs (5000×) of the surface of a TiO_2 microporous membrane. The left picture shows an untreated membrane. The middle picture shows the membrane after an acid shock in combination with a thermoshock. The right picture shows the membrane subjected to an ultrasound treatment in an acid bath.

long time to stabilize: phenomena such as solvent adsorption on ceramic surfaces cannot be excluded [23, 24].

3. Thermal Stability

3.1. Background

The thermal stability of ceramics in general, and of porous ceramic membranes especially, is better studied than their chemical stability. However, the definition of thermal stability of a porous ceramic membrane is again not unambiguous, since different characteristics of such membranes can change with temperature, for example, the pore size, the pore-size distribution, the grain size, and the crystallographic phase(s). These parameters may be closely related to each other, but this is not necessarily so. The effect of temperature on all these parameters will be discussed.

In determining the thermal stability of a porous ceramic membrane, three characteristics of the membrane are important:

1. the sinter activity of the membrane material,
2. the phase transformation of the membrane material, and
3. the support.

3.1.1. The sintering process

Calcination, the final step during the preparation of a ceramic membrane, is the first thermal treatment that the material undergoes before use. Its aim is to stabilize the crystallographic and morphological structure of the dried layer. It leads to a specific grain size and related pore size of the porous ceramic membrane (layer). Calcination is taking place in standard practice at temperatures of several hundred degrees centigrade, depending on the membrane material used. When the material is kept long enough at the calcination temperature, it can be assumed that the material approaches its equilibrium structure [25]. Therefore, the structure of the membrane will not vary when it is exposed to temperatures of 100–150 °C below its calcination temperature, since the reaction rates and driving forces for the sintering process will be too low there. However, when the material is heated to higher temperatures, it acts as every other ceramic material, meaning that sintering and grain-growth processes start again, leading to a new equilibrium. Therefore, in order to prevent the membrane from changing its pore structure during application, sintering has to be avoided or suppressed.

During calcination/sintering, the ceramic material tends to minimize its Gibbs free energy, mainly by decreasing its surface energy [6,25]. This decrease in surface energy is realized by diffusion of material from places with high energy to places of low energy, that is, from places with high curvature to places with low curvature. In a initial step of the sintering process, this leads to neck formation between the contacting grains of the dried membrane layers (see Section 1.1), as shown in Fig. 1.14. When sintering at relatively low temperatures (low compared to the melting temperature), this surface diffusion is the only effect of the thermal treatment.

When sintering at temperatures closer to the melting temperature [(0.5–0.9) × the melting temperature, depending on the binding strength of material], grain-boundary diffusion takes place, and hence grain growth. In this second step of the sintering process, small particles (strong curvature and so high energy) are swallowed by larger ones (smaller curvature and so lower energy). This grain growth leads to the simultaneous growth of the pore size, and a decrease in the porosity. Consequently, the surface area of the porous material decreases. Finally, for sintering at temperatures near the melting temperature of the material, bulk diffusion occurs, leading to complete densification of the porous system.

In order to obtain porous ceramic membranes with a high thermal stability, the sinter activity of the material should be as low as possible. This is the case for materials with high melting temperature, and correlated high thermodynamic stability. Burggraaf has discussed a few other strategies to inhibit sintering in porous materials [26]. One of his suggestions was to lower the number of contact points between the particles in the porous materials, for example, by producing highly asymmetric particles as needles or plates and using them in a low-density packing,

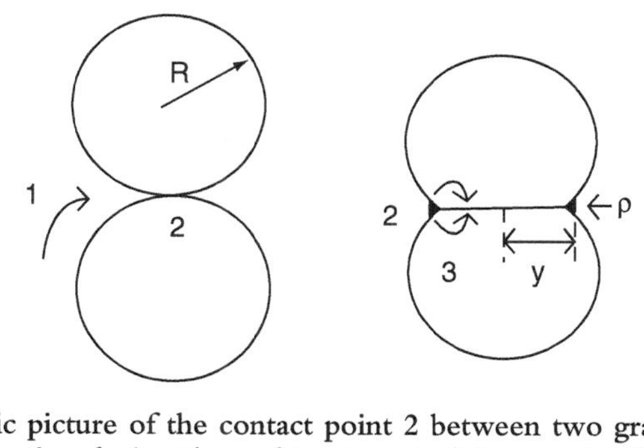

Figure 1.14 Schematic picture of the contact point 2 between two grains with radius R (left picture) and neck formation during thermal treatment of these grains (right picture) Pathways 1 and 3 represent the diffusion of material over typical distances y. Reprinted from [6] with permission from Hofman-Züter.

instead of the usual close packing spheres. Another possibility to decrease the sinter activity mentioned is to block the diffusion along the diffusion path. This can be realized by suitable doping of the membrane material.

3.1.2. Phase transformations

Another issue that is of major importance for the thermal stability is the phase-transformation behavior of the membrane material. In materials like Al_2O_3, ZrO_2, or TiO_2, different phase transformations occur. Al_2O_3 material is at low temperature in the boehmite phase (AlOOH) and transforms subsequently in its γ-Al_2O_3 phase, δ-Al_2O_3 phase, and finally, the most stable α-Al_2O_3 phase. TiO_2 evolves from an amorphous phase into the anatase phase, and finally, to the rutile phase. ZrO_2 changes from the amorphous phase to its monocline and tetragonal phases, and finally, to its cubic phase. SiO_2 is a glass-forming oxide and shows a phase transformation from the usual amorphous phase to the crystalline cristobalite phase at relatively high temperature compared to the other oxides (above 900 °C). For granular material with sufficient grain size, the temperatures of these phase transitions are fixed. But for porous sol–gel material, typically with grain sizes on the nanometer scale, the exact transition temperatures depend on the precursors used, and on the details of the sol–gel process [6].

Phase transformations cause large volume and energy changes, leading to large variations in the particle size and therefore in the membrane pore size and porosity [27]. For the thermal stability of a ceramic membrane, it is therefore very important that the material is in a thermodynamically stable phase and does not show any phase transformations below its working temperature.

In order to shift the phase transformations to higher temperature, and so enhance the thermal stability, specific doping can be used. Metal ions such as the popular La^{3+} are especially used [28]. Section 3.2 will point out other possible dopants.

3.1.3. Support

The support also has an important effect on the thermal stability of a porous ceramic membrane. The support of the membrane acts as a constraint and decreases the diffusion rate of the membrane-layer material and therefore retards the sintering process [8,28]. Therefore, the support has a negative effect on the densification rate of materials. This is consistent with the lower pore size of supported membranes compared to unsupported membrane flakes [8]. For thick membrane layers, the sintering stresses are relatively high, inducing tensile stresses in the film, which can result in cracks and even delamination of the top layer from the support. When the membrane layer is thin enough, these tensile stresses only lead to a decrease in the sintering rate, and finally, an end to the sintering process after some pore change.

The support not only retards the sintering but can also increase the temperature of the phase transitions, as will be seen in Section 3.2.

As the support clearly delays the rate of sintering, and can increase the phase transitions of membrane material, it definitely increases the thermal stability for sufficiently thin layers.

3.2. Experiments

As already mentioned in the previous section, a good definition of the thermal stability of porous ceramic membranes is missing. As a consequence, there is no standard experimental procedure for the determination of thermal stability. The papers published [6,8,16,28–38] focus on specific aspects of the thermal stability of ceramic membranes. Some concentrate on the pore size and porosity and some on the crystallographic phase stability. In case the phase stability is followed, especially the phase-transformation temperatures are followed, which is a badly defined parameter, as mentioned in Section 3.1.2. , and, moreover, depends on the heating treatment used during experiments. Chang [29] presented one of the most detailed and complete papers on the thermal and hydrothermal stability of Al_2O_3, TiO_2, and ZrO_2 membranes, providing information on the change in pore size, pore volume, surface area, gas permeability, and phase stability of the membranes. Common to all the published papers are the short times ($\leq$30 h) used to study the thermal stability with the exception of the work of van Veen [31]. This section is meant to give a summary of the main results found in the thermal-stability studies described in the literature.

3.2.1. Effect of the sintering process

As explained in Section 3.1, the calcination temperature defines the pore size of the membrane material. All experiments show that higher sintering temperatures cause the membrane grains and pores to grow, and the pore volume and the surface area to decrease. Moreover, in the vicinity of phase transitions, the change in the pore size and porosity is extra large. This is nicely illustrated for Al_2O_3, TiO_2, and mixed Al_2O_3–TiO_2 membrane flakes in [16]. Figure 1.15 and Table 1.9. show their results of colloidal sol material (with binder) and of polymeric sol material (without binder). The use of a binder such as poly-vinyl-alcohol (PVA) or hydroxy propyl-cellulose (HPC) in colloidal sol–gel layers and not in polymeric sol–gel layers is a common membrane manufacturing practice. The use of a binder increases the pore size of the formed layers and is therefore avoided in membrane layers with intended microporosity, as the ones made by the polymeric sol–gel method.

To retard the pore growth and other sintering processes, doping of the porous ceramic material can be used. The idea of the dopant is to block the diffusion path of the material during the sintering process. This effect of doping has been shown among others by Lin [27,28,36] for boehmite, γ-Al_2O_3 or TiO_2 doped with lanthanum, and for ZrO_2 doped with yttrium. In comparison with the pure materials, the effect of stabilization of the pore structure was larger for the ZrO_2 than for the TiO_2 and the Al_2O_3 material. The main gap in this work is the unknown distribution of the dopants throughout the samples; more specifically, it is unclear whether the second oxide is adsorbed on the surface of the first oxide or whether the two oxides form a solid solution or a real composite. More recently, Sekulic [32] has shown the same structural effects of ZrO_2 doping into TiO_2 material. The pore growth is clearly retarded by the doping, and the pore size decreased from 4.5 nm down to 3.6 nm by addition of 20 mol% of ZrO_2, while the surface area and the porosity increased. These results are summarized in Table 1.10. Sekulic explains these results by the existence of a small cluster of

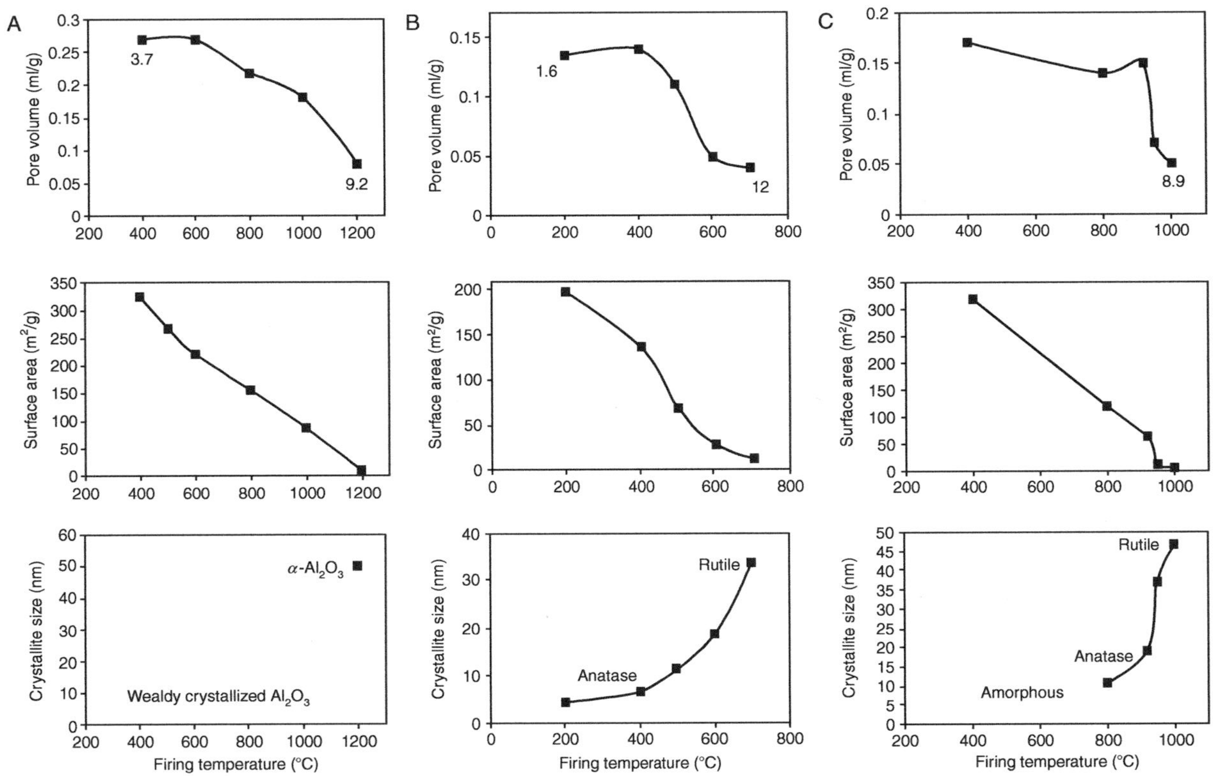

Figure 1.15 Structural development as a function of temperature for Al_2O_3, TiO_2, and mixed Al_2O_3/TiO_2 membrane flakes made by the colloidal sol–gel method with binder. The numbers mentioned in the pore volume graphs represent the pore sizes in nanometers. Reprinted from [16] with permission from Elsevier.

Table 1.9 Pore structure characteristics of TiO_2 membrane flakes made by the polymeric sol-gel method without binder [16]

TiO_2 phase	Firing T (°C)	Mean pore size (nm)	Surface area (m^2/g)	Pore volume (cm^3/g)
Amorphous	200	1.6	384	0.21
Anatase	300	2.4	307	0.19
Anatase	400	3.4	130	0.13

Table 1.10 Properties of unsupported TiO_2 material doped with various ZrO_2 contents and calcined at 450 °C

ZrO_2 content (mol%)	Surface area (m^2/g)	Porosity (%)	Pore size (nm)
0	73.0	28	4.5
5	120.0	38	4.1
10	142.0	38	3.9
20	165.1	40	3.6
30	151.0	47	5.4

ZrO_2 particles between the necks of neighboring TiO_2 particles, blocking the normal sintering of the material. When too large clusters of ZrO_2 particles are present, at 30 mol% of ZrO_2, the pore size increases again. The same effects of retarded pore growth by doping ZrO_2 into TiO_2 have also recently been shown for microporous ZrO_2/TiO_2 membranes [33].

3.2.2. Effect of phase transformations

Other limiting factors for the thermal stability of porous ceramic membranes are the crystallographic phase transformations. Doping of the ceramic material is again the common method to retard the phase transformations. However, literature shows that although some dopants retard the phase transformation, others promote it. For example, Yang and Ferreira [38] have studied different doping ions for TiO_2 in order to shift the anatase to rutile transformation. They reported that additives such as WO_3 and P_2O_5 retard the phase transformation process, whereas others such as CuO and Fe_2O_3 accelerate the process. It is still not clear what characteristics of the dopant play a role [32].

Successful enhancement of the thermal stability by an increase in transformation temperatures has been shown, for example, in the work of Lin [36]. The phase transformation temperature from γ-Al_2O_3 to α-Al_2O_3 was shifted by 200 °C using lanthanum doping; the transformation temperature from tetragonal to monoclinic ZrO_2 was shifted by 300 °C using yttrium doping; and the anatase to rutile transformation temperature was shifted by 150 °C using lanthanum doping. Also, mixed metal oxide membranes such as Al_2O_3–ZrO_2 [6], SiO_2–ZrO_2 [34], Al_2O_3–TiO_2 [16], or ZrO_2–TiO_2 membranes [30] show an increased thermal stability. This can, for example, be seen in Fig. 1.15, where the anatase–rutile transition of TiO_2 is clearly shifted by mixing with Al_2O_3. However, it is important to mention that the chemical stability of the TiO_2/Al_2O_3 composition is decreased by the intrinsically more unstable Al_2O_3 [17]. The effect of doping of the chemically more stable ZrO_2 in mesoporous TiO_2 has also recently been studied by Sekulic [32]. In their pure TiO_2 membranes, they see the start of the anatase to rutile transformation around 350–400 °C. After 3 h at 450 °C, their TiO_2 membrane layer consists of 10% rutile particles, causing severe cracking of the layer. However, when 10% ZrO_2 is mixed in the TiO_2 material, the phase transformation can be shifted above 700 °C, and crack-free membrane layers can be made up to sintering temperatures at 700 °C. This can be clearly seen in their XRD measurements, shown in Fig. 1.16. Kumar [37] has shown that the phase transformation from anatase to rutile for TiO_2 can even be completely avoided by doping with SnO_2, and directly forming rutile membranes during the synthesis.

The temperature stability of microporous sol–gel-derived SiO_2 is limited by the phase transition to the crystalline cristobalite phase (between 900 and 1400 °C). However, below this temperature, at about 800 °C, strong sintering effects occur in the structure, leading to very strong shrinkage of the material. For more details, see Chapter 2.

3.2.3. Effect of the support

Effects of the support on the thermal stability of porous ceramic membranes have also been studied. Kumar [8] measured the difference between supported and unsupported membranes. They found that supported membranes show a

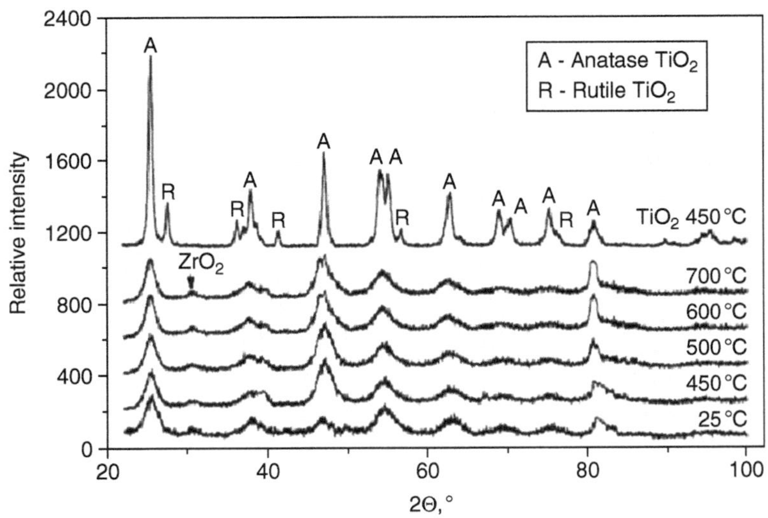

Figure 1.16 *In situ* high temperature X-ray diffraction (XRD) diagrams for the composite powder 90 mol% TiO_2/10 mol% ZrO_2, and XRD spectrum of pure TiO_2 calcined at 450 °C. Reprinted from [32] with permission from Elsevier.

considerably higher anatase to rutile transformation temperature than do unsupported membranes. Moreover, they found that the supported membranes have a smaller pore size and a higher porosity when sintered at a given temperature than do unsupported membranes. All this is explained by the fact that the support acts as a constraint to the sintering process and the phase transformation, as already mentioned in Section 3.1.3. This is also confirmed by other measurements [6,30,32].

3.2.4. Hydrothermal stability

The hydrothermal stability of porous ceramic membranes has first been studied by Chang [29]. They studied Al_2O_3 and TiO_2 as well as ZrO_2 membranes under calcination in dry air, and in steam/air atmosphere. They found that the pore sizes of samples fired in steam/air are larger than that of samples fired in dry air. Simultaneously, the surface area is smaller. This can be explained by an accelerated dissolution–redeposition process and correlated densification in the presence of steam. For TiO_2 membranes, steam had the largest effect. The effect is less important for Al_2O_3 membranes, and the lowest for ZrO_2 membranes.

However, the metal oxide most vulnerable to steam is SiO_2. SiO_2 is known to have a very limited hydrothermal stability (see also Chapter 2). This is, for example, evident by the fast decreasing flux (during the first 400 h) and simultaneous selectivity decrease of SiO_2 membranes in pervaporation at 95 °C [39]. The origin of this effect is not completely clear, but it is believed that adsorption of water on the pore surface area of SiO_2 plays an important role. Another sign of the hydrothermal instability of SiO_2 is the decreasing gas-separation performance with hot steam >100 °C [40]. This is explained by enhanced densification under the influence of steam. Incorporation of methyl groups in the SiO_2 clearly increases the hydrothermal stability [39,40]. As a consequence, methylated SiO_2 shows a high flux stability in pervaporation at 95 °C for up to 18 months. This is attributed to the lower water adsorption on the more hydrophobic material. The stability of the methylated SiO_2 pervaporation membrane at temperatures above 100 °C is severely limited (to about 10 days) because of reorganization of the separating layer. However, methylated SiO_2 does show long-term stable gas-separation performance in the presence of hot steam >100 °C. This is due to the decreased densification effects of the modified SiO_2.

In order to improve the stability of SiO_2 membranes in steam, doping of other inorganic oxides such as TiO_2, ZrO_2, Fe_2O_3, Al_2O_3, and NiO, was also successfully used [41–44]. The dopants cause a retardation in the densification effects due to steam, similar to their effect in case of an increase in the thermal stability (see Section 3.2.1). The dispersion of the dopant in the SiO_2 matrix is important for the efficiency of the stabilization [44].

The same positive effect of doping on the hydrothermal stability has also been shown recently for mesoporous γ-Al_2O_3 doped with La_2O_3 and Ga_2O_3 [45]. Double doping has the best effect here, and avoids the large pore changes occurring within the first few hours of the hydrothermal treatment [46]. Both for doped γ-Al_2O_3 and for doped SiO_2, the increased hydrothermal stability has been shown by their gas permeation fluxes before and after hydrothermal treatment, at temperatures up to 650 °C, and steam contents of up to 90%.

4. Resuming

Ceramic materials are known for their high thermal and chemical stability. But this stability is not ideal, and especially not for porous ceramic membranes. The membrane material, its multilayer structure, and the membrane morphology all play an important role in the final real thermal and chemical stability.

The chemical stability of porous ceramic membranes is a complex process and can be defined as the inertness of the membrane against corrosive attack by aggressive liquids. In this process, both thermodynamic and kinetic aspects play a role. From a thermodynamic point of view, materials with a strongly negative free energy and total energy of formation such as Al_2O_3, TiO_2, and ZrO_2 are all recognized for their potential chemical stability. From a kinetic point of view, aspects such as contact time, porosity, temperature, working pressure, and flow velocity determine the supply of corrosive reactants and the discharge of corrosion products. Consequently, the design of corrosion experiments influences the corrosion measured. The corrosion of a porous membrane starts at the surface of the membrane grains and at the grain boundaries. These locations contain more energetically favorable positions for material dissolution, compared to the bulk of the material. Other preferred dissolution spots are positions with high roughness or high curvature. In general, we can draw the following conclusions for the chemical stability of porous ceramic membranes, as confirmed by experiments:

- The chemical stability depends on the pore size, and is lower for smaller pore sizes.
- The chemical stability depends on the crystallographic phase of the material, and is higher for crystalline phases than for amorphous phases. The chemical stability therefore depends on calcination temperature.
- Considering the existing ceramic membrane materials, γ-Al_2O_3 and SiO_2 are the most unstable. The stability of ZrO_2 and α-Al_2O_3 is good. The stability of TiO_2 is outstanding, also for fine-porous material in the crystalline anatase phase.
- The solvent stability is very good, but adsorption effects are repeatedly seen.
- The combination of corrosive chemicals and thermoshock can cause cracks.

The chemical stability of a specific metal oxide membrane can be increased by doping with another metal oxide with a higher chemical stability, if the preparation is such that the stable metal oxide forms a protective coating on the unstable metal oxide grains.

The thermal stability of porous ceramic membranes is better studied than their chemical stability, but is badly defined, since different characteristics of the membrane, such as the pore size, the pore-size distribution, the grain size, and the crystallographic phase(s), can all change. Calcination, the final step during the preparation of a ceramic membrane, and the calcination temperature greatly determine the thermal stability of a membrane. During calcination, sintering of the ceramic material occurs, leading to the diffusion of material from places with high energy to places with low energy. This leads to neck formation between membrane grains, and grain and pore growth, before complete densification. Phase transformations are another major factor determining the thermal stability of a membrane. As experimentally confirmed, doping of membrane material can enhance the thermal stability of membranes by blocking the material diffusion during sintering and by shifting phase transformations to higher temperatures. The support has a positive effect on the thermal stability of membranes, as long as the membrane layers are thin enough. Steam has a negative effect on the thermal stability because it leads to accelerated sintering. The hydrothermal stability of the metal oxides decreases in the following order: $ZrO_2 > Al_2O_3 > TiO_2 > SiO_2$. Similar to the thermal stability, the hydrothermal stability can be increased by doping.

REFERENCES

[1] Gillot, J. (1991). *In* "Inorganic Membranes: Synthesis, Characterization and Applications" (R. Bhave, ed.). Van Nostrand Reinhoud, New York.

[2] Burggraaf, A. J., and Keizer, K. (1991). *In* "Inorganic Membranes: Synthesis, Characterization and Applications" (R. Bhave, ed.). Van Nostrand Reinhoud, New York.

[3] Hsieh, H. (1991). *In* "Inorganic Membranes: Synthesis, Characterization and Applications" (R. Bhave, ed.). Van Nostrand Reinhoud, New York.

[4] Holstein, T. (1999). Corrosion of advanced ceramics by liquid media: A review. *In* "Proceedings of the Ninth CIMTEC World Ceramic Congress", Florence, Italy, 14–19 June, pp. 433–444.

[5] Westerheide, R., Holstein, T., and Füting, M. (1999). Aqueous corrosion of advanced ceramics. *In* "Proceedings of the Euromat 1999", Munich, Germany, 27–30 september, pp. 392–397.

[6] Hofman-Züter, J. M. (1995). "Chemical and Thermal Stability of Mesoporous Ceramic Membranes", Ph.D Thesis, Universiteit Twente, Enschede, The Netherlands.

[7] Franke, M. D., Ernst, W. R., and Myerson, A. S. (1987). Kinetics of dissolution of alumina in acidic solution. *AIChE J.* **33,** 267–273.

[8] Kumar, K. N. P., Keizer, K., Burggraaf, A. J., Okubo, T., and Nagamoto, H. (1993). Textural evolution and phase transformations in titania membranes. Part 2. Supported membranes. *J. Mater. Chem.* **3,** 1151–1159.

[9] Moosemiller, M. D., Hill, C. G., Jr., and Anderson, M. (1989). Physicochemical properties of supported gamma-Al_2O_3 and TiO_2 ceramic membranes. *Sep. Sci. Technol.* **24,** 641–657.

[10] Nevant, M., and Klein, L. C. (1992), Durability of porous zerconia powder in caustic, *Key Eng. Mater.* **61/62,** 137–142.

[11] Mardilovich, P. P., Paterson, R., and Govyadinov, A. N. (1992). Preparation and properties of new and modified anodic alumina membranes. *In* "Proceedings of Euromembrane 92, Récents Progrès en Génie des Proceédés" (P. Aimar and P. Aptel, eds.), Vol. 6, No. 22, pp. 33–38.

[12] Thoraval, V., Castelas, B., Joulin, J. P., Larbot, A., and Cot, L. (1991). Preparation and characterisation of microporous ceramic media,. *Key Eng. Mater.* **61/62,** 213–220.

[13] Wieserman, L. F., Wefers, K., Cross, K., Martin, E. S., Hsieh, H. P. L., and Quayle, W. H. (1990). US patent application nr. 4,957,890.

[14] Egushi, K., Tanaka, H., Yazawa, T., and Yamaguro, T., (1986). European patent application nr 86201756.3, publication nr. 0220764 A1.

[15] Schaep, J., Vandecasteele, C., Peeters, B., Luyten, J., Dotremont, C., and Roels, D. (1999). Characteristics and retention properties of a mesoporous γ-Al_2O_3 membrane for nanofiltration. *J. Membr. Sci.* **163,** 229–237.

[16] Van Gestel, T., Vandecasteele, C., Buekenhoudt, A., Dotremont, C., Luyten, J., Leysen, R., Van der Bruggen, B., and Maes, G. (2002). Alumina and titania multilayer membranes for nanofiltration: preparation, characterisation and chemical stability. *J. Membr. Sci.* **207,** 73–89.

[17] Van Gestel, T., Vandecasteele, C., Buekenhoudt, A., Dotremont, C., Luyten, J., Van der Bruggen, B., and Maes, G. (2003). Corrosion properties of alumina and titania NF membranes. *J. Membr. Sci.* **214,** 21–29.

[18] Voigt, I. (2007). Private communication. Inocermic GmbH, Hermsdorf, Germany.

[19] Voigt, I., Fischer, G., Puhlfürss, P., and Seifert, D. (1998). New filtration ceramics—from support to NF-membrane completely of TiO_2. *In* "Proceedings of the 5th International Conference on Inorganic Membranes", June 22–26, pp. 42–45. Nagoya, Japan.
[20] Lescoche, P., Bergel, J.-Y. (2004). FILTANIUMTM: The 100% titanium dioxide ceramic membrane. *In* "Proceedings of the 8th International Conference on Inorganic Membranes, Cincinnati," pp. 430-433. USA, july 18-22.
[21] Bolduan, P., and Flörke, J., Private Communication. Atech Innovations GmbH. Gladbeck, Germany.
[22] Wang, Y. H., Tian, T. F., Liu, X. Q., and Meng, G. Y. (2006). Titania membrane preparation with chemical stability for very harsh environments applications. *J. Membr. Sci.* **280,** 261–269.
[23] Tsuru, T., Sudoh, T., Yoshioka, T., and Asaeda, M. (2001). Nanofiltration in non-aqueous solutions by porous silica-zirconia membranes. *J. Membr. Sci.* **185,** 253–261.
[24] Verrecht, B., Leysen, R., Buekenhoudt, A., Vandecasteele, C., and Van der Bruggen, B. (2006). Solvent flux behaviour of hydrophilic and hydrophobic mesoporous—Al_2O_3, TiO_2, and ZrO_2 membranes. *In* "Proceedings of the 9th International Conference on Inorganic Membranes", June 25–29, pp. 135–138. Lillehammer, Norway.
[25] Burggraaf, A. J. (1996). "Fundamentals of Inorganic Membrane Science and Technology" (A. J. Burggraaf and L. Cot, eds.). Elsevier, Amsterdam.
[26] Burggraaf, A. J. (1993). Internal Report CT93/402/31. University of Twente.
[27] Lin, Y. S., de Vries, K. J., and Burggraaf, A. J. (1991). Thermal stability and its improvement of the alumina membrane top layer prepared by sol-gel methods. *J. Mater. Sci.* **26,** 715–720.
[28] Lin, Y. S., and Burggraaf, A. J. (1991). Preparation and characterisation of high temperature thermally stable alumina composite membrane. *J. Am. Ceram. Soc.* **74,** 219–225.
[29] Chang, C. H., Gopalan, R., and Lin, Y. S. (1994). A comparative study on thermal and hydrothermal stability of alumina, titania and zirconia membranes. *J. Membr. Sci.* **91,** 27–45.
[30] Xu, Q., and Anderson, M. A. (1993). Sol-gel route to synthesis of microporous ceramic membranes—thermal stability of TiO_2–ZrO_2 mixed oxides. *J. Am. Ceram. Soc.* **76,** 2093–2097.
[31] van Veen, H. M., Top, J. P. M., Terpstra, R. A., Engel, A. J. G., and Veringa, H. J. (1989). The calcination and thermal stability of ceramic Al_2O_3 membranes for gas separation. *In* "Euroceramies I" (G. de With, R. A. Terpstra, and R. Metselaar, eds.), Vol. 3, pp. 585–589. Elsevier, Amsterdam.
[32] Sekulic, J., Magraso, A., ten Elshof, J. E., and Blank, D. H. A. (2004). Influence of ZrO_2 addition on microstructure and liquid permeability of mesoporous TiO_2 membranes. *Microporous Mesoporous Mater.* **72,** 49–57.
[33] Aust, U., Benfer, S., Dietze, M., Rost, A., and Tomandl, G. (2006). Development of microporous ceramic membranes in the system TiO_2/ZrO_2. *J. Membr. Sci.* **281,** 463–471.
[34] Liu, W., Zhang, B. Q., Liu, X. F., and Xu, L. M. (2006). Thermal stability of silica-zirconia membranes. *Chin. J. Chem. Eng.* **14,** 31–36.
[35] Larbot, A., Fabre, J. P., Guizard, C., and Cot, L. (1989). New inorganic ultrafiltration membranes: Tiania and zirconia membranes. *J. Am. Ceram. Soc.* **72,** 257–261.
[36] Lin, Y. S., Chang, C. H., and Gopalan, R. (1994). Improvement of thermal-stability of porous nanostructured ceramic membranes. *Ind. Eng. Chem. Res.* **33,** 860–870.
[37] Kumar, K. N. P., Keizer, K., Burggraaf, A. J., Okubo, T., and Nagamoto, H. (1993b). Synthesis and textural properties of unsupported and supported rutile membranes. *J. Mater. Chem.* **3,** 923–929.
[38] Yang, J., and Ferreira, J. M. F. (1998). On the titania phase transition by zirconia additive in a sol-gel powder. *Mater. Res. Bull.* **33,** 389–394.
[39] Vente, J. F., Campaniello, J., Engelen, C. W. R., Haije, W. G., and Pex, P. P. A. C. (2006). Long term pervaporation performance of microporous methylated silica membranes. *In* "Proceedings of the 9th International Conference on Inorganic Membranes", pp. 81–84. June 25–29, Lillehammer, Norway.
[40] De Vos, R. (1998). High-Selectivity, High-Flux Silica Membranes for Gas-Separation. Ph.D Thesis, Universiteit Twente, Enschede, The Netherlands.

[41] Yoshida, K., Hirano, Y., Fujii, H., Tsuru, T., and Asaeda, M. (2001). Hydrothermal stability and performance of silica-zirconia membranes for hydrogen separation in hydrothermal conditions. *J. Chem. Eng. Jpn.* **34,** 523–530.

[42] Asaeda, M., Kanezashi, M., Yoshioka, T., and Tsuru, T. (2003). Gas permeation characteristics and stability of composite silica-metal oxide membranes. *Mater. Res. Soc. Symp. Proc.* **752,** 213–218.

[43] Kanezashi, M., Yoshioka, T., Tsuru, T., and Asaeda, M. (2004). Stability of Ni-doped silica membranes for H_2 separation at high temperature. *Trans. Mater. Res. Soc. Jpn.* **29,** 3267–3270.

[44] Kanezashi, M., and Asaeda, M. (2006). Hydrogen permeation characteristics and stability of Ni-doped silica membranes in steam at high temperature. *J. Membr. Sci.* **271,** 86–93.

[45] Zahir, M. H., Sato, K., and Iwamoto, Y. (2005). Development of hydrothermally stable sol-gel derived La_2O_3-doped Ga_2O_3 composite mesoporous membrane. *J. Membr. Sci.* **247,** 95–101.

[46] Gallaher, G. R., and Liu, P. K. T. (1994). Characterization of ceramic membranes. I. Thermal and hydrothermal stabilities of commercial 40 Å membranes. *J. Membr. Sci.* **92,** 29–44.

CHAPTER TWO

Microporous Silica Membrane: Basic Principles and Recent Advances

André Ayral, Anne Julbe, Vincent Rouessac, Stéphanie Roualdes, *and* Jean Durand

Contents

1. Introduction

Silica (SiO_2) exhibits unique properties related to the ability of its elemental bricks, that is, SiO_4 tetrahedra, to be connected together to give rise to a large number of different amorphous or crystallized solids which can be microporous, mesoporous, or macroporous (Table 2.1). In comparison with other common single oxides like alumina (Al_2O_3), titania (TiO_2), or zirconia (ZrO_2), silica can be more easily prepared as ultra- or super-microporous amorphous thin layer and thus can be used for molecular sieving applications. Zeolites, which are mainly silica or silica-based crystallized solids with structural ultramicroporosity, are presented in another chapter of this book.

From Table 2.2 reporting the main separative processes using ceramic membranes, it appears that the microporous membranes are convenient for

Institut Européen des Membranes, UMR no 5635 CNRS-ENSCM-UMII, CC047, Université Montpellier II, Place Eugène Bataillon, 34095 Montpellier cedex 5, France

Membrane Science and Technology, Volume 13
ISSN 0927-5193, DOI: 10.1016/S0927-5193(07)13002-3

Table 2.1 IUPAC classification of pores as a function of their size

Micropores (<2 nm)			
Ultramicropores	Supermicropores	Mesopores	Macropores
<0.7 nm	>0.7 nm	2–50 nm	>50 nm

nanofiltration, pervaporation, and gas separation applications. In the case of gas separation, the highest selectivity values are reached with microporous membranes (Table 2.3). The molar permeance π of a gas going through the membrane is the stationary flux J, normalized for transmembrane pressure ($\pi = J/\Delta P$). The permselectivity (or ideal selectivity) is the ratio of the single gas permeances, measured under the same condition ($\alpha^*(i/j) = \pi_i/\pi_j$). Finally, the mixture selectivity (or separation factor) is the ratio of the gas permeances, measured for mixtures.

The performance of membranes is related to their permselectivity/separation factor and also to their permeability. For the microporous membranes, the developed pore surface area is very large and the permeability is very low. Thus, the thickness of the separative layer must be thin enough to reach attractive fluxes with experimentally acceptable transmembrane pressure. On the contrary, the mechanical strength of the membrane must be large enough to withstand the applied pressure. These considerations led to the concept of an asymmetric structure based on a macroporous support and successive layers with decreasing thickness and pore size (Table 2.4).

Several old and recent reviews on inorganic membranes are available with parts devoted to the synthesis and applications of microporous amorphous silica membranes [1–4]. Microporous amorphous silica exhibits a weak stability in aqueous solutions or wet streams which limits its use as separative membranes to specific applications. The main current developments deal with gas separation applications, in particular hydrogen separation. The thermal stability of silica membranes compared with organic and hybrid layers and their interconnected porosity inside a stiff oxide skeleton enable to reach very good permeability–selectivity balances. For instance, H_2 permeance of the order of 10^{-6} mol/m^2/s/Pa with a H_2/N_2 selectivity up to 100 has been recently reported in [4]. Other potential applications are pervaporation or nanofiltration of the organic mixtures. In the case of nanofilters, it must be noted that requirements in terms of pore size are not so much restrictive and it enables to take advantage of the preparation of silica-based mixed oxide membranes to adapt the separative properties to the treated solvents [5].

After an introduction on the silica properties, this chapter will present the synthesis methods used to prepare supported amorphous silica membranes. Finally, silica membrane designs and associated separative properties will be detailed.

Table 2.2 Main separative processes using ceramic membranes

Process	Nature of feed/strip	Pore size	Origin of selectivity	Pressure gradient	Elemental operation
Microfiltration, MF	Liquid/liquid	0.1–10 μm	Sieving effect	1–3 bars	Clarification, debacterization, separation
Ultrafiltration, UF		2 nm–0.1 μm		3–10 bars	Clarification, purification, concentration
Nanofiltration, NF		<2 nm	Sieving + specific interactions with the membrane	10–40 bars	Purification, water softening, separation, concentration
Pervaporation, PV	Liquid/gas	<2 nm	Sieving + additional specific interactions	1 bar	Separation
Gas filtration, GF	Gas/gas	100 μm–0.01 μm	Sieving effect	0.1–5 bars	Separation, dusting
Gas separation, GS		50 nm–<2 nm	Sieving + additional specific interactions	0.1–50 bars	Separation, extraction, purification
Gas separation, GS		Dense	Ionic conduction of O^{2-} or protons by oxides	$\Delta P(O_2)$ or $\Delta P(H_2)$	Air separation or H_2 purification transport of O_2 or H_2

Table 2.3 Gas transport mechanisms inside porous materials and related selectivity

Type of membrane	Transport mechanism	Selectivity
Macroporous, >50 nm	Viscous flow 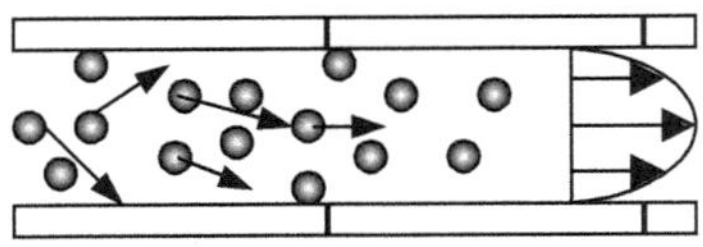	–
	Molecular diffusion 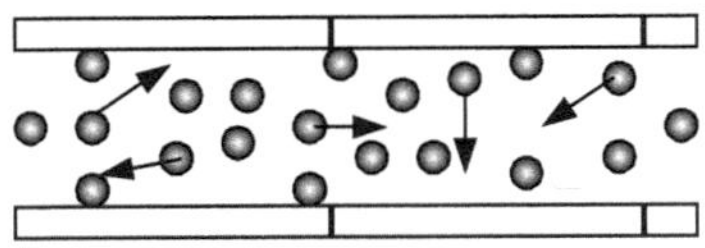	–
Mesoporous, 2 nm–50 nm	Knudsen diffusion 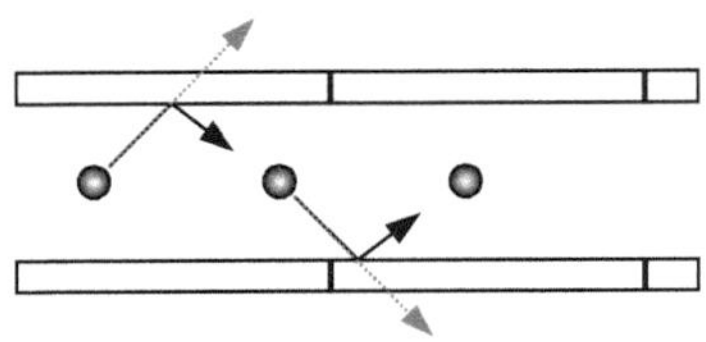	$\alpha^*_{A/B} = \sqrt{\frac{M_B}{M_A}}$
	Surface diffusion 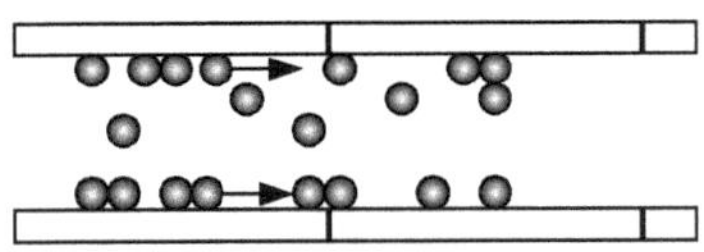	+
	Capillary condensation, pore filling 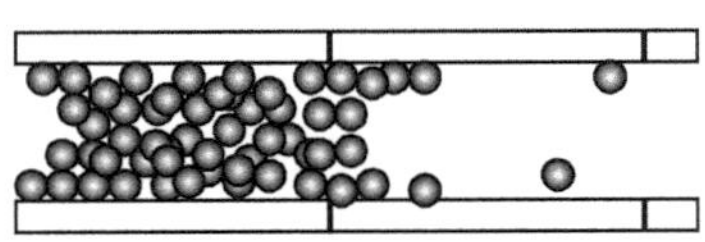	++
Microporous, <2 nm	Micropore diffusion 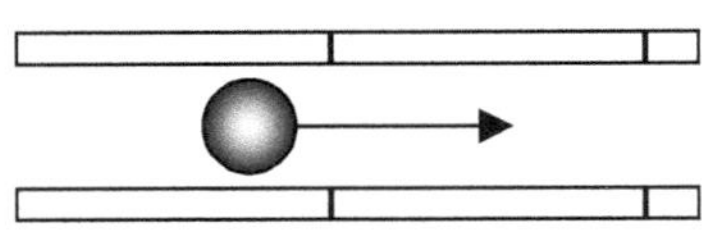	++

Table 2.4 Characteristics of the intermediate and of the separative top layers (from [1])

Process	Average number of layers	Thickness of the separative layer	Pore size in the separative layer	Nature of the porosity
Microfiltration	1–3	Few ten micrometers	5–0.1 μm	Macroporosity
Ultrafiltration	3–4	Few micrometers	5 nm	Mesoporosity
Nanofiltration/ gas separation	4–5	<1 μm	1 nm	Microporosity

2. Specific Properties of Amorphous Silica: Comparison with Other Oxides

Silica is a well-known glass-forming oxide whereas Al_2O_3, TiO_2, and ZrO_2 are intermediate oxides in the classification of Zachariasen [6] (Table 2.5). The latter cannot form glasses by themselves. The structure of silica glass consists of corner-sharing SiO_4 tetrahedra covalently connected in a continuous infinite three-dimensional (3-D) network having no long-range order (Fig. 2.1).

Conventional preparation of glasses is performed by cooling melted solids, but amorphous phases can also be obtained at lower temperature from liquid solutions or gaseous mixtures. Amorphous phases are intrinsically metastable phases (Table 2.6). Thermal crystallization of intermediate oxides occurs at a lower temperature compared with that of silica (Table 2.6). Crystallization is systematically associated with a microstructural change. In the case of microporous solids, it usually results in a significant enlargement of pore sizes. For high-purity silica, crystallization can be delayed at a very high temperature (close to the domain of the thermodynamic stability of cristobalite). The thermal stability of porous silica is in fact limited by viscous flow sintering occurring at around 800 °C in the case of microporous silica (Fig. 2.2). At lower temperature (from 200 °C), low shrinkage phenomena can be observed due to condensation reactions between silanol groups (≡SiOH).

The strong effect of temperature and pH on the solubility of amorphous silica in water (Fig. 2.3) clearly explains the limited use of silica membranes in aqueous

Table 2.5 Classification of the oxides versus glass formation ability

Former	Modifier	Intermediate
SiO_2	Li_2O	Al_2O_3
GeO_2	Na_2O	PbO
B_2O_3	K_2O	ZnO
P_2O_5	CaO	CdO
As_2O_5	BaO	TiO_2
V_2O_5		ZrO_2

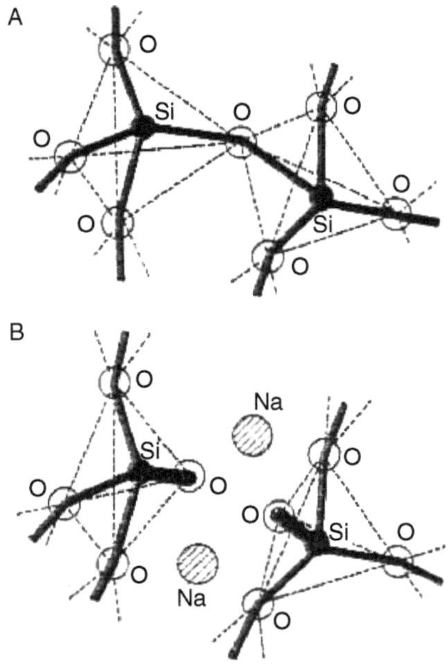

Figure 2.1 Silica structure, (A) corner-sharing SiO_4 tetrahedra, (B) nonbridging oxygen pair associated with the presence of modifier Na_2O (from [7]).

solutions or wet atmospheres. In addition, dissolution–redeposition phenomena are enhanced in the case of microporous materials associated with surface nanometric radii of curvature (Fig. 2.4). The solubility strongly decreases for methanol-rich water–methanol mixtures, whereas for pure alcohols it also decreases with the length of the alkyl chain (Table 2.7). This explains the preservation of the silica microporosity in organic media.

3. Synthesis Methods

Homogeneous and defect-free amorphous silica films can be deposited on porous substrates using sol–gel routes or chemical vapor deposition (CVD) methods to produce asymmetric membranes. We will discuss the synthesis conditions required to obtain microporous amorphous thin layers.

3.1. Sol–gel routes

The sol–gel route is a very convenient method to prepare porous thin layers of ceramic oxides. Fluid starting sols can be deposited on dense or porous substrates leading to defect-free submicrometer thin films. However, the porosity in the final layers depends drastically on the various synthesis parameters. The requirements

Table 2.6 Structural data for common single oxides

	SiO_2	Al_2O_3	TiO_2	ZrO_2
Crystalline phase thermodynamically stable at room temperature	Quartz α	Corundum	Rutile	Baddeleyite
Thermal crystallization of amorphous phases	In the range of 900–1400 °C to cristobalite	In the range of 400–700 °C to transition aluminas	In the range of 150–400 °C to anatase or rutile	In the range of 200–400 °C to metastable cubic or tetragonal phases
Aging of amorphous phases in aqueous solution	No structural evolution	Crystallization of hydroxides $AlO(OH)$ or $Al(OH)_3$	Crystallization of anatase or rutile	Formation of hydrated oxyhydroxides

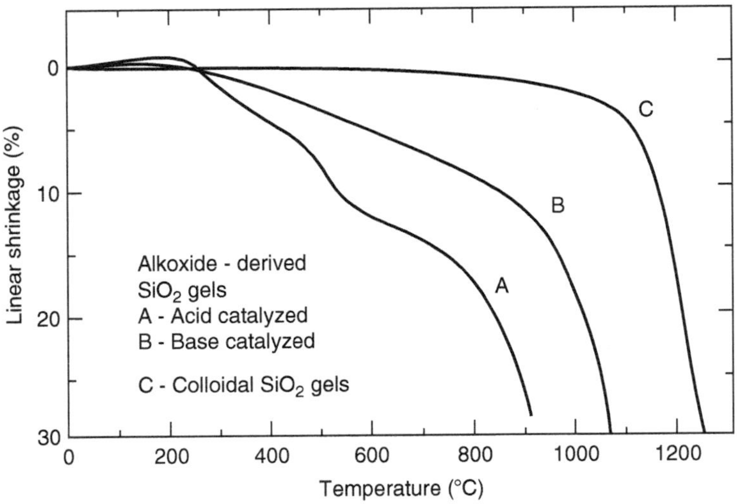

Figure 2.2 Percent linear shrinkage for silica xerogels prepared by two-step acid-catalyzed hydrolysis of tetraethoxysilane, TEOS (water/TEOS molar ratio, $r = 5$) (A); two-step acid–base catalyzed ($r = 3.8$) (B); or destabilization of a particulate sol from Aerosil® (C). Heating treatments were performed at 2 °C/min in air (from [8]).

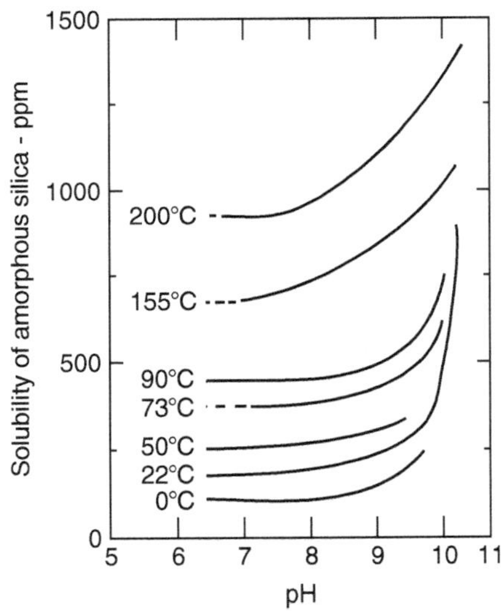

Figure 2.3 Solubility of amorphous silica versus pH at different temperatures (from [9]).

associated with the final application of the layers allow to define an ideal porosity in terms of total porosity, pore size, pore size distribution, connectivity, and tortuosity of the porous network.

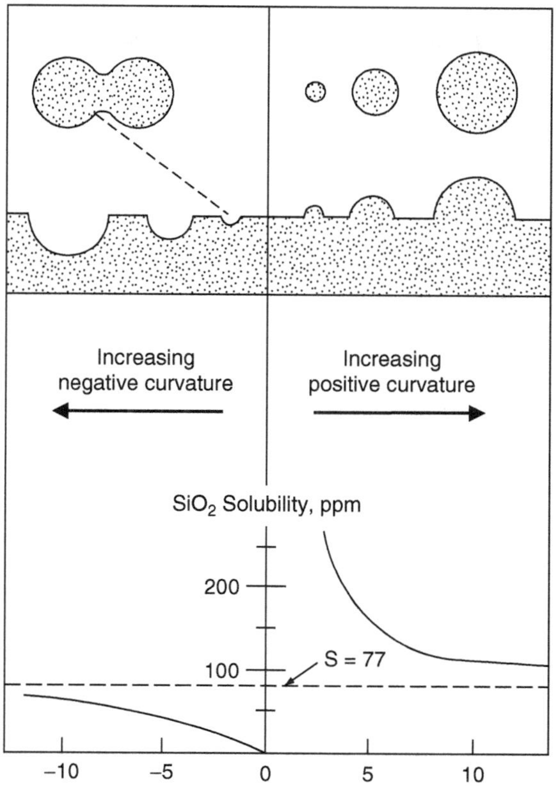

Figure 2.4 Variation in the solubility of silica with radius of curvature of surface. The positive radii of curvature are shown in cross section as particles and projection from a silica surface. Negative radii are shown as depressions or holes in the silica surface, and in the crevice between two particles (from [9]).

Table 2.7 Solubility of amorphous silica in alcoholic solutions (from [9])

Wt. % of methanol in water–methanol mixtures	Solubility at 25 °C (mg/liter)	Anhydrous alcohol	Solubility at 500 °C (ppm)
0	140	CH_3OH	1890
25	75	C_2H_5OH	164
50	40	n-C_3H_7OH	8
75	15		
90	5		

3.1.1. Conventional sol–gel routes

The initial choice is related to the nature of the precursor of the silica network. Two main types of sol–gel routes can be distinguished. In the polymeric route, the elemental unit which is initially used is a molecular precursor, usually a silicon alkoxide, $Si(OR)_4$. The most common precursors are tetramethoxysilane

$Si(OCH_3)_4$ (TMOS) and tetraethoxysilane $Si(OC_2H_5)_4$ (TEOS) in solution in their associated alcohols. In the colloidal route, the elemental unit is a solid nanoparticle dispersed as stable sol in a liquid. The sol stability is related to electrostatic and steric repulsive interactions. Various silica hydrosols are commercially available with an average particle size ranging from 10 to 40 nm.

These elemental units or clusters of elemental units are the bricks from which the oxide network is built. The choice of the size of the brick will first define the size of the pores in the layer. For spheres with a single particle size (radius R), an ordered close-packed arrangement (face-centered cubic or hexagonal) gives rise to a porosity between the bricks of 26% and to two types of pores with a radius equal to $0.225R$ (tetrahedral sites) and $0.414R$ (octahedral sites), respectively. For a dense random-packed arrangement, the lowest limit for the porosity is 36% with a pore size distribution ranging from 0.2 to $0.6R$ [10]. In the case of a distribution of particle sizes, denser packing can be obtained [11] and the pore size distribution depends strongly on the composition of the mixture of spheres [12] (Fig. 2.5).

From this first set of data, it appears clearly that the preparation of mesoporous layers (pore size between 2 and 50 nm) can be considered using colloidal dispersions. For the preparation of microporous layers (pore size <2 nm), the most convenient way is the polymeric route from solutions of molecular precursors like alkoxides (Table 2.8).

The formation of the oxide network results in the above case from the polymerization of the molecular precursor. The hydrolysis of the alkoxide (a) produces activated species and the condensation by alcoxolation (b) or oxolation (c) leads to a reticulation by formation of siloxane bridges ≡Si-O-Si≡.

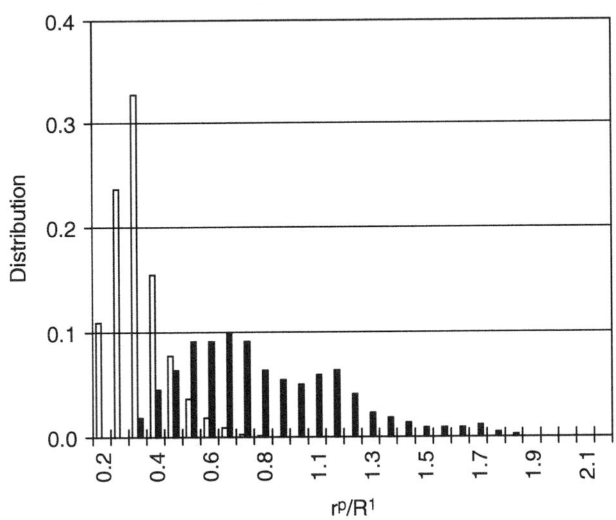

Figure 2.5 Relative pore radius distribution (r_p/R_1) for two random close-packed arrangements produced from a bimodal mixture of small spheres (radius R_1) and large spheres (radius $R_2 = 4R_1$) and exhibiting the same total porosity (30%); black bars: 10% of small spheres; white bars: 75% of small spheres (from [12]).

Table 2.8 Sol–gel process of silica from colloidal dispersions or molecular solutions

Sol–gel route	Elemental unit	Accessible porosity
Colloidal	Oxide nanoparticle	Mesopores
Polymeric	Alkoxide molecule $Si(OR)_4$	Micropores and/or mesopores

$$Si(OR)_4 + H_2O \rightarrow Si(OR)_3OH + ROH \quad (a)$$

$$\equiv Si - OH + RO - Si \equiv \rightarrow \equiv Si - O - Si \equiv + ROH \quad (b)$$

$$\equiv Si - OH + HO - Si \equiv \rightarrow \equiv Si - O - Si \equiv + H_2O \quad (c)$$

Because of the fact that the silicon alkoxide are not soluble in water, a common solvent is required which is usually the corresponding alcohol of the silicon alkoxide (Fig. 2.6).

A lot of different chemical environments for the silicon atoms can be simultaneously present in the alkoxide-derived solutions: $Si(OR)_x(OH)_y(OSi)_z$ with $x + y + z = 4$. Powerful analytical methods like ^{29}Si NMR (Table 2.9) or Raman spectroscopy can be used to follow the kinetics of hydrolysis and condensation in the gelling solutions [8].

The silica network is generated by polymerization of alkoxides and/or aggregation of silica clusters. It depends on the probability of aggregation during a contact between two species (reactivity) and on the diffusion of the elemental species and of the growing clusters (effects of dilution, viscosity, temperature). Various models were proposed which are associated with a limiting mechanism for the growth:

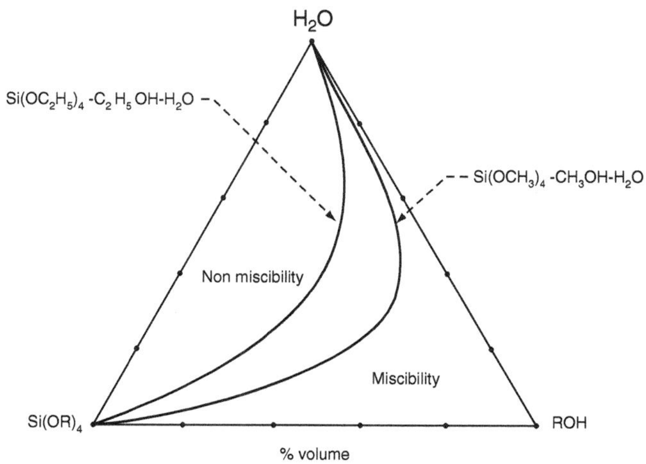

Figure 2.6 Miscibility ternary diagrams for $Si(OR)_4$–ROH–H_2O with R=CH_3, C_2H_5 (from [13]).

Table 2.9 ^{29}Si NMR chemical shift for silicon atoms in silica polymers; *Q* refers to silicon atoms bonded with four oxygen atom and the index corresponds to the number of siloxane bridges around Si (from [8])

	Q^0	Q^1	Q^2	Q^3	Q^4
δ (ppm/TMS)	$-73.1 < \delta < -82.0$	$-79.5 < \delta < -89.0$	$-82.0 < \delta < -96.2$	$-90.4 < \delta < -102.1$	-110

diffusion-limited monomer–cluster or cluster–cluster growth and reaction-limited monomer–cluster or cluster–cluster growth (Fig. 2.7). The resulting clusters or aggregates usually exhibit a fractal (or hierarchical) structure for scales higher than that of the elemental unit size (Fig. 2.7). Small angle X-ray scattering (SAXS) measurements are convenient to study *in situ* the mechanisms of growth and aggregation of the clusters (cluster size from the Guinier radius, cluster structure from the fractal dimension).

The hydrolysis and condensation reactions are catalyzed in acidic and basic media, respectively. The hydrolysis ratio *h* (water/alkoxide molar ratio) is another important parameter. Acid catalysis induces long gelation times, t_G. Coupled with a low hydrolysis ratio, it promotes the formation of linear polymers (Fig. 2.8). The formation of dense clusters and short gelation times t_G are favored by base catalysis. For a given alkoxide, differences in the synthesis conditions give rise to important differences for the porous texture of the final material as shown in Tables 2.10 and 2.11.

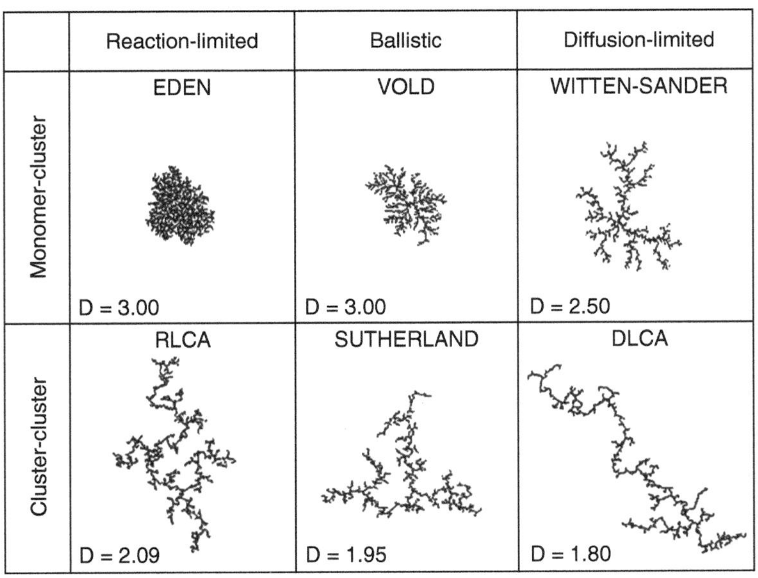

Figure 2.7 Aggregation models; *D* is the associated fractal dimension (from [14]).

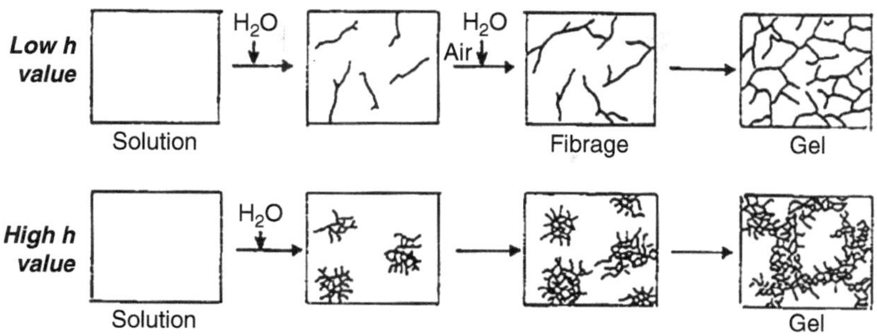

Figure 2.8 Influence of the hydrolysis ratio on the polymerization process (from [15]).

Table 2.10 Influence of the hydrolysis method on the porous texture of polymeric silica gels synthesized using TEOS (from [8])

Hydrolysis	Porosity (%)	Pore diameter (nm)	Specific surface area (m^2/g)
Two-step acid catalyzed	54	1–5	740
Two-step acid–base catalyzed	67	1–10	910
One-step base catalyzed	70	1–20	515

Table 2.11 Textural characteristics determined by thermoporometry on wet and dried silica gels prepared with various hydrolysis ratio, *h* (molar ratio water/TEOS) (from [16])

	Porosity (%)		Average pore diameter (nm)		Specific surface area (m^2/g)	
h	Wet	Dried	Wet	Dried	Wet	Dried
4	78	42	7	6.5	1012	296
6	78	42	9	6.5	720	337
10	82	47	10	6	812	518
15	77	45	11	5	618	452

In the case of thin layers, the deposition is carried out, preferably under clean room conditions, by dip-coating, slip-casting, or spin-coating (flat substrates only) with a fluid sol prior to the gelation (at $t < t_G$, gelation time). The dispersed clusters form a dense-packed layer on the substrate (suction capillary forces), while the smallest clusters, the solvent and the dissolved additives infiltrate the support. The reticulation occurs simultaneously with the drying of the layer which induces a concentration of the sol. The final porous texture is then strongly dependent on the relative kinetics of polymerization and solvent evaporation/evacuation.

An important parameter is the aging of the sol prior to deposition. This parameter defines the advancement of the polymerization process (cluster size) before the deposition of the sol and consequently the porous texture of the resulting layer (Table 2.12).

The porosity of the wet gels is filled by the liquid phase. Capillary stresses are applied on the solid network which shrinks. The porous texture is strongly modified by the drying step (Table 2.13). The monolithic structure of the bulk gels is usually not preserved as cracking occurs. Sol–gel films thinner than 0.5 μm can be dried without crack whereas for a thickness higher than 1 μm, cracking is usually observed [8].

The evaporation process can be reduced or stopped during and after the deposition in order to promote a reinforcement of the inorganic network prior to the drying stage. Table 2.13 illustrates the effect of wet gel aging on the porous texture of the final material.

Table 2.12 Aging of polymeric silica sols prior to film deposition; effect on the porous texture of the resulting layers (from [17])

Aging before deposition, t/t_G	Porosity (%)	Average pore diameter (nm)	Specific surface area (m^2/g)
0–0.15	–	<0.4	1–2
0.15	16	3.0	146
0.33	24	3.2	220
0.66	33	3.8	263
1	52	6.0	245

Table 2.13 Effect of aging prior to drying for wet silica gels synthesized by two-step acid–base hydrolysis of TMOS

Sample	Aging	Pore volume (cm^3/g)	Porosity (%)	μ-porosity (% of porosity)	Specific surface area (m^2/g)
a	Dried after deposition at $t \sim 0.05t_G$	0.68	60	86	1264
b	Dried immediately after the gelation	0.74	62	72	1106
c	Dried after 3 days of aging at 50 °C	1.23	73	42	1056

Acidic conditions with low HNO_3/TEOS or HCl/TEOS ratios, diluted sols, short aging times before deposition (in comparison to the gelation time in bulk solution), and low drying rates are required to get ultramicroporous silica layers.

3.1.2. Tailoring of the porosity in sol–gel-derived membranes

Templating strategies can also be advantageously applied to control the microporosity of the silica layers. The first way consists of using templating molecules (incorporated in the gelation medium) which are inert toward the chemical process leading to the inorganic network [route (c) and (d) in Fig. 2.9] [18]. An example illustrating such strategy deals with hydrolysis and condensation of silicon alkoxide in the presence of nonionic surfactants and more precisely octyl aryl polyether alcohols with a general formula C_8H_{17}–C_6H_4–$(OCH_2CH_2)_x$–OH where $x = 1$, 3, 9, 10, or 30 [19]. The porous characteristics of the materials after a thermal treatment up to 450 °C are reported in Table 2.14. While the materials synthesized without surfactant additives do not exhibit any measurable porosity by nitrogen adsorption, the results of Table 2.6 clearly show that both the porous volume and the micropore size can be tailored by the added amount of surfactant and by the length of the polyoxyethylene chain.

A second strategy consists of the use of modified alkoxides. A molecular group acting as a template is covalently bonded to the Si atoms. This synthesis method is also schematically represented in Fig. 2.9 [route (a)–(b)–(d)]. The synthesis conditions are always selected in order to directly associate the porosity of the final material with the thermal elimination of the templating species during stage (d). Several papers were published on the improvement of microporous silica layers starting from mixtures of TEOS with a modified alkoxide-like methyltriethoxysilane [20], methacryloxypropyltrimethoxysilane [21], phenyltrimethoxysilane, or methyltrimethoxysilane [22].

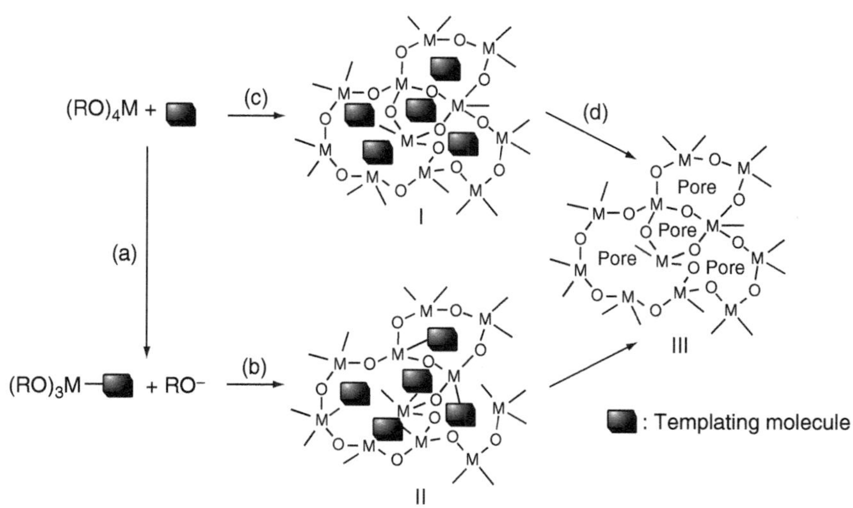

Figure 2.9 Schematic illustration of the organic/inorganic template strategies (from [18]).

Table 2.14 Effect of addition of nonionic surfactants (C_8H_{17}–C_6H_4–$(OCH_2CH_2)_x$–OH) on the porous texture of unsupported silica layers measured from N_2 adsorption–desorption isotherms (from [19])

			Porous volume (cm^3/g)			Hydraulic pore radius (Å)	
x	Surfactant/ TEOS molar ratio	S_{BET} (m^2/g)	Total	Microporous	Porosity (%)	Range	Mean
1	0.55	400	0.217	0.210	32.2	3–7	3.9
3	0.16	250	0.121	0.105	20.9	3–6	3.2
3	0.35	470	0.197	0.190	30.1	3–6	3.4
3	0.55	500	0.249	0.239	35.5	3–6	3.7
10	0.55	728	0.510	0.480	52.7	5–9	6.3
30	0.55	800	0.541	0.514	54.2	4–10	6.0

Potential increase of the permeability in such microporous layers can be derived from hierarchical porous structures with noninterconnected extraporosity at a larger scale. In such conditions, the permselectivity of the membrane is still defined by the microporosity of the continuous phase. This strategy is illustrated with mesophase-templated silica membranes resulting in ordered but not interconnected mesopores separated by microporous silica walls (Fig. 2.10) [23, 24].

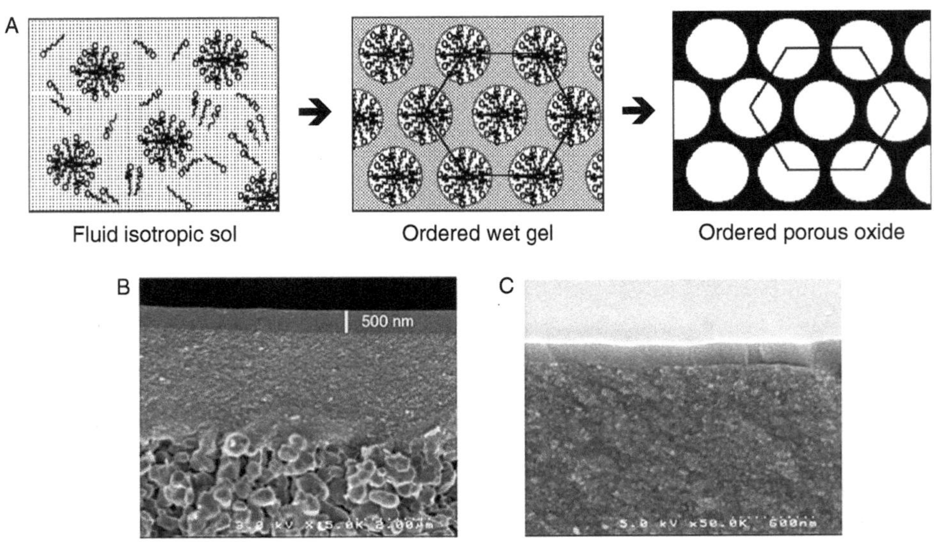

Figure 2.10 (A) Schematic representation of the sol–gel process using a hexagonal mesophase as structure directing template; (B and C) SEM images of (B) 2D hexagonal and (C) 3D hexagonal mesoporous layers deposited on an asymmetric substrate with a γ-Al_2O_3 top layer (from [23] and [24]).

Macroporous spherical pores can also be generated by the addition of spherical polystyrene latex in the silica sols (Fig. 2.11) [25]. The two additional levels of porosity can also be combined as recently demonstrated with titania layers (Fig. 2.12) [26].

It must be finally noted that hydrophobic sol–gel-derived silica membranes can be prepared by one-pot methods from mixtures of TEOS and modified alkoxides (for instance, see [27]) avoiding further heat treatment at temperatures higher than the thermal degradation of the organic groups directly bonded to Si (usually around 250 °C). Another route involves surface grafting from liquids or gaseous phases, using experimental procedures derived from the preparation of chromatography phases [28].

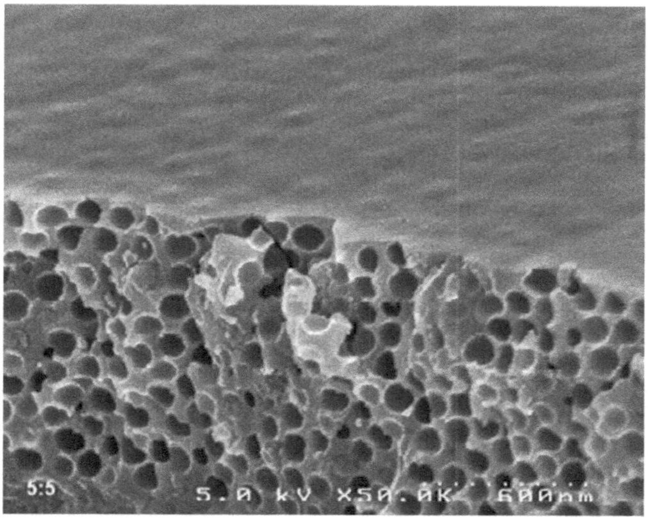

Figure 2.11 Silica layer with spherical macropores generated by templating with polystyrene latex (from [25]).

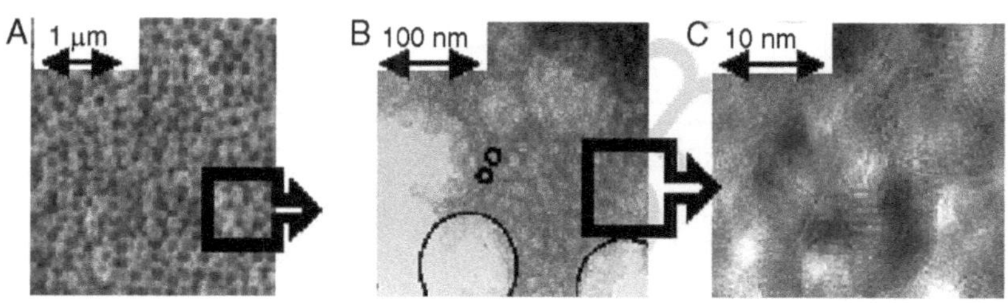

Figure 2.12 Hierarchical anatase-based membrane consisting of isolated spherical macropores (A), isolated mesopores (B), and nanocrystalline anatase walls with interconnected micropores (C) (from [26]).

3.2. CVD routes

In parallel to the sol–gel routes, CVD methods can be used and have been intensively investigated for preparing homogeneous and defect-free amorphous silica films and membranes.

3.2.1. Thermal CVD

CVD allows to deposit nonporous or microporous inorganic thin films. Hence, CVD can be used to modify the external surface of a substrate and also its mesopores or macropores. Several CVD techniques, detailed below, appear to be ideal techniques to control the pore size reduction of an inorganic membrane in order to control the permeance of the species going through. Steric obstruction of the pores covered by a CVD thin film enhances the selectivity for the small species.

CVD reaction, to obtain a silica film, consists of the thermal decomposition of a silicon-based precursor and the chemical reaction with an oxidant gas in contact with the hot substrate where the growth of the film takes place. For CVD silica membranes, the gaseous precursors are mainly TEOS and TMOS fed to the hot surface by argon or nitrogen as carrier gas. Other possible precursors are silicon tetrachloride ($SiCl_4$) or silane (SiH_4), whereas oxidant gases are air, pure O_2, O_3, N_2O, or water vapor.

CVD methods, to prepare a membrane on a porous substrate, can be classified into two categories, based on the supplying configuration of the precursors: co-current CVD or counter-diffusion CVD, also mentioned as chemical vapor infiltration (CVI). In co-current CVD, the precursors are provided from one side of the substrate [29–32], while the other side of the substrate is usually vacuumed to obtain a pinhole free membrane [33–36] (Fig. 2.13).

Lee *et al.* [32] used an argon flow as carrier gas bubbled in TEOS at 25 °C for the CVD reaction at the inner side of a porous alumina support tube (average pore size 5 nm) heated at 600 °C. The CVD process was interrupted at various times and the permeance of H_2, CH_4, CO, and CO_2 were measured. The hydrogen selectivities over such gases are shown as a function of hydrogen permeance in Fig. 2.14.

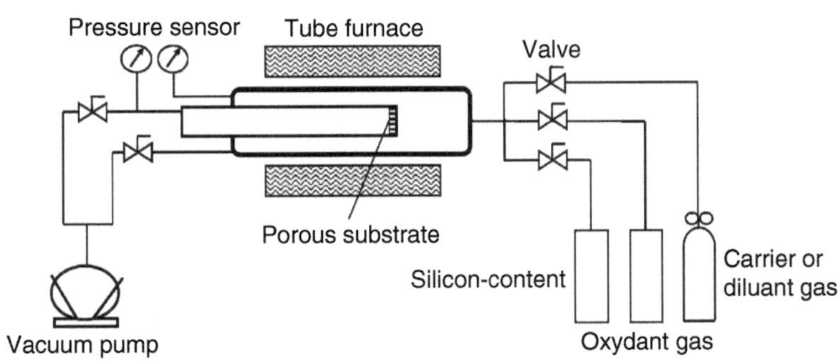

Figure 2.13 Experimental setup for chemical vapor deposition (CVD) of silica membrane on a porous substrate (from [36]).

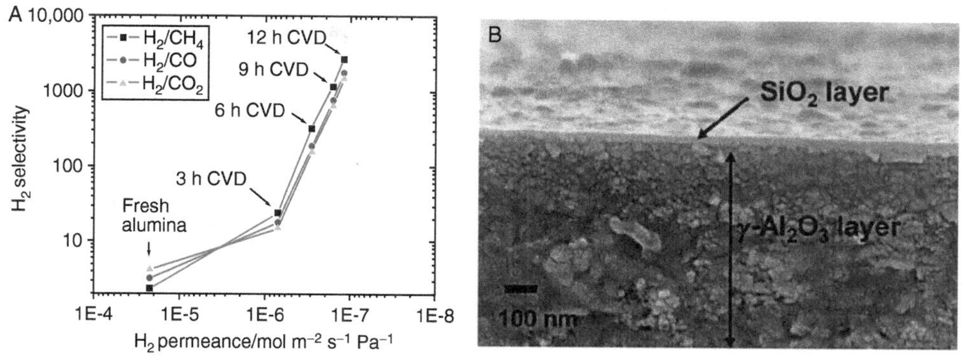

Figure 2.14 (A) Dependence of the deposition time (CVD, 600 °C) on the hydrogen selectivity and permeance. (B) SEM cross-sectional photograph of the same silica membrane (from [32]).

By changing the flow of the reactants (pure or diluted in argon or helium carrier gases), it is possible to control the deposition on the poremouth of the substrate. Depending on the pore size, the reduction of the permeation during CVD reaction can be drastically different. Figure 2.15A presents the slow decrease of He and N_2 permeances as a function of the deposition time on a symmetric α-Al_2O_3 porous support with an average pore size of 100 nm (membrane M1; [34]). Figure 2.15B shows the reduction of permeance as a function of the deposition time on an asymmetric support composed of three superimposed layers with an average pore size of 1 μm, 80 nm, and 10 nm (membrane M2). In this case, a drastic permeance reduction suddenly occurs corresponding to the mesopore blocking and leading to a high He/N_2 selectivity.

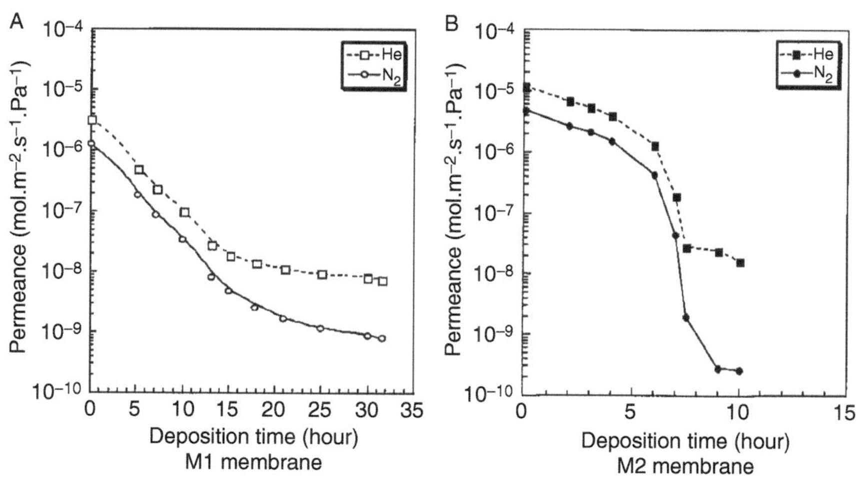

Figure 2.15 The relationship between permeances of He and N_2 and deposition time (CVD, 600 °C) depending on the average pore size of the support (from [34]). (A) Symmetric α-Al_2O_3, 100 nm pore size; (B) Asymmetric with 3 layers, 1 μm, 80 nm, 10nm pore sizes.

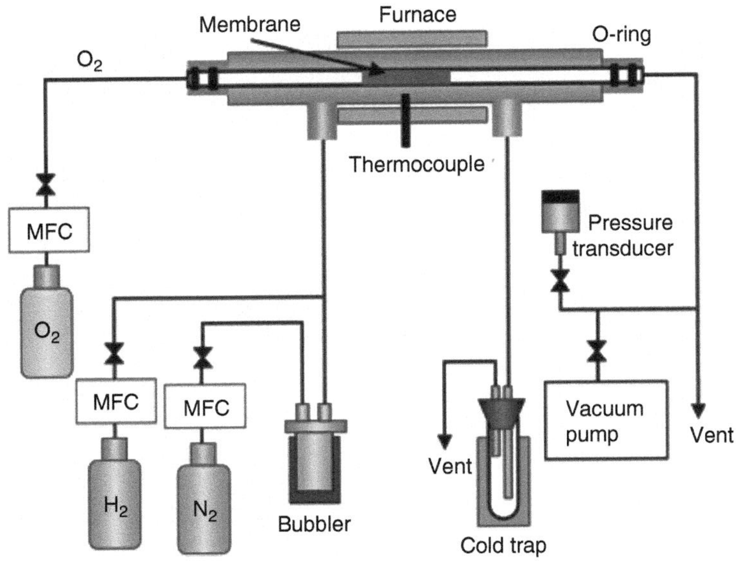

Figure 2.16 Experimental setup for tubular silica membrane deposition by counter-diffusion chemical vapor deposition (CVD) (from [43]).

The second method is counter-diffusion CVD (Fig. 2.16) where the two reactants are supplied from the opposite sides of the substrate [37–45]. These reactants diffuse into a porous substrate and the silica deposition occurs in the pores of the substrate. The deposition finishes by the limitation of supply of the reactants due to the reduction of the pore diameters.

Stable silica membranes can be prepared by counter-diffusion CVD treatment at 600 °C inside α-Al_2O_3 tube [42] with an average pore diameter of 100 nm or inside γ-Al_2O_3 mesoporous tube with an average pore diameter of 4 nm [43] from TMOS and O_2. Pore sizes and effective membrane thickness can be controlled by changing reactants nature and reaction conditions: pressures, flows, gas dilutions, temperature of decomposition, time of deposition.

Vycor glass tubes have also been extensively used as substrates to synthesize H_2 permselective silica membrane using TEOS/O_3 [40]. Reaction procedures were optimized by varying the CVD temperature (150–250 °C) and time and also the bubbling temperature of TEOS (40–100 °C) to increase its partial pressure carried by argon up to the CVD chamber. Other groups deal with the synthesis of silica membranes inside a γ-Al_2O_3 top layer deposited on a flat support from SiH_4 and O_2 at 250 °C [45].

Some low-temperature CVD techniques have been investigated to reduce the cost due to the use of a high-temperature CVD reactor. CVI of silicon tetra-acetate ($SiAc_4$) at 250–275 °C carried by a helium stream has been used by Nijmeijer *et al.* [45] to modify a γ-Al_2O_3 membrane. Catalyzed atomic layer deposition (ALD) at room temperature has been also employed to progressively reduce mesoporous silica membranes by successively cycling $SiCl_4$ and H_2O, both bubbled in nitrogen flows

[46]. Each precursor is provided to the reaction surface at a high pressure and then purged before providing the other precursor. The ALD technique, involving a layer-by-layer growth, allows to better control the growth which is very slow compared with that of CVD.

The two successive surface reactions are:

$$\mathrm{Si-OH_{/surface} + SiCl_4 \rightarrow SiOSiCl_{3/surface} + HCl} \quad \text{(a)}$$

$$\mathrm{SiCl_{/surface} + H_2O \rightarrow SiOH_{/surface} + HCl} \quad \text{(b)}$$

Pyridine allows the reaction (a) to take place at near-room temperature. One ALD cycle [reactions (a) + (b)] is estimated to deposit ~0.21 nm of silica. Twenty cycles are needed to reduce the mesopore size of the substrate down to diameters where reactant exclusion takes place. A higher number of cycles is needed to reduce larger pores and defects.

Coupled deposition methods have also been tested. Catalytic composite silica membrane can be prepared by dispersing catalytic nanoparticles (rhodium) in the macropores of the support before a deposition of a silica top layer by counter-diffusion CVD [44]. This method is adapted to membrane preparation but also to the post-treatment [47] or improvement of ceramic membrane properties as in the case of zeolite membranes [23].

3.2.2. Plasma-enhanced CVD

Compared with thermal CVD, plasma-enhanced CVD works at low temperature which allows depositing inorganic or hybrid thin films on nonrefractory materials. Until now, the PECVD technique cannot deposit a thin film on the pore walls inside a porous substrate, like it can be done by counter-diffusion CVD. The main advantage of this surface deposition technique is to be able to control the decomposition of gaseous precursors in order to deposit hybrid thin films with high thermal and chemical stabilities.

Plasmas can be created at laboratory and industrial scales by applying direct or alternative high voltage to a gas. Depending upon whether the gas is at reduced pressure or at atmospheric pressure, it is possible to obtain a low-temperature nonequilibrium glow discharge plasma or equilibrium thermal plasma. The non-equilibrium plasma is of particular importance for technological applications. The highly energetic electrons are responsible for initiating chemical reactions by breaking covalent bonds of gas molecules. Through selection of gas or gas mixture, method for energy input, and reactor geometry, it is possible to optimize the conditions for a wide range of materials and processes for technological applications particularly in microelectronics and optics and also for membrane processes.

Plasma-enhanced chemical vapor deposition (PECVD) presents several possible advantages. The process is "dry," that is, does not produce liquid effluents. The deposition rate can be high (several micrometers per minute) [48, 49], making the process ideal for industrial applications. Plasma generates highly reactive species like ions, free radicals, and metastable species allowing reactions to occur at much lower temperatures than when using conventional methods. Finally, PECVD is a convenient technique for layer fabrications since one can have easy control over factors like thickness, refractive index, and roughness of the layers. Hence, PECVD technique was advantageously employed for the low-temperature deposition of hybrid silicon-based membranes [50–54] up to silica ones [55–57].

Numerous parameters can be controlled by PECVD. The nature of the deposits depends on the reactor shape, the power W, and frequency of the electric current applied to form the plasma phase; on the precursor nature and its flow F and pressure; on possible oxidant or nonreactive dilutions at the substrate temperature where the deposition occurs, and on the process duration. Yasuda [58] has introduced a simple parameter W/FM, where F is the precursor flow, M is its molar weight, and W is the radiofrequency power, which helps to compare PECVD deposits issued from the same reactor and the same power frequency.

Figure 2.17 shows two extreme PECVD conditions. At low Yasuda's parameter, that is, in "soft conditions," the precursor is slightly fragmented, as the plasma energy is low. Most of the fragments, whose chemical architecture is close to that of the precursor (here called "monomer"), stick the substrate and form the amorphous deposit due to their low mobility at low temperature. In that case, the deposit

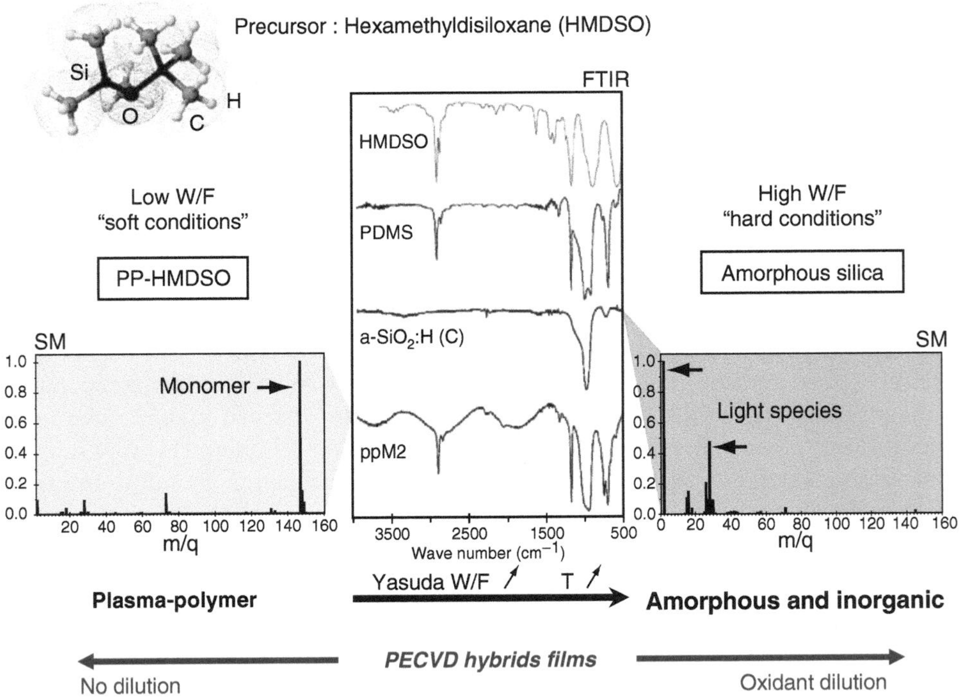

Figure 2.17 Extreme conditions to obtain a plasma polymer (PP-M2) and an amorphous silica film from the same precursor hexamethyldisiloxane (HMDSO).

is a plasma polymer (PP), mainly different from a conventional polymer due to its high cross-linking and its 3-D growth. Figure 2.18 is a representation of the microstructure of such PP-M2 material obtained from a vapor of hexamethyldisiloxane (M2). Fourier transform infrared (FTIR) spectroscopy analysis (Fig. 2.17) shows that the deposit is close to PDMS, according to their chemical bonds. *In situ* mass spectrometry spectra of the stable fragments of the plasma reveal the low fragmentation of the precursor.

At high Yasuda's parameter, that is, in hard conditions, the precursor is strongly fragmented. The resulting deposit is inorganic and generally amorphous. If the highly fragmented precursor is diluted with a reactive gas like e.g. O_2, NH_3, CH_4, the deposits are oxides, nitrides, or carbides, respectively. In the case of any silicon-based precursor diluted with O_2 or N_2O, the deposit is amorphous silica, as its chemical architecture is completely lost by high fragmentation. Carbon-based fragments coming from the condensable precursor are generally evacuated to the exit of the reactor in the form of gaseous species (hydrocarbons, CO_2,...).

Thermal stability can be high, even if the deposit has been formed at low temperature, due to the nonthermodynamical equilibrium of low temperature plasmas. Finally, silicon-based material density ranges from 1.1 for plasma-polymerized a-SiOC:H films up to 2.2 for amorphous silica; these values have been obtained by the grazing incidence X-ray reflectometry (XRR) technique.

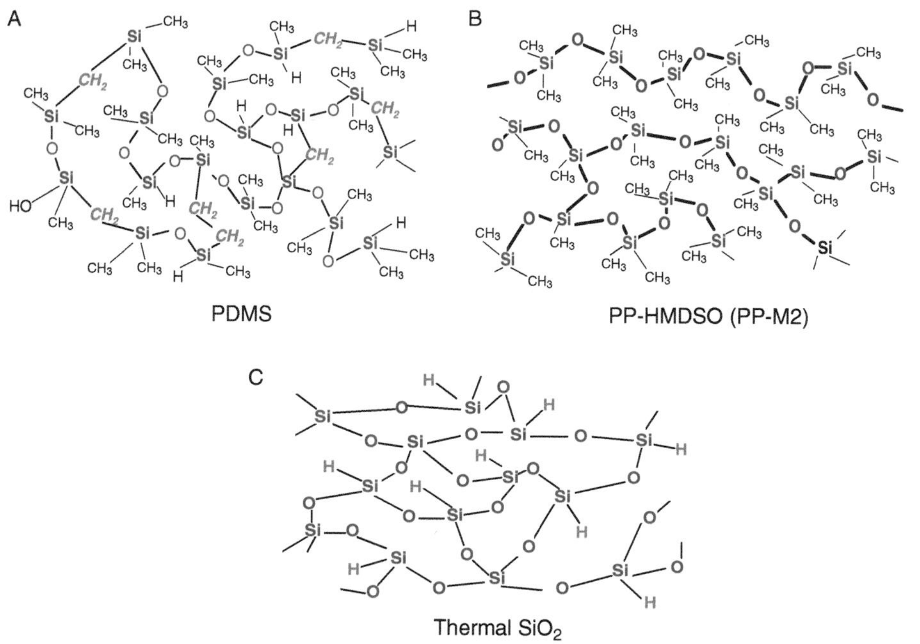

Figure 2.18 Schematic representations showing the differences between the chemical bond networks of a silicon-based plasma polymer (B) compared with that of the PDMS conventional polymer (A) and that of an amorphous hydrogenated silica (C).

According to the deposition conditions, permeabilities of plasma membranes are presented in Fig. 2.19. It is clear that in plasma polymers, the permeance properties of the deposited material are strongly dependent on the starting precursor vapor, especially starting from cyclic siloxanes which are the origin of the free volume. For drastic plasma conditions, the gas diffusion coefficients drop with the Yasuda's parameter (*W*/*FM*) because of the important densification of materials: the chain cross-linking is the limiting factor in this region. Whatever the monomers, H_2 logically shows the best mobility.

The growth of a silica membrane by PECVD inside a long tubular support is possible with the setup shown in Fig. 2.20A [59]. The setup consists of a full concentric reactor with a noncondensable/reactive plasma top zone. The oxygen plasma is moved from this zone toward the inner surface of the support up to the deposition zone where the condensable precursor (organosilicon) exits from a capillary. The capillary diameter is small enough to avoid any deposition inside, even if the capillary is crossing the full plasma top zone. Figure 2.20B shows the local deposition profile of the PECVD layer at the bottom end of the capillary. To obtain a regular thin PECVD layer (Fig. 2.20C), the support is moved along its axis.

New methods to produce porous silica or hybrid silica membranes by PECVD have been recently studied. They are based on the use of two vapor precursors: one to form the silica-based skeleton and the other one, generally a hydrocarbon with a low thermal stability will form the porosity or will initiate a high roughness leading to a porous film.

In the first case, the two precursors are co-deposited by PECVD. A further thermal or UV treatment is applied to remove most of the organic part issued from the hydrocarbon plasma polymerization, often called the porogen [60]. Open mesoporosity is then created (Fig. 2.21) [61].

In the second case, depositions of the two precursors are sequentially and cyclically applied, changing the energetic plasma conditions at each successive layer: an inorganic a-SiO_2 layer and a plasma-polymer a-C:H one. Both deposition

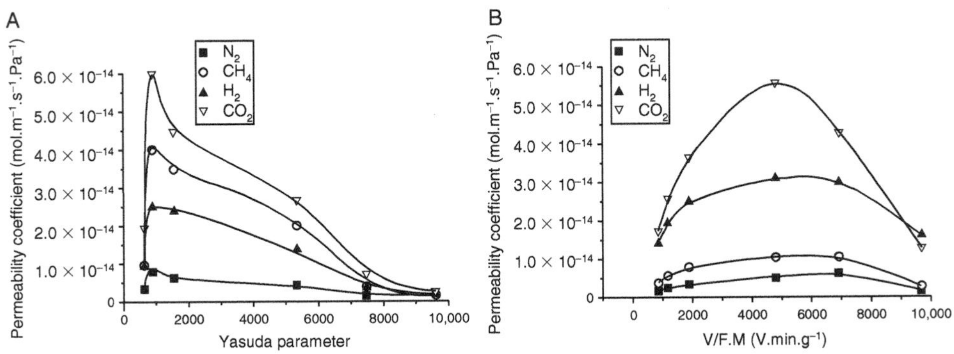

Figure 2.19 Permeability coefficient of different plasma-enhanced chemical vapor deposition (PECVD) membranes from plasma polymer like a-SiOC:H up to amorphous silica a-SiO_2:H, (A) starting from hexamethyldisiloxane precursor, (B) starting from a cyclic precursor: octamethylcyclotetrasilaxane $[OSi(CH_3)_2]_4$ (from [53]).

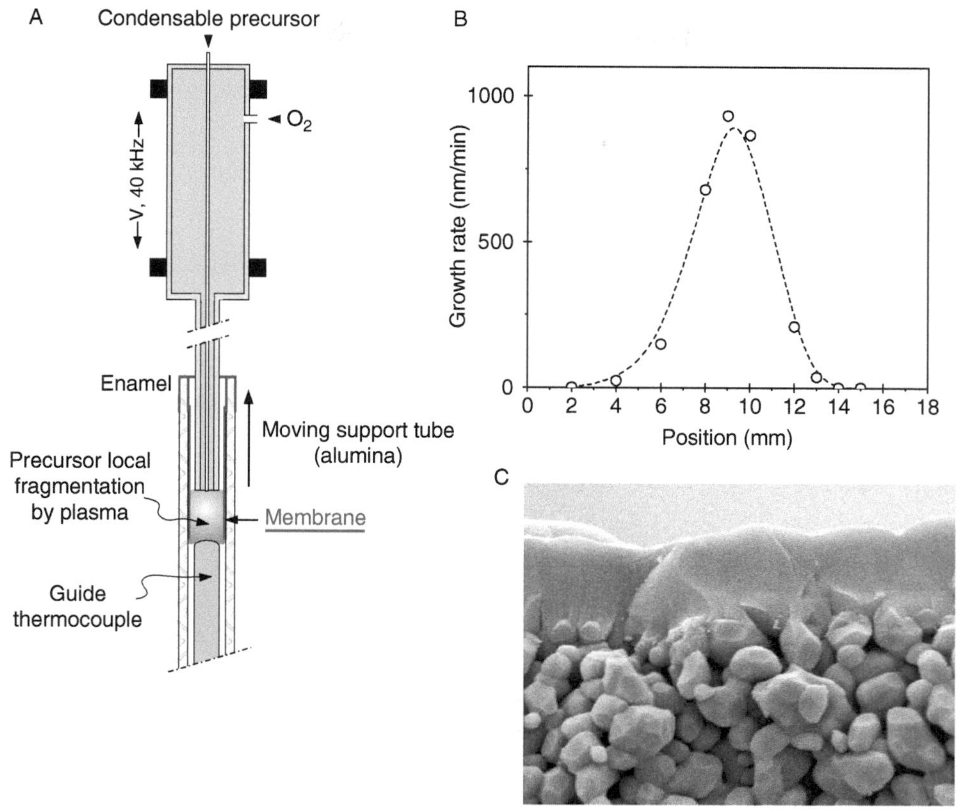

Figure 2.20 Plasma-enhanced chemical vapor deposition (PECVD) setup to deposit thin films or membranes inside a tube (A). Growth profile of the PECVD layer (B). Tubular support is regularly moved to obtain a homogeneous membrane (C).

of a-SiO_2 and etching of a-C:H simultaneously take place. Figure 2.22 shows the deposition process used to obtain such porous PECVD membranes. Porosity and pore size can be controlled by the PECVD deposition times during each cycle [62].

Surface-modified silica membrane by plasma is a way to graft specific groups on the surface in order to modify the chemical affinity to fluid in contact and also to modify surface density or reticulation of hybrid silica membranes. Some groups have studied several attempts using different kinds of plasmas of noncondensable gas. Hydrophilic or hydrophobic surface behavior can be rendered to a silica membrane surface with plasmas of H_2O, NH_3, or fluorocarbons vapor, respectively. For instance, a CF_4/Ar plasma treatment can further increase hydrophobic properties of PECVD hybrid silica changing both the surface morphology and the surface chemical terminations [63]. Figure 2.23 shows Atomic Force Microscopy (AFM) images of the surface and local adhesion forces measured also by AFM, compared with water contact angles. X-ray Photoelectron Spectroscopy (XPS) analyses show the surface chemical grafting of CF groups.

Recently, surface modification by ammonia-based plasma has been shown to drastically reduce the permeance of ethanol through a hybrid mesoporous silica

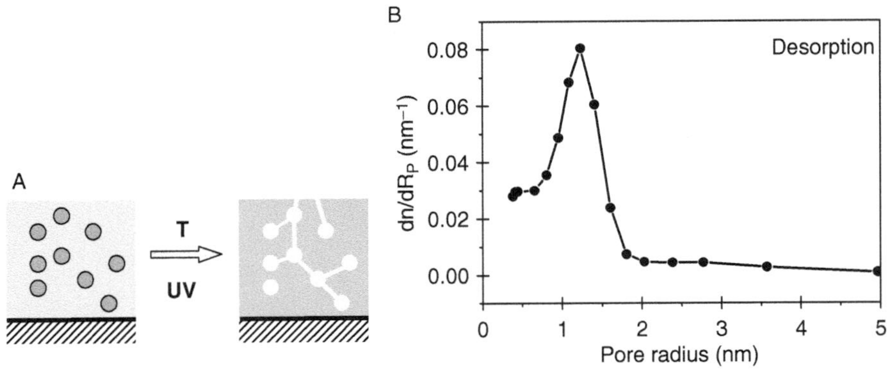

Figure 2.21 Plasma copolymerization (A). Pore size distribution of such plasma-enhanced chemical vapor deposition (PECVD) membrane (B) (from [61]).

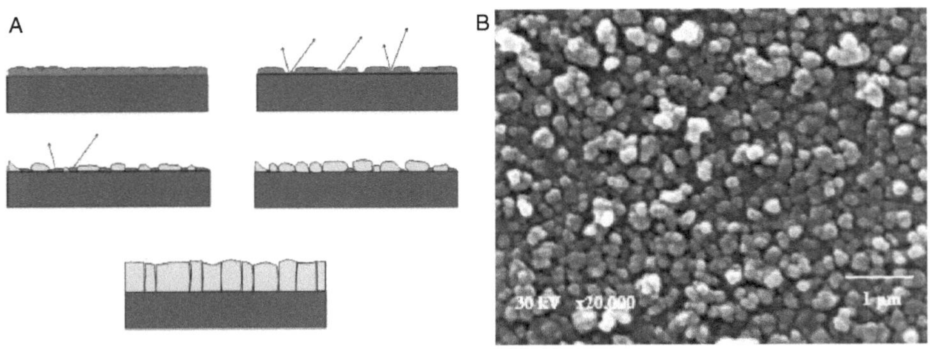

Figure 2.22 Sequential and cyclic plasma-enhanced chemical vapor deposition (PECVD) of two precursors (A) to obtain porous silica membranes (B) (from [62]).

membrane by creating a thin surface barrier [61]: $\pi = 7.8 \times 10^{-11}$ mol/m^2/s/Pa, much lower than ethanol permeance of an unmodified mesoporous (2 nm) membrane.

4. Design and Performance of Microporous Silica Membranes

The characteristics, performance, stability, and industrial potentialities of microporous silica membranes are directly related to the selection of an appropriate synthesis methodology, to the selection of relevant reactants, and to the membrane design. For the preparation of high-quality membranes, high-quality supports are needed and the thermochemical stability of both the membranes and the support have to be adapted to the process conditions used. Another key issue for membrane

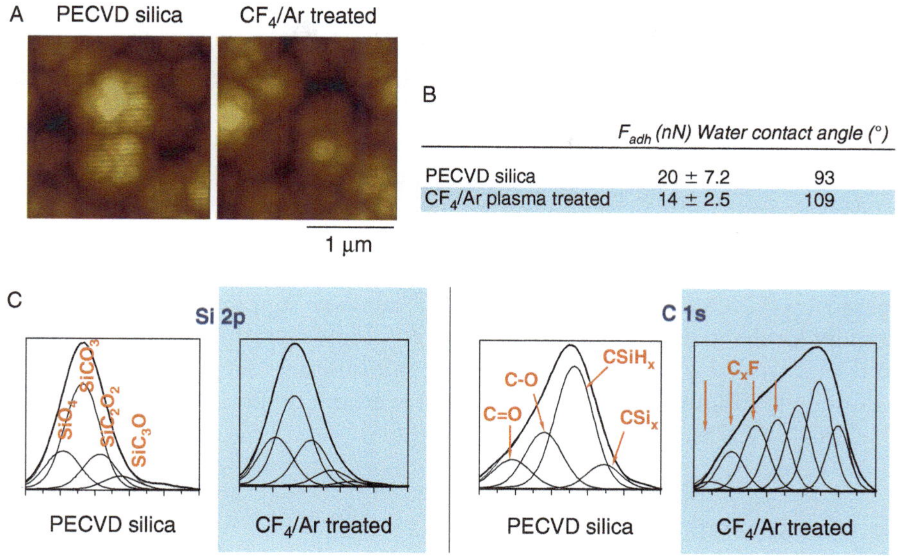

	F_{adh} (nN)	Water contact angle (°)
PECVD silica	20 ± 7.2	93
CF_4/Ar plasma treated	14 ± 2.5	109

Figure 2.23 (A) AFM images of the surface of plasma-enhanced chemical vapor deposition (PECVD) hybrid silica before and after a CF_4/Ar plasma treatment. (B) Water contact angles compared to adhesion forces measured by AFM. (C) XPS analysis in the silicon 2p and carbon 1s areas (from [63]).

development is the engineering of compact modules that are able to operate in harsh conditions.

A number of reviews partially or entirely dedicated to the design and performance of microporous silica membranes can be found in the recent literature [64–68]. On the basis of the state of the art, the present paragraph will highlight recent developments and related strategies.

4.1. Silica membrane applications

Ultramicroporous amorphous silica membranes, with their almost-dense structure deposited as thin films on strong and permeable porous ceramic supports, are able to sieve molecules bigger than CO_2 (kinetic diameter, 3.3 Å), provided that crack-free membranes can be prepared. Both pervaporation and gas separation applications have been considered for these membranes.

4.1.1. Pervaporation

Since their early synthesis, the potentialities of sol–gel-derived microporous silica for pervaporation applications have been investigated [66, 69], for example, for the dehydration of organic or organic acid solutions and for the separation of alcohol/aromatic organic solvents mixtures.

A number of groups tried to develop silica membranes at the semi-industrial and industrial level. Pervatech BV was founded in 1999 by F. Velterop in order to provide membranes and modules for laboratory and pilot testing, and to serve small and medium size industrial plants with up to several hundred square meters of membrane surface [70–72]. Thanks to the work of the Netherlands Energy Research Corporation (ECN, The Netherlands), sol–gel-derived silica microporous membranes became commercially available in 2000 for large-scale pervaporation applications (Sulzer Chemtech, Achema) [73].

Silica membranes have shown good performance in the dewatering of solvents like methylethylketone, dimethylformamide, tetrahydrofuran, and ethylacetate [74]. However, in spite of their attractive high flux, their selectivity hardly competes with those of hydrophilic NaA zeolite membranes provided by SMART Chemical Company Ltd. (UK) or Mitsui (Japan) [75]. The sol–gel-derived membranes prepared by TNO on the outer surface of capillary supports revealed very high selectivity for the dehydration of *n*-butanol [76]. However, the poor stability of silica membranes in water above 100 °C is a strong limiting factor to their large-scale development, although stability can be improved in composite membranes such as SiO_2–ZrO_2 [77, 78].

4.1.2. Gas separation

During the last 15 years, much work in relation with ultramicroporous silica membranes concentrated on their hydrogen separation properties. Considerable progress was made very recently for improving their permeability, permselectivity, hydrothermal stability/hydrophobicity, reproducibility, and integration in high temperature modules.

The first almost dense silica membranes which were revealed to be quite effective for H_2 separation were prepared by CVD on a porous (Vycor®) glass substrate by Gavalas *et al.* 1989 [37]. Nowadays, silica membranes with attractive performance can be obtained either by CVD and CVI or by the sol–gel process, although both types of membranes show rather poor stability in steam at high temperature.

Typical H_2 permeance and H_2/N_2 permselectivity values reported in the literature for CVD and sol–gel-derived silica membranes are compared in Table 2.15, in the range of 500–600 °C. The performance of Pd-based and Ni-doped silica membranes are also reported for comparison. Clearly, the CVD method results in membranes with lower permeance, but higher selectivity in comparison with sol–gel-derived membranes (Fig. 2.24). The objective of much research work is to develop silica membranes with performance above the dotted line in Fig. 2.24, that is, approaching those of Pd-based membranes which are more expensive, and sensitive to both cycling and poisoning. Although the transport mechanisms and poisoning resistance of Ni-doped silica membranes are not yet clarified, the performance of these membranes clearly compete with those of Pd membranes.

Ultramicroporous silica membranes are of particular interest in energy-efficient H_2 separation from other gases because of their high flux, good selectivity, and their relatively good stability in harsh conditions. They are namely considered for selective removal of H_2 in equilibrium-restricted processes like dehydrogenation of alkanes, dry reforming of methane [31], steam reforming of methane [88, 89], ethane, propane [89], ethanol [91] or water-gas shift [92], and also for partial oxidation of methane [93], coal gasification, and decomposition of H_2S or thermochemical cycles involving HI [94, 95] or HBr [96].

Table 2.15 Comparison of H_2 permeances and α^*H_2/N_2 ideal selectivities at 500–600 °C through sol–gel and CVD-derived silica membranes from literature

	Temperature (°C)	Permeance, π 10^{-6} (mol/m^2/s/Pa)	Ideal selectivity, α^*H_2/N_2	Author, year	Reference
CVD silica	500	0.05	1300	Nomura, 2006	[43]
	600	0.004	72	Wu, 1994	[79]
	600	0.006	160	Hwang, 1999	[34]
	600	0.05	800	Nomura, 2005	[42]
	600	0.07	450	Nomura, 2006	[43]
	600	0.1	1000	Lee, 2004	[32]
Sol–gel silica	500	1.08	32	Collins, 1996	[80]
	600	0.7	102	Yoshino, 2005	[81]
	500	0.26	87	Tsuru, 2004	[82]
	500	0.45	100	Asaeda, 1998	[83]
Ni-doped	500	4.6	340	Kanezashi, 2006	[84]
Palladium	500	2.7	350	Su, 2004	[85]
	500	1.17	1600	Jun, 2000	[86]
	500	0.2	1000	Yan, 1994	[87]

4.2. Gas transport in almost dense silica membranes

The transport in silica membranes (often referred as almost dense membranes) occurs through the network of connected micropores; thanks to a gradient of chemical potential of species in the gas phase (driving force) [97]. The separation is mainly based on a sieving effect through small ultramicropores (~3 Å) and a favorable combination of sorption selectivity (interactions gas molecules/membrane surface) and diffusion mobility (rate of gas passage through the membrane) produces reasonably high values for flux and selectivity.

Until recently, the mechanism of H_2 permeation through silica membranes has not been studied in detail. The resistance of gas transport of small molecules like H_2 through the thin silica layer is such that the resistance in the supporting layers could even dominate the transport properties of the final membrane. This is clearly demonstrated in Fig. 2.25 for a Nanosil membrane [98]. The gas permeance through

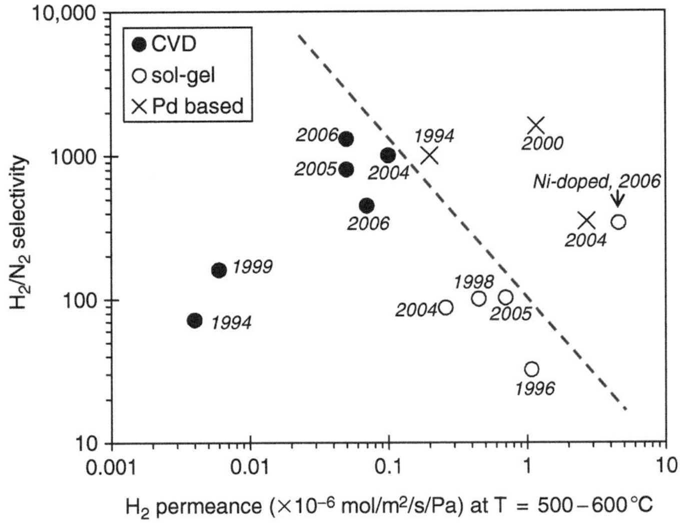

Figure 2.24 Comparison of H_2 permeances and H_2/N_2 ideal selectivities at 500–600 °C through sol–gel and chemical vapor deposition (CVD)-derived silica membranes from literature, and compared to palladium-based membranes.

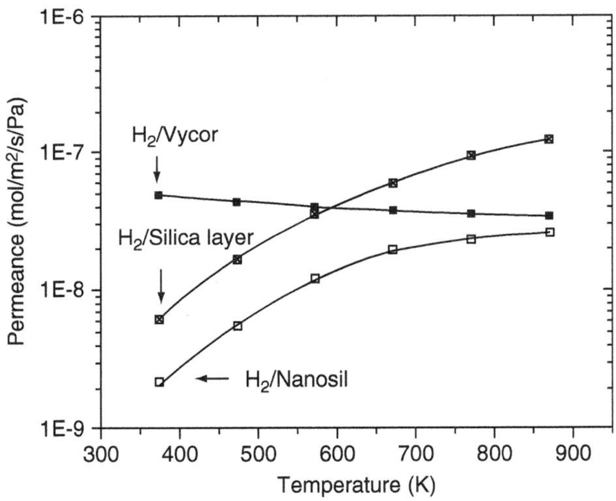

Figure 2.25 Model-calculated Ne permeance on the silica layer, and experimental permeance on the Vycor® and Nanosil membrane (from [98]).

the alumina or Vycor® support decreased with the inverse square root of temperature in accordance with a Knudsen permeation mechanism (Table 2.3). Initial designs with mesoporous Vycor® were abandoned because of the too low permeance of these supports.

In [99], it was shown that the gas permeation characteristics through the silica membrane layers obtained by CVD on a porous support (membranes referred as

Nanosil when the silica is outside the pores; [100]) were similar to those for vitreous silica glass, with the presence of solubility sites for the permeating species (Fig. 2.26) [101]. Indeed, silica glass had been described as a disordered form of β-cristobalite that contains 5, 6, 7, and 8 membered rings (Fig. 2.26 from [99]), in which solubility sites are ~0.3 nm in diameter [102]. The restricted size of the solubility sites in the silica membrane matrix structure allows sorption and permeation of the small gas molecules (He = 0.26 nm, Ne = 0.275 nm, H_2 = 0.289 nm) while restricting sorption of the large gas molecules (CO_2 = 0.33 nm, CO = 0.376 nm, CH_4 = 0.38 nm) by size differentiation. This result is clearly evidenced after 6 h deposition time, resulting in a defect-free and uniform silica molecular sieve layer (Fig. 2.27A from [98]). After 12 h of silica deposition, the resulting composite membrane showed a high H_2 permeance of $\sim 10^{-7}$ mol/m^2/s/Pa at 600 °C with H_2 selectivities over CH_4, CO, and CO_2 in excess of 2000.

The mechanism of small gas transport through the silica membrane is molecular diffusion, in which gas molecules enter into the solubility sites and diffuse through the silica structure by jumping to adjacent solubility sites under the driving force of a

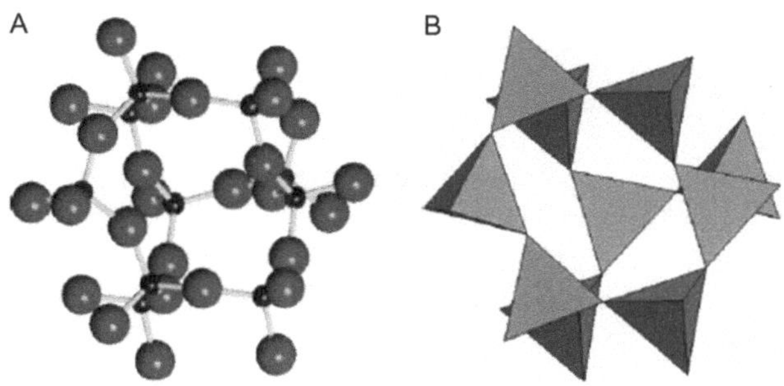

Figure 2.26 Solubility site in β-cristobalite. (A) Ball-and-stick model (large spheres: oxygen; small spheres: silicon); (B) polyhedral representation of the same site (from [99]).

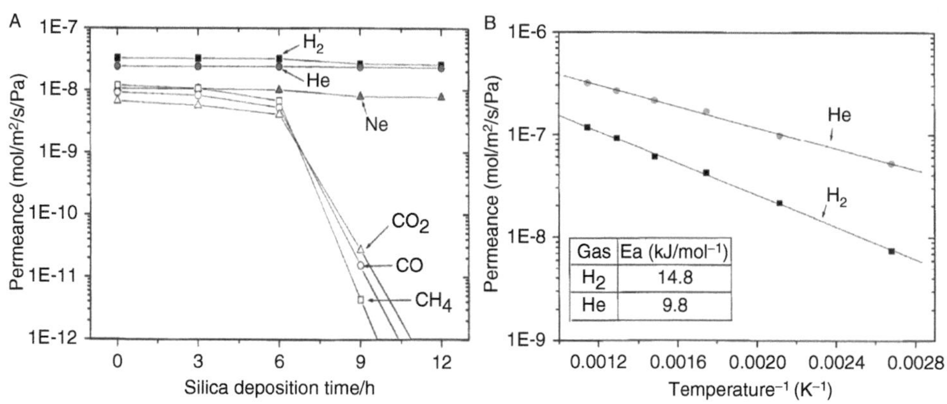

Figure 2.27 Evolution of single gas permeance for chemical vapor deposition (CVD)-derived silica membrane (A) versus deposition time on Vycor glass (from [98]) (B) Arrhenius plot for a membrane deposited on γ-Al_2O_3 support (from [32]).

concentration gradient [99]. Gas transport was successfully described by a statistical solid state diffusion model, which was able to account for the unusual permeation order He > H_2 > Ne, which did not follow mass or atomic size. This behavior was explained by a statistical jump mechanism [99].

As shown in Fig. 2.27B from [32], the permeance of small gases (e.g., He, H_2, and Ne) through the silica membrane is activated, and increases with temperature. The activation energies of permeation E_a are obtained by fitting the experimental gas permeance data to an Arrhenius law:

$$\pi = \pi_0 \exp\left(\frac{-E_a}{RT}\right)$$

The measured activation energies for CVD membranes on either alumina supports (H_2 = 14.8 kJ/mol, He = 9.8 kJ/mol in [32]; H_2 = 18–26 kJ/mol in [103]) or on Vycor® glass support (H_2 = 16.4 kJ/mol, He = 12.4 kJ/mol, Ne = 15.7 kJ/mol in [98]) are much lower than those for the vitreous silica glass reported in the literature (H_2 = 37.2–38.3 kJ/mol in [104, 105]; He = 17.8–21.1 kJ/mol in [106, 107]; Ne = 33.8–39.5 kJ/mol in [108, 109]), indicating that the interstitial structure of silica obtained by CVD is more open than in vitreous glass. The larger permeability of He over H_2 can be understood from the larger number of solubility sites for He, coupled with a lower jump barrier [99].

The activation energies measured for H_2 or He in sol–gel-derived membranes are usually lower than those measured for CVD ones. Typically, a value of 9.5 kJ/mol was reported for H_2 [81] and values in the range 3.2–14.9 kJ/mol were obtained in [110]. These results confirm that the structure of sol–gel-derived membranes is less compact than the CVD-derived ones. This explains why sol–gel-derived membranes are more permeable but less selective.

The He/H_2 selectivity could also be considered as a good indication of the membrane structure and quality. Because of the differences in activation energy, the ideal selectivity α^*He/H_2 decreases when the temperature increases and values can be as high as 5 [111]. It is also interesting to note that the H_2/H_2O permeance ratio is also relatively low in sol–gel-derived membranes (typically 10 at 500 °C according to [112]), although it can reach 290 for CVD-derived membranes [43].

A bimodal structure is proposed in [113] for CVD silica membranes, in which He (and also H_2, Ne, and H_2O) is transported mainly through siloxane rings (containing at least six Si atoms – Fig. 2.28A), while N_2 transport should occur only through the silica cluster pores (Fig. 2.28B). The decrease of He/N_2 separation over thermal treatment (from 500 to 580 °C) seems to result from the rearrangement of the silica morphology leading to either opening up or increasing the number of larger pores that contribute mostly to N_2 flux (which increases with temperature). The silica microstructure could then be schematically represented as shown in Fig. 2.28.

However, the structure of sol–gel-derived membranes is not as compact as the CVD one and sol–gel-derived membrane does not seem to behave identically. Indeed, according to [67] when the firing temperature of sol–gel silica membranes

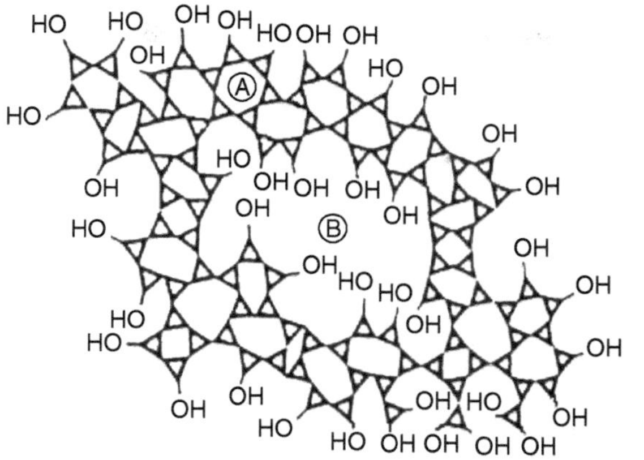

Figure 2.28 Schematic representation of the gas transport sites in amorphous almost dense silica membranes. (A) Siloxane rings accessible only to He, H_2, Ne, and H_2O; (B) pores between silica clusters accessible to bigger molecules. Adapted from [113].

is increased up to 600 °C, a significant decrease of the CO_2 permeance was observed and attributed to a decrease of the hydroxyl group concentration on the pore surface in the silica layer. In this case, the H_2/CO_2 permselectivity was improved by a factor of more than 10, while the H_2 permeance remained high (6×10^{-7} mol/m^2/s/Pa at 300 °C). Further studies comparing the CVD and sol–gel membrane structure, performance, and behavior as a function of temperature should be helpful.

4.3. Membrane supports and intermediate layers

Most of the silica membranes reported in the literature are deposited either on alumina-based supports (macroporous α-Al_2O_3 with a mesoporous final layer) or on Vycor® glass (7930 glass, Corning Inc.). The latter has a tubular geometry with an outside diameter of 10 mm, a wall thickness of 1 mm, and 4 nm nominal average pore size. Alumina is a more desirable support than Vycor® glass because of its higher permeability and resistance to high pressures. Stainless steel is also used as a support in several studies [92]. Metals simplify the sealing problem at high temperature but their high expansion coefficients hardly match with that of silica.

Silica membranes reported in literature have either a flat plate (disc) or tubular geometry. The first one is mainly used for academic studies, although tubular supports (mono- or multichannel) are preferred when large surface areas have to be developed. Commercially available pervaporation silica membranes have a tubular geometry [114]. The classical tubular geometry leads to surface area to volume ratio typically lower than 500 m^2/m^3. Recently, TNO and CEPAration developed attractive low-cost capillary supports (OD/ID = 4/3, 3/2, 2/1.4, 2/1 mm) and modules, allowing surface area to volume ratio higher than 1000 m^2/m^3 [76]. In parallel, hollow ceramic fibers (OD $\leq$ 2 mm) are also under development [115, 116]. A micrograph of the capillary structure with the silica layer deposited on their surface is shown in Fig. 2.29.

Figure 2.29 SEM micrographs of the various layers constituting the membrane: (A) cross-section substrate (magnification ×100), (B) cross-section intermediate γ-Al_2O_3 layers on top of substrate (magnification ×3300), (C) cross-section silica membrane on the intermediate γ-Al_2O_3 layers (magnification ×75,000) (from [76]).

In the preparation of ultramicroporous ceramic membranes, the quality of the support is of crucial importance to the membrane layer integrity. However, compared with the huge efforts to develop ever better membranes for gas separation, relatively little is published in the open literature on support development. The surface roughness and homogeneity of the support determines not only the integrity of the membrane layer but also the minimal thickness of the membrane layer for complete surface coverage. It is consequently very important to favor the development of low cost, high S/V ratio, highly permeable supports, and good surface quality ensuring the right conditions for coating high-quality membrane layers. The support wall thickness has to be decreased while maintaining a sufficient strength value, of ~30 MPa according to [81]. On the contrary, multilayer or graded supports whose top-layer has an ordered or oriented porosity (similar to anodic alumina capillary; [111]) are highly desirable. Colloidal filtration and centrifugal injection casting are able to provide supports with a higher degree of homogeneity than conventional die-pressed, tape-cast, and extruded supports. The methods can be used to prepare multilayer or graded membrane supports in just one production step [117].

In order to obtain high fluxes, the permselective silica top-layer has to be extremely thin. The thickness is limited to ~20–30 nm for CVD membranes (Fig. 2.14B). In the case of sol–gel-derived membranes, the exact top-layer thickness (usually 100–200 nm) is much difficult to evidence because of sol infiltration (Fig. 2.30).

The support surface quality can of course be improved by polishing. However, this method, adapted to flat supports, is tedious and repetitive. The use of thin intermediate layers is a more attractive alternative which can be used to generate a smooth surface, to improve the chemical adhesion of the silica layer to the support (e.g., with phosphates; [119]), to limit the effect of differential thermal expansion coefficients, and finally, to limit the infiltration of the silica layer in the macroporous support. Typical composition for the intermediate layers include transition Al_2O_3, TiO_2, and ZrO_2. A mesoporous γ-Al_2O_3 layer with pores sizes of ~5 nm is often used, although recent developments tend to favor silica-based mesoporous intermediate layers such as SiO_2–ZrO_2 on α-Al_2O_3 (Fig. 2.31) [84] or Ni-SiO_2 on stainless steel [120] supports.

A more sophisticated approach, involving a four-layer configuration, can also be adopted. The first example is reported in [121], where improvement of both flux

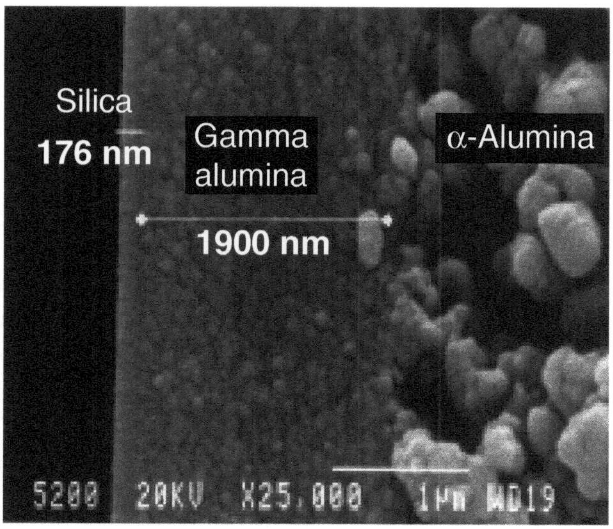

Figure 2.30 FE-SEM observation of an ultramicroporous sol–gel-derived silica membrane deposited on γ-Al_2O_3 support (from [118]).

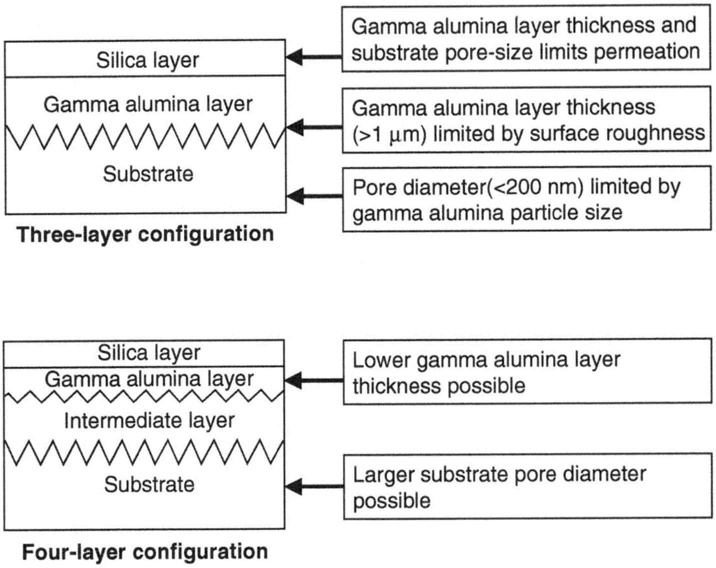

Figure 2.31 Schematic of three-layered and four-layered silica membrane configuration (from [81]).

and selectivity can be reached by deposition (on the γ-Al_2O_3 layer) of a first microporous silica layer with ~1 nm pore size (supermicropores obtained by surfactant-assisted self-organization) and then the final 30 nm thick ultramicroporous silica top layer (Fig. 2.32). A second example is reported in [81], where the objective is to limit strongly the thickness of the mesoporous γ-Al_2O_3 layer.

Carbon-based barriers can also be impregnated in the supports in order to limit the infiltration of the silica layer. Furfuryl alcohol was typically used in [122] for CVD layers. The "hot coating method" proposed in [123], involving sol disposition on a hot support, is also a possible way to limit infiltration, although it is probably difficult to have a good layer thickness uniformity on large-scale samples.

Most of the recently developed membranes are deposited on the external side of tubes, whatever the selected synthesis method, that is, sol–gel [76, 118] or CVD [31]. This strategy is probably related to an easier upscaling but could also influence the mechanical behavior of the membrane upon heating.

Processing of large area membrane films is seldom reported in the literature. An example is given in [81] with porous tubes ID/OD = 4/6 mm, 400–500 mm long. Modules (substrate bundle) containing six alumina tubes, 400 mm long (with intermediate layer) are sealed on dense alumina end-plates using glass-based sealant materials. A comparison of the H_2 permeance and H_2/N_2 separation performances of a silica membrane single tube and module evidence the problems related to upscaling in the case of sol–gel-derived membranes (Fig. 2.33A and B). This problem seems to be more easily overcome in the case of CVD-derived membranes [103, 124] (Fig. 2.33C).

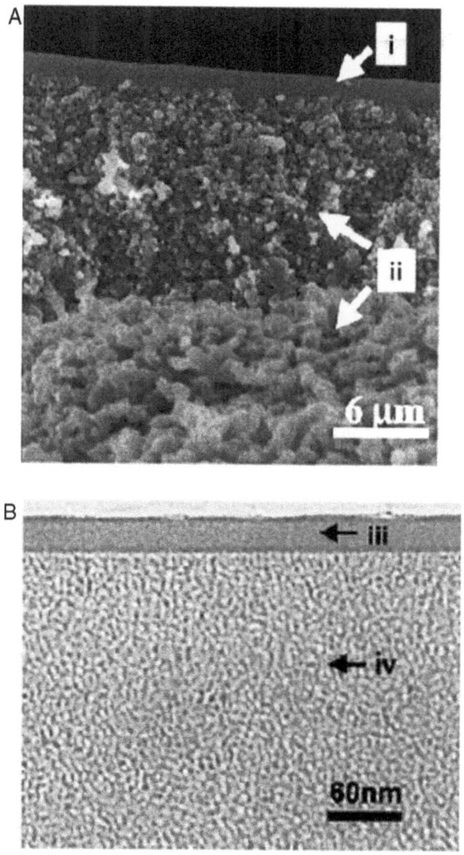

Figure 2.32 Cross-sectional electron micrographs of an asymmetric membrane: (A) SEM overview of (i) silica membrane on top of γ-Al_2O_3 layer, (ii) asymmetric α-Al_2O_3 support and (B) TEM micrograph revealing the dual-layer silica membrane (iii) top-layer and (iv) sublayer (from [121]).

4.4. Thermal stability of silica membranes on steam

The main concern regarding the silica membranes is that they are unstable under hydrothermal conditions. Indeed, most of the microporous silica membranes suffer from loss of permeability (which can reach 50% or greater in the first 12 h) on exposure to moisture [83]. This loss of permeability is generally coupled with the loss of selectivity which is highly detrimental to long-term application of silica membranes in the presence of steam [118].

This behavior has been attributed to the removal of Si–OH groups leading to the formation of Si–O–Si bonds which closes pore channels [9]. This phenomenon, termed densification, is catalyzed by moisture at high temperatures. Densification not only leads to lower permeability but can also cause embrittlement of the silica film that compromises selectivity.

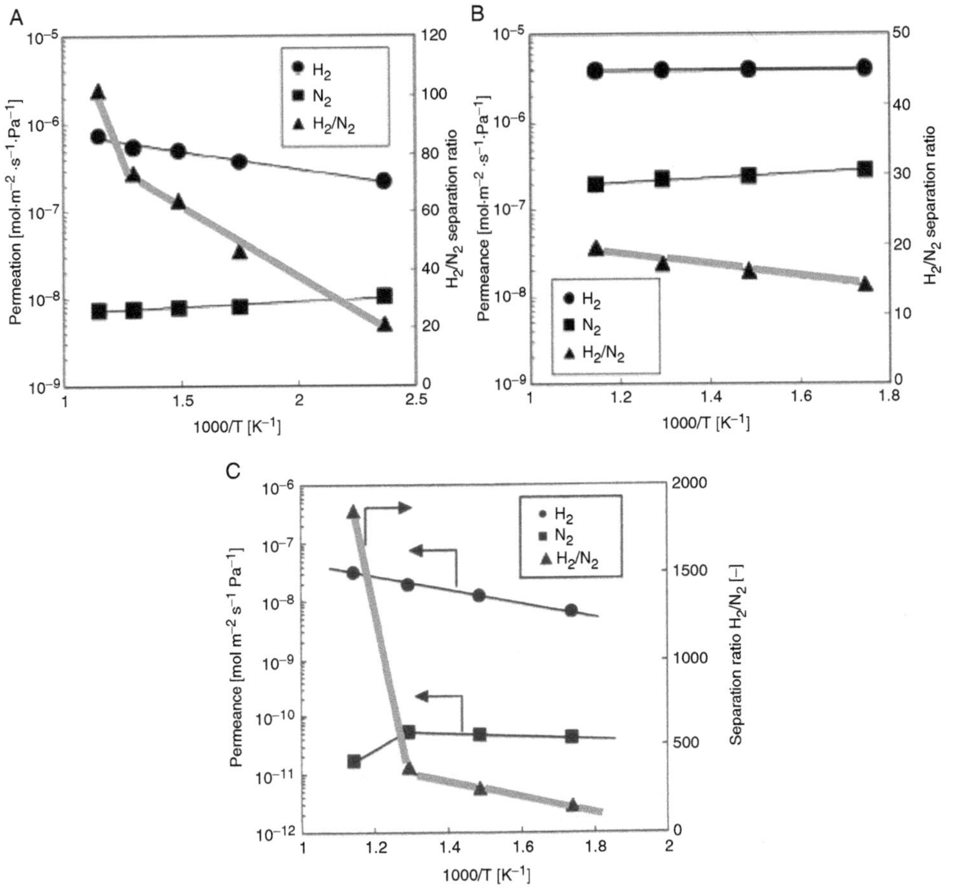

Figure 2.33 Arrhenius plots of H_2 permeance and H_2/N_2 separation ratio for (A) a sol–gel-derived single membrane tube [81], (B) a sol–gel-derived silica membrane bundle (from [81]), and (C) a chemical vapor deposition (CVD)-derived silica membrane module prepared at 600 °C (from [103]).

Although the mechanisms are not fully understood [113], silica membrane hydrothermal densification occurs mostly by restructuring siloxane rings (accessible only to H_2, He, Ne, and H_2O) at the surface of the amorphous silica clusters.

Sol–gel-derived silica membranes may strongly change their pore structure and separation performance due to adsorption of water molecules and successive rearrangements while left in the atmosphere [123]. Exposition of a standard silica membrane to steam at 350 °C during 15 days yields a decline of both the H_2/CO_2 selectivity from 29.7 to 20.9 and of the hydrogen permeance, by a factor 3.

On the contrary, CVD membranes seem to be more stable as far as their exposure to moisture (10% H_2O in Ar) at 873 K resulted in only a 4% drop in permeability in the first 12 h with no further drop in permeability for up to 120 h [31]. In [43], the H_2/N_2 permeance ratio of silica membranes obtained by counter-diffusion CVD was kept over 800 for 21 h at 773 K steam treatment. A quite stable behavior was

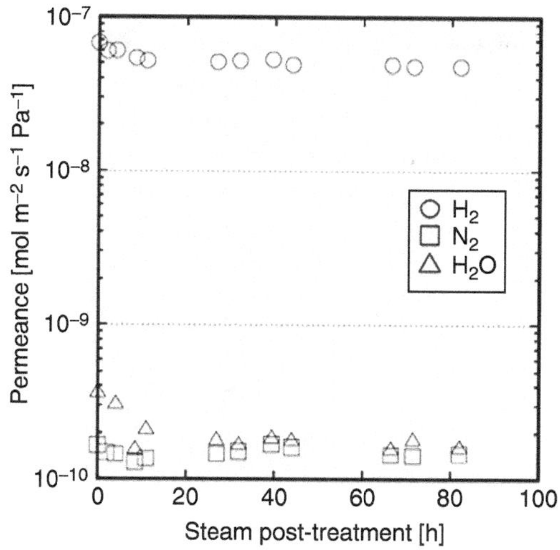

Figure 2.34 Hydrogen, nitrogen, and water permeations at 600 °C of counter-diffusion chemical vapor deposition (counter-diffusion CVD) silica membranes as a function of a steam posttreatment at 500 °C (from [43]).

obtained after nearly 100 h under steam at 500 °C. Figure 2.34 shows the evolution of H_2, N_2, and H_2O permeances at 600 °C, with a final H_2/H_2O permeation ratio of 290. In spite of the relatively short aging time, it was concluded that the high steam stability and high H_2/H_2O permeance ratio of the membrane prepared by the counter-diffusion CVD method were sufficiently attractive for application in steam reforming reactor. The stability of the Nanosil membranes is attributed to a smaller quantity of hydroxyl groups which tend to render the membrane hydrophobic [31].

It is clear that only long-term treatment under realistic conditions is the only way to validate the stability of silica membranes in steam atmosphere. According to [118], the formation of cracks in sol–gel-derived membranes was only evidenced after 1150 h on steam at 350 °C. This explains the drastic conditions used in [81] for studying the membrane stability: 0.1 MPa, $H_2O{:}H_2 = 1{:}1$, at 500 °C for 1000 h. The demonstrated instability of porous sol–gel-derived silica membranes during long-term operation in steam led to the conclusion that they are not adapted to reactions such as water gas shift [118].

A number of strategies have been recently developed in order to improve the hydrothermal stability of sol–gel-derived membranes. These strategies concern not only modifications of the silica membrane material itself but also stabilization of the membrane substrate. The hydrothermal stability of the substrate is strongly related to the selected support and intermediate-layer compositions (e.g., La-stabilized γ-Al_2O_3; [125], SiO_2–ZrO_2 [84]). On the contrary, it is also suggested in [84] that a pretreatment of the support in steam at 550–650 °C can improve the hydrothermal stability of the final membrane.

The development of the hydrophobic silica membrane material including carbon species is another attractive strategy for obtaining steam-stable silica membranes. Incorporation of hydrophobic methyl moieties was proposed in [126] by using a silica alkoxide precursor containing methyl groups (e.g., by replacing the classical TEOS precursor by methyltriethoxysilane). Compared with the conventional TEOS-derived membranes, the "methylated" silica membranes have an improved hydrothermal stability, but a systematically lower H_2/CO_2 selectivity [118].

Two other approaches have been proposed in the recent literature in order to insert carbon precursors in the silica matrix. The first one involves the pyrolysis at 800—1000 °C of a two-phase copolymer (imide-siloxane) in order to generate a pyrolytic C-SiO_2 composite membrane with inclusions of carbon rich domains with the silica-rich matrix [127]. The initial precursor morphology strongly influences pores sizes and shapes, and consequently membrane performance. A schematic representation of the structure of the composite membrane is shown in Fig. 2.35. The second approach consists of preparing functional *carbonised template molecular sieve silica* (CTMSS) using a novel surfactant templating procedure [128]. Although the membrane performance is still rather low (H_2 permeance is 2.1×10^{-8} mol/m^2/s/Pa and $\alpha^*H_2/CO = 13$ at 200 °C), their relative hydrostability and possible regeneration of the initial values by thermal treatment at 500 °C is promising.

The multilayer strategy, that is, the protection of the silica membrane with an hydrophobic layer can also be proposed, for example, with the PECVD derived a-SiOC:H layers studied in [63]. Finally, the insertion of inorganic additives in the silica structure has been proposed a number of years ago. Typical additives are Al_2O_3, ZrO_2, or TiO_2 [5] but Mg/Al or Pt [129], cobalt [130], and nickel [84, 111, 131] are also used. Highly attractive performances were reported for Ni-silica membranes [84], although the transport mechanisms and poisoning behavior of this type of membrane with sulfur and CO species are still to be studied in details. An SEM observation of a thin Ni-SiO_2 membrane deposited on a stabilized SiO_2–ZrO_2 intermediate layer is shown in Fig. 2.36.

The final thermal treatment of the membrane is also strongly important for obtaining good performance and hydrothermal stability. Silica densification typically occurs around 800 °C; it is consequently important to select the optimum temperature to guarantee both high selectivity and permeability. According to [84], a thermal treatment in steam in the range of 550–650 °C allows to stabilize efficiently the Ni-silica membrane.

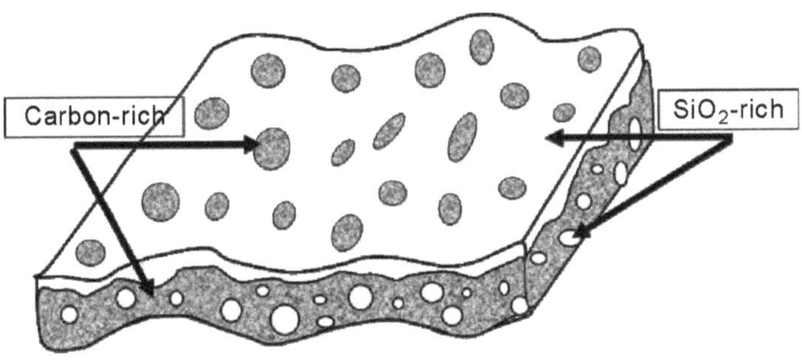

Figure 2.35 Schematic representation of the structure of a C-SiO_2 composite membrane prepared by pyrolysis of a two-phase copolymer (from [127]).

Figure 2.36 Ni-doped silica membranes (Si/Ni = 2/1) fired in the steamed atmosphere at 650 °C (from [84]).

5. Conclusion

Taking into account the specific properties of silica and using convenient deposition methods with well-defined synthesis parameters, it is possible to prepare microporous silica membranes of particular interest for specific applications in high temperature gas separation, pervaporation, or nanofiltration in organic media. Sol–gel and CVD routes are versatile methods which can be efficiently applied to prepare ultra- or supermicroporous amorphous silica membranes with tunable composition (pure silica, doped silica, surface-modified silica) and design.

Since their first synthesis in 1989, many improvements have been achieved in the occurrence of membrane defects, reproducibility, operational stability, and manufacturing costs which make membranes a viable option for a number of energy-efficient H_2 separations.

The sol–gel method suffers from a lack of reproducibility and great sensitivity to dust although CVD methods require complicated and expensive equipments, but lead to highly selective membranes. In spite of their limited permeance, counter-diffusion CVD membranes and small-scale modules have been recently developed by the Nakao Laboratory in Tokyo [132]. The higher flux but lower selectivity of sol–gel-derived membranes is also attractive for various applications, namely in liquid and vapor phase. Composite sol–gel-derived membranes such as Ni-SiO_2 or C-SiO_2 could also be considered attractive for high temperature gas phase separations.

In all cases, the selection of the ideal membrane composition and supporting structure has to be derived from an application-based design and a concerted effort is needed for understanding membrane transport, developing reproducible low-cost synthesis routes for modules, improving operational stability, and developing relevant quality estimator methods for membrane units.

On this basis, silica-based membrane separations will obviously play a key role in future power production systems and industrial chemical production processes.

REFERENCES

[1] Burgraff, A. J., and Cot, L. (eds.) (1996). *Fundamentals of Inorganic Membrane Science and Technology*. Membrane Science and Technology Series, 4, Elsevier, Amsterdam.
[2] Cot, L., Ayral, A., Durand, J., Guizard, C., Hovnanian, N., Julbe, A., and Larbot, A. (2000). Inorganic membranes and solid state sciences. *Solid State Sci.* **3,** 313–334.
[3] Meinema, H. A., Dirrix, R. W. J., Brinkman, H. W., Terpstra, R. A., Jekerle, J., and Kösters, P. H. (2005). Ceramic membranes for gas separation—recent developments and state of the art. *Interceram* **54,** 86–91.
[4] Sommer, S., and Melin, T. (2005). Performance evaluation of microporous inorganic membranes in the dehydration of industrial solvents. *Chem. Eng. Process.* **44,** 1138–1156.
[5] Guizard, C, Ayral, A., and Julbe, A. (2002). Potentiality of organic solvents filtration with ceramic membranes. A comparison with polymer membranes. *Desalination* **147,** 275–280.
[6] Zachariasen, W. H. (1932). The atomic arrangement in glass. *J. Am. Chem. Soc.* **54,** 3841–3851.
[7] Zarzycki, J. (1982). *Les Verres et l'état Vitreux.* Masson, Paris.
[8] Brinker, C. J., and Scherer, G. W. (1990). *Sol-Gel Science, The Physics and Chemistry of Sol-Gel Processing* Academic Press, Boston.
[9] Iler, R. K. (1979). *The Chemistry of Silica.* Interscience, New York.
[10] Frost, H. J., and Raj, R. (1982). Limiting densities for dense random packing of spheres. *J. Am. Ceram. Soc.* **65,** C19.
[11] Ben Aim, R. (1970). *Etude de la Texture des Empilements de Grains: Application à la Détermination de la Perméabilité des Mélanges Binaires en Régime Moléculaire, Intermédiaire, Laminaire.* Thesis, University of Nancy, France.
[12] Troadec, J. P., and Bideau, D. (1992). Milieux granulaires: structure, distribution de taille. *In* "Proceedings of the Journée sur la préparation, la caractérisation et la manipulation des poudres céramiques, Paris, France," 12 February, pp. 33–36. Groupe Français de la Céramique, Paris.
[13] Prassas, M., and Hench, L. L. (1984). Physical chemical factors in sol-gel processing. *In* "Ultrastructure Processing of Ceramic, Glasses, Composites" (L. L. Hench and D. R. Ulrich, eds.), pp. 100–125. Wiley, New York.
[14] Schaefer, D. W. (1988). Fractal models and the structure of materials. *MRS Bull.* **8,** 22–27.
[15] Mizuno, T. (1984). *Procédé sol-gel et génie des matériaux,* Ph.D. Thesis, University of Montpellier, France.
[16] Quinson, J. F. (1987). La thermoporométrie: Une méthode de caractérisation texturale des gels. *In*"Proceedings of the Ecole d'été Procédés sol-gel, Bombannes, France," 28 September–2 October. Vol. II-10, pp. 1–9. CNRS, Paris1987.
[17] Brinker, C. J., Frye, G. C., Hurd, A. J., and Ashley, C. S. (1991). Fundamentals of sol-gel dip coating. *Thin Solid Films* **201,** 97–108.
[18] Roger, C., and Hampden-Smith, M. J. (1994). General routes to porous metal oxides via inorganic and organic templates. *J. Sol-Gel Sci. Technol.* **2,** 67–72.
[19] Ayral, A., Balzer, C., Dabadie, T., Guizard, C., and Julbe, A. (1995). Sol-gel derived silica membranes with tailored microporous structures. *Catal. Today* **25,** 219–224.
[20] Raman, N. K., and Brinker, C. J. (1995). Organic "template" approach to molecular sieving silica membranes. *J. Membr. Sci.* **105,** 273–279.
[21] Kim, Y. S., Kusakabe, K., Moroka, S., and Yang, S. M. (2001). Preparation of microporous silica membranes for gas separation. *Korean J. Chem. Eng.* **18,** 106–112.
[22] West, G. D., Diamond, G. G., Holland, D., Smith, M. E., and Lewis, M. H. (2002). Gas transport mechanisms through sol-gel derived templated membranes. *J. Membr. Sci.* **203,** 53–69.
[23] Klotz, M., Ayral, A., Guizard, C., and Cot, L. (2001). Synthesis and characterization of silica membranes exhibiting an ordered mesoporosity—control of the texture and effect on the membrane permeability. *Sep. Purif. Technol.* **25,** 71–78.
[24] Klotz, M., Besson, S., Ricolleau, C., Bosc, F., and Ayral, A. (2003). Ordered 3D hexagonal mesoporous silica membranes: synthesis and characterization. *Mater. Res. Soc. Symp. Proc.* **752,** 123–128.

[25] Ayral, A., Guizard, C., and Cot, L. (1994). Synthesis and characterization of hybrid organic-inorganic colloidal gels. *J. Mater. Sci. Lett.* **13,** 1538–1539.
[26] Bosc, F., Lacroix-Desmazes, P., and Ayral, A. (2006). TiO_2 anatase-based membranes with hierarchical porosity and photocatalytic properties. *J. Colloid Interf. Sci.* **304,** 545–548.
[27] Campaniello, J., Engeln, C. W. R., Haije, W. G., Pex, P. P. A. C., and Vente, J. F. (2004). Long-term pervaporation performance of microporous methylated silica membranes. *Chem. Commun.* **7,** 834–835.
[28] Unger, K. K. (1979). *Porous Silica: Its Properties and Use as Support in Column Liquid Chromatography.* Elsevier, Amsterdam.
[29] Okubo, T., and Inoue, H. (1989). Introduction of specific gas selectivity to porous glass membranes by treatment with tetraethoxysilane. *J. Membr. Sci.* **42,** 109–110.
[30] Ha, H. Y., Nam, S. W., Hong, S. A., and Lee, W. K. (1993). CVD of hydrogen permselective silica films on porous glass supports from TEOS. *J. Membr. Sci.* **85,** 279–290.
[31] Prabhu, A. K., and Oyama, S. T. (2000). Highly hydrogen selective ceramic membranes: Application to the transformation of greenhouse gases. *J. Membr. Sci.* **176,** 233–248.
[32] Lee, D., Zhang, L., Oyama, S., Niu, S., and Saraf, R. (2004). Synthesis, characterization, and gas permeation properties of a hydrogen permeable silica membrane supported on porous alumina. *J. Membr. Sci.* **231,** 117–126.
[33] Morooka, S., Yan, S. C., Kusakabe, K., and Akiyama, Y. (1995). Formation of hydrogen-permselective SiO_2 membrane in macropores of a-alumina support tube by thermal decomposition of TEOS. *J. Membr. Sci.* **101,** 89–98.
[34] Hwang, G. J., Onuki, K., Shimizu, S., and Ohya, H. (1999). Hydrogen separation in H_2-H_2O-HI gaseous mixture using the silica membrane prepared by CVD. *J. Membr. Sci.* **162,** 83–90.
[35] Kuraoka, K., Kakitani, T., Suetsugu, T., and Yazawa, T. (2001). Methanol vapor separation through the silica membrane prepared by the CVD method with the aid of evacuation. *Sep. Purif. Technol.* **25,** 161–166.
[36] Alsyouri, H., Langheinrich, C., Lin, Y., Ye, Z., and Zhu, S. (2003). Cyclic CVD modification of straight pore alumina membranes. *Langmuir* **19,** 7307–7314.
[37] Gavalas, G. R., Megiris, C. E., and Nam, S. W. (1989). Deposition of H_2 permselective SiO_2 films. *Chem. Eng. Sci.* **44,** 1829–1835.
[38] Megiris, C. E., and Glezer, J. H. E. (1992). Synthesis of H_2 permselective membranes by modified CVD. Microstructure and permselectivity of SiO_2/C/Vycor membranes. *Ind. Eng. Chem. Res.* **31,** 1293–1299.
[39] Tsapatsis, M., and Gavalas, G. R. (1992). A kinetic model of membrane formation by CVD of silica and alumina. *AIChE J.* **38,** 847–856.
[40] Nakao, S. I., Suzuki, T., Sugawara, T., Tsuru, T., and Kimura, S. (2000). Preparation of microporous membranes by TEOS/O_3 CVD in the opposing reactants geometry. *Microporous Mesoporous Mater.* **37,** 145–152.
[41] Yamaguchi, T., Ying, X., Tokimasa, Y., Nair, B. N., Sugawara, T., and Nakao, S. I. (2000). Reaction control of tetraethyl orthosilicate (TEOS)/O_3 and tetramethyl orthosilicate (TMOS)/O_3 counter diffusion chemical vapor deposition for preparation of molecular-sieve membranes. *Phys. Chem. Chem. Phys.* **2,** 4465–4469.
[42] Nomura, M., Ono, K., Gopalakrishnan, S., Sugawara, T., and Nakao, S. I. (2005). Preparation of a stable silica membrane by a counter diffusion chemical vapor deposition method. *J. Membr. Sci.* **251,** 151–158.
[43] Nomura, M., Aida, H., Gopalakrishnan, S., Sugawara, T., Nakao, S. I., Yamazaki, S., Inada, T., and Iwamoto, Y. (2006). Steam stability of a silica membrane prepared by counter-diffusion chemical vapor deposition. *Desalination* **193,** 1–7.
[44] Nomura, M., Seshimo, M., Aida, H., Nakatani, K., Gopalakrishnan, S., Sugawara, T., Ishikawa, T., Kawamura, M., and Nakao, S. I. (2006). Preparation of a catalyst composite silica membrane reactor for steam reforming reaction by using a counterdiffusion CVD method. *Ind. Eng. Chem. Res.* **45,** 3950–3954.
[45] Nijmeijer, A., Bladergroen, B., and Verweij, H. (1998). Low-temperature CVI modification of gamma-alumina membranes. *Microporous Mesoporous Mater.* **25,** 179–184.

[46] McCool, B., and DeSisto, W. (2004). Self-limited size reduction of mesoporous silica membranes via pyridine-catalyzed silicon dioxide ALD. *Chem. Vap. Deposit.* **10,** 190–194.
[47] Nomura, M., Yamaguchi, T., and Nakao, S. (1997). Silicalite membranes modified by counterdiffusion CVD technique. *Ind. Eng. Chem. Res.* **36,** 4217–4223.
[48] Kim, Y., Cho, S., Lee, H., Yoon, H., and Yoon, D. (2003). N_2 doped SiO_2–SiON planar waveguides deposited by PECVD method. *Surf. Coat. Technol.* **174/175,** 166–169.
[49] Storgaard-Larsen, T., and Leistiko, O. (1997). Plasma-enhanced chemical vapor deposited silicon oxynitride films for optical waveguide bridges for use in mechanical sensors. *J. Electrochem. Soc.* **144,** 1505–1513.
[50] Bankovic, P., Demarquette, N., and da Silva, M. (2004). Obtention of selective membranes for water and hydrophobic liquids by PECVD on porous substrates. *Mater. Sci. Eng. B Solid State Mater. Adv. Technol.* **112,** 165–170.
[51] Rouessac, V., Tuluc, P., and Durand, J. (2004). Composite plasma-polymerized organosiloxane-based material for hydrocarbon vapor selectivity. *J. Membr. Sci.* **230,** 49–59.
[52] Bosc, F., Sanchez, J., Rouessac, V., and Durand, J. (2003). Sorption and permeation characteristics of hybrid organosilicon thin films deposited by PECVD. *Sep. Purif. Technol.* **32,** 371–376.
[53] Roualdès, S., Sanchez, J., and Durand, J. (2002). Gas diffusion and sorption properties of polysiloxane membranes by PECVD. *J. Membr. Sci.* **198,** 299–310.
[54] Roualdès, S., and Durand, J. (2004). Experimental design and modelling in the investigation of PECVD parameters on the structural and gas transport properties of plasma polysiloxane membranes. *J. Membr. Sci.* **230,** 39–48.
[55] Inoue, Y., Sugimura, H., and Takai, O. (1999). *In situ* observation of behavior of organosilicon molecules in low-temperature plasma enhanced CVD. *Thin Solid Films* **345,** 90–93.
[56] Teshima, K., Inoue, Y., Sugimura, H., and Takai, O. (2001). Room-temperature deposition of high-purity silicon oxide films by RF plasma-enhanced CVD. *Surf. Coat. Technol.* **146/147,** 451–456.
[57] Rouessac, V., Ferreira, P., and Durand, J. (2003). Preparation of silica membranes inside macroporous alumina tubes by PECVD for hydrogen selectivity. *Sep. Purif. Technol.* **32,** 37–43.
[58] Yasuda, H. (1985). *Plasma Polymerisation.* Academic Press Inc, New York.
[59] Pages, X., Rouessac, V., Cot, D., Nabias, G., and Durand, J. (2001). Gas permeation of PECVD membranes inside alumina substrate tubes. *Sep. Purif. Technol.* **25,** 399–406.
[60] Favennec, L., Jousseaume, V., Rouessac, V., Fusalba, F., Durand, J., and Passemard, G. (2004). Porous extreme low k dielectrics using a PECVD porogen approach. *Mater. Sci. Semicond. Process.* **7,** 277–282.
[61] Rouessac, V., Puyrenier, W., Broussous, L., Rebiscoul, D., and Ayral, A. (2006). How ellipsometry can characterize a gas diffusion barrier. *In* "Proceedings of the 9th International Conference on Inorganic Membranes, ICIM" (R. Bredesen and H. Raeder, eds.), pp. 484–487. Lillehammer, Norway.
[62] Barranco, A., Cotrino, J., Yubero, F., Espinos, J. P., Contreras, L., and Gonzales-Elipe, A. R. (2004). A novel PECVD procedure for the room-temperature synthesis of SiO_2 thin films with controlled porosity. *Chem. Vap. Deposition* **10,** 17–20.
[63] Finot, E., Roualdes, S., Kirchner, M., Rouessac, V., Berjoan, R., Durand, J., Goudonnet, J. P., and Cot, L. (2002). Surface investigation of plasma HMDSO membranes post-treated by CF_4/Ar plasma. *Appl. Surf. Sci.* **187,** 326–338.
[64] Bredesen, R., Jordal, K., and Bollan, O. (2004). High-temperature membranes in power generation with CO_2 capture. *Chem. Eng. Process.* **43,** 1129–1158.
[65] Verweij, H., Lin, J. Y. S., and Dong, J. (2006). Micro-porous silica and zeolite membranes for hydrogen purification. *MRS Bull.* **31(10),** 756–764.
[66] Asaeda, M. (2006). Microporous silica-based membranes—present status and prospect. *In* "Proceedings of the 9th International Conference on Inorganic Membranes, $ICIM_9$" (R. Bredesen and H. Raeder, eds.), pp. 38–47. Lillehammer, Norway.
[67] Benes, N. E., Nijmeijer, A., and Verweij, H. (2000). Microporous silica membranes. *In* "Recent Advances in Gas Separations by Microporous Membranes, Membrane Science and Technologies Series, 6" (N. Kanellopoulos, ed.), pp. 335–372. Elsevier, Amsterdam.
[68] M. L. Mottern, K. Shqau, J. Y. Shi, D. Yu, H. Verweij, Thin supported inorganic membranes for energy-related gas and water purification, *Int. J. Hydrogen Energy* (2006), doi:10.1016/j.ijhydene.2006.08.042

[69] Asaeda, M, and Kitao, S. (July 1–4, 1991). Separation of molecular mixtures by inorganic porous membrane of ultrafine pores. *In* "Proceedings of the 2nd International Conference on Inorganic Membranes, $ICIM_2$" (A. J. Burggraaff, J. Charpin, and L. Cot, eds.) , pp. 295–300. Montpellier, FranceKey Eng. Mater. **61/62,** 295–300.
[70] http://www.mamut.com/pervatechbv.
[71] Casado, C., Urtiaga, A., Gorri, D., and Ortiz, I. (2005). Pervaporative dehydration of organic mixtures using a commercial silica membrane: determination of kinetic parameters. *Sep. Purif. Technol.* **42,** 39–45.
[72] ten Elshof, J. E., Abadal, C. R., Sekuli, J., Chowdhury, S. R., and Blank, D. H. A. (2003). Transport mechanisms of water and organic solvents through microporous silica in the pervaporation of binary liquids. *Microporous Mesoporous Mater.* **65,** 197–208.
[73] Wynn, N. (2000). Dehydration with silica pervaporation membranes. *Sulzer Tech. Rev.* **82,** 10–11.
[74] van Veen, H. M., van Delft, Y. C., Engelen, C. W. R., and Pex, P. P. A. C. (2001). Dewatering of organics by pervaporation with silica membranes. *Sep. Purif. Technol.* **22–23,** 361–366.
[75] Gallego-Lizon, T., Edwards, E., Lobiundo, G., and Freitas dos Santos, L. (2002). Dehydration of water/*t*-butanol mixtures by pervaporation: Comparative study of commercially available polymeric, microporous silica and zeolite membranes. *J. Membr. Sci.* **197,** 309–319.
[76] Peters, T. A., Fontalvo, J., Vorstman, M. A. G., Benes, N. E., van Dam, R. A., Vroon, Z. A. E. P., van Soest-Vercammen, E. L. J., and Keurentjes, J. T. F. (2005). Hollow fibre microporous silica membranes for gas separation and pervaporation—synthesis, performance and stability. *J. Membr. Sci.* **248,** 73–80.
[77] Asaeda, M., Sakou, Y., Yang, J., and Shimazaki, K. (2002). Stability and performance of porous silica–zirconia composite membranes for pervaporation of aqueous organic solutions. *J. Membr. Sci.* **209,** 163–175.
[78] Urtiaga, A., Casado, C., Asaeda, M., and Ortiz, I. (2006). Comparison of SiO_2-ZrO_2–50% and commercial SiO_2 membranes on the pervaporative dehydration of organic solvents. *Desalination* **193,** 97–102.
[79] Wu, J. C. S., Sabol, H., Smith, G. W., Flowers, D. L., and Liu, P. K. T. (1994). Characterization of hydrogen-permselective microporous ceramic membranes. *J. Membr. Sci.* **96,** 275–287.
[80] Collins, J. P., Schwartz, R. W., Sehgal, R., Ward, T. L., Brinker, C. J., Hagen, G. P., and Udovich, C. A. (1996). Catalytic dehydrogenation of propane in hydrogen permselective membrane reactors. *Ind. Eng. Chem. Res.* **35,** 4398–4405.
[81] Yoshino, Y., Suzuki, T., Nair, B. N., Taguchi, H., and Itoh, N. (2005). Development of tubular substrates, silica based membranes and membrane modules for hydrogen separation at high temperature. *J. Membr. Sci.* **267,** 8–17.
[82] Tsuru, T., Yamaguchi, K., and Yoshioka, T. (2004). Methane steam reforming by microporous catalytic membrane reactors. *AIChE J.* **50**(11), 2794–2805.
[83] Asaeda, M., and Kashimoto, M. (1998). Sol-gel silica membranes for separation of hydrogen at high temperature—separation performance and stability against steam. *In* "Proceedings of the 5th International Conference on Inorganic Membranes, $ICIM_5$" (S.-I. Nakao, M. Asaeda, and S. Morooka, eds.), pp. 172–175. Nagoya, Japan.
[84] Kanezashi, M., and Asaeda, M. (2006). Hydrogen permeation characteristics and stability of Ni-doped silica membranes in steam at high temperature. *J. Membr. Sci.* **271,** 86–93.
[85] Su, C., Jin, T., Kuraoka, K., Matsumura, Y., and Yazawa, T. (2005). Thin palladium film supported on SiO_2-modified porous stainless steel for high hydrogen flux membrane. *Ind. Eng. Chem. Res.* **44,** 3053–3058.
[86] Jun, C. S., and Lee, K. H. (2000). Palladium and palladium alloy composite membranes prepared by metal-organic chemical vapor deposition method (cold-wall). *J. Membr. Sci.* **176,** 121–130.
[87] Yan, S., Maeda, H., Kusakabe, K., and Morooka, S. (1994). Thin palladium membrane formed in support pores by metal-organic chemical vapor deposition method and application to hydrogen separation. *Ind. Eng. Chem. Res.* **33,** 616–622.
[88] Hacarlioglu, P., Gu, Y., and Oyama, S. T. (2006). Studies of the methane steam reforming reaction at high pressure in a ceramic membrane reactor. *J. Nat. Gas Chem.* **15,** 73–81.

[89] Tsuru, T., Shintani, H., Yoshioka, T., and Asaeda, M. (2006). A bimodal catalytic membrane having a hydrogen-permselective silica layer on a bimodal catalytic support: Preparation and application to the steam reforming of methane. *Appl. Catal. A Gen.* **302,** 78–85.
[90] Z. D. Ziaka, S. Vasileiadis, (1996). Reactor-Membrane Permeator Process for Hydrocarbon Reforming and Water Gas-Shift Reactions, US Patent 6090312
[91] Haryanto, A., Fernando, S., Murali, N., and Adhikari, S. (2005). Current status of hydrogen production techniques by steam reforming of ethanol: A review. *Energy Fuels* **19,** 2098–2106.
[92] Brunetti, A., Barbieri, G., Drioli, E., Lee, K. H., Sea, B., and Lee, D. W. (2007). WGS reaction in a membrane reactor using a porous stainless steel supported silica membrane. *Chem. Eng. Process.* **46,** 119–126.
[93] Ioannides, T., and Verykios, X. E. (1996). Application of a dense silica membrane reactor in the reactions of dry reforming and partial oxidation of methane. *Catal. Lett.* **36(3–4),** (165–169.
[94] Hwang, G. J., and Onuki, K. (2001). Simulation study on the catalytic decomposition of hydrogen iodide in a membrane reactor with a silica membrane for the thermochemical water-splitting IS process. *J. Membr. Sci.* **194,** 207–215.
[95] Hwang, G. J., Kim, J. W., Choi, H. S., and Onuki, K. (2003). Stability of a silica membrane prepared by CVD using γ- and α -alumina tube as the support tube in the HI–H_2O gaseous mixture. *J. Membr. Sci.* **215,** 293–302.
[96] Morooka, S., Kim, S. S., Yan, S., Kusakabe, K., and Watanabe, M. (1996). Separation of hydrogen from an H_2-H_2O-HBr system with an SiO_2 membrane formed in macropores of an α-alumina support tube. *Int. J. Hydrogen Energy* **21,** 183.
[97] Verveij, H. (2003). Ceramic membranes: Morphology and transport. *J. Mater. Sci.* **38**(23), 4677–4695.
[98] Lee, D., and Oyama, S. T. (2002). Gas permeation characteristics of a hydrogen selective supported silica membrane. *J. Membr. Sci.* **210,** 291–306.
[99] Oyama, S. T., Lee, D., Hacarlioglu, P., and Saraf, R. F. (2004). Theory of hydrogen permeability in nonporous silica membranes. *J. Membr. Sci.* **244,** 45–53.
[100] S. T. Oyama, A. Prabhu, (2003). Hydrogen-Selective Silica Based Membrane, US Patent 6527833.
[101] Masaryk, J. S., and Fulrath, R. M. (1973). Diffusivity of helium in fused silica. *J. Chem. Phys.* **59** (3), 1198–1202.
[102] Barrer, R. M., and Vaughan, D. E. W. (1967). Solution and diffusion of helium and neon in tridymite and cristobalite. *Trans. Faraday Soc.* **63,** 2275–2290.
[103] Yoshino, Y., Suzuki, T., Nair, B. N., Taguchi, H., Nomura, M., Nakao, S., and Itoh, N. (2006). Silica membrane module for hydrogen separation at high temperatures. *In* "Proceedings of the 9th International Conference on Inorganic Membranes, $ICIM_9$" (R. Bredesen and H. Raeder, eds.), pp. 143–148. Lillehammer, Norway.
[104] Lee, R. W., Frank, R. C., and Swets, D. E. (1962). Diffusion of hydrogen and deuterium in fused quartz. *J. Chem. Phys.* **36(4),** 1062–1071.
[105] Lee, R. W. (1963). Diffusion of hydrogen in natural and synthetic fused quartz. *J. Chem. Phys.* **38,** 448–455.
[106] Shelby, J. E. (1972). Helium migration in natural and synthetic vitreous silica. *J. Am. Ceram. Soc.* **55,** 61–64.
[107] Shelby, J. E. (1971). Temperature dependence of He diffusion in vitreous SiO_2. *J. Am. Ceram. Soc.* **54,** 125–126.
[108] Shelby, J. E. (1972). Neon migration in vitreous silica. *J. Am. Ceram. Soc.* **55,** 167–170.
[109] Perkins, W. G., and Begeal, D. R. (1971). Diffusion and permeation of He, Ne, Ar, Kr and D_2 through silicon oxide thin films. *J. Chem. Phys.* **54,** 1683–1694.
[110] de Lange, R. S. A., Keizer, K., and Burggraaf, A. J. (1995). Analysis and theory of gas transport in microporous sol-gel derived ceramic membranes. *J. Membr. Sci.* **104,** 81–100.
[111] Iwamoto, Y., Ikuhara, Y. H., Mori, H., and Fujisaki, S. (2006). Novel amorphous silica-based nanocomposite membranes for high temperature separation of hydrogen. *In* "Proceedings of the 9th International Conference on Inorganic Membranes, $ICIM_9$" (R. Bredesen and H. Raeder, eds.), pp. 143–148. Lillehammer, Norway.
[112] Tsuru, T., Yamaguchi, K., Yoshioka, T., and Asaeda, M. (2004). Methane steam reforming by microporous catalytic membrane reactors. *AIChE J.* **50,** 2794–2805.

[113] Galuska, J., Giddings, T., and Clelland, I. (2006). Silica H-membrane thermal and hydrothermal stability. *In* "Proceedings of the 9th International Conference on Inorganic Membranes, ICIM$_9$" (R. Bredesen and H. Raeder, eds.), pp. 124–126. Lillehammer, Norway.
[114] Wynn, N. (2001). Dehydration with silica pervaporation membranes. *Membr. Technol.* **2001** (129), 10–11.
[115] Pan, X. L., Stroh, N., Brunner, H., Xiong, G. X., and Sheng, S. S. (2003). Deposition of sol–gel derived membranes on α-Al_2O_3 hollow fibers by a vacuum-assisted dip-coating process. *J. Membr. Sci.* **226,** 111–118.
[116] de Jong, J., Benes, N. E., Koops, G. H., and Wessling, M. (2004). Towards single step production of multi-layer inorganic hollow fibers. *J. Membr. Sci.* **239,** 265–269.
[117] Biesheuvel, P. M., Breedveld, V., Higler, A. P., and Verweij, H. (2001). Graded membrane supports produced by centrifugal casting of a slightly polydisperse suspension. *Chem. Eng. Sci.* **56**(11), 3517–3525.
[118] Pex, P. P. A. C., and van Delft, Y. C. (2005). Silica membranes for hydrogen fuel production by membrane water gas shift reaction and development of a mathematical model for a membrane reactor. *In* "Carbon Dioxide Capture for Storage in Deep Geologic Formations—Results from the CO_2 Capture Project: Capture and Separation of Carbon Dioxide from Combustion Sources." Vol. 1. ECN report RX-05-123. (Chapter 17).
[119] Nijmeijer, A., Kruidhof, H., Bredesen, R., and Verweij, H. (2001). Preparation and properties of hydrothermally stable γ-alumina membranes. *J. Am. Ceram. Soc.* **84(1),** 136–140.
[120] Lee, D. W., Lee, Y. G., Nam, S. E., Sea, B., and Lee, K. H. (2003). Preparation and characterization of SiO_2 composite membrane for purification of hydrogen from methanol steam reforming as an energy carrier system for PEMFC. *Sep. Purif. Technol.* **32,** 45–50.
[121] Tsai, C. Y., Tam, S. Y., Lu, Y., and Brinker, C. J. (2000). Dual-layer asymmetric microporous silica membranes. *J. Membr. Sci.* **169,** 255–268.
[122] Jiang, S., Yan, Y., and Gavalas, G. R. (1995). Temporary carbon barriers in the preparation of H_2-permselective silica membranes. *J. Membr. Sci.* **103,** 211–218.
[123] Asaeda, M., and Yamasaki, S. (2001). Separation of inorganic/organic gas mixtures by porous silica membranes. *Sep. Purif. Technol.* **25,** 151–159.
[124] Gopalakrishnan, S., Nomura, M., Sugawara, T., and Nakao, S. (2006). Preparation of a multi-membrane module for high-temperature hydrogen separation. *Desalination* **193,** 230–235.
[125] Lin, Y. S., de Vries, K. J., and Burggraaf, A. J. (1991). Thermal stability and its improvement of the alumina membrane top-layers prepared by sol-gel methods. *J. Mater. Sci.* **26(3),** 715–720.
[126] de Vos, R. M., Maier, W. F., and Verweij, H. (1999). Hydrophobic silica membranes for gas separation. *J. Membr. Sci.* **158,** 277–288.
[127] Park, H. B., and Moo Lee, Y. (2003). Pyrolytic carbon–silica membrane: a promising membrane material for improved gas separation. *J. Membr. Sci.* **213,** 263–272.
[128] Duke, M. C., Diniz da Costa, J. C., Lu, G. Q. M., Petch, M., and Gray, P. (2004). Carbonised template molecular sieve silica membranes in fuel processing systems: permeation, hydrostability and regeneration. *J. Membr. Sci.* **241,** 325–333.
[129] Nijmeyer, A. (1999). Hydrogen-selective silica membranes for use in membrane steam reforming. Ph.D. Thesis Twente University, The Netherlands.
[130] Mori, H., Fujisaki, S., Siato, T., Sumino, T., and Iwamoto, Y. (2006). Characterization and hydrogen interaction studies on cobalt-doped amorphous silica composite materials for high temperature hydrogen separation membranes. *In* "Proceedings of the 9th International Conference on Inorganic Membranes, ICIM$_9$" (R. Bredesen and H. Raeder, eds.), pp. 382–385. Lillehammer, Norway.
[131] Ikuhara, Y., Saito, T., Mori, H., and Iwamoto, Y. (2006). Hydrogen adsorption property and permselective properties of Ni nanoparticle-dispersed amorphous Si-O membranes. *In* "Proceedings of the 9th International Conference on Inorganic Membranes, ICIM$_9$" (R. Bredesen and H. Raeder, eds.), pp. 488–491. Lillehammer, Norway.
[132] http://www.nedo.go.jp/english/activities/2_sinenergy/1/p02010e.html

CHAPTER THREE

From Polymeric Precursors to Hollow Fiber Carbon and Ceramic Membranes

Ahmad Fauzi Ismail* *and* Kang Li†

Contents

1. General Introduction

Polymeric membranes are widely used in industrial applications because they are reproducible in manufacturing and offer acceptable separation performance. However, current polymeric membrane materials have reached a limit in their permeability/selectivity trade-off. To overcome this limitation, many researchers have investigated carbon molecular sieve and ceramic materials in view of developing advanced membrane materials for fluid separation processes.

In recent years, inorganic membranes particularly carbon and ceramic membranes emerged as the important membranes for fluid separation because of their

* Membrane Research Unit, Faculty of Chemical and Natural Resources Engineering, Universiti Teknologi Malaysia, 81310 Skudai, Johor, Malaysia
† Department of Chemical Engineering, Imperial College London, South Kensington Campus, London SW7 2AZ, United Kingdom

Membrane Science and Technology, Volume 13
ISSN 0927-5193, DOI: 10.1016/S0927-5193(07)13003-5

superior thermal resistance, stability in corrosive environment, and excellent selectivity and permeability. It is, therefore, the aim of this chapter to elaborate the preparation of carbon and ceramic membranes from its polymeric precursor. The discussion will be based on hollow fiber configuration because it provides large membrane area per unit volume. This chapter has been written in two parts which represent the two membrane materials. This is because of the differences in the preparation of polymeric precursors and the treatment processes involved.

2. Part 1: Polymeric Precursors of Hollow Fiber Carbon Membranes

2.1. Introduction

The development of porous inorganic membranes commenced before 1945, long before the development of today's synthetic organic membranes. Not much technical information was available in the early development of inorganic membranes because the first porous inorganic membranes were developed for the separation of uranium isotopes, which were mainly used for military purposes or nuclear applications.

However, the concept of carbon membrane or film for gas separation can be traced back to the early 1970s when carbon membranes were prepared by compressed nonporous graphited carbon into a plug, called as carbon membrane [1]. At present, carbon membrane materials become more important in the new era of membrane technology for gas separation and liquid-liquid phase separation due to their higher selectivity and permeability, stability and high temperature operations, and well-defined stable pore structure. In addition, carbon membranes have showed attractive characteristics among molecular sieving materials such as excellent shape selectivity for planar molecule and high hydrophobicity, and it is more feasible to form carbon molecular sieve membranes (CMSMs) [2–4]. These characteristics had stimulated interest among researchers in the early 1980s to further investigate the separation characteristics of these membranes. Recently, carbon membranes were also found to be a very good membrane material in the preparation of nanofiltration [5], pervaporation [6], and microfiltration membrane [7].

The interest in developing carbon membranes grew rapidly after a successful preparation of apparently crack-free molecular sieving hollow fiber membranes by carbonizing cellulose hollow fiber [8–11]. The dependence of permeabilities and selectivities on temperature, pressure, and extent of pore for both adsorbing and nonadsorbing permeates has been reported. However, those membranes would be lack of mechanical strength for practical applications.

The preparation of carbon membranes is not an easy task because it involves a number of important steps that must be carefully controlled and optimized. The success of preparing carbon membranes lies in the experience and knowledge of the preparation of polymeric precursor or membranes in particular because good quality membranes will ensure good quality carbon membranes. The preparation of carbon membranes can be divided into five steps: precursor material selection, polymeric membrane preparation, pretreatment of the precursor, pyrolysis process, and posttreatment of pyrolyzed membranes.

Among the most important factors that influence the preparation of carbon membranes are: pretreatment variables, pyrolysis process parameters, and posttreatment conditions. Thus, understanding and manipulation of such variables will provide an opportunity to further enhance the performance of carbon membranes. All of these aspects will be discussed in detail below.

2.2. Preparation of carbon membranes: A general process

This section briefly discusses the process taken in the preparation of carbon membrane. Generally, the preparation of carbon membranes involves five important processes, as shown in Fig. 3.1. Simultaneously, there are several factors that need to be considered in order to ensure the success of each process.

Among these processes, the pyrolysis is the most important stage and can be regarded as the heart of the carbon membrane preparation process. At this stage, the pore structure of the carbon membrane is formed, and this determines the ability of a carbon membrane to activate separation processes [12]. Various methods and parameters that are involved in each step are discussed and compared in the next section. However, the choice of the most suitable method depends mainly on the intended application of the carbon membrane.

2.3. Precursor selection

Most of the carbon membranes available today are prepared from polymeric precursors, therefore, the choice of the polymeric precursors is the first important factor that needs consideration. This is because pyrolysis of different precursors may result

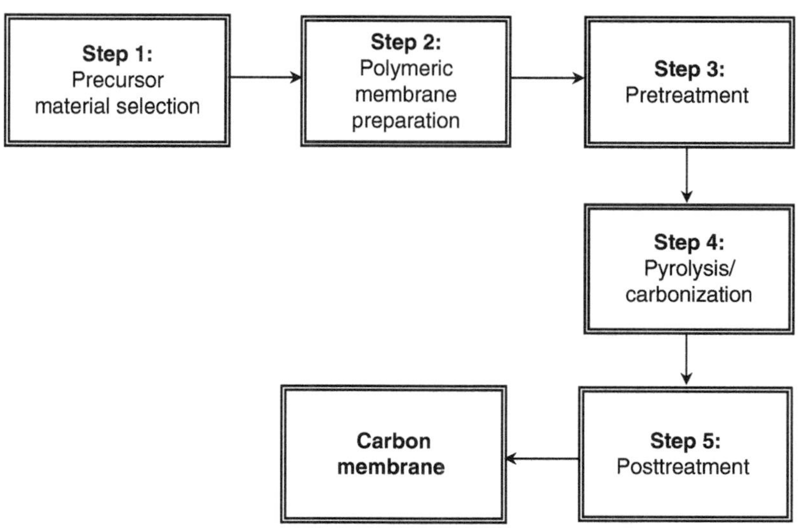

Figure 3.1 Carbon membrane preparation process [11].

in different types of carbon membranes [13]. The currently available carbon membranes can be produced through the carbonization or pyrolysis process of suitable carbon-containing materials such as thermosetting resin, graphite, coal, pitch and plants, under inert atmosphere, or vacuum [14]. At present, there are numerous synthetic precursors being used in the preparation of carbon membranes; among the most widely used are polyimide and derivatives, polyacrylonitrile (PAN), phenolic resin, polyfurfuryl alcohol (PFA), polyvinylidene chloride-acrylate terpolymer (PVDC-AC), phenol formaldehyde, cellulose, and others.

Because of the variety of polymeric precursors available for carbon membrane preparation, the most suitable characteristics for carbon membrane must be identified in producing successful carbon membranes. Through practical experience and knowledge gathered during carbon membrane preparation exercises, two most dominant characteristics have been identified that are thermoresistance and good arrangements of polymer molecules at molecular levels. These characteristics allow further fine tuning of membrane morphology as well as separation performance optimization.

In addition to high temperature resistance, a thermosetting polymer neither liquefies nor softens during any stage of pyrolysis [15]. Suitable precursor materials for carbon membrane preparation will not cause any pore holes or cracks to appear after the pyrolysis step [16]. Based on a practical experience, another important reason why polymer precursor is preferred is because the resultant membrane prepared has fewer impurities as compared with activated carbon precursor [17]. Based on these rationales, therefore, thermosetting polymeric precursors have been primarily used in manufacturing carbon membranes, as shown in Table 3.1. Detailed descriptions for each selected precursor are given below.

2.3.1. Polyimide and derivatives

At present, aromatic polyimide is one of the most systematically studied polymer precursor and has been successfully used for the preparation of carbon membranes (Table 3.1). Aromatic polyimides, either used as it is or blended with other polymers like polyethylene glycol (PEG), are also known as the best materials for the preparation of carbon membranes due to their good separation and mechanical properties [18, 19]. This is because polyimides possess the required properties such as rigidity, high melting point, high glass transition temperature (T_g), and thermally stable polymers formed by the condensation reactions of dianhydrides with diamines as depicted in Fig. 3.2 [20, 21]. Polyimide is normally synthesized on laboratory scale with different types of dianhydrides and diamines in order to tailor its separation properties when being used as a membrane material as shown in Fig. 3.2.

The most common commercial polyimide used for preparing carbon membrane is Kapton (DuPont™) polyimide. Carbon molecular sieves film with homogeneous fine pores and without cracks or large pores can normally be produced by carbonizing Kapton film at 800°C [22]. Kapton polyimide had also been carbonized in the form of a capillary membrane, which can retain high gas permselectivity even at high temperature [23]. The membrane exhibited a permselectivity of 2000 for He/N_2 at 0°C and a permselectivity of 170 at 250°C. Recent Kapton polyimide-based carbon

Table 3.1 Examples of carbon membrane precursors and pyrolysis conditions [11]

Precursor	Connuration	Pyrolysis condition (temperature; heating rates; soak times; atmosphere)
Acrylonitrile	Hollow fiber	800–1600°C; –; 5–60 min; N_2
Acrylonitrile	Hollow fiber	600–1200°C; –; 10 min; N_2
Polyacrylonitrile (PAN)	Hollow fiber	250°C–800°C, 9°C/min; 10–180 min; N_2
PAN	Hollow fiber	600 and 950°C; 1°C/min; –; N_2
PAN	Hollow fiber	900°C; 5°C/min; –; N_2
Cellulose	Hollow fiber or supported film	400–900°C; 1–10°C/min; –; Ar
Cellulose	Hollow fiber or supported film	500–800°C; –; 12 h; Ar
Cellulose	Hollow fiber	120–400°C; 0.1–0.6°C/min; –; Ar
Coal tar pitch	Plate	600–900°C; 1°C/min; –; N_2
Kapton and Matrimid	Supported film	450–700°C; 0.5°C/min; 1 h; vacuum
Kapton and Matrimid	Film	500–800°C; 1 or 4°C/min; 2 or 8 h; vacuum
Kapton Polyimide	Film	600–1000°C; 10°C/min; 2 h; vacuum or Ar
Kapton Polyimide	Capillary	950°C; 5°C/min; 120 min; –
Phenol formaldehyde	Supported film	900°C; 0.5°C/min; 60 min; Ar
Phenolic resin	Supported film	500–1000°C; 5°C/min; –; vacuum
Phenolic resin	Supported film	700°C; 0.5°C/min; –; vacuum;
Phenolic resin	Supported film	700°C; –; –; vacuum
Poly (*para*/phenylene pyromellitimide) blend with PEG	Film	600°C; 3°C/min; 1 h; Ar

membrane showed that the permselectivities for O_2/N_2, CO_2/CH_4, and CO_2/N_2 are 4, 16, and 9, respectively, at 25°C [24].

Another form of polyimide used to form carbon membrane is polyetherimide (PEI), which is manufactured by General Electric under the trade name of Ultem® 1000. This material has been used to prepare polymeric membranes due to its superior strength and chemical resistance [25, 26]. Among the first application of PEI as carbon membrane precursor was in the preparation of carbon film through

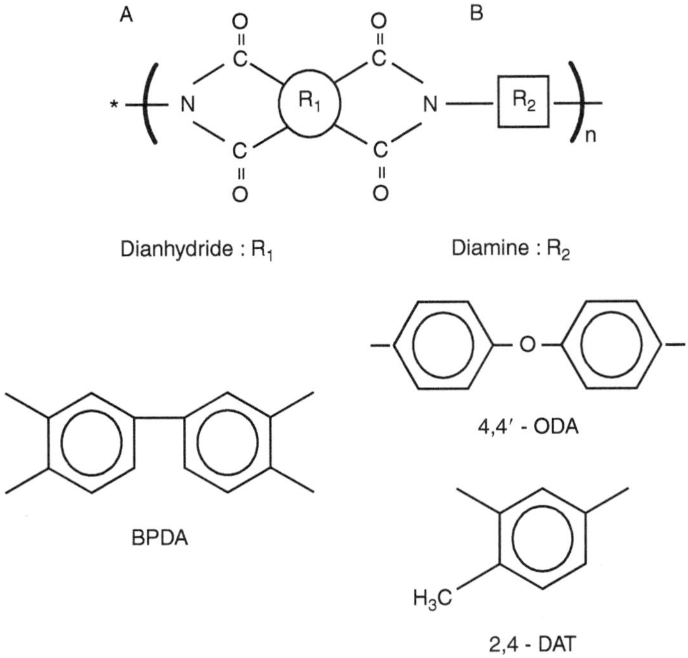

Figure 3.2 Basic molecular structure of polyimide [21].

one step of casting resulting in the permselectivity value of 7.4, 121, 25, and 15 for O_2/N_2, He/N_2, CO_2/CH_4, and CO_2/N_2, respectively, at 25°C [26].

More recently, PEI carbon hollow fiber membrane has been prepared to be used as a catalyst support in a membrane reactor. The pyrolysis process parameters and stabilization parameters were found to be the dominant parameters in the preparation of PEI-based carbon membranes [27].

2.3.2. Phenolic resin

Phenolic resins are very popular and inexpensive polymers, which are employed in a wide range of applications, spanning from commodity construction to high technology materials [28]. Phenolic resins present suitable features to be applied as carbon membrane precursors, such as:

1. The pyrolysis of these materials provides carbon films with molecular sieve properties [29].
2. They provide a high carbon yield after carbonization [28].
3. They are low-cost polymers with numerous applications in different fields [29].

Intensive study on phenolic resin precursor in the preparation of carbon membranes was done by coating the phenolic resin film on a porous carbon with a spin-coating technique and then the supported film was cured by air at 150°C for 2 h, and followed by carbonization step under vacuum. The phenolic resin that was first carbonized at 700°C showed an O_2/N_2 selectivity around 10, and separation factors of 160 and 45 for CO_2/CH_4 and CO_2/N_2, respectively, in gas mixture separation experiments [28]. The phenolic resin was also used to prepare adsorption-selective carbon membranes (ASCMs) [30], which can separate nonadsorbable or weakly adsorbable gases (e.g., He, H_2, air, O_2, CH_4, CO_2) from adsorbable gases such as hydrocarbons (C_{2+}), NH_3, SO_2, H_2S, etc.

2.3.3. Polyfurfuryl alcohol

PFA is an amorphous polymer with a nongraphitizable structure, and potentially a good precursor to prepare CMSM [31]. Because of its simple molecular structure and formation mechanism, it is an appropriate material for fundamental experimental and atomistic simulation studies. Besides these characteristics, PFA has been chosen based on the knowledge that PFA-derived carbon could have desirable properties such as a narrow pore size distribution and chemical stability [32]. PFA is in liquid state at room temperature; it can be used only in supported films [33].

PFA has been extensively used as a precursor during the preparation of nanoporous carbon (NPC) membrane for gas separation [34–40]. Spray-coating technique was employed in order to produce a thin layer of NPC on the porous surface of a stainless steel disk support [32, 36]. Selectivity for O_2/N_2 separation of up to 4 was achieved for the supported membrane carbonized at 600°C in flowing helium. Further improvement in the preparation technique by using an ultrasonic deposition system, which uniformly distribute the polymeric solution onto the support, has successfully produced reproducible NPC membranes which improved O_2/N_2 selectivity within the range of 2–30, and provide large separation factors for H_2/CH_4, CO_2/CH_4, N_2O/N_2, and H_2/CO_2 of 600, 45, 17, and 14 respectively [34]. These successes recently stimulated systematic research in using PFA precursor for carbon membrane synthesis.

2.3.4. Polyacrylonitrile

PAN primarily has been recognized as the most important and promising precursor for producing high-performance carbon fiber. PAN possesses numerous advantages including a high degree of molecular orientation, higher melting point (PAN fiber tends to decompose below its melting point, T_m, of 317–330°C) and a greater yield of the carbon fiber [41].

The first work involving the preparation of PAN carbon membrane was found in 1990 [12]. Porous capillary carbon membranes have been prepared by nitrogen gas pyrolysis system at pyrolysis temperatures ranging from 800 to 1600°C. Preoxidation and carbonization, the precursor membranes were subjected to pretreatment with hydrazine solution. Later in 1992, researchers from Mitshubishi Rayon, Japan, patented their invention on the preparation of porous hollow fiber carbon membranes [42].

PAN-based carbon membranes were widely studied in hollow fiber configuration during the period of 1994–1995 by blending with other polymers, such as methyl methacrylate, PEG, and polyvinylpyrrolidone (PVP) [43, 44], in order to alter the final pore size distribution of the carbon membranes. In addition to blending, coating of PAN fibers with zeolites [45], polyimide [46], and perfluoro-sulfonated ionomer (PSI) [47], in order to produce composite membranes, was also sought.

However, after this period, there was not much research involving the use of PAN as a precursor for carbon membranes conducted until late 2002, when David and Ismail [48–49], prepared their carbon hollow fiber membrane from PAN precursor.

Although significant efforts have been made to develop asymmetric carbon hollow fiber membranes from PAN, it seems that PAN did not work well as a precursor for carbon membrane preparation due to its low separation performance. PAN, however, should be considered as one of the most interesting polymer precursors for carbon membranes in future due to its impressive mechanical properties. Therefore, intensive efforts need to be made in order to improve its intrinsic properties to be the future polymer precursor for the carbon membranes.

2.4. Preparation of carbon hollow fiber membrane

The first and foremost important step in achieving a good carbon hollow fiber membrane from polymeric precursors lies in the preparation of polymeric hollow fiber membrane. Polymeric precursors for preparation of carbon membranes must be prepared at optimum conditions in order to produce a carbon membrane of high quality. In the case of poor quality polymeric membranes, a pyrolysis process cannot be expected to satisfactorily produce good carbon membrane properties. Therefore, precursor membranes must be prepared in relatively defect-free form in order to minimize problems in subsequent processing during the preparation of carbon membranes.

For polymeric hollow fiber membranes, spinning parameters are crucial factors that must be controlled during the preparation of the membranes. These parameters include the amount and type of polymers, solvents, and additives mixed into the spinning dope solution, the dope and bore fluid flow rates, the fiber take-up velocity, the air gap distance (unless wet spinning is used), and the coagulant bath temperature [50]. The main challenges in hollow fiber spinning are as follows [51]:

- Synthesizing polymeric melts and solutions having appropriate thermodynamic and rheological properties.
- Determining the kinetic and thermodynamic conditions involved in transforming the solution/melt to a solid.
- Annealing and crystallizing (and orienting) the formed bodies.
- Posttreating to alter surface properties and stabilize the products.

2.4.1. Preparation of hollow fiber membrane

The dry/wet phase separation process has been known as a suitable process for preparing hollow fiber membranes as shown schematically in Fig. 3.3. In the spinning process, the dope was fed to the tube-in-orifice spinneret via a gear pump. The gear pump smoothly delivered the polymer solution to the spinneret with a few milliliters per minute dope extrusion rate. Syringe pump provided an accurate and pulse-free supply of water to the capillary in the spinneret. Spinning solution and bore fluids were extruded simultaneously through a tube-in-orifice spinneret to form a nascent hollow fiber at an ambient temperature. The inside surface of hollow fiber was contacted with a bore fluid and experienced wet phase

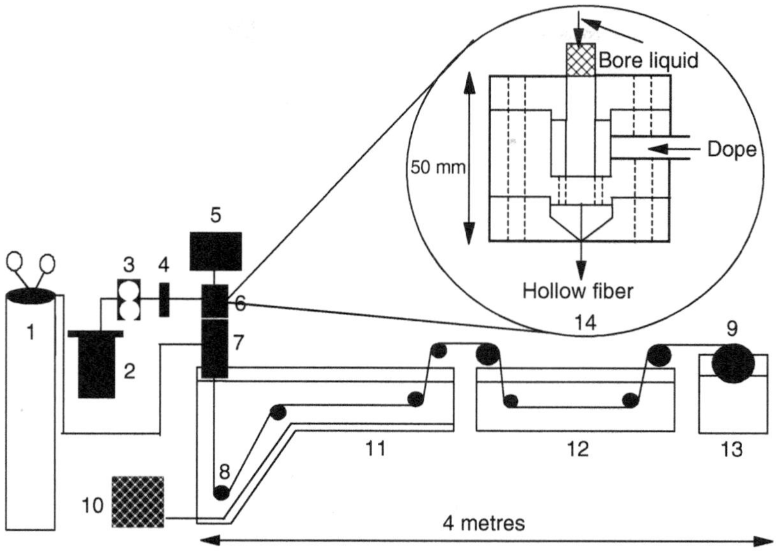

Figure 3.3 Schematic diagram of hollow fiber spinning system: (1) nitrogen cylinder; (2) dope reservoir; (3) gear pump; (4) on-line filter, 7 mm; (5) syringe pump; (6) spinneret; (7) forced convective tube; (8) roller; (9) wind-up drum; (10) refrigeration/heating unit; (11) coagulation bath; (12) washing/treatment bath; (13) wind-up bath; and (14) schematic spinneret.

separation in order to form a circular hollow lumen. Water can be chosen as the bore fluid.

The fiber was then passed through an air gap which was subjected to a controlled force convective environment for dry phase separation. The inert gas used was nitrogen. Subsequently, the fiber was immersed into the nonsolvent coagulation bath and the washing or treatment bath. The coagulation conditions ensure rapid solidification when the spinning dope comes in contact with coagulant. Wet phase separation process occurs in the coagulation bath.

After collecting the hollow fibers on a wind up drum, the fibers were washed in water and then, subjected to solvent exchange treatment using methanol and hexane. It is necessary to replace the water contained within the pores of the membrane structure with a volatile organic liquid having a much lower surface tension. The purpose of the solvent exchange is to reduce the drying stress. The hollow fibers were hung vertically and air dried at room temperature prior to pyrolysis process. Figure 3.4 shows a bundle of typical polymeric hollow fiber membranes.

2.4.2. Stabilization process

Before carbonization process, the hollow fibers were treated in an oxidizing atmosphere at a temperature range of 200–300 °C. The stabilization process is necessary to cross-link polymer chains and to prepare a structure that can withstand the rigors of high temperature processing. Stabilization process can ensure that both the molecules and the fibrillar orientation will not be lost during final heat treatment.

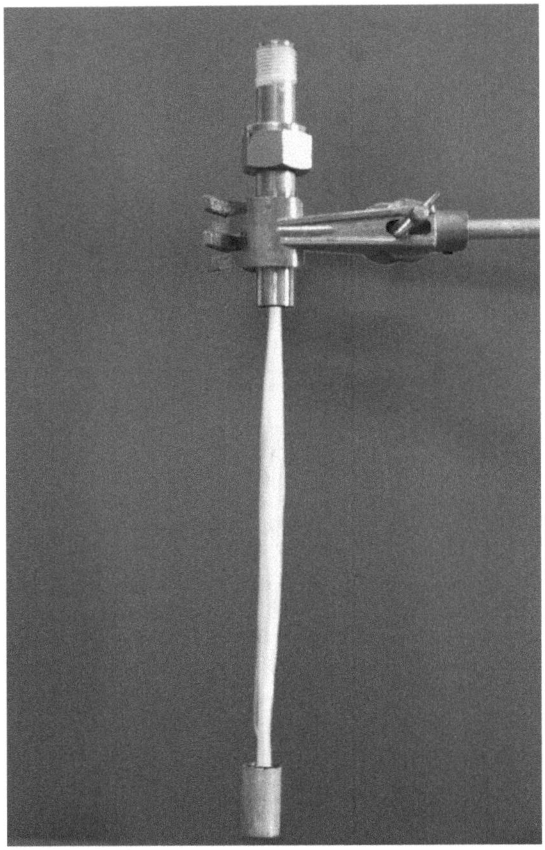

Figure 3.4 Typical polymeric hollow fiber membranes.

Therefore, either the inherent stiffness of the polymer molecules must be increased or the molecules must be "tied" together in order to eliminate or at least limit the relaxation and chain scission during the final carbonization step. The reaction is highly exothermic and if not controlled, can lead to run away of autocatalytic processes resulting in "melting" down and coalescing of fibers [52, 53]. The treatment converts polymeric membrane into ladder compound capable of withstanding the high temperature present in the subsequent process and increases the carbonization yield [41].

2.5. Pyrolysis process

The heart of the carbon hollow fiber membrane fabrication process is the pyrolysis process or carbonization process. Pyrolysis is a process in which a suitable carbon precursor is heated in a controlled atmosphere (vacuum or inert) to the pyrolysis temperature at a specific heating rate for a sufficiently long thermal soak time. The pyrolysis process is conventionally used for the production of porous carbon fibers, and causes the product to have a certain microporosity of molecular dimensions that is responsible for the molecular sieve properties of the carbon [12].

A pyrolysis system consists of a wire-wound tube furnace, flow meter, Pyrex tube or quartz tube, thermostabilization gases, and inert gas. Tube furnace was chosen rather than box furnace because it allows the purging of the inert gas during heating process. Quartz tube should be used in the pyrolysis system not only because it can withstand high temperature, but it also will not react with the membrane sample. Long quartz tube must be used to prevent the melting problem of the connecting tube at the end of the quartz tube. However, Pyrex tubes that are more economical can be used for heating temperature of not more than 600 °C.

Before the carbonization process started, the inert gas was purged into the pyrolysis system to remove unwanted air or oxygen. This is to prevent the oxidation process during high temperature process. The heating rate for the entire pyrolysis process should be maintained and precaution is required when approaching final pyrolysis temperature to avoid an overshoot problem [54]. Then, the precursor was heated to a required pyrolysis temperature and held for certain duration. After that, the resulting carbon membrane was cooled down to ambient temperature in the inert gas atmosphere. Finally, the precursor was completely carbonized to form carbon membrane. Figure 3.5 shows the resulting typical carbon hollow fiber membranes that were obtained.

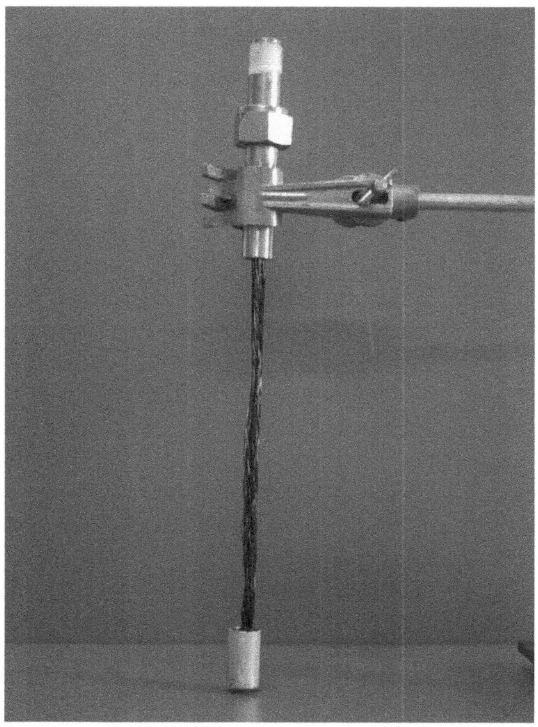

Figure 3.5 Typical carbon hollow fiber membranes.

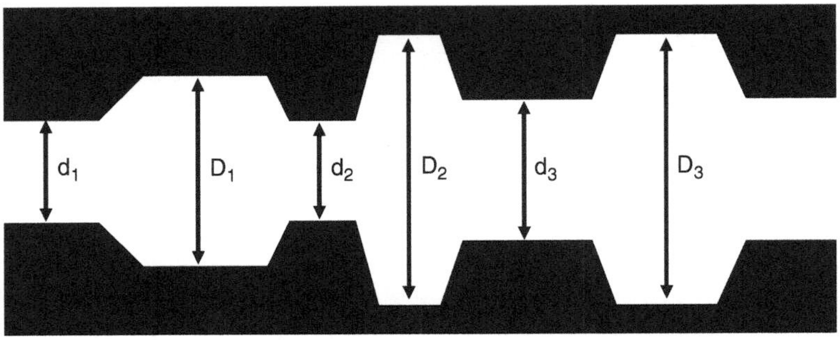

Figure 3.6 Idealized structure of a pore in a carbon material [41].

Pyrolysis creates an amorphous carbon material that exhibits a distribution of micropore dimension with only short-range order of specific pore sizes, and also pores larger than the ultramicropores required to exhibit molecular sieving properties. These larger pores connect the ultramicropores that perform the molecular sieving process, and therefore allow high gas separation productivities to be attained [20].

Generally, the pore system of carbon membranes is nonhomogeneous, as it is comprised of relatively wide openings with a few constrictions [55]. The pores vary in size, shape, and degree of connectivity depending on the morphology of the organic precursor and the chemistry of its pyrolysis. An idealized structure of a pore in a carbon material is shown in Fig. 3.6. The poremouth "*d*" often referred to as an ultramicropore (<10 Å) [29, 56] is thought to allow molecular sieving of the penetrating molecules. The larger micropore "*D*" of the material (6–20 Å) may allow the diffusion of gas molecules to occur through the carbon material. Therefore, properly prepared carbon materials not only have the ability to perform molecular sieving, but also may allow a considerably high flux of the penetrating molecules through the material [57].

The pores of the starting material do not change uncontrollably when the pyrolysis process is being carried out. Instead, the pore structure is essentially retained and can be controlled selectively by adjusting the various process parameters [12]. A number of different variables can affect the pyrolysis process, and small changes in pyrolysis parameters have been found to have a significant impact on the final properties of a carbon membrane [4, 18]. Among the factors that influence the pyrolysis process are the pyrolysis temperature, heating rates, thermal soak times, type of pyrolysis atmosphere and, most noticeably during inert gas pyrolysis, the gas flow rates, pressure, and concentration.

The pyrolysis process can be conducted within a wide range of temperatures, in between the decomposition temperature of the carbonaceous precursor and its graphitization temperature (generally about 3000°C). Usually, pyrolysis is carried out in the range between 500 and ~1000°C [13, 58]. Figures 3.7 and 3.8 show the typical hollow fiber membranes with a skin layer structure and substructure morphology produced under different pyrolysis temperatures.

The thermal soak time can be different depending on the final pyrolysis temperature [12]. This parameter may be used to fine-tune the transport properties of a carbon membrane using a particular final pyrolysis temperature [59]. Previous

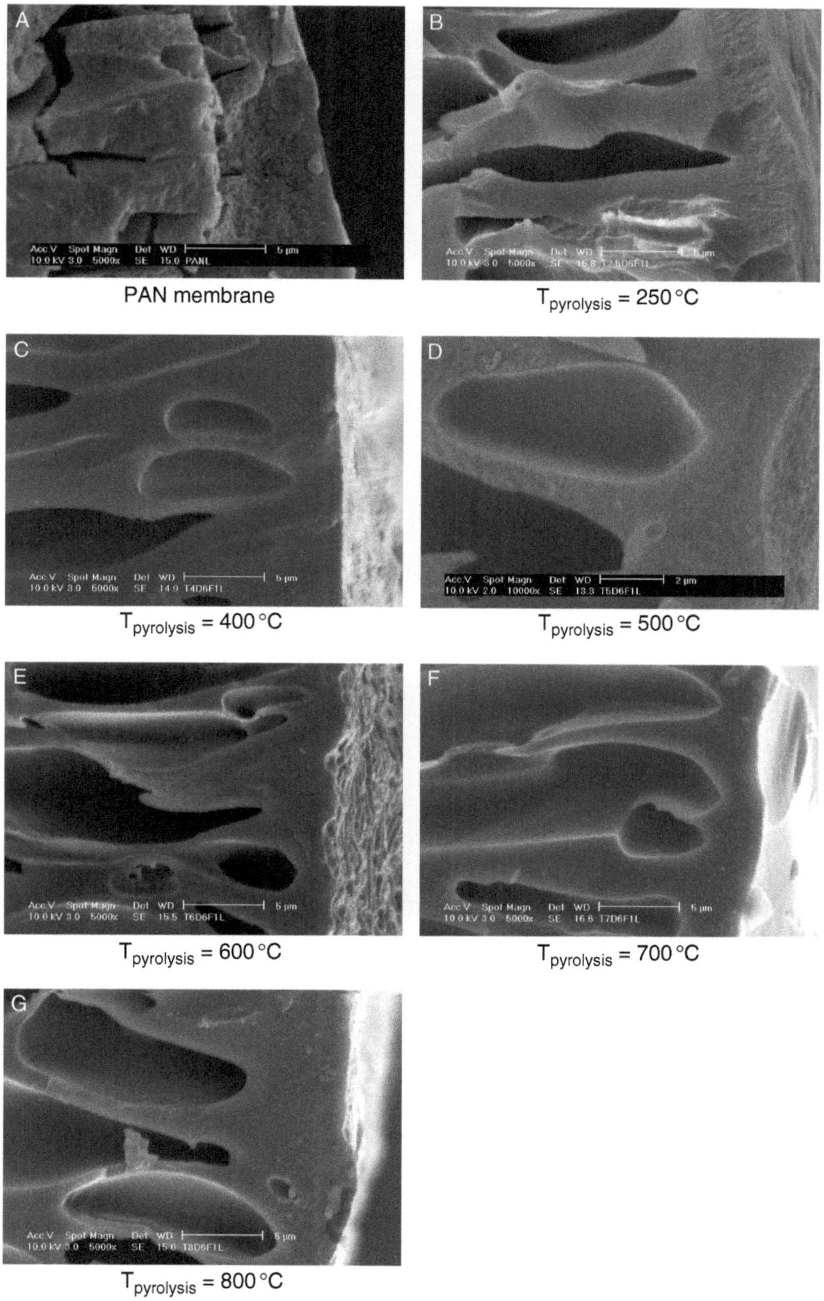

Figure 3.7 Structural changes of the typical hollow fiber membrane skin layer at different pyrolysis temperatures.

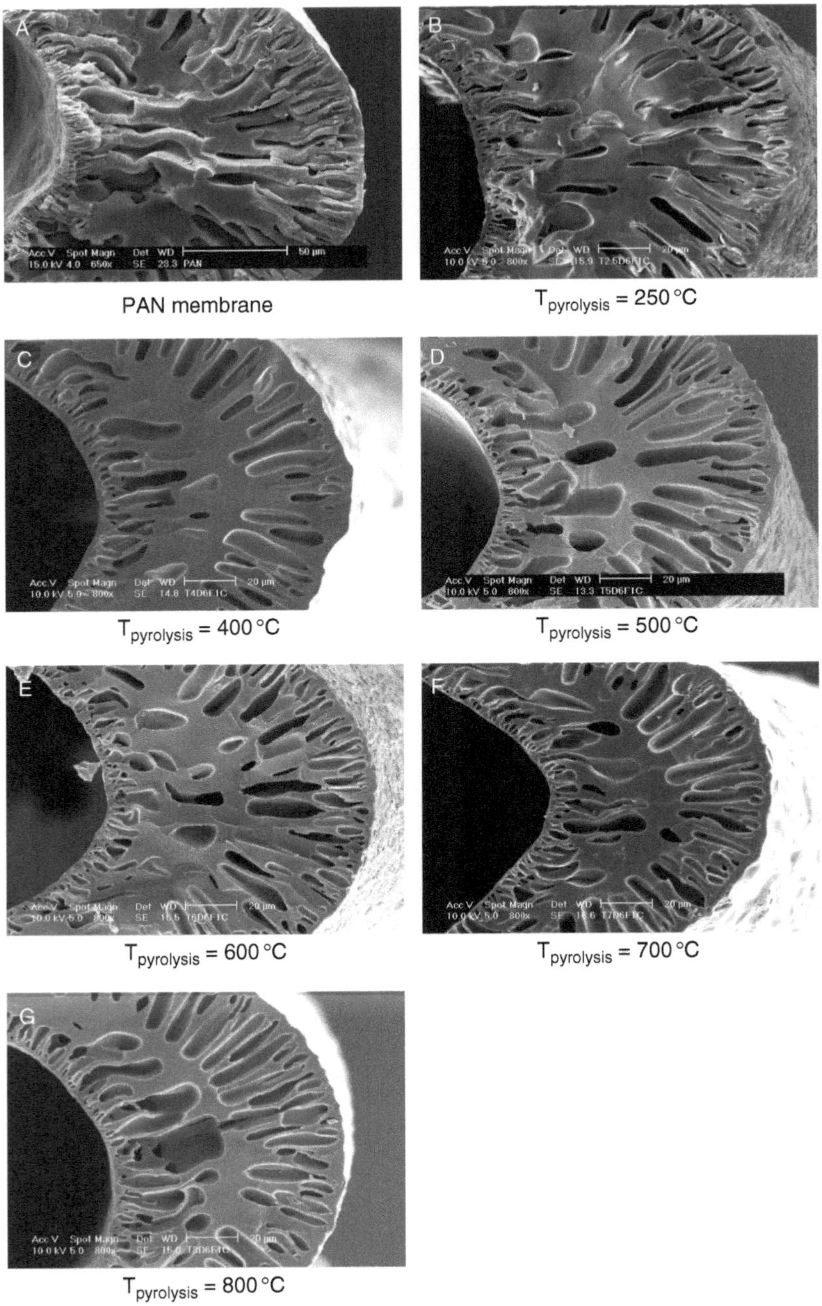

Figure 3.8 Structural change of the typical hollow fiber membrane substructure at different pyrolysis temperatures.

studies [24, 56, 57, 59, 60] showed that increments in thermal soak time would increase the selectivity of carbon membranes. It is believed that only microstructures rearrangement occurs during the thermal soak time, thus affecting the pore size distribution and average porosity of carbon membranes [20].

Heating rates will determine the evolution rate of volatile component from a polymeric membrane during its pyrolysis, and consequently affect the formation of pores in carbon membranes [16]. Widely different heating rates can be used, ranging from 1 to >10°C/min [13]. However, lower heating rates are preferable in order to produce small pores, to increase carbon crystallinity, and consequently to produce carbon membranes of higher selectivity [13, 24]. Higher heating rates may lead to the formation of pinholes, microscopic crack, blisters and distortions, which in extreme cases may render membranes useless (i.e., of low selectivity) for gas separation [16].

The pyrolysis atmosphere must be controlled in order to prevent undesired burn off and chemical damage of the membrane precursor during pyrolysis. Therefore, the pyrolysis can be carried out either in vacuum or in an inert atmosphere. Vacuum pyrolysis was reported to yield more selective but less permeable carbon membranes (from a polyimide precursor) than an inert gas pyrolysis system [20, 58]. When dealing with the inert gas pyrolysis system, one must consider how the inert gas flow rate will affect the performance of the resulting carbon membranes. Generally, an increase in gas flow rate will improve the permeability of carbon membranes without significantly interfering with the selectivity [58, 59].

2.6. Posttreatment

As a result of the pyrolysis process, polymeric membranes are transformed into carbon membranes with varying degrees of porosity, structure, and separation properties that depend to an extent on the carbonization conditions employed. In most cases, it is found to be an advantage that the pore dimensions and distribution in a carbon membrane can be finely adjusted by a simple thermochemical treatment to meet different separation needs and objectives [3]. At present, various posttreatment methods have been applied to meet the desired pore structure and separation properties of carbon membranes, including postoxidation, chemical vapor deposition (CVD), postpyrolysis, and coating. Discussions on each method are given below.

2.6.1. Postoxidation

Postoxidation or activation is the favorite posttreatment used by researchers to alter the pore structure of the carbon membranes. Typically, when a membrane is exposed to an oxidizing atmosphere after the pyrolysis step, the ensuing oxidation increases the average pore size [12, 16, 60, 61]. The oxidation of carbon membranes can be performed using pure oxygen, oxygen admixed with other gases, air, or other oxidizing agents such as steam, carbon dioxide, nitrogen oxides and chlorine oxides, or solutions of oxidizing agents such as nitric acid, mixtures of nitric and sulfuric acids, chromic acid, and peroxide solutions at elevated temperatures [61].

Kusakabe *et al.* [62] showed in their study that the postoxidation of a polyimide carbon membrane in a mixture of O_2–N_2 increased its permeability to CO_2 and O_2 without damaging its permselectivity significantly. Due to the oxidation process, the micropore volume increased significantly but the pore size distribution was not broadened.

As the oxidation temperature was raised, there was an increase in gas permeance for all gases (permanent and hydrocarbon) and a reduction in permselectivity of permanent gas pairs such as O_2/N_2 or CO_2/N_2 [31]. However, hydrocarbon gases (e.g., n-C_4H_{10}, C_2H_4, C_3H_6) exhibited higher permeances than permanent gases (e.g., He, N_2), which is a consequence of enhanced transport via hydrocarbon adsorption in the micropores [31]. It is believed that the activation process increases the permeability by expelling carbon atoms from the pore walls, thus enlarging them [63]. The variation observed in permselectivities suggests that for permanent gases (e.g., O_2, N_2), the transport mechanism through the membrane changes as oxidation progresses, from a molecular sieving mechanism to surface diffusion for highly oxidized samples [31].

2.6.2. Chemical vapor deposition

The selectivity of a carbon membrane may be increased through the introduction of organic species into the pore system of the carbon membrane and their pyrolytic decomposition (i.e., CVD) [61, 63–65]. Generally, to manufacture carbon molecular sieves, the inherent pore structure of the carbonaceous precursor is initially tailored into a suitable pore size range by controlling the thermal pretreatment, followed by a final adjustment of the pore apertures by CVD [61].

The organic species used for cracking and the conditions of their pyrolysis should be carefully chosen to produce sufficient and selective deposition of carbon at the pore apertures [66]. The organic molecules which can be used for this purpose are ethane, propane, ethylene, benzene, and other hydrocarbons [67]. A few additional organic sources for CVD and their application conditions are summarized in Table 3.2.

The CVD of carbon onto a carbon membrane may bring about three distinct results, namely homogeneous deposition, adlayer deposition, and in-layer

Table 3.2 Chemical vapor deposition (CVD) of carbon membrane

Precursor	CVD conditions	Reference
Cellulose	1–1–1-Trichloroethane saturated in Ar; 620°C; 5 min	[67]
Phenolic resin	N_2 saturated with trichloroethane (TCE); 500°C; 1–10 min	[68]
Polyfurfuryl alcohol	20% propylene in He; 600°C; 2 h	[52]
Polyimide	0.05 mol fraction propylene in carrier gas; 650°C; 1 h	[8]

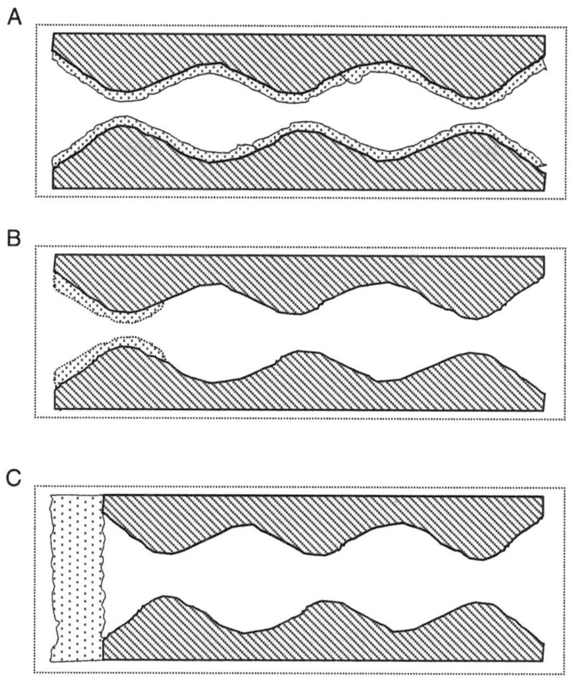

Figure 3.9 Mechanism of a carbon deposition on the pore system of carbon membrane: (A) homogeneous carbon deposition on membrane pore walls; (B) in-layer carbon deposition on membrane pore wall entrances; and (C) adlayer carbon deposition outside membrane pores [63].

deposition, as depicted in Fig. 3.9(A)–(C) [63]. The preferred mode is homogeneous deposition, but in practice, the three modes need not necessarily be completely distinct. For instance, there may be an adlayer on top of the in-layer or some deposition in depth in addition to the in-layer.

2.6.3. Postpyrolysis

Besides CVD, postpyrolysis is another treatment that can be used to decrease membrane pore size. However, this treatment is rarely used, because the first pyrolysis step at high temperature produces small pores efficiently due to the shrinkage of the carbon structure. Typically, postpyrolysis is applied after postoxidation in order to recover from an excessive pore enlargement.

The process of sintering is one way to decrease the pore volume. Sintering involves pore collapse, which proceeds more readily on the smaller pores. Heating the carbon membranes in an inert or reducing atmosphere to above 400°C (typically, 800°C) is one way to induce pore sintering. The more facile collapse of smaller pores is due to the fact that the surface energy of high curvature (small pore size) surfaces is higher. Thus, the contractive surface forces acting to diminish the pore wall surface areas and to close the pores are greater for smaller pores [63].

2.6.4. Coating

There are at least two methods of coating being employed to repair defects in carbon membranes: (1) dip coating and (2) surface coating. For dip coating, a carbonized membrane is treated by coating a thin layer of polydimethylsiloxane (PDMS) to prevent gas flow through defect structures. The membrane was immersed in a PDMS usually with 2% (wt/wt) solution in heptane, and then withdrawn to harden the coated layer [24]. For surface coating method, an alcohol solution of 60% B rank phenolformaldehyde resin (PFR) and dispersant, followed by recarbonization (700°C, 60 min), was used to prepare carbon-carbon composite membranes [14]. This method tends to decrease gas permeation rates by a factor of 10, and the ideal separation factor is increased, probably due to the PFR's dispersing effect [14].

In addition to the above-mentioned chemicals, poly(4-methyl-1-pentene) (PMP) was also used as a coating agent in order to counteract the decrease in carbon membrane performance when exposed to water vapor or in environments having high humidity [69, 70].

3. Part 2: Polymeric Precursors of Hollow Fiber Ceramic Membranes

3.1. Introduction

Polymeric hollow fiber membranes are currently used for a number of process applications, including filtration, desalination, gas separation, and biochemical reactions. However, this membrane configuration is seldom found in ceramic membranes; possibly due to the lack of methods available in producing the ceramic membranes in hollow fiber geometry. It would be interesting to develop a method for preparation of hollow fiber ceramic membranes, as they would unfold their applications to areas where high alkaline (or acid) feed solution or high temperature operation precludes the use of existing polymeric hollow fiber membranes. A typical example is distillation where hollow fibers would give better, more productive separations than distillation in random or structured packing [71]. Although some attempts have been made in fabricating a silica glass hollow fiber membrane for gas separation [72], the membrane was too fragile to be used in large industrial scales. Recently, Tan *et al.* [73] and Liu and coworkers [74–77] have successfully produced hollow fiber ceramic membranes using materials such as Al_2O_3 and perovskites by employing a phase inversion method which is a three-step process including (1) preparation of a spinning suspension, (2) spinning of ceramic hollow fiber precursors, and (3) final sintering.

Preparation of ceramic membranes using the phase inversion method is a quite complex process. Many factors may affect the final membrane performance. Therefore, a detailed discussion on the preparation of ceramic hollow fiber membrane is presented according to the sequence of the three-step process.

3.2. Preparation of spinning suspension

Preparation of a spinning suspension may consist of several steps in sequence: (1) mixing a dispersant and a solvent of choice, (2) adding preconditioned particles for deflocculant and dispersion, (3) adding polymer binders and plasticizers, and (4) degassing the suspension prior to the spinning of the membrane precursor. The deflocculation is a very important step, as soft agglomerates of the particles in the spinning suspension would have an effect on pore size and permeability of the final membrane product [78]. A more detailed preparation procedure will be given at the end of this section.

The main ingredients in the spinning suspension for the fabrication of hollow fiber ceramic membranes are the ceramic powders, additives, and solvents. In selection of the ceramic powders, one of the important factor is the particle size and its distribution as well as the shape of the particles. These factors affect the porosity, pore size, and pore size distribution of the final membrane product [78].

Additives such as deflocculants/dispersants, binders, plasticizers, antifoaming agents, pore formers, etc., are used to provide the spinning suspension with the required property. In preparing the ceramic hollow fiber membrane using the phase inversion technique, the requirement of the organic binders is that they should not only be invertible, but also must be burnt out without leaving ash and tar during the calcinations. The quantity of organic binders should be as low as possible, but without affecting the inversion capability.

The solvent(s) must dissolve the additives used and must show a high exchange rate with nonsolvent (coagulants). The rates of solvent outflow and coagulant inflow have an effect on the cross-sectional structures of the membrane precursor, hence the structures of the final membrane products [73].

As preparation of the spinning suspension is still in the development stage, the following few general rules are, so far, considered to be useful in preparation of the spinning suspension:

1. The amount of dispersant must be maintained to ensure the stability of the suspension.
2. The amount of solvent must be fixed at a minimum to maintain a homogeneous suspension.
3. The ratio between organic components and ceramic powders must be as low as possible.
4. The plasticizer to binder ratio must be adjusted to make the membrane precursor flexible, resistant, and easy to release.

3.2.1. Particles and packing of particles

In ceramic membrane fabrication processes, the most important component in the spinning suspension is the ceramic powder. After binder removal and final sintering, the ceramic powders are the only component left and eventually would determine the properties of the ceramic membrane produced. The other components in the spinning suspension, such as solvents, plasticizers, binders, dispersant, etc., are used simply to predesign the membrane morphology (i.e., pore size, pore size distribution, porosity, cross-sectional structures) and to facilitate the spinning and phase inversion which is, in general, a process in obtaining and holding the powder particles in the hollow fiber configuration.

During the phase inversion (consolidation process), the particles in the spinning suspension undergo a process of particle packing. It is, therefore, possible to adjust the particles to attain high packing densities or to attain specific compact properties. The most widely used parameters of the particle packing are the packing density (or bulk density) and the coordination number. The bulk density (defined as weight of the particle bed/volume of the bed) is directly proportional to the density of the individual particles and a high bulk density may be obtained simply because the particles have a high density rather than because the particles are closely packed. In order to avoid this difficulty, an expression involving the ratio of the volume of particles composing the bed (V_p) to the volume of the bed (V_b) may be adopted. This ratio, that is, packing fraction (defined as bulk density/particle density) is usually referred to as the packing fraction or the fractional solids content. The fractional voidage ($\varepsilon = 1 - V_p/V_b$), expressed as a percentage, is often known as the volumetric porosity.

The packing density can be improved by introducing smaller sized particles. Small particles are introduced into a packing of large particles, to fill its voids without increasing the overall volume of the bed. Several studies have been conducted to view the two-dimensional structure using disks and spheres [79–81]. Two dominant factors are the size ratio and composition of the mixture.

When two sizes of spheres are mixed, the smaller particles fill the unoccupied voids among the large particles. So as long as the proportion of the smaller sphere is sufficiently great, the proportion of large spheres could reach the stage at which they will touch each other. The resulting packing is optimal. When the quantity of small particles becomes too great since all of the voids are filled, further addition will force the large particles apart and no longer improve the packing density. The basic behavior for random dense packing is sketched in Fig. 3.10, which shows that the packing volume, termed the specific volume (volume to mass ratio), is a

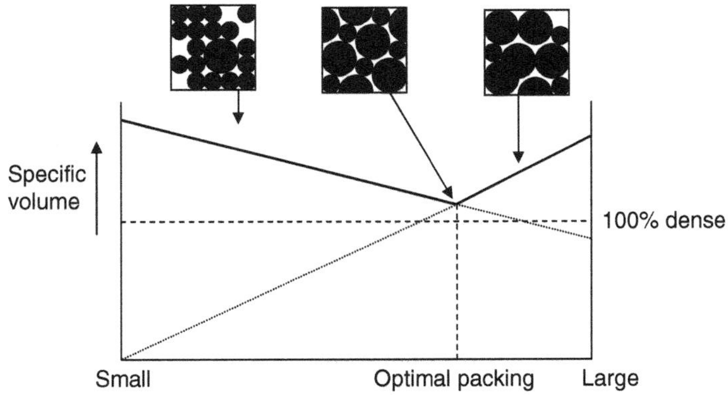

Figure 3.10 The reduction in specific volume for mixed large and small spheres, showing the condition of optimal packing where the small spheres fill all voids in the large sphere packing.

function of the composition for a mixture composed of large and small spheres. There is a composition of maximum packing density [82]. The relative improvement in packing density depends on the particle size ratio of the large and small particles. Detailed discussion on the packing of different particle sizes was given by Rahaman [83].

3.2.2. Dispersion of particles

As mentioned above, in preparation of ceramic membranes, one of the problems to be tackled is the dispersion of each individual ceramic particle in a solvent. When individual primary particles are in proximity, they often have a tendency to form loosely bound groups usually known as soft agglomerates. In a fluid suspension of particles, particularly when the powder has little affinity for wetting, the soft agglomerates can be formed to lower the free energy of the suspension by reducing the solid-liquid interface area. There are many types of surface forces acting between particles; among them electrostatic forces, van der Waals forces, and steric forces are the three most important ones. Descriptions of these three forces can be found elsewhere [83].

In order to deflocculate the soft agglomerate as well as to disperse the primary particles in a solvent, addition of a dispersant is usually required. The purpose of the dispersant is to disperse the primary particles and to hold them in a homogeneous suspension. As the dispersant separates primary particles and holds them in suspension, which allows the solvent to form a separating layer between particles, the suspension usually shows a lower viscosity due to the particle mobility offered by the solvent in the interparticle layer. There are basically two different mechanisms, that is, ionic repulsion and steric hindrance in describing the deflocculation and dispersion. The term, ionic repulsion, is used to describe the charging of particle surfaces so that they magnetically repel each other. In aqueous systems, this can be achieved by changing the pH. The surface charge of the particles can also be accomplished by introducing polyelectrolyte solutions, which coat the particle surfaces and provide the surface charge. The steric hindrance term is used to describe the separation of particles by coating a layer on one particle that would physically prohibit another particle to be in contact with it. Organic polymers are often used in both aqueous and nonaqueous systems to accomplish this particle-particle separation.

There are several reasons for a spinning suspension to be fully dispersed. First, the ceramic powders being added into the spinning suspension are usually in the form of soft agglomerates because they are normally produced through a sol/gel process and are high surface area powders. The soft agglomerates tend to trap air in the interstitial space between the primary particles. If the soft agglomerates are not broken up, the trapped air would produce bubbles in the membrane precursor, resulting pin holes and unwanted surface porosity in the final sintered membranes. Second, when the polymeric binder is introduced, it will envelop the soft agglomerates instead of the individual (primary) particles. This will make the soft agglomerates a permanent group throughout the rest of the process, which produces less-density membrane precursors, leading to high shrinkage during the final consolidation process.

The dispersion of particles in a solvent takes place in several stages. When the powder is first put in proximity to the solvent, the particle surfaces (or agglomerates) get coated with the solvent. Therefore, the stage 1 of suspension is the wetting of the particle surface. After the particle surface is wetted, soft agglomerates held together by electrostatic forces or by the van der Waals forces need to be separated, that is, the stage 2 of suspension is the mechanical separation of primary particles. After the completion of the stage 2, the particles are evenly distributed throughout the solvent. The final stage of suspension consists of protecting and stabilizing what has been accomplished, that is, prohibiting soft agglomerates from reforming in the spinning suspension.

Dispersion mechanisms have been fairly well studied in nonaqueous solvent systems, primarily with oxides. Although the main particle separation mechanism has been found to be steric hindrance, Lee and Rives [84] concluded that both steric and ionic mechanisms are active in the organic solvent-based suspension. The ionic repulsion dispersion mechanism may come into organic solvent systems with certain additives. However, since the spinning suspension generally contains a high solid loading and the particles are in close proximity to each other, the steric hindrance is still an important stabilization method [85].

One of the measurable effects of using a dispersant is a higher packed-bed density. This phenomenon has been used to determine the proper dispersant level. The level or extent of dispersion can be measured by a number of techniques such as microelectrophoresis, optical microscopy, and laser scattering. Targeted variables of these tests include zeta potential, isoelectric point, and equivalent spherical diameter. A settling experiment is perhaps another means for the characterization of the spinning suspension, as one of the main reasons for dispersing the solution is to form a densely packed bed of particles. It follows that a screening experiment should be performed with the densest packed bed as the goal. Therefore, an experiment in which different concentrations of dispersant are used to disperse identical suspensions of powder and solvent can be carried out. After a proper screening, the extent of dispersion in powder suspensions can be observed.

3.2.3. General preparation procedure

Based on each of the above individual discussion, a general procedure in preparation of the spinning suspension can now be drawn. The first step is the dispersion of particles in the presence of a dispersant in the solvent(s). The usual procedure would be to dissolve a predetermined amount of dispersant in terms of weight or volume by stirring or rolling. The powder(s) is then added.

Percentage of the solids loading during dispersion is dependent on the powder density, particle size, and dispersant effectiveness. It is beneficial to optimize the solids loading during this step, since any excess solvent will have to be removed during the phase inversion (consolidation step). The solid loading is often adjusted with experience, because the solvent volume also affects the final viscosity after the polymer binder and/or plasticizer(s) are added to the mixture. The viscosity of the dispersion is not a primary concern, as long as deagglomeration occurs. Viscosity of the dispersions at this stage is extremely low, but will increase when the binder is added.

The purpose of the dispersion procedure is to accomplish three goals: (1) to break apart any soft agglomerates, (2) to coat the primary particles with a dispersant, which (3) keeps the particles apart by steric effects, electrostatic effects, or both.

After the dispersion process is complete, the plasticizers (if any) are poured onto the top of the mixture, and followed by the weighed binder, either in powder form or dissolved in a solvent.

Once the plasticizer and binder are added, the mixture is stirred or rolled to make sure that the binder is not in large clumps. Usually a few hours are required for the complete dissolution of the binder.

Usually, the suspension becomes more viscous with the addition of a long-chain polymeric binder. The only time that this does not occur is when the binder is a more powerful dispersant and replaces or overrides the chosen dispersant. This is also a good reason to use the suspension in the phase inversion process as soon as possible after the binder is mixed into the suspension. Rheological properties of the suspension can change with time. Once the plasticizer(s) and binder(s) have been mixed into the suspension formulation, it is ready for prespinning conditioning and characterization.

The next step is to degas the suspension to remove any air that may have been entrained during the mixing process. Air bubbles cause defects in the membrane precursor. Pinholes (small holes left by air bubbles) are the most common defect, and these can lead to cracking which radiates from the pinhole during the phase inversion (consolidation step). It is therefore essential that entrained air be removed before the spinning. Several techniques are used for the degassing process. The simplest technique is the partial vacuum accompanied by gentle stirring or agitation. Once the spinning suspension is properly degassed, it is ready to be spun into hollow fiber precursors and this is discussed below.

3.3. Spinning of ceramic hollow fiber precursors

Spinning of ceramic hollow fiber precursors can be carried out in a spinning apparatus, which is identical to the system used in spinning the polymeric membranes. Detailed procedures and illustrations have been discussed in Section 2.4.1 and Fig. 3.3. Figure 3.11 shows a typical fiber precursor spun from a suspension containing $La_{0.6}Sr_{0.4}Co_{0.2}Fe_{0.8}O_{3-\alpha}$ (LSCF) powders. It can be seen that the precursors are very flexible because of polymeric binder and dispersant used. In spinning of ceramic hollow fiber precursors, there are two important parameters, that is, spinnability and invertability of the powder suspension, which are discussed next.

3.3.1. Rheology and spinnability

In any polymer solutions or particle suspensions, the rheological properties play a key role in controlling the shape-forming behavior [86]. A general observed phenomenon is that the viscosity increases with the increase of polymer contents in polymer solutions and increases with volume fraction of solids and finally ceases to flow at some critical maximum volume fraction of solids in particle suspensions. The rheological behavior of a concentrated suspension of ceramic particles is affected by a

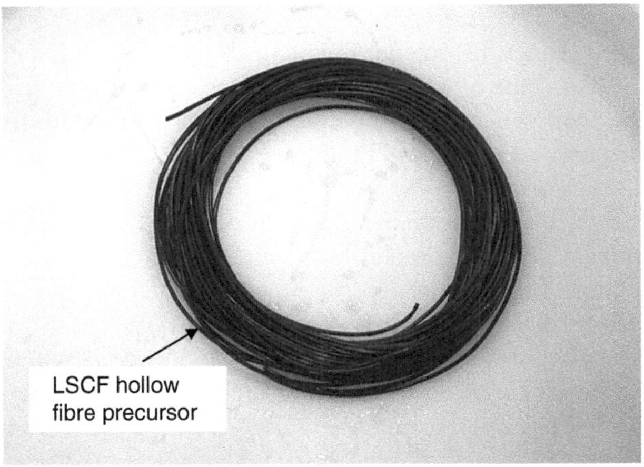

Figure 3.11 Ceramic hollow fiber precursors.

number of factors including the particle size distribution, particle shape, volume fraction of solids, and the range and magnitude of the interparticle forces [87–89].

When the viscosity reaches about 10 poise for a polymer solution, the solution becomes spinnable and fibers can be drawn through an orifice [90, 91]. For a suspension containing spherical particles, the relation between the reduced viscosity η_{sp}/c and the particles concentration c, where η_{sp} is the specific viscosity, can be examined. In principle, solutions containing spherical particles follow Einstein's formula [90]:

$$\frac{\eta_{sp}}{c} = \frac{k}{\rho} \tag{3.1}$$

where k is the constant and ρ is the density of the particles.

For a concentrated particle suspension, the best spinnability should be observed when the viscosity behavior is highly shear-thinning but not thixotropic. A very high viscosity suspension is necessary to prevent the drawn fiber from breaking up into droplets. High viscosity is required for a stable fiber formation and can generally be achieved by increasing the concentration of the suspension or adding some additives.

3.3.2. Phase inversion

In the 1960s, Loeb and Sourirajan [92] first developed an immersion-induced phase inversion method and produced an asymmetric polymeric membrane, which exhibited a high selectivity and high water flux. This membrane shows an asymmetric structure: a dense and thin skin (about 25 μm) integrated on porous substrate of the same material. The dense skin layer shows permselective properties and suitable to separate salt from seawater, while it is also thin enough for a high water flux. The porous substrate provides the mechanical stability for the membrane. Since then, this method has been one of the most useful methods for the fabrication of polymeric membranes. Conceptually, membrane formation by the phase inversion can be described in terms of a three-component system: nonsolvent, solvent, and polymer. After immersion of a polymeric solution into a nonsolvent bath, the solvent diffuses out of the polymer solution, while nonsolvent diffuses into the solution. This exchange of solvent and nonsolvent, in general, leads to a change of the polymer solution from a thermodynamically stable state to a metastable or unstable state. Then, the liquid-liquid demixing takes place in the solution. In case the solvent flux is higher than the nonsolvent flux, the polymer concentration at the interface would increase. The highly concentrated polymer layer at the surface reduces the exchange speed of the solvent and nonsolvent hence, the polymer concentration in the sublayer is lower than the polymer concentration in the top layer. The further replacement of solvent by nonsolvent will result in solidification of the polymer-rich phase and the asymmetric structure of the membrane is, thus, obtained. Figure 3.12 shows SEM photographs of a typical asymmetric structure with a finger-like support layer of a hollow fiber membrane produced by the phase inversion method.

The fact that should be noted in the phase inversion process is the concentration of polymer in the top layer, which is increased with the exchange of the solvent and nonsolvent. When the out-diffusion of solvent is faster that the in-diffusion of nonsolvent, the total volume of the original polymer solution is reduced. The polymer molecules move from bulk to the interface. For the polymer solution system, this move will only result in an increase in the polymer concentration. But for a suspension system containing different sized ceramic powders, the situation may be different. When the particles are well covered by polymer molecules in the suspension, the particles may be considered as polymer nodules. The rate of movement of different sized nodules may be different when exchange of solvent and nonsolvent takes place in the phase inversion process. It has been postulated that smaller particles may move faster than the larger particles [93]. Therefore,

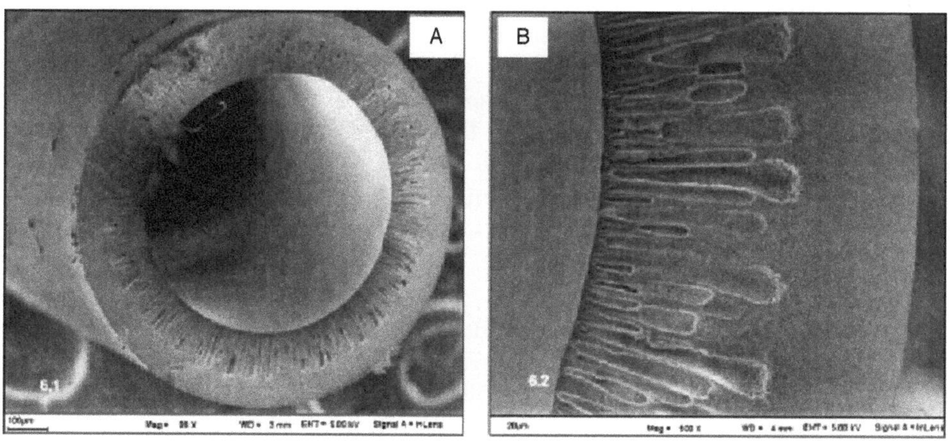

Figure 3.12 The SEM photograph of a typical asymmetric structure with finger-like support layer of a hollow fiber membrane produced by the phase inversion method: (A) cross section, (B) wall.

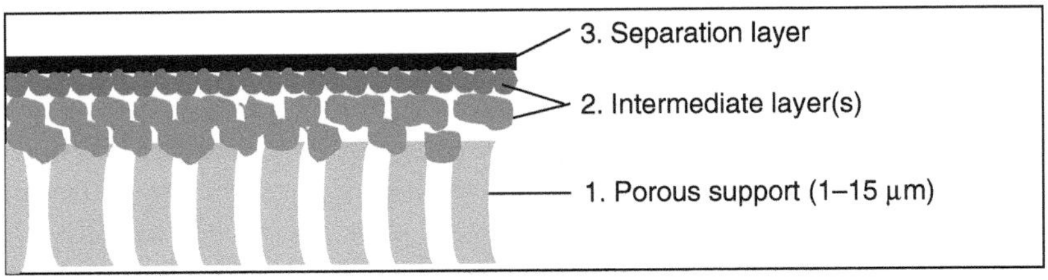

Figure 3.13 A general structure of ceramic membranes.

membranes with a graded structure as shown in Fig. 3.13 can be prepared. It, thus, follows that a good control of the ratio of small and large particles in the suspension and the exchange rate of the solvent and nonsolvent may tailor the membrane structure, which can only be prepared with multiple steps if conventional techniques are used.

3.4. Sintering

Sintering processes have been studied for long time and considerable breakthrough in understanding the fundamental aspects of the basic mechanism of sintering of ceramics have been achieved [92, 94]. In sintering the hollow fiber membrane precursors mentioned above, three main steps including presintering, thermolysis, and final sintering are required and described separately in the following sections.

3.4.1. Presintering

Sintering does not commonly begin until the temperature exceeds one-half to two-thirds of the melting temperature of the material. This temperature will be sufficient to cause a significant atomic diffusion for solid-state sintering or a significant diffusion and viscous flow when a liquid phase is present or produced by a chemical reaction. In the presintering process, the material changes due to heating prior to sintering mainly include vaporization of chemically combined water from the surface of particles or formed within inorganic phases containing water of crystallization. Therefore, care must be taken to prevent the precursor from cracks or fractures because of the stresses from the pressure of the vapor evolved or from differential thermal expansion of phases. The initial heating may remove any liquid remaining after forming and drying the membrane precursor and any moisture adsorbed from the atmosphere during transporting and setting. The adsorbed moisture may persist in the membrane precursor up to a temperature exceeding 200°C.

3.4.2. Thermolysis

Thermolysis is a process, which burns out the organic components such as binder, dispersant, etc. It is an important step prior to the densification on sintering. Incomplete binder removal and uncontrolled thermolysis may introduce membrane defects. These defects may impair the performance of the membrane. With the proper choice of binders for the membrane and controlled heating in an appropriate atmosphere, the membrane precursors should survive theremolysis without deformation, distortion, formation of cracks, or expanded pores. Binder burnt out is very dependent on the composition of the binder material and the composition and flow of the gas surrounding the precursor. It is also highly dependent on (a) the microstructure of the organic, powder, and porosity phases and (b) dynamic changes in the microstructure as the binder is eliminated. The thermochemistry of the binder and additives, binder concentration, precursor dimension, precursor's configuration, heating rate, and furnace atmosphere may influence the thermolysis behavior. Precursors containing low amounts of a binder will open pore channels sufficiently to allow the transport of vapors and gases between the reaction zone and the membrane surface. The time for thermolysis is controlled by the diffusion length of vapor phase transport rather than the characteristic dimension of the binder phase. Decomposition and vaporization of the organics will cause an internal gas pressure, which depends on the gas evolution rate, gas permeability, and the precursor size. The gas permeation rate is much lower in dense precursors of very fine particles. In precursors containing a higher organic content, the weight loss is much higher and the time for thermolysis without bubbling is relatively long. Thermolysis of an organic binder in an inert atmosphere such as nitrogen, as it is used when calcining silicon nitride and aluminum nitride ceramics, proceeds differently. Some residual carbon may be expected in the absence of oxygen. The residual carbon commonly removes adsorbed oxygen from surfaces of particles and serves as a sintering aid in carbide nitride ceramics.

3.4.3. Final sintering

Sintering is normally thought to occur in three stages, that is: (i) the initial stage, (ii) the intermediate stage, and (iii) the final stage [95]. The movements of particles have different features at each stage and attributed to full density, grain coarsening, and pores closing. The temperature program of sintering process would be different for different materials. Rapid densification with limited grain growth occurs at lower temperature followed by rapid grain growth with little densification at higher temperature as shown in Fig. 3.14.

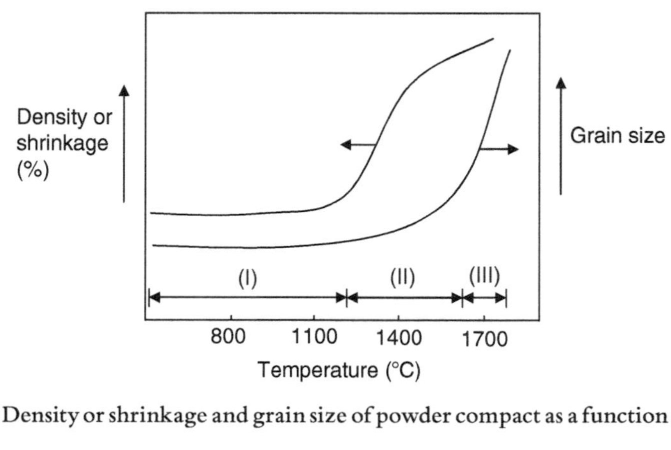

Figure 3.14 Density or shrinkage and grain size of powder compact as a function of the sintering temperature.

Table 3.3 Mass transport mechanisms in sintering

Mechanism	Densification
Surface diffusion	No
Evaporation-condensation	No
Grain boundary diffusion	Yes
Lattice diffusion	Yes
Viscous flow	Yes
Plastic flow	Yes

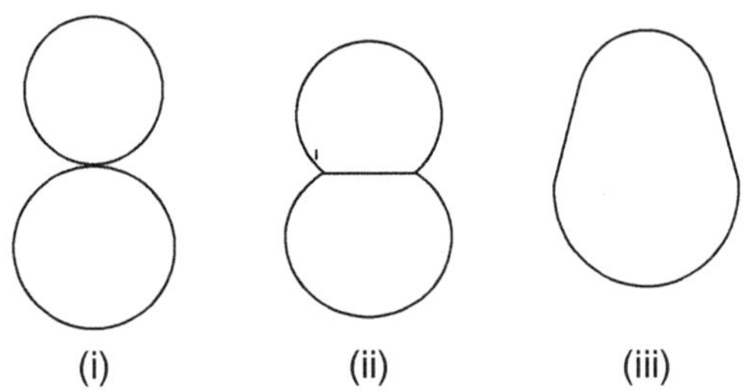

Figure 3.15 A two-sphere sintering model for a qualitative mechanism for grain growth in porous powder compacts: (i) particles of slightly different size in contact; (ii) neck growth by surface diffusion between particles; and (iii) grain growth.

There are at least six different mechanisms for mass transport during sintering as listed in Table 3.3. All of these will lead to growth of necks between particles. As shown in Fig. 3.15, the neck growth produces bonding between the particles, so the strength of the consolidated powders increases during sintering. Only certain mechanisms, however, lead to shrinkage and densification. Surface diffusion is a general transport mechanism that can produce surface smoothing, particle joining, and pore rounding, but it does not produce volume shrinkage. In materials where the vapor pressure is relatively high, sublimation and vapor transport to surfaces of lower vapor pressure also produce these effects. Diffusion along the grain boundaries and diffusion through the lattice of the grains produce both neck growth and volume shrinkage. The mechanisms of bulk viscous flow and plastic deformation may be effective when a wetting liquid is present and when a mechanical pressure is applied, respectively.

3.4.4. Grain growth

The properties of ceramic membrane are strongly dependent on its microstructure. The important microstructural features are the size and shape of the grains, the porosity, the pore size, and the distribution of the pores in the structure. In order to achieve the desired properties during the fabrication of ceramic membranes, the microstructural features must be controlled. For most applications, microstructural control usually means the achievement of high density, small grain size, and homogeneous microstructure as much as possible. There are a few factors that should be understood such as how the grains grow, the interaction between the pores and the moving grain boundaries, and the coarsening process.

Grain growth is the term used to describe the increase in the grain size of a single-phase solid or in the matrix grain size of a solid containing second-phase particles at a sufficiently high temperature. For the conservation of matter, the sum of the individual grain sizes must remain constant; therefore, the increase in the average size of the grains is always accompanied by the disappearance of some grains, usually the smaller ones. In porous solids, both the grains and the pores normally increase in size while decreasing in number. Frequently, the term coarsening is used to describe the progress by which grains and pores grow. The driving force for grain growth is the decrease in grain boundary energy that results from a decrease in the grain boundary area. Grain growth in ceramic membranes is generally divided into two types: normal and abnormal. In the normal grain growth, the grain sizes and shapes fall within a fairly narrow range and the distribution in grain sizes at a later time is fairly similar to that at an early time. In the abnormal grain growth, a few large grains develop and grow fairly rapidly at the expense of the smaller ones. The grain size distribution may change significantly.

For stable structures, the grain boundary energy of all grain boundaries should be the same. The edges must meet at an angle of 120°. Taking N as the number of edges, when $N = 6$, grain boundaries are straight. Grains with $N > 6$ have concave boundaries, and those with $N < 6$ have convex boundaries. Since the boundaries migrate toward their center of curvature, grains with $N > 6$ tend to grow while those with $N < 6$ tend to shrink.

3.4.5. Pore elimination or evolution

Following the above description of the grain growth, pore changing in the sintering should be discussed, because it is the direct reflection of membrane performance. For the case of a grain surrounded by other grains, a pore is surrounded by N number of grains. The N is normally called the pore coordination number. The pore has straight sides if $N = 6$, convex sides for $N < 6$, and concave sides for $N > 6$ as shown in Fig. 3.16. Since the surface of the pore moves toward to its center of curvature, the pore with $N < 6$ will shrink whereas the one with $N > 6$ will grow. The pore is metastable for $N = 6$, and this number is called the critical pore coordination number, N_c. For making a dense membrane, the shrinkable pores are favorable. These pores can be obtained by careful control of the particle packing and the grain growth in the sintering process.

3.5. Example: Preparation of porous Al_2O_3 hollow fiber membranes

Studies on the formation principles were first started with a spinning suspension containing only 1 μm of Al_2O_3 powders because of the complicity of the fabrication process. It was further explored using spinning suspensions containing Al_2O_3 with different sizes. By addition of different weight ratios of 1, 0.3, and 0.01 μm Al_2O_3 particles in the spinning suspension and sintering at an appropriate temperature, the Al_2O_3 hollow fiber membranes with excellent properties had been prepared.

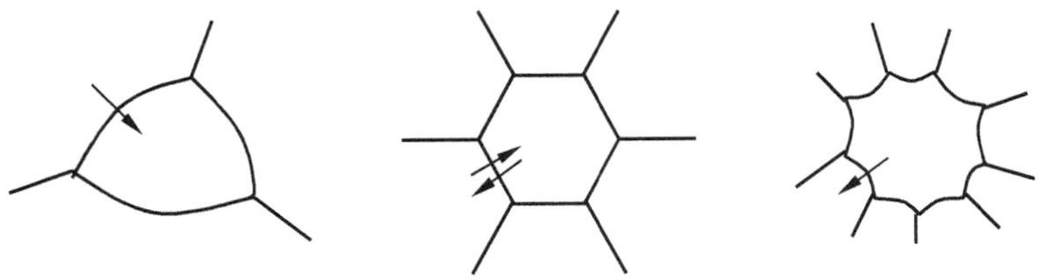

Figure 3.16 Pore stability in two dimensions for a dihedral of 120 °.

Materials used in preparation of Al_2O_3 hollow fiber membranes were obtained commercially. Al_2O_3 powders with three different particle diameters of 0.01 μm (gamma/alpha, surface area 100 m^2/g), 0.3 μm (gamma/alpha, surface area 15 m^2/g), and 1 μm (alpha, surface area 10 m^2/g), polyethersulfone (PESf), *N*-methyl-2-pyrrolidone (NMP), and PVP (K90) were used as membrane materials, a binder, a solvent, and an additive, respectively. Tap water was used as both the internal and external coagulant.

3.5.1. Morphology

Morphologic study of Al_2O_3 hollow fibers has been performed using SEM. Figures 3.17 and 3.18 show the SEM micrographs of the Al_2O_3 hollow fiber precursors and their sintered fibers, spun from two different spinning suspensions with one containing PESf 10%, PVP 0.5%, NMP 39.5%, and 1 μm Al_2O_3 50% and the other containing PESf 10%, PVP 0.5%, NMP 39.5%, and 0.3 μm Al_2O_3 50% (wt.%), respectively. The sintering process was carried out in air at temperature of 1500 °C. It can be seen from the micrograph of Fig. 3.17(AI) that the OD and ID of the fiber precursor prepared from 1 μm particles were 1287 and 847 μm shrunk to 1044 and 726 μm (Fig. 3.17(BI)), respectively, after sintering. Similarly, Fig. 3.18 (AI) and (BI) depicts that the OD and ID of the fiber precursor and the sintered fiber prepared from 0.3 μm particles were shrunk from 1705 and 1118 μm to 1333 and 911 μm, respectively. Further comparing the fiber dimensions of the precursors and the sintered ones revealed that the fibers prepared from 0.3 μm particles have the highest shrinkage during the sintering process. This is probably due to the fact that for the same weight of Al_2O_3 adding into a given volume of the binder solution, the smaller particle size (0.3 μm) would leave higher voids compared to the larger particle size (1 μm). As the polymer binder, which is ultimately removed during the sintering process, generally occupies the voids, the higher shrinkage of the resulting fibers is, thus, expected to be 0.3 μm particles.

Cross-sectional structures of the fiber precursor and the sintered fiber prepared from 1 μm particles are shown in Fig. 3.17(AII) and (BII). It can be seen from Fig. 3.17(AII) that near the outer and inner walls of the fiber precursor, long finger-like structures are present and that at center of the hollow fiber precursor, sponge-like structures are possessed. The appearance of the fiber structures shown in the figure can be attributed to the rapid precipitation occurred at both the inner and outer fiber walls resulting in long fingers and to the slow precipitation giving the sponge-like structure at the center of the fiber. It is interesting to note that the

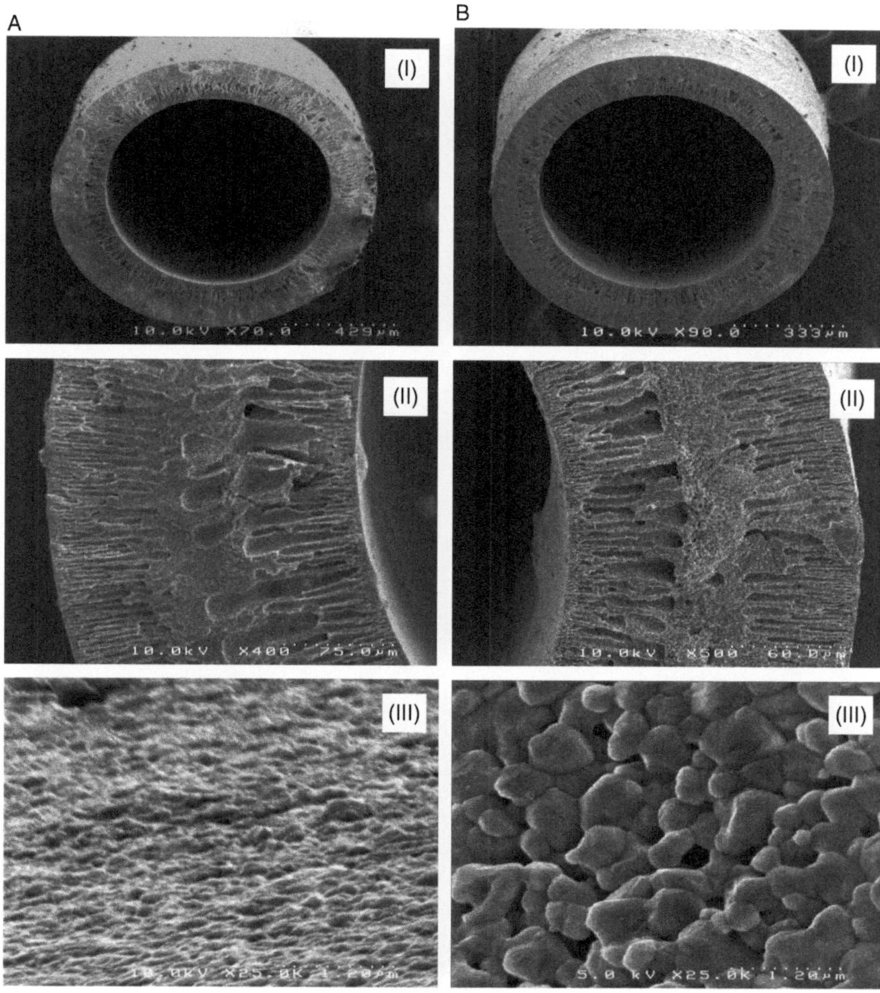

Figure 3.17 SEM diagrams of the hollow fibers fabricated from the 1 μm Al_2O_3 particles: (A) before sintering; (B) after sintering; (I) overall view; (II) membrane walls; (III) membrane surfaces [96].

cross-sectional structure of the fiber precursor is well maintained after the sintering process as shown in Fig. 3.17(BII). Figure 3.18(AII) and (BII) depicts the cross-sectional structures of the fiber precursor and the sintered fiber prepared from 0.3 μm particles. As compared to the fibers prepared from 1 μm particles, it clearly shows that after the sintering, the fiber prepared from 0.3 μm particles is much denser, which probably resulted from higher shrinkage during the sintering process. It is known that as compared with the traditional flat and tubular membranes, the Al_2O_3 hollow fiber membranes have the advantage of higher surface area/volume ratios. For example, the sintered hollow fiber membranes from Figs. 3.17(B) and 3.18(B) have the surface area per membrane volume ratios of 3831 and 3000 m^2/m^3, respectively. When these hollow fiber membranes are assembled in

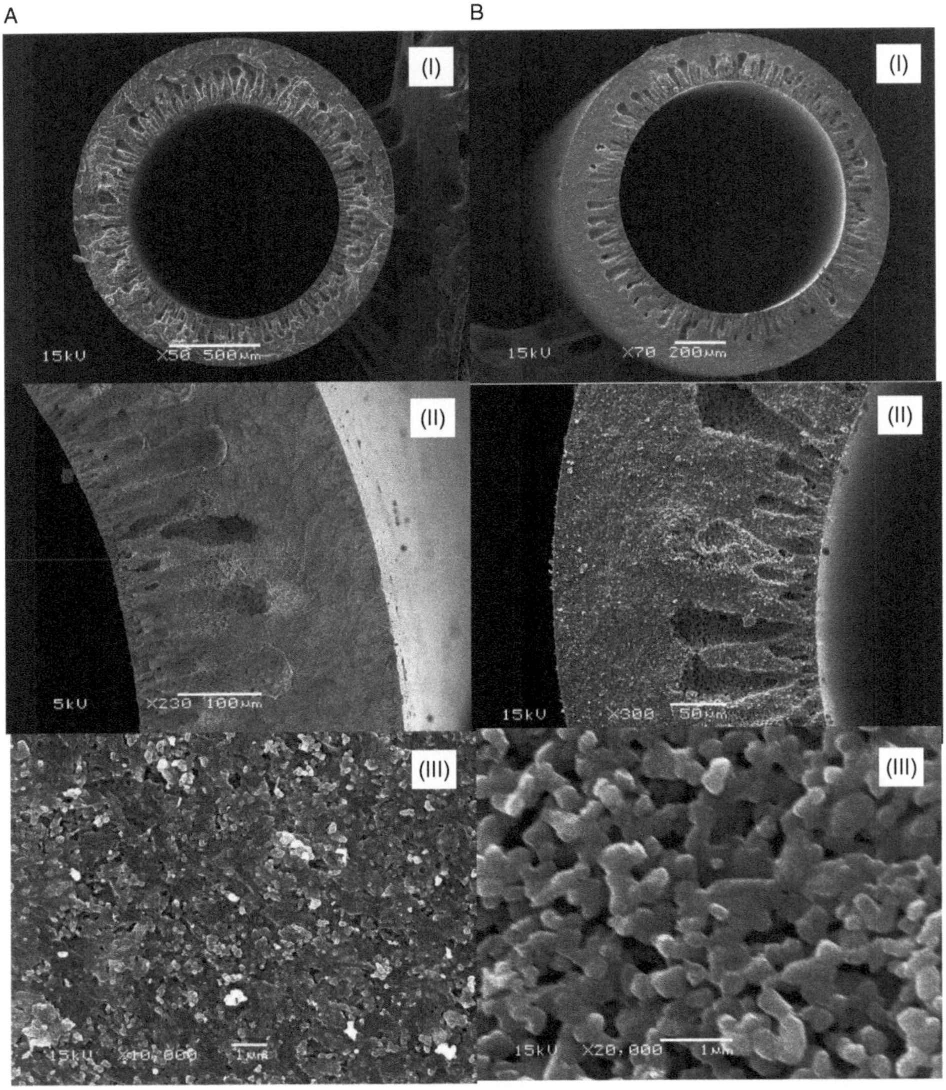

Figure 3.18 SEM diagrams of the hollow fibers fabricated from the 0.3 µm Al_2O_3 particles: (A) before sintering; (B) after sintering; (I) overall view; (II) membrane walls; (III) membrane surfaces [96].

the modules, they are very compact and have a practical surface area/volume ratio of >3000 m^2/m^3.

Figure 3.17(AIII) shows the surface of the hollow fiber precursor. It can be seen that the surface of the precursor is similar to the conventional polymeric membranes prepared via the phase inversion technique. Further a comparison of the SEM photographs, especially for the surfaces between the precursors (Figs. 3.17(AIII) and 3.18(AIII)) and the sintered fiber (Figs. 3.17(BIII) and 3.18(BIII)), reveals that the pore quantity and pore size may have changed after the sintering process although the general structure is maintained. Such structure changes depend on the particle sizes, composition of the dope solution, and the sintering temperature, which are discussed below.

3.5.2. Mechanical strength

Mechanical strength of the Al_2O_3 hollow fiber membranes is one of the important issues in preparation of ceramic hollow fiber membranes. It depends on several preparation conditions such as Al_2O_3 particle size and its size distributions, the Al_2O_3 content in the spinning suspension and sintering environment (i.e., sintering temperature and time), etc. Experimental data on the hollow fiber mechanical strength and permeation property is shown in Fig. 3.19 where the hollow fiber membrane was prepared from 0.3 μm particles. It can be seen that increase of sintering temperature would enhance the mechanical strength. For example, at temperatures of 1300 and 1550°C sintered for 10 h, the three-point (3P) values are 20.9 and 80.9 MPa, respectively. When the sintering is less than 1500°C, the 3P value is proportional to the sintering temperature. As the sintering temperature is further increased and greater than 1550°C, the 3P value increases sharply. It, thus, suggests that preparation of the Al_2O_3 hollow fiber membranes with high mechanical strength are possible at the sintering temperature of 1550°C or higher, as a slight increase in temperature would dramatically increase the mechanical strength. Sintering of the hollow fiber at the temperature of 1600°C shows, of course, the increase in the mechanical strength; however, gas permeability decreases considerably as shown in Fig. 3.19. Therefore, there is a trade-off between the mechanical strength and gas permeability. The 3P values at various sintering temperatures suggest that the

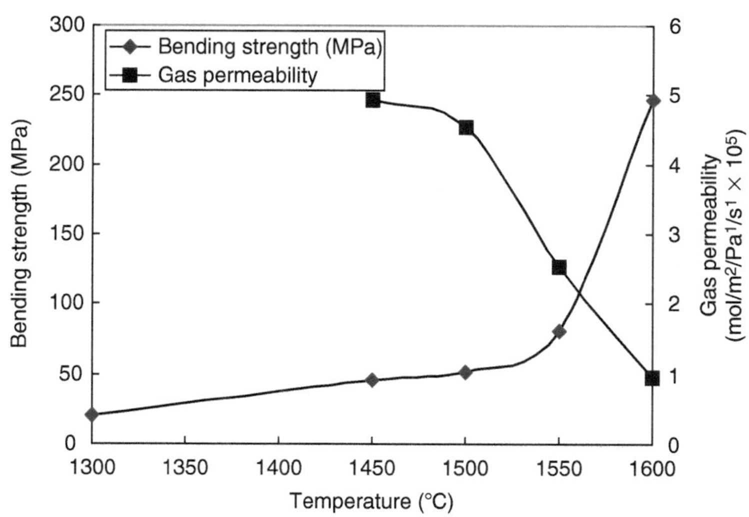

Figure 3.19 Effect of sintering temperature on the mechanical strength and gas permeability of hollow fiber membranes prepared from 0.3 μm Al_2O_3 powder at Al_2O_3/PESf ratio of 5; sintered for 10 h [96].

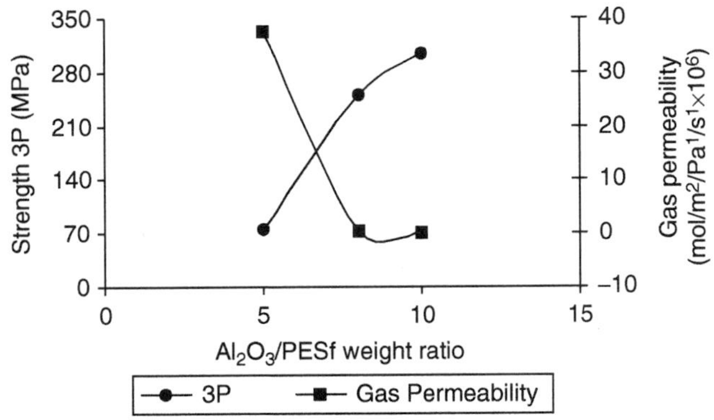

Figure 3.20 Effect of Al_2O_3 content on the mechanical strength and gas permeability of the hollow fiber membranes prepared from 1.0 μm Al_2O_3 particles, sinter temperature of 1550 °C [96].

Table 3.4 Experimental results for preparation of Al_2O_3 hollow fiber membranes [96]

No.	Al_2O_3 powders 1/0.3/0.01 (μm)	Sintering temperature (°C)	Gas permeability (mol/m²/Pa/s) (N_2, 1 atm)	Bending strength, σ_F (MPa)
AI	100/0/0	1500	6.50E−5	48.6
BI	50/50/0	1500	3.90E−5	59.3
CI	0/100/0	1500	4.53E−5	51.3
DI	46.5/46.5/7	1500	2.78E−5	72.8
AII	100/0/0	1550	3.75E−5	76.1
BII	50/50/0	1550	3.59E−5	93.6
CII	0/100/0	1550	2.54E−5	80.9
DII	46.5/46.5/7	1550	1.60E−5	107.0
AIII	100/0/0	1600	1.15E−5	182.4
BIII	50/50/0	1600	0.98E−5	253.08
CIII	0/100/0	1600	0.95E−5	246.2
DIII	46.5/46.5/7	1600	0.22E−5	341.7

sintering temperature at 1550°C will give sufficient strength for the fiber to be fabricated into a module without breaking.

The hollow fiber precursor formed through the phase inversion techniques contains the Al_2O_3 powder and the PESf binder. During the sintering process, the PESf is removed and the Al_2O_3 hollow fiber is ultimately formed. Therefore, the Al_2O_3 content in the spinning suspension plays the important role in determining its mechanical strength. Figure 3.20 illustrates the effect of Al_2O_3 content on the fiber mechanical strength and its gas permeability. It can be seen that the 3P value enhances greatly, as the Al_2O_3/PESf ratio is increased. Compared to the sintering temperature, the increase of the Al_2O_3 powder content in the spinning dope would result in a much obvious effect on the fiber mechanical strength. Therefore, in order to produce a Al_2O_3 hollow fiber membrane with higher mechanical strength, the higher Al_2O_3 content in the solution dope must be maintained. At Al_2O_3/PESf ratio of 7 or greater, reduction in gas permeability is tailed, indicating that the membrane is transformed to a much denser structure.

3.5.3. Effect of particle size and size distribution

Pores in the sintered ceramic hollow fiber membranes are voids left between packed particles, having neither a regular shape, nor a regular size. The particle size and size distribution of the Al_2O_3 powders in the spinning dopes plays an important role in determining the membrane wall density and gas permeability. It has been shown by [73] that a higher Al_2O_3 content in the spinning suspension would produce a denser membrane with a lower porosity and smaller pore size. When the Al_2O_3 content is too high, it is difficult to form a hollow fiber precursor using the phase inversion process. The Al_2O_3/PESf ratio of 5 has, therefore, been suggested and employed for preparation of a Al_2O_3 hollow fiber membrane with a meaningful mechanical strength and appreciable permeation characteristics using the spinning suspension containing different sizes of the Al_2O_3 powders. The experimental result is given in Table 3.4. It indicates that the gas permeability decreases and mechanical strength of the hollow fiber increases as the average particle size of the Al_2O_3 powders is decreased. The results given in Table 3.4 clearly correspond with the SEM graphs presented earlier. Further comparison of the experimental results given in Fig. 3.20 and Table 3.4 reveal that blending the smaller Al_2O_3 particles in the spinning suspension is a more effective way in producing the Al_2O_3 hollow fiber membrane with increased mechanical strength without considerably losing its permeation characteristics.

REFERENCES

[1] Ash, R., Barrer, R. M., and Lowson, R. T. (1973). Transport of single gases and of binary mixtures in a microporous carbon membrane. *J. Chem. Soc. Faraday Trans. I* **69,** 2166–2178.

[2] Hayashi, J., Mizuta, H., Yamamoto, M., Kusakabe, K., and Morooka, S. (1997). Pore size control of carbonized BPTA-*pp*′ODA polyimide membrane by chemical vapor deposition of carbon. *J. Membr. Sci.* **124,** 243–251.

[3] Kyotani, T. (2000). Control pore in carbon. *Carbon* **38,** 269–286.

[4] Jones, C. W., and Koros, W. J. (1995). Characterization of ultramicroporous carbon membranes with humidified feeds. *Ind. Eng. Chem. Res.* **34,** 158–163.

[5] Kishore, N., Sachan, S., Rai, K. N., and Kumar, A. (2003). Synthesis and characterization of a nanofiltration carbon membrane derived from phenol–formaldehyde resin. *Carbon* **41**(15), 2961–2972.

[6] Sakata, Y., Muto, A., Uddin, M. A., and Suga, H. (1999). Preparation of porous carbon membrane plates for pervaporation separation applications. *Sep. Purif. Technol.* **17**(2), 97–100.

[7] Song, C., Wang, T., Pan, Y., and Qiu, J. (2006). Preparation of coal-based microfiltration carbon membrane and application in oily wastewater treatment. *Sep. Purif. Technol.* **51**(1), 80–84.

[8] Koresh, J. E., and Soffer, A. (1987). The carbon molecular sieve membranes: general properties and the permeability of CH_4/H_2 mixture. *Sep. Sci. Technol.* **22,** 973–982.

[9] Koresh, J. E., and Soffer, A. (1986). Mechanism of permeation through molecular sieve carbon membrane. Part 1—The effect of adsorption and the dependence on pressure. *J. Chem. Soc. Faraday Trans. I* **82,** 2057–2063.

[10] Koresh, J. E., and Soffer, A. (1983). Molecular sieve carbon permselective membrane. Part I: Presentation of a new device for gas mixture separation. *Sep. Sci. Technol.* **18,** 723–734.
[11] Saufi, S. M., and Ismail, A. F. (2004). Fabrication of carbon membranes for gas separation—a review. *Carbon* **42,** 241–259.
[12] Schindler, E., and Maier, F. (1990). *Manufacture of Porous Carbon Membranes,* US Patent 4,919,860.
[13] Suda, H., and Haraya, K. (1997). Gas permeation through micropores of carbon molecular sieve membranes derived from Kapton polyimide. *J. Phys. Chem. B* **101,** 3988–3994.
[14] Liang, G., Sha, G., and Guo, S. (1999). Carbon membrane for gas separation derived from coal tar pitch. *Carbon* **37,** 1391–1397.
[15] Morthon-Jones, D. H. (1989). *Polymer Processing,* Chapman and Hall, London.
[16] Soffer, A., Rosen, D., Saguee, S., and Koresh, J. (1989). *Carbon Membranes,* GB Patent 2,207,666.
[17] Laszlo, K., Bota, A., and Nagy, L. G. (2000). Comparative adsorption study on carbons from polymer precursors. *Carbon* **38,** 1965–1976.
[18] Jones, C. W., and Koros, W. J. (1994). Carbon molecular sieve gas separation membranes—I Preparation and characterization based on polyimide precursors. *Carbon* **32,** 1419–1425.
[19] Hatori, H., Kobayashi, T., Hanzawa, Y., Yamada, Y., Iimura, Y., Kimura, T., and Shiraishi, M. (2001). Mesoporous carbon membranes from polyimide blended with poly(ethyleneglycol). *J. Appl. Polym. Sci.* **79,** 836–841.
[20] Vu, D. Q., Koros, W. J., and Miller, S. J. (2002). High pressure CO_2/CH_4 separation using carbon molecular sieve hollow fibre membranes. *Ind. Eng. Chem. Res.* **41,** 367–380.
[21] Yamamoto, M., Kusakabe, K., Hayashi, J., and Morooka, S. (1997). Carbon molecular sieve membrane formed by oxidative carbonization of a copolyimide film coated on a porous support tube. *J. Membr. Sci.* **133,** 195–205.
[22] Hatori, H., Shiraishi, M., Nakata, H., and Yoshitomi, S. (1992). Carbon molecular sieve films from polyimide. *Carbon* **30,** 719–720.
[23] Petersen, J., Matsuda, M., and Haraya, K. (1997). Capillary carbon molecular sieve membranes derived from Kapton for high temperature gas separation. *J. Membr. Sci.* **131,** 85–94.
[24] Fuertes, A. B., Nevskaia, D. M., and Centeno, T. A. (1999). Carbon composite membranes from Matrimid and Kapton polyimide. *Microporous Mesoporous Mater.* **33,** 115–125.
[25] Fuertes, A. B., and Centeno, T. A. (1998). Carbon molecular sieve membranes from polyether-imide. *Microporous Mesoporous Mater.* **26,** 23–26.
[26] Sedigh, M. G., Xu, L., Tsotsis, T. T., and Sahimi, M. (1999). Transport and morphological characteristics of polyetherimide-based carbon molecular sieve membranes. *Ind. Eng. Chem. Res.* **38,** 3367–3380.
[27] Coutinho, M. B., Salim, V. M. M., and Borgees, C. P. (2003). Preparation of carbon hollow fibre membranes by pyrolysis of polyetherimide. *Carbon* **41,** 1707–1714.
[28] Centeno, T. A., and Fuertes, A. B. (1999). Supported carbon molecular sieve membranes based on phenolic resin. *J. Membr. Sci.* **160,** 201–211.
[29] Centeno, T. A., and Fuertes, A. B. (2001). Carbon molecular sieve membranes derived from a phenolic resin supported on porous ceramic tubes. *Sep. Purif. Technol.* **25,** 379–384.
[30] Fuertes, A. B. (2001). Effect of air oxidation on gas separation properties of adsorption-selective carbon membranes. *Carbon* **39,** 697–706.
[31] Sedigh, M. G., Onstot, W. J., Xu, L., Peng, W. L., Tsotsis, T. T., and Sahimi, M. (1998). Experiments and simulation of transport and separation of gas mixtures in carbon molecular sieves membranes. *J. Phys. Chem. A* **102,** 8580–8589.
[32] Acharya, M. (1999). *Engineering Design and Theoretical Analysis of Nanoporous Carbon Membranes for Gas Separation,* Ph.D. Thesis, University of Delaware.
[33] Chen, Y. D., and Yang, R. T. (1994). Preparation of carbon molecular sieve membrane and diffusion of binary mixtures in the membrane. *Ind. Eng. Chem. Res* **33,** 3146–3153.
[34] Shiflett, M. B., and Foley, H. C. (2000). On the preparation of supported nanoporous carbon membranes. *J. Membr. Sci.* **179,** 275–282.
[35] Shiflett, M. B., and Foley, H. C. (1999). Ultrasonic deposition of high selectivity nanoporous carbon membranes. *Science* **285,** 1902–1905.

[36] Acharya, M., and Foley, H. C. (1999). Spray-coating of nanoporous carbon membranes for air separation. *J. Membr. Sci.* **161,** 1–5.

[37] Corbin, D. R., Foley, H. C., and Shiflett, M. B. (2001). *Mixed Matrix Nanoporous Carbon Membranes.* WO Patent 01/97956.

[38] Shiflett, M. B., and Foley, H. C. (2001). Reproducible production of nanoporous carbon membranes. *Carbon* **39,** 1421–1446.

[39] Strano, M. S., and Foley, H. C. (2001). Synthesis and characterization of catalytic nanoporous carbon membranes. *AIChE J.* **47,** 66–78.

[40] Strano, M. S., and Foley, H. C. (2002). Temperature- and pressure-dependent transient analysis of single component permeation through nanoporous carbon membranes. *Carbon* **40,** 1029–1041.

[41] Donnet, J. B., and Bansal, R. C. (1984). *Carbon Fibre.* Marcel Dekker, New York.

[42] Yoneyama, H., and Nishihara, Y. Y. (1990). *Porous Hollow Carbon Fibre Film and Method of Manufacturing the Same.* EP Patent 0,394,449A1.

[43] Linkov, V. M., Sanderson, R. D., and Jacobs, E. P. (1994). Highly asymmetrical carbon membranes. *J. Membr. Sci.* **95,** 93–99.

[44] Linkov, V. M., Sanderson, R. D., and Jacobs, E. P. (1994). Carbon membranes from precursors containing low carbon residual polymers. *Polym. Int.* **35,** 239–242.

[45] Smith, S. P. J., Linkov, V. M., Sanderson, R. D., Petrik, L. F., O'Conner, C. T., and Keiser, K. (1995). Preparation of hollow-fibre composite carbon-zeolite membranes. *Microporous Mater.* **4,** 385–390.

[46] Linkov, V. M., Sanderson, R. D., and Rychkov, B. A. (1994). Composite carbon-polyimide membranes. *Mater. Lett.* **20,** 43–46.

[47] Linkov, V. M., Bobrova, L. P., Tomifeev, S. V., and Sanderson, R. D. (1995). Composites membranes from perfluoro-sulfonated ionomer on carbon and ceramic supports. *Mater. Lett.* **24,** 147–151.

[48] David, L. I. B., and Ismail, A. F. (2003). Influence of the thermastabilization process and soak time during pyrolysis process on the polyacrylonitrile carbon membranes for O_2/N_2 separation. *J. Membr. Sci.* **213,** 285–291.

[49] Ismail, A. F., and David, L. I. B. (2001). A review on the latest development of carbon membranes for gas separation. *J. Membr. Sci.* **193,** 1–18.

[50] Geiszler, V. C. (1997). *Polyimide Precursors for Carbon Membranes,* Ph.D. Thesis, University of Texas.

[51] Puri, P. S. (1990). Fabrication of hollow fibre gas separation membranes. *Gas Sep. Purif.* **4,** 29–36.

[52] Edie, D. D. (1998). The effect of processing on the structure and properties of carbon fibres. *Carbon* **36**(4), 354–362.

[53] Gupta, A., and Harisson, I. R. (1996). New aspects in the oxidative stabilization of PAN-based carbon fibres. *Carbon* **34**(11), 1427–1445.

[54] Ghosal, A. S., and Koros, W. J. (2000). Air separation properties of flat sheet homogeneous pyrolytic carbon membranes. *J. Membr. Sci.* **174,** 177–188.

[55] Koresh, J. E., and Soffer, A. (1980). Study of molecular sieve carbon. Part 1—Pore structure, gradual pore opening and mechanism of molecular sieving. *J. Chem. Soc. Faraday Trans. I* **76,** 2457–2471.

[56] Steel, K. M., and Koros, W. J. (2005). An investigation of the effect of pyrolysis parameters on gas separation properties of carbon membranes. *Carbon* **43**(9), 1843–1856.

[57] Steel, K. M. (2000). *Carbon Membranes for Challenging Gas Separations,* Ph.D. Thesis, University of Texas.

[58] Geiszler, V. C., and Koros, W. J. (1996). Effect of polyimide pyrolysis conditions on carbon molecular sieve membrane properties. *Ind. Eng. Chem. Res.* **35,** 2999–3003.

[59] David, L. I. B. (2001). *Development of Asymmetric Polyacrylonitrile Carbon Hollow Fibre Membrane for Oxygen/Nitrogen Gas Separation,* M.Sc. Thesis, Universiti Teknologi Malaysia.

[60] Suda, H., and Haraya, K. (1997). Alkene/alkane permselectivities of a carbon molecular sieve membrane. *J. Chem. Soc. Chem. Commun.* 93–94.

[61] Soffer, A., Koresh, J., and Saggy, S. (1987). *Separation Device,* US Patent 4685940.

[62] Kusakabe, K., Yamamoto, M., and Morooka, S. (1998). Gas permeation and micropore structure of carbon molecular sieving membranes modified by oxidation. *J. Membr. Sci.* **149,** 59–67.
[63] Soffer, A., Azariah, M., Amar, A., Cohen, H., Golub, D., Saguee, S., and Tobias, H. (1997). *Method of Improving the Selectivity of Carbon Membranes by Chemical Vapor Deposition*, US Patent 5,695,818.
[64] Hayashi, J., Yamamoto, M., Kusakabe, K., and Morooka, S. (1995). Simultaneous improvement of permeance and permselectivity of 3,3′4,4′-biphenyltetracarboxylic dianhydride-4,4′-oxydianiline polyimide membrane by carbonization. *Ind. Eng. Chem. Res.* **34,** 4364–4370.
[65] Cabrera, A. L., Zehnar, J. E., Coe, C. G., Gaffney, T. R., and Farris, T. S. (1993). Preparation of carbon molecular sieve I. Two steps hydrocarbon deposition with a single hydrocarbon. *Carbon* **31,** 969–976.
[66] Chihara, K., and Suzuki, M. (1979). Control of micropore diffusivities of molecular sieving carbon by deposition of hydrocarbons. *Carbon* **17,** 339–343.
[67] Verma, S. K., and Walker, P. L., Jr. (1992). Carbon molecular sieves with stable hydrophobic surfaces. *Carbon* **30,** 837–844.
[68] Fuertes, A. B., and Mendez, I. (2002). Separation of hydrocarbon gas mixtures using phenolic resin-based carbon membranes. *Sep. Purif. Tech.* **28,** 29–41.
[69] Koros, W. J., and Jones, C. W. (1994). *Composite Carbon Fluid Membranes*, US Patent 5,288,304.
[70] Jones, C. W., and Koros, W. J. (1995). Carbon composite membranes: A solution to adverse humidity effects. *Ind. Eng. Chem. Res.* **34,** 164–167.
[71] Zhang, G. L., and Cussler, E. L. (2003). Distillation in hollow fibers. *AIChE J.* **49**(9), 2344–2351.
[72] Way, J. D., and Roberts, D. L. (1992). Hollow fiber inorganic membranes for gas separations. *Sep. Sci. Technol.* **27**(1), 29–41.
[73] Tan, X. Y., Liu, S. M., and Li, K. (2001). Preparation and characterization of inorganic hollow fiber membranes. *J. Membr. Sci.* **188**(1), 87–95.
[74] Liu, S. M., Li, k., and Hughes, R. (2004). Preparation of $SrCe_{0.95}Yb_{0.05}O_{3-a}$ perovskite for use as a membrane material in hollow fibre fabrication. *Mater. Res. Bull.* **39**(1), 119–133.
[75] Liu, S. M., Li, K., and Hughes, R. (2003). Preparation of porous aluminium oxide (Al_2O_3) hollow fibre membranes by a combined phase-inversion and sintering method. *Ceram. Int.* **29**(8), 875–881.
[76] Liu, S. M., and Li, K. (2003). Preparation TiO_2/Al_2O_3 composite hollow fibre membranes. *J. Membr. Sci.* **218**(1–2), 269–277.
[77] Liu, S. M., Tan, X. Y., Li, K., and Hughes, R. (2001). Preparation and characterisation of $SrCe_{0.95}Yb_{0.05}O_{2.975}$ hollow fibre membranes. *J. Membr. Sci.* **193**(2), 249–260.
[78] Hsieh, H. P. (1996). *In* "organic Membranes for Separation and Reaction." Elsevier Science B. V., Amsterdam, The Netherlands, pp. 23–86
[79] Bideau, D., and Troadec, J. P. (1984). Compacity and mean coordination-number of dense packings of hard disks. *J. Phys. C Solid State Phys.* **17**(28), L731–L735.
[80] Visscher, W. M., and Bolsterl, M. (1972). Random packing of equal and unequal spheres in 2 and 3 dimensions. *Nature* **239**(5374), 504.
[81] Kausch, H. H., Fesko, H. H., and Tschoegl, N. W. (1971). Random packing of circles in a plane. *J. Coll. Interf. Sci.* **37**(3), 603.
[82] Starr, T. L. (1986). Packing density of fiber powder blends. *Am. Ceram. Soc. Bull.* **65**(9), 1293–1296.
[83] Rahaman, M. N. (1995). *In* "Ceramic Processing and Sintering." Marcel Dekker Inc., New York, pp. 265–279.
[84] Lee, B. I., and Rives, B. I. (1991). Dispersion of alumina powders in nonaqueous media. *Colloids Surf.* **56,** 25–43.
[85] Mikeska, K. R., and Cannon, K. R. (1984). Dispersants for tape casting barium titanate. *In* "Advances in Ceramics" (J. A. Mangels and G. L. Messing, eds.), pp. 164–183. The American Ceramic Society, Westerville.
[86] Bergstrom, W. R. (1996). Rheological properties of concentrated, nonaqueous silicon nitride suspensions. *J. Am. Ceram. Soc.* **79**(12), 3033–3040.
[87] Bergstrom, L. (1994). Rheology of concentrated suspensions. *In* "Surface and Colloid Chemistry" (R. J. Pugh and L. Bergstorm, eds.). Marcel Dekker, New York.

[88] Bergstrom, L. (1994). Surface and colloid chemistry. *In* "Advanced Ceramic Processing" (P. J. Pugh and L. Bergstrom, eds.), pp. 193–224. Marcel Dekker, New York.
[89] Kitano, T., Kataoka, T., and Shirota, T. (1981). An empirical-equation of the relative viscosity of polymer melts filled with various inorganic fillers. *Rheol. Acta* **20**(2), 207–209.
[90] Sakka, S. (1989). Rheology of sols in sol-gel processing. *In* "Sol-Gel Science and Technology" (M.A. Aegerter *et al.*), pp. 76–103. World Scientific Publisher, Singapore.
[91] Deshmukh, S. P., and Li, K. (1998). Effect of ethanol composition in water coagulation bath on morphology of PVDF hollow fibre membranes. *J. Membr. Sci.* **150**(1), 75–85.
[92] Richardson, D. W. (1992). *Morden Ceramic Engineering*. Marcel Dekker Inc., New York.
[93] Li, K., Xiaoyao, K., and Liu, Y. (2006). Single-step fabrication of ceramic hollow fibers for oxygen permeation. *J. Membr. Sci.* **272**(1–2), 1–5.
[94] Binner, J. G. P. (1990). *Advanced Ceramic Processing and Technology*. Noyes Publications, New Jersey.
[95] Reed, J. S. (1995). Principles of Ceramic Processing. second ed. John Wiley & Sons, New York.
[96] Liu, S., Li, K., and Hughes, R. (2003). Preparation of porous aluminium oxide (Al_2O_3) hollow fibre membranes by a combined phase-inversion and sintering method. *Ceram. Int.* **29,** 875–881.

CHAPTER FOUR

Organic-Inorganic Membranes

Suzana Pereira Nunes

Contents

1. Introduction

Polymeric membranes are established now in several large-scale applications, varying from water desalination, food, chemical, and pharmaceutical industry, and play a decisive role in the medical field. Inorganic membranes emerged later and are complementing the polymeric materials, being in many cases the only alternative, particularly in processes, which operate at high temperature and in the presence of organic solvents. Although the membrane development had a huge progress in the last decades, the use of pure homopolymers for membrane manufacture leads to a stagnation of perspectives when exploring the membrane application in new fields like energy and nanofiltration in nonaqueous medium. The introduction of new materials would expand the membrane application and solve problems of stability, swelling, etc. The possibility of combining polymers and inorganic components in one material has been explored from the beginning of the polymer industrialization, particularly with the objective of improving mechanical and optical properties. A new and elegant perspective appeared in the last decades with the birth of soft inorganic chemistry and the introduction of the concept of "hybrid organic-inorganic" materials, which allows tailoring innovative materials with organic and inorganic phases coexisting in a nanoscale and with multifunctionalization, which make them interesting for applications such as electronics, sensors, nonlinear optics, and catalysis. Comprehensive reviews have been published on this topic [1, 2].

Institute of Polymer Research, GKSS Research Centre, Geesthacht, Germany

Membrane Science and Technology, Volume 13
ISSN 0927-5193, DOI: 10.1016/S0927-5193(07)13004-7

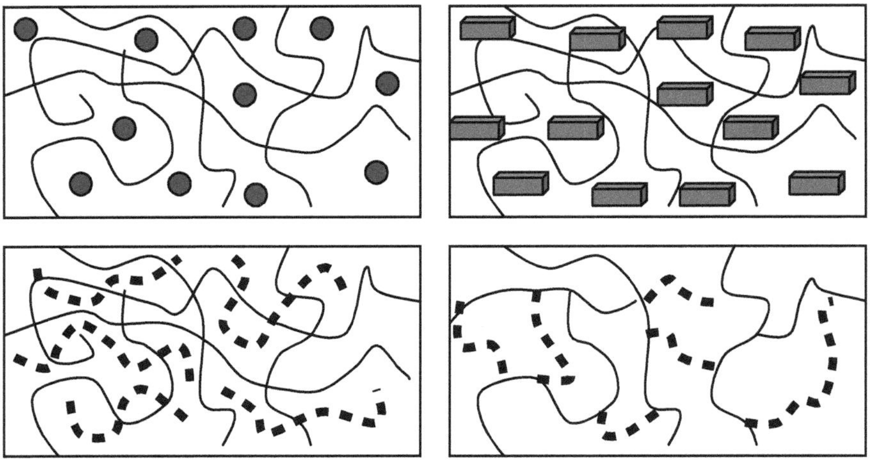

Figure 4.1 Organic-inorganic hybrids for membranes: isotropic and anisotropic inorganic phases dispersed in a polymer matrix; interpenetrated organic and inorganic networks and covalently bonded organic and inorganic segments in a common network.

The use of organic-inorganic materials for membranes is very attractive, having in mind the possibility of achieving synergetic effects on permeability and selectivity, as well as introducing new functionalizations and improving mechanical and thermal stability. Figure 4.1 summarizes different possibilities of combining organic and inorganic phases or segments for membrane preparation.

2. Polymers with Impermeable Fillers

2.1. Effect of the aspect ratio

The simplest case of composite materials for membranes is that of a polymeric matrix with impermeable fillers. In the case of spherical fillers, it is possible to have an estimation of the permeability of the final membrane, P, by taking into account the permeability of the matrix, P_0, and the volume fraction of the filler, ϕ, using the Eq. (4.1) [3, 4].

$$P = \frac{P_0(1 - \phi)}{[1 + \phi/2]} \tag{4.1}$$

If the fillers have a higher aspect ratio, for instance in the form of flakes (Fig. 4.2), the equation for permeability must include an additional factor. Equation (4.2) considers the system depicted in Fig. 4.2, containing flakes with the length L and the width W [3, 4]:

$$P = \frac{P_0(1 - \phi)}{[1 + (L/2W)\phi]} \tag{4.2}$$

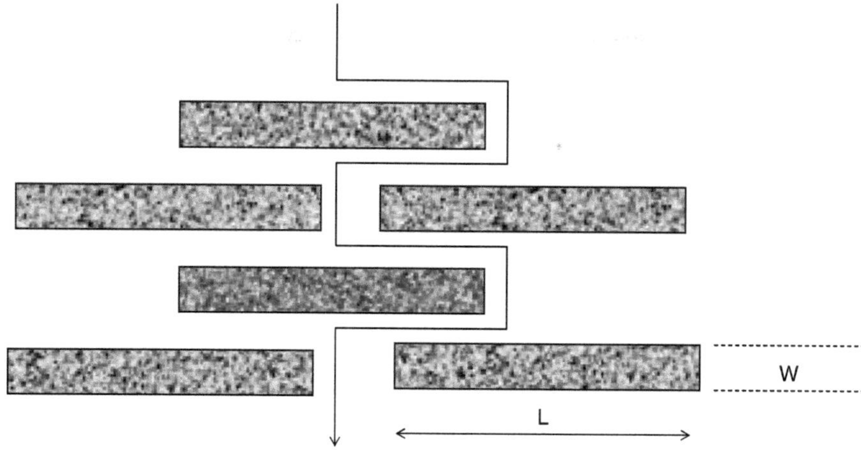

Figure 4.2 Fillers in a polymeric/inorganic composite.

For isotropic fillers, Eqs. (4.1) and (4.2) are the same. This simple consideration has been taken into account for instance to decrease the methanol permeability of membranes to be used for direct methanol fuel cell (DMFC) [5–7] by adding silica or zirconium oxide to an ionomeric matrix. DMFC is a promising fuel cell technology for powering small devices due to the relatively high-energy density. It uses liquid methanol to feed the electrochemical cell. Protons are produced in the anode and transported through a polymer electrolyte membrane to the cathode to react with oxygen. One of the main drawbacks is the excessive alcohol crossover, which reduces the cell performance because of competitive reactions in the cathode. Also from Eq. (4.1) it is clear that the use of fillers with high aspect ratio ($L/W \gg 1$) like layer silicates is an effective way to decrease the permeability of a film. An immediate application for that can be found in the sector of packaging and bottles for soft drinks containing carbon dioxide. Many biological tissues like enamel of tooth, vertebral bone, and shell are made of a soft matrix with hard platelets [8]. This has an important effect not only on mechanical toughness but also on permeability like in the case of human skin [9].

More recently zeolites like mordenite [12, 13] and layer silicates [10, 11] such as laponite [14] and montmorillonite [15] have been effectively used by different groups also to decrease the methanol permeability of fuel cell membranes. Figure 4.3 summarizes the effect of fillers of different aspect ratios on the permeability and proton conductivity of membranes for DMFC. While aerosil is a spherical filler, laponite is a layered silicate. $R\text{-}SiO_{3/2}$ is generated inside the membrane and its geometry can vary from spherical irregular particles to a dispersed network. The effective dispersion of layered fillers requires their exfoliation (Fig. 4.4), which can be achieved by treatment with surfactants like hexadecyltrimethylammonium [15].

The intercalation of perfluorosulfonylfluoride copolymer resin into the montmorillonite was achieved after treatment with dodecylamine. Membranes were then produced by direct melt [16].

Another important issue in fuel cell is the search for polymeric materials with high proton conductivities even above 100 °C, a temperature at which the

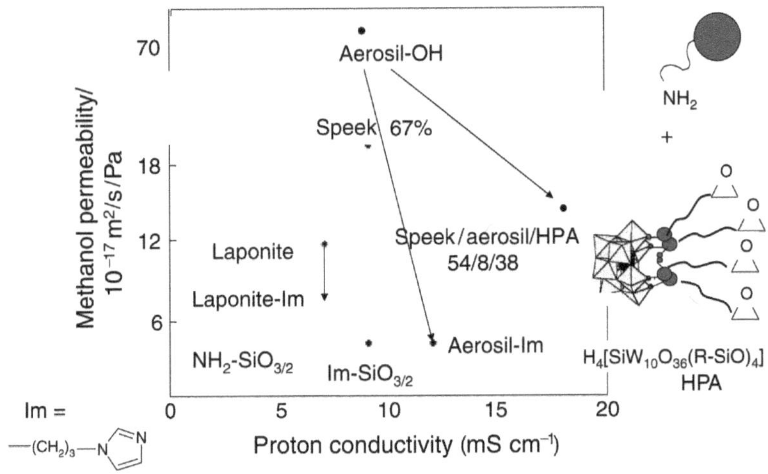

Figure 4.3 Comparison between fillers of different aspect ratio and different surface chemistry incorporated in membranes for direct methanol fuel cell (adapted from [14]).

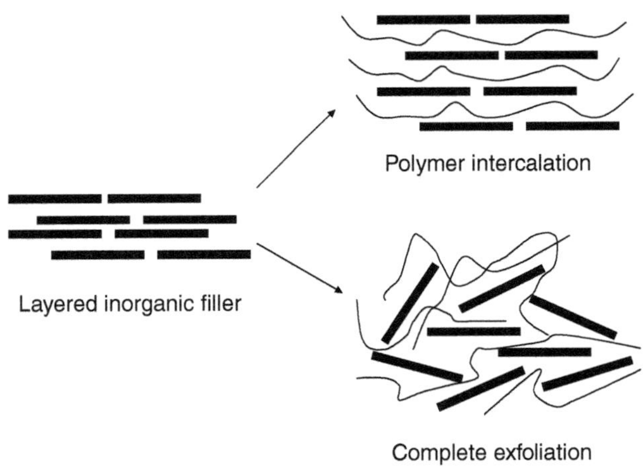

Figure 4.4 Polymer intercalation and filler exfoliation.

sulfonated polymers dehydrate. Stonehart and Watanabe [17] and Antonucci and Arico [18] were the first to claim that the introduction of small silica particles to Nafion ® increase the water retention at higher temperatures. The effect is believed to be a result of the water adsorption on the oxide surface. A layered filler, which has been well explored for fuel cell membranes in different forms, is zirconium phosphates. Alberti [19] was one of the pioneers in this field and has investigated the proton conductivity of zirconium phosphate for decades. In a recent review, Alberti and Casciola summarized their results and commented on other currently investigated organic-inorganic fuel cell membranes for application at temperatures above 100 °C [19]. Their most successful results were obtained with zirconium sulfoaryl phosphonates, which was investigated pure or in composites with sulfonated poly(ether ether ketone). Other groups have been investigating modified approaches for proton conductive membranes also based on zirconium phosphate [20–24]. In our group, colloidal dispersions of zirconium oxide in water and organic solvents was first obtained by the neutralization reaction of zirconyl chloride and converted to zirconium phosphate by the treatment with phosphoric acid [5, 25, 26]. The partial exfoliation of the zirconium phosphate was obtained by further treatment with *n*-propylamine and polybenzimidazole could be followed by using anomalous small angle X-ray scattering, as is shown in Fig. 4.5.

2.2. Effect of the surface chemistry

2.2.1. Acidity

Much more important than the filler aspect ratio or geometry is its surface chemistry. Aricò *et al.* [28] analyzed the effect of filler acidity on the water retention and the performance of fuel cell membranes. The pH varies according to the following series: SiO_2 with adsorbed heteropolyacid $< SiO_2 < ZrO_2 < n - Al_2O_3 < b{-}Al_2O_3$. They recorded a very clear correlation to the electrochemical properties (proton conductivity and power density) of this series, Nafion® membranes with $\beta - Al_2O_3$ having the lowest performance. They claimed that both strength and amount of acidic surface functional groups in the fillers would enhance the water retention inside the composite membranes through an electrostatic interaction, in the presence of humidification constraints, in the same way as it occurs for the adsorption of hydroxyl ions in solution. More recently Gomes *et al.* [29] reported the modification of silica particles modified with sulfonated oligomeric segments. The particles were reacted with bromine containing silanes, which were later reacted with sulfonated polyoxadiazole telechelics, which were hydroxyl terminated. The functionalized particles were then incorporated in sulfonated poly(ether ether ketone), as is shown in Fig. 4.6. The increase of proton conductivity was much higher than when the silica surface was modified only with short acid segments (Fig. 4.7).

This procedure allows the introduction of highly sulfonated segments into the membrane without affecting water solubility.

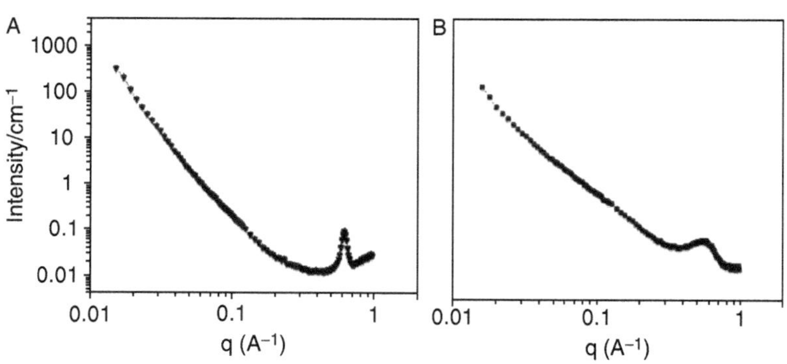

Figure 4.5 Anomalous small angle X-ray scattering (Zr-separated scattering) of membranes containing zirconium phosphate (A) without and (B) after treatment with *n*-propyl amine and polybenzimidazole (adapted from [27]).

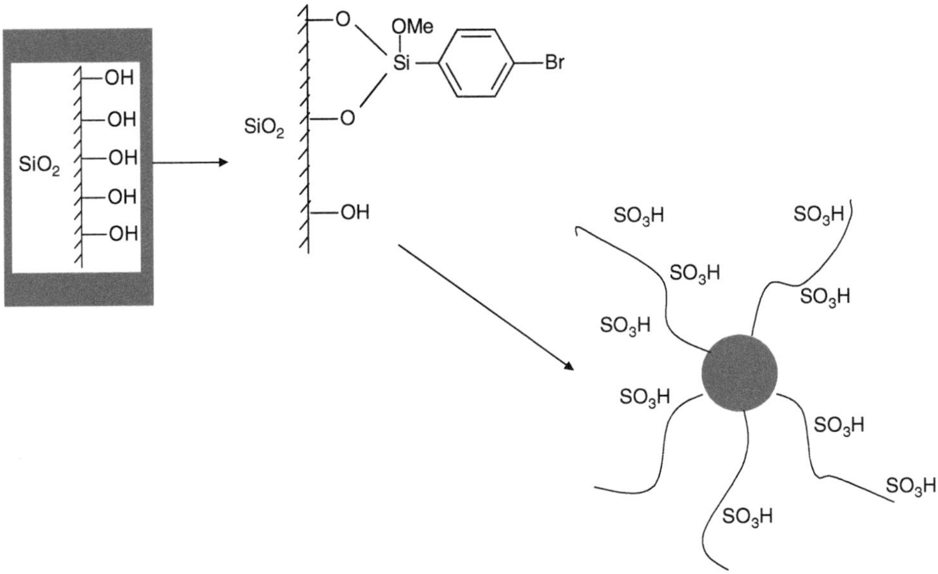

Figure 4.6 Functionalization of silica particles with sulfonated telechelics for incorporation in fuel cell membranes.

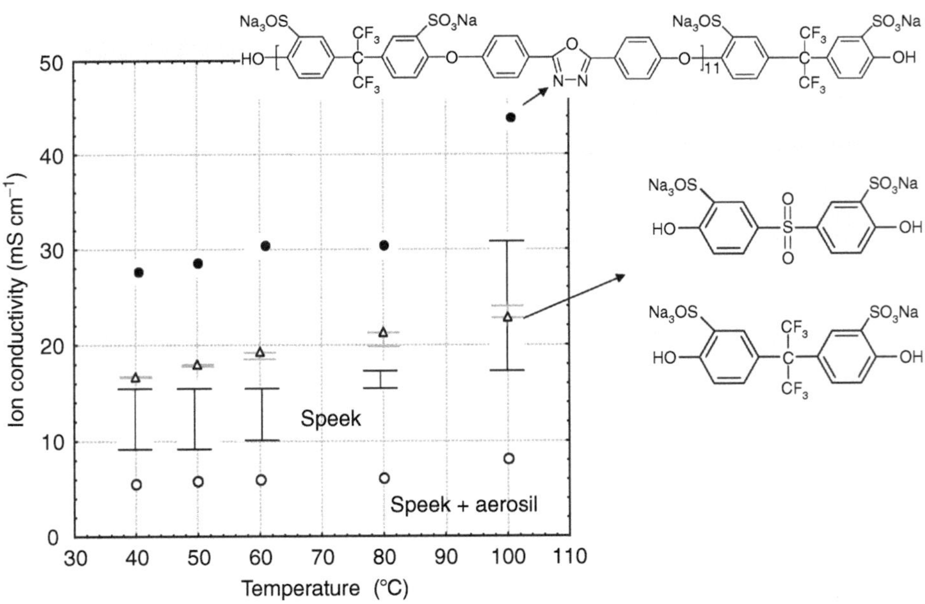

Figure 4.7 Proton conductivity of sulfonated poly(ether ether ketone) membranes containing 5% of aerosil, which was previously functionalized with sulfonated monomers or oligomers (adapted from [29]).

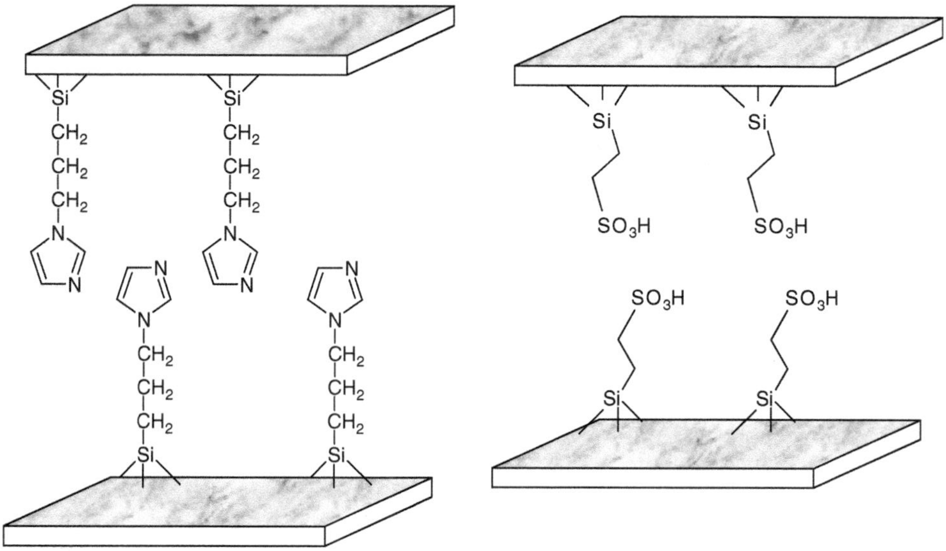

Figure 4.8 Functionalization of layered silicates (adapted from [14, 30]).

Figure 4.8 shows two surface modifications of layered silicates for use in DMFC. The first one contains basic groups [13] and improves the adhesion between the filler and the polymer matrix. The second one was reported by Rhee *et al.* [30]. They grafted organic species bearing an organic sulfonic acid (HSO_3-) onto the surface of montmorillonite by the condensation of 3-mercaptopropyltrimethoxysilane. The fillers were cast together with Nafion® to prepare composite membranes for DMFC. The methanol permeability of the composite membrane decreased dramatically with increasing content of functionalized filler in the composite membrane, without changing much the conductivity.

2.2.2. Solubility/leaching out

An insoluble inorganic phase can be very effective to incorporate rather soluble active components. Heteropolyacids have been described in the literature as good proton conductors [31–40] and investigated for hydrogen fuel cell, but their main disadvantage is the high solubility in water. To overcome the problem of the electrolyte dissolution, there are many works focused on the entrapping of the heteropolyacid in silica oxide networks. Another approach was the modification of the anion structure of the heteropolyacid [39–40]. A more effective procedure was the modification of heteropolyacids by covalent bonding to an insoluble inorganic phase dispersed in a polymer matrix. The insoluble inorganic phase can decrease the water and methanol crossover, besides fixing the heteropolyacid to the membrane. For this reason, organosilyl derivatives of divacant tungstosilicate were prepared using 3-glycidoxypropyltrimethoxysilane. The vacant sites of the heteropolyacid allow the covalent grafting of the epoxy group. Later the epoxy group could react with amino groups attached to the surface of silica particles. A quite insoluble phase with immobilized heteropolyacid is obtained leading to membranes with high proton conductivity with minimized leaching out of acid components. The procedure is not only useful for proton conductive membrane preparation, but could also be relevant for the fixation of heteropolyacids as catalysts in the membrane.

Again commenting on results obtained by Alberti and Casciola [19], using sulfoaryl zirconium phosphates, a careful investigation was performed using pendant groups with different acidities attached to the inorganic layer. They observed that the acid strength has an appreciable influence on the proton conduction at low relative humidity values. However at very high-humidity values, the increasing hydration tends to level the acid strength of the sulfonic group, thus, determining its full dissociation and the difference in the protonic conduction becomes smaller.

2.2.3. Adhesion

Besides the acidity, another important issue related to the filler surface chemistry is its adhesion to the polymer matrix. This is clearly shown in Fig. 4.3[14]. The introduction of untreated fillers may lead to the formation of cavities due to poor adhesion between the filler and the matrix. These cavities create a path of low resistance for methanol and water across the membrane and lead to high methanol permeability values in the case for instance of Aerosil-OH, whose surface is dominated by OH groups. The simple surface treatment with basic (imidazole- or amino-functionalized) silanes (Aerosil-Im) leads to strong acid-base interaction between the basic groups of the filler and the acid sulfonic groups of the polymer, increasing their adhesion and decreasing the permeability. The effect was independent of filler geometry. Both membranes with aerosil and layered silicates (Laponite), which were previously modified with silanes containing imidazole groups (Laponite-Im), had a much lower permeability than those with the unmodified filler.

Also in the case of membranes using zirconium phosphates for direct methanol fuel cells (DMFC), improving the adhesion between particles and the matrix is essential for a good performance. The simple dispersion of phosphate in the membrane might not be critical fuel cell application using hydrogen as feed. However for DMFC the hydrophilic phosphate favors the undesired transport of methanol and water. The treatment with polybenzimidazole (PBI) promotes an acid-base interaction between the filler, the PBI, and the matrix and reduces the liquid transport across the membrane [26, 27].

2.3. Effect of the free volume

Equation (4.1) predicts always a permeability decrease when impermeable fillers are added to the polymer matrix. However, a surprising result was recently reported by Merkel *et al.* [41, 42] for nanocomposites containing fumed silica particles in glassy amorphous poly(4-methyl-2-pentyne) (PMP). They observed a simultaneous increase both of membrane permeability and of selectivity for large organic molecules over small permanent gases. PMP is known for its high-free volume. It is believed that the small inorganic particles disrupt the packing of rigid bulky PMP chains, increasing the accessible free volume in the polymer matrix without introducing cavities large enough to promote weakly selective or nonselective free-phase flow mechanisms. Significant increases in permeability were observed only for particles <50 nm.

Gomes *et al.* [43] investigated nanocomposites of poly(1-trimethylsilyl-1-propyne) (PTMSP), practically the polymer with the highest-known free volume and silica particles, which were *in-situ* generated by a sol-gel process, using organically modified silanes as precursors. Different organic segments were investigated, which led to silica particles not only with different surfaces but also with different sizes. Here again it seems that the smaller particles had a predominant effect on the permeability and selectivity. For instance membranes prepared with a mixture of tetraethoxysilane (TEOS) and methyltriethoxy silane had a higher butane/methane selectivity and higher permeability than the plain PTMSP membrane. Membranes prepared with pure TEOS as inorganic precursor had larger inorganic domains and both lower selectivity and permeability.

3. Polymers with Permeable Filler: Mixed Matrix Membranes

The most representative examples of introducing permeable inorganic fillers in a polymeric matrix for separation purposes are the so-called mixed matrix membranes. The term "mixed matrix membranes" was introduced in 1988 by Kulprathipanja *et al.* [44], who proposed the introduction of silicalites into a matrix of cellulose acetate, increasing the CO_2/H_2 selectivity from 0.77 to 5.15. From that time the concept of mixed matrix membranes has been considered by several groups [45–47] mainly for gas separation. Molecular sieves such as zeolites have a much higher selectivity for many gas mixtures than polymeric membranes due to their well-defined pore sizes. The preparation of defect-free zeolite layers on a large scale is extremely difficult. The possibility of incorporating zeolites into a flexible organic polymer matrix enables the combination of the superior gas selectivities of these molecular sieves with the processibility of polymeric membranes.

One of the early attempts to improve the selectivity of composite membranes by the introduction of zeolites was published by Jia *et al.* [48]. Particularly in the case of composite membranes with very thin selective layers an important requirement is the availability of small zeolite particles. But even with well-dispersed small particles, an essential consideration is the choice of the polymer matrix with the right permeability. The estimation of the permeability and selectivity of membranes with permeable fillers needs a more detailed equation than in the previous section. This can be done using the equation derived by Maxwell to calculate the electric conductivity of a metal in which small spheres of a second metal are dispersed [49]. The permeability of the membrane to a defined gas is

$$P = \frac{P_c[P_d + 2P_c - 2\phi_d(P_c - P_d)]}{[P_d + 2P_c + \phi_d(P_c - P_d)]} \tag{4.3}$$

The selectivity for gases 1 and 2 of a membrane prepared with a continuous phase of permeabilities P_{c1} and P_{c2} and a dispersed phase with permeabilities P_{d1} and P_{d2} can be calculated therefore by Eq. (4.4):

$$\alpha_{\text{eff}} = \alpha_c \frac{1 + 2P_{\text{rel}} - 2\phi(P_{\text{rel}} - 1)}{1 + 2P_{\text{rel}} + \phi(P_{\text{rel}} - 1)} \times \frac{1/\alpha_d + 2P_{\text{rel}}/\alpha_c + \phi[(P_{\text{rel}}/\alpha_c) - (1/\alpha_d)]}{1/\alpha_d + 2P_{\text{rel}}/\alpha_c - 2\phi[(P_{\text{rel}}/\alpha_c) - (1/\alpha_d)]} \tag{4.4}$$

where α_c and α_d are the selectivity of the continuous and dispersed phases for the gas pair 1 and 2; P_{rel} is the permeability ratio between continuous and dispersed phase.

From this equation it can be seen that the selectivity is highly dependent on the permeability of the filler and the matrix. When the polymer permeability is too high the selectivity of the mixed matrix membrane approaches the polymer selectivity. However if the permeability of the filler and the matrix are not too far apart, the properties of the membrane will have a large contribution from the highly selective filler. As a matter of fact a selectivity maximum is observed when the polymer permeability more or less matches the permeability of the filler. The practical challenge is again here to improve the compatibility between inorganic molecular sieves and polymers in order to eliminate gas diffusion pathways at the interface between them. A very informative analysis of the influence of the interface between the filler and the polymer on the membrane properties has been published by Moore and Koros [50]. The Maxwell model correctly describes only the ideal case, for which the filler and the matrix are perfectly attached to each other. If the adhesion is poor and cavities surround the filler, the system is seen as a "sieve in a cage." The permeability increases. However, the selectivity values might be much lower than those predicted by the Maxwell equation in the ideal case. Another possibility is that the filler surface is modified (e.g., with organosilanes) to effectively improve the adhesion to the matrix and an interface region of low permeability is generated. When fillers in the nanoscale are dispersed this interface might constitute a large volume of the sample with a consequent permeability decrease. The selectivity could follow the predicted value or even have a positive deviation. This effect could be observed for instance in the case of fillers, which are originally much more permeable than the matrix. The surface modification may reduce the filler pores, leading to permeability values nearer to that of the matrix, a situation for which the contribution of the filler selectivity is higher. Instead of using zeolites as highly selective filler, a promising alternative is the use of carbon molecular sieves (CMS). Like zeolites, CMS are quite rigid. This inherent brittleness creates difficulties in forming high-surface area modules and makes pure CMS membranes economically impractical for industrial-scale implementation. On the other hand, when used as filler in a mixed matrix membrane, they provide a very attractive alternative strategy [51]. Membranes were prepared using Matrimid® 5218 and

Ultem 1000 as polymer matrix. Selectivity values for CO_2/CH_4 higher than 50 were achieved with CMS loading up to 36 vol %, while the selectivity of the plain membranes is 35–39.

A recent strategy, which has been rather seldom explored, is the use of three-dimensional (3D)-network nanoporous transition metal complexes or metal organic frameworks (MOF) as filler. Good reviews on MOFs have been published by Kitagawa *et al.* [52] and Rowsell and Yaghi [53]. MOFs, also mentioned in the literature as coordination polymers, are organic-inorganic hybrids themselves with organic linker coordinated around transition metals. A huge number of architecture have been published in the last years. The possibility of tailoring the structures and choosing the right linkers to build networks with sieve dimensions able to discriminate gas permeants is particularly attractive for membrane development. Issues such as stability and reproducibility must however to be controlled. Some early results on membranes using MOFs have been reported recently. Won *et al.* recently reported membranes made of polysulfone and $[Cu_2(PF_6)(NO_3)(4,4'bpy)4]2PF_6{\cdot}2H_2O]_n$ and tested them for hydrogen/methane separation [54].

4. Organic-Inorganic Covalent Network

Some of the examples mentioned above already consisted of an inorganic phase generated in a polymer matrix or solution. This strategy was early used by our group in order to decrease the swelling of membranes during the permeation of hydrocarbon mixtures [55] or to decrease the compaction of polyetherimide porous supports in operation in the presence of hydrocarbons [56]. A network of organic and inorganic segments was prepared by reacting epoxy-functionalized silanes and diamines containing long ethylene oxide segments [57]. The selectivity for CO_2 could be enhanced by the ethylene oxide segments. On the other hand, the addition

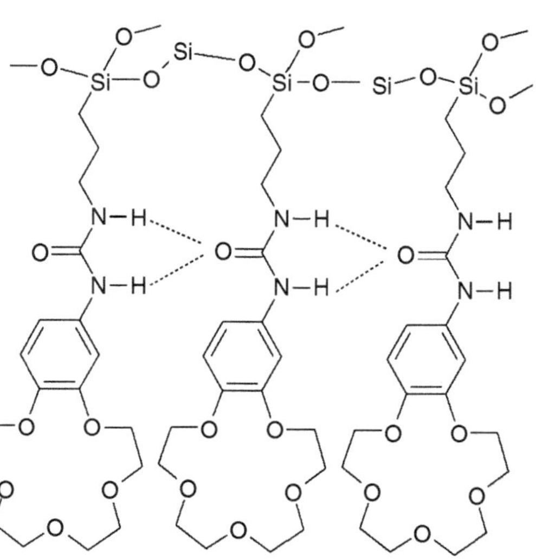

Figure 4.9 Hybrid superstructure with crown ether as a receptor group (adapted from [60]).

of tetraethoxy silane to the reaction mixture increased the content of the inorganic phase in the membrane and minimized excessive swelling.

Recently Steininger *et al.* [58] reported the synthesis of phosphonic acid functionalized oligosiloxanes for application as proton conductive membranes.

Practically all the examples commented up to know were based on inorganic phases dispersed in a polymeric matrix or hybrid organic-inorganic networks. Another perspective is the functionalization of inorganic porous sieves with organic segments. The early efforts to graft organic groups into the micropores of zeolites generally failed, resulting in the modification of primarily the external surface of the crystals. The first microporous, crystalline silicates with organic groups covalently bound within the micropores [59] were made by adding organosilanes $(CH_3O)_3SiR$ to the silicate synthesis gel and incorporating into the zeolites during synthesis, yielding organic-functionalized molecular sieves with potential application for catalysis. Similar approaches could also be very useful for membrane application.

Barboiu *et al.* [60] used the self-organization of organic-inorganic molecules for the construction of new functional nanomaterials in the form of tubular super-structures with macrocyclic "receptors" (Fig. 4.9). The receptor, in this case a crown ether group, allows the transport of specific molecules by molecular recognition. This system has been employed successfully to design a membrane, functioning as an ion-powered adenosine triphosphate (ATP2-) pump.

REFERENCES

[1] Sanchez, C., de, G. J., Soler-Illia, A. A., Ribot, F., Lalot, T., Mayer, C. R., and Cabuil, V. (2001). Designed hybrid organic-inorganic nanocomposites from functional nanobuilding blocks. *Chem. Mater.* **13,** 3061–3083.

[2] Kickelbick, G. (2003). Concepts for the incorporation of inorganic building blocks into organic polymers on a nanoscale. *Prog. Polym. Sci.* **28,** 83–114.

[3] Nielsen, L. E. (1967). Models for the permeability of filled polymer systems. *J. Macromol. Sci. Part A* **1,** 929–942.

[4] Falla, R., Mulski, M., and Cussler, E. (1996). Estimating diffusion through flake-filled membranes. *J. Membrane Sci.* **119,** 129.

[5] Nunes, S. P., Ruffmann, B., Rikowski, E., Vetter, S., and Richau, K. (2002). Inorganic modification of conductive polymer membranes for direct methanol fuel cell. *J. Membr. Sci.* **203,** 215–225.

[6] Silva, V. S., Schirmer, J., Reissner, R., Ruffmann, B., Silva, H., Mendes, A., *et al.* (2005). Proton electrolyte membrane properties and direct methanol fuel cell performance II. Fuel cell performance and membrane properties effects. *J. Power Sources* **140,** 41–49.

[7] Silva, V. S., Ruffmann, B., Silva, H., Gallego, Y. A., Mendes, A., Madeira, L. M., *et al.* (2005). Proton electrolyte membrane properties and direct methanol fuel cell performance. I. Characterization of hybrid sulfonated poly(ether ether ketone)/zirconium oxide membranes. *J. Power Sources* **140,** 34–40.

[8] Gao, H., Ji, B., Jäger, I. L., Arzt, E., and Fratz, P. (2003). Materials become insensitive to flaws at nanoscale: Lessons from nature. *Proc. Natl. Acad. Sci. USA (PNAS)* **100,** 5597–5600.

[9] Talreja, P. S., Kasting, G. B., Kleene, N. K., Pickens, W. L., and Wang, T.-F. (2001). Visualization of the lipid barrier and measurements of lipid path length in human stratum corneum. *AAPS J.-PharmSci.* **3**(2), Article 13.

[10] Prado, L. A. S., de, A., Karthikeyan, C. S., Schulte, K., Nunes, S. P., and de Toriani, I. L. (2005). Organic modification of layered silicates: Structural and thermal characterisations. *J. Non-Cryst. Solids* **351,** 970–975.

[11] Karthikeyan, C. S., Nunes, S. P., and Schulte, K. (2005). Ionomer-silicates composite membranes: Permeability and conductivity studies. *Eur. Polym. J.* **41,** 1350–1356.
[12] Kwaka, S.-H., Yang, T.-H., Kim, C.-S., and Yoon, K. H. (2003). Nafion/mordenite hybrid membrane for high-temperature operation of polymer electrolyte membrane fuel cell. *Solid State Ionics* **160,** 309–315.
[13] Won, J., and Kang, Y. S. (2003). Proton-conducting polymer membranes for direct methanol fuel cells. *Macromol. Symp.* **204,** 79–92.
[14] Karthikeyan, C. S., Nunes, S. P., Prado, L. A. S. A., Ponce, M. L., Silva, H., Ruffmann, B., and Schulte, K. (2005). Polymer nanocomposite membranes for DMFC application. *J. Membr. Sci.* **254,** 139–146.
[15] Gaowen, Z., and Zhentao, Z. (2005). Organic/inorganic composite membranes for application in DMFC. *J. Membr. Sci.* **261,** 107–113.
[16] Jung, D. H., Cho, S. Y., Peck, D. H., Shin, D. R., and Kim, J. S. (2003). Preparation and performance of a Nafion/montmorillonite nanocomposite membrane for direct methanol fuel cell. *J. Power Sources* **118,** 205–211.
[17] Stonehart, P., and Watanabe, M. (June 4, 1996). US Patent 5523181.
[18] Antonucci, V., and Aricò, A. S. (June 30, 1999). EP 0926754A1.
[19] Alberti, G., and Casciola, M. (2001). Solid state protonic conductors, present main applications and future prospects. *Solid State Ionics* **145,** 3–16.
[20] Yang, C., Srinivasan, S., Bocarsly, A. B., Tulyani, S., and Benziger, J. B. (2004). A comparison of physical properties and fuel cell performance of Nafion and zirconium phosphate/Nafion composite membranes. *J. Membr. Sci.* **237,** 145–161.
[21] Costamagna, P., Yang, C., Bocarsly, A. B., and Srinivasan, S. (2002). Nafion® 115/zirconium phosphate composite membranes for operation of PEMFCs above 100 °C. *Electrochim. Acta* **47,** 1023–1033.
[22] Sue, H. J., Gam, K. T., Bestaoui, N., Spurr, N., and Clearfield, A. (2004). Epoxy nanocomposites based on the synthetic alpha-zirconium phosphate layer structure. *Chem. Mater.* **16,** 242–249.
[23] Tchicaya-Bouckary, L., Jones, D. J., and Roiziere, J. (2002). Hybrid polyaryletherketone membranes for fuel cell applications. *Fuel Cells* **2,** 40–45.
[24] Hogarth, W. H. J., Diniz da Costa, J. C., and Lu, G. Q.(Max) (2005). Solid acid membranes for high temperature proton exchange membrane fuel cells. *J. Power Sources* **142,** 223–237.
[25] Ruffmann, B., Silva, H., Schulte, B., and Nunes, S. P. (2003). Organic/inorganic composite membranes for application in DMFC. *Solid State Ionics* **162–163,** 269–275.
[26] Silva, V. S., Ruffmann, B., Vetter, S., Mendes, A., Madeira, L. M., and Nunes, S. P. (2005). Characterization and application of composite membranes in DMFC. *Catal. Today* **104,** 205–212.
[27] Prado, L. A. S. A., Wittich, H., Schulte, K., Goerigk, G., Garamus, V. M., Willumeit, R., *et al.* (2003). Anomalous small-angle X-ray scattering characterization of composites based on sulfonated poly(ether ether ketone), zirconium phosphates, and zirconium oxide. *J. Polym. Sci., Part B: Polym. Phys.* **42,** 567–575.
[28] Aricò, A. S., Baglio, V., Di Blasi, A., Modica, E., Antonucci, P. L., and Antonucci, V. (2004). Surface properties of inorganic fillers for application in composite membranes-direct methanol fuel cells. *J. Power Sources* **128,** 113–118.
[29] Gomes, D., Buder, I., and Nunes, S. P. (2006). Sulfonated silica-based electrolyte nanocomposite membranes. *J. Polym. Sci., Part B: Polym. Phys.* **44,** 2278–2298.
[30] Rhee, C. H., Kim, H. K., Chang, H., and Lee, J. S. (2005). Nafion/Sulfonated montmorillonite composite: A new concept electrolyte membrane for direct methanol fuel cells. *Chem. Mater.* **17,** 1691–1697.
[31] Saiti, P., Aricò, A. S., Hocevar, S., and Antonucci, V. (1998). Physico-chemical characteristics of phosphotungstic acid solutions and electrolyte evaluation for application in fuel cell. *J. New Mater. Electrochem. Syst.* **1,** 1.
[32] Nakamura, O., Kodama, T., Ogino, I., and Miyake, Y. (1979). High conductivity solid proton conductors: Dodecamolybdophosphoric acid and dodecatungstophosphoric acid crystals. *Chem. Lett.* **1,** 17–18.
[33] Katsoulis, D. E. (1998). A survey of applications of polyoxometalates. *Chem. Rev.* **98,** 359–387.

[34] Saiti, E., Hocevar, S., and Giordano, N. (1997). Fuel cells with $H_3PW_{12}O_{40}$•$29H_2O$ as solid electrolyte. *Int. J. Hydrogen Energy* **22,** 809.
[35] Giordano, N., Saiti, E., Hocevar, S., and Aricò, A. S. (1996). High performance fuel cell based on phosphotungstic acid as proton conducting electrolyte. *Electrochim. Acta* **41,** 397.
[36] Zaidi, S., Mikhailenko, S., Robertson, G., Guiver, M., and Kaliaguine, S. (2000). Proton conducting composite membranes from polyether ether ketone and heteropolyacids for fuel cell applications. *J. Membr. Sci.* **173,** 17–34.
[37] Tazi, B., and Savadogo, O. (2000). Parameters of PEM fuel-cells based on new membranes fabricated from Nafion (R), silicotungstic acid and thiophene. *Electrochim. Acta* **45,** 4329.
[38] Dimitrova, R., Friedrich, K., and Stimming, U. (2002). Modified Nafion®-based membranes for use in direct methanol fuel cells. *Solid State Ionics* **150,** 115–122.
[39] Ponce, M. L., Prado, L., Richau, K., Mohr, R., and Nunes, S. P. (2003). Reduction of methanol permeability in polyetherketone-heteropolyacid membranes. *J. Membr. Sci.* **217,** 5–15.
[40] Ponce, M. L., de, L. A. S., Prado, A., Silva, V., and Nunes, S. P. (2004). Membranes for direct methanol fuel cell based on modified heteropolyacids. *Desalination* **162,** 383–391.
[41] Hill, A. J., Freeman, B. D., Jaffe, M., Merkel, T. C., and Pinnau, I. (2005). Tailoring nanospace. *J. Mol. Struct.* **739,** 173–178.
[42] Merkel, T. C., Freeman, B. D., Spontak, R. J., He, Z., Pinnau, I., Meakin, P., and Hill, A. J. (2002). Ultrapermeable, reverse-selective nanocomposite membranes. *Science* **296,** 519–522.
[43] Gomes, D., Nunes, S. P., and Peinemann, K. V. (2005). Membranes for gas separation based on poly(1-trimethylsilyl-1-propyne)-silica nanocomposites. *J. Membr. Sci.* **246,** 13–25.
[44] Kulprathipanja, S., Neuzil, R. W., and Li, N. N. (April 26, 1988). US Patent 4740219.
[45] Nunes, S. P., and Peinemann, K. V. (2006). *In* "Membrane Technology in the Chemical Industry," 2nd edn, Wiley-VCH.
[46] Mahajan, R., and Koros, W. J. (2000). Factors controlling successful formation of mixed-matrix gas separation materials. *Ind. Eng. Chem. Res.* **39,** 2692–2696.
[47] Zimmermann, C. M., Singh, A., and Koros, W. J. (1997). Tailoring mixed matrix composite membranes for gas separations. *J. Membr. Sci.* **137,** 145.
[48] Jia, M., Peinemann, K.-V., and Behling, R.-D. (1991). Molecular sieving effect of the zeolite-filled silicone rubber membranes in gas permeation. *J. Membr. Sci.* **57,** 289–292.
[49] Maxwell, J. C. (1891). A Treatise on Electricity and Magnetism. Vol. 1, p. 440. Dover Publication.
[50] Moore, T. T., and Koros, W. J. (2005). Non-ideal effects in organic–inorganic materials for gas separation membranes. *J. Mol. Struct.* **739,** 87–98.
[51] Vu, D. Q., Koros, W. J., and Miller, S. J. (2003). Mixed matrix membranes using carbon molecular sieves: II. Modeling permeation behavior. *J. Membr. Sci.* **211,** 335–348.
[52] Kitagawa, S., Kitaura, R., and Noro, S. (2004). Functional porous coordination polymers. *Angew. Chem. Int. Ed.* **43,** 2334–2375.
[53] Rowsell, J. L. C., and Yaghi, O. M. (2004). Metal-organic frameworks: A new class of porous materials. *Microporous Mesoporous Mater.* **73,** 3–14.
[54] Won, J., Seo, J. S., Kim, J. H., Kim, H. S., Kang, Y. S., Kim, S. J., Kim, Y., and Jegal, J. (2005). Coordination Compound Molecular Sieve Membrances. *Adv. Mater.* **17,** 80.
[55] Nunes, S. P., Schultz, J., and Peinemann, K. V. (1996). Silicone membranes with silica nanoparticles. *J. Mater. Sci. Lett.* **15,** 1139–1141.
[56] Nunes, S. P., Peinemann, K. V., Ohlrogge, K., Alpers, A., Keller, M., and Pires, A. T. N. (1999). Membranes of poly(ether imide) and nanodispersed silica. *J. Membr. Sci.* **157,** 219.
[57] Sforca, M. L., Yoshida, I. V. P., and Nunes, S. P. (1999). Organic-inorganic membranes prepared by hydrolysis and condensation of polyether-diamine/epoxy silane. *J. Membr. Sci.* **159,** 197.
[58] Steininger, H., Schuster, M., Kreuer, K. D., and Maier, J. (2006). Intermediate temperature proton conductors based on phosphonic acid functionalized oligosiloxanes. *Solid State Ionics* **177,** 2457–2462.
[59] Jones, C. W. (2003). Zeolites go organic. *Science* **300,** 439.
[60] Barboiu, M., Cerneaux, S., van der Lee, A., and Vaughan, G. (2004). Ion-driven ATP pump by self-organized hybrid membrane materials. *J. Am. Chem. Soc.* **126,** 3545–3550.

CHAPTER FIVE

Preparation and Characterization of Zeolite Membranes

Patricia Gorgojo, Óscar de la Iglesia, *and* Joaquín Coronas

Contents

1. Introduction

Zeolites are crystalline, hydrated aluminosilicates with microporous regular structures. The zeolite micropores are of molecular size, which give them adsorption, catalytic properties [1], and ion exchange properties [2], of paramount importance

Department of Chemical and Environmental Engineering, University of Zaragoza, 50018 Zaragoza, Spain
Nanoscience Institute of Aragon, University of Zaragoza, 50009 Zaragoza, Spain

Membrane Science and Technology, Volume 13
ISSN 0927-5193, DOI: 10.1016/S0927-5193(07)13005-9

in both the industrial chemical field and for the study of new applications related to, *inter alia*, process intensification [3], green chemistry [4], hybrid materials [5], and medicine [6]. The ability to grow zeolites on porous supports can produce devices with very specific properties such as membranes [7], membrane reactors [8], microreactors [9], reactive [10] and nonreactive gas sensors such as quartz crystal microbalance (QCM; [11]) or zeolite-based capacitors [12], corrosion protection coatings [13], antimicrobial coatings [14], zeolite-coated catalyst particles [15], and hollow zeolite particles [16].

In spite of the large number of zeolite and zeolite-like structures described in the literature, only a few such materials have been studied as membranes or even membrane reactors. For instance, membranes with interesting separation properties have been reported for the following zeolite-type materials [8, 17, 18]: silicalite, ZSM-5, mordenite, zeolite A, zeolite Y, zeolite beta, ETS-4, ETS-10, and MCM-48. Figure 5.1 shows the structures of the LTA-, MFI-, FAU-, and MOR-type zeolites, which can be considered as among the most studied membrane zeolitic materials. These structures were obtained from the Atlas of zeolite framework types [19].

Table 5.1 lists some milestones in the development of zeolite membranes. The first membranes having zeolites in their composition were mixed matrix membranes (MMMs; [20]). Thus in this case, there was no continuous layer of intergrown zeolite crystals. This concept was introduced by Suzuki [21], who in 1987 described a large array of methods to prepare zeolite (6-, 8-, 10-, or 12-membered ring) membranes on porous metal, Vycor glass, or ceramic supports. The first zeolite membranes prepared by hydrothermal synthesis and showing separation performance were of zeolite A (1989) and silicalite (1991) on a glass tube [22] and Teflon slab [23], respectively. In 1992, a layer of zeolite crystals was coupled to a QCM to produce a physical sensor device [24]. In 1994, the dry gel method [in its vapor-phase transport (VPT) version] was used for the first time to prepare a ferrierite/ZSM-5 membrane [25]. In 1995, the first zeolite (silicalite) membrane reactor results related to the dehydrogenation of isobutene were reported [26]. The secondary (seeded) growth method, decoupling nucleation and membrane growth, was presented in detail in 1996 for silicalite membranes [27]. Soon after the year 2000 zeolite membranes were introduced into microelectronic fabrication [28], the first industrial-scale pervaporation plant using zeolite A membranes became operational [29], while the concept of the zeolite membrane was extended to coat particles [15] and zeolite crystals [30] of different structures in an attempt to provide shape and size

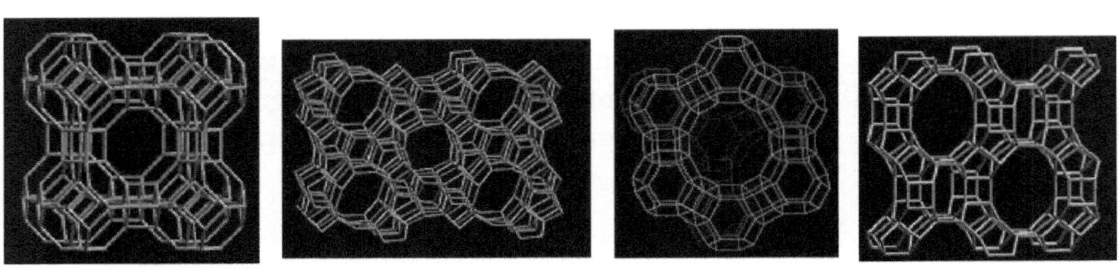

Figure 5.1 Some zeolite structure types [19].

Table 5.1 Some milestones in the development of zeolite membranes

1986	Mixed matrix membranes (Kulprathipanja)
1987	Suxuki's patent
1989	Zeolite A membrane (Ishikawa *et al.*)
1991	Silicalite membrane (Haag and Tisikoyannis)
1992	Coupling of zeolite crystals to QCM (Yan and Bein)
1994	VPT method (Metsukata *et al.*)
1995	ZM reactor (Casanave *et al.*)
1996	Secondary growth method (Lovallo *et al.*)
2001	Microelectronic fabrication (Wan *et al.*)
2001	Mitsui plant (Morigami *et al.*)
2003	Silicalite b-oriented membranes (Lai *et al.*)
2004	Silicalite-coated catalyst particles (Nishiyama *et al.*)
2006	Core-shell zeolite microcomposites (Bouizi *et al.*)

selectivity to catalysts and adsorbents. Finally, the most spectacular progress in synthesis was achieved by Tsapatsis and coworkers in 2003 [31] when b-oriented, 1-μm-thick silicalite membranes were prepared, establishing that the zeolite membrane performance depended strongly on membrane microstructure. This was expected but unobserved until that moment because of the composite membrane meso- or macrostructure.

1.1. How is a zeolite membrane different?

Table 5.2 summarizes the advantages of inorganic membranes (zeolite membranes belong to the larger family of inorganic membranes) when compared to others (polymeric or organic membranes). From the point of view of transport and performance applications, a zeolite membrane, because of its very microporous character, has a special permeation pattern that a mesoporous or a defective zeolite membrane does not possess, as illustrated in (Fig. 5.2). This pattern will be explained in depth in the characterization section. Besides this unique permeation pattern, zeolite membranes also differ from other membranes because (1) they can separate difficult mixtures, such as those of isomers and azeotropic compounds, trace gas separations (volatile organic compounds from air), and permanent gases; (2) they can simultaneously perform two functions, separation and reaction, as zeolite membrane reactors (ZMRs).

1.2. New zeolitic membrane materials

Since the early 1990s, there has been an increasing interest in the synthesis of new ordered microporous materials such as titanosilicates and their isomorphs ETS-10, ETS-4, and umbite, and other heteropolyhedral structures [32]; mesoporous templated silicates, M41S (MCM-41 and MCM-48), SBA-15, SBA-16 [6, 33]; and metal-organic frameworks (MOF; [34]). In the field of membranes and membrane reactors only a few such materials have been studied. Besides the already mentioned

Table 5.2 Advantages of zeolite membranes, adapted from Caro *et al.* [18]

Advantages	Disadvantages
Stability at high temperature	High cost
Resistance to harsh environments	Brittleness
Resistance to high-pressure drops	Poor intensification
Inertness to microbiological degradation	Difficulty in industrial scale-up
Easy cleaning after fouling	Low permeability of the high selective (dense) membranes
Easy catalytic activation	Sealing problems

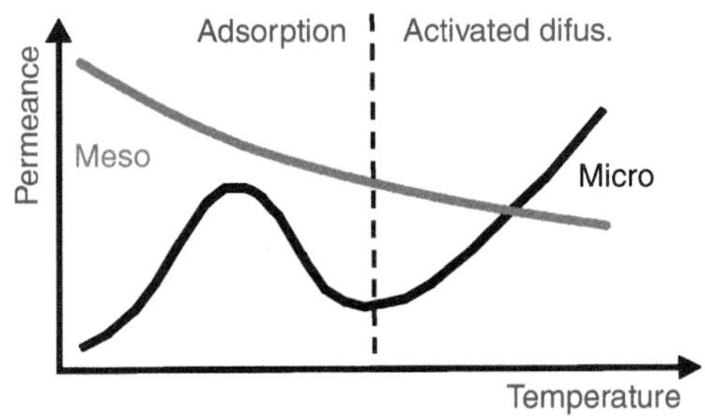

Figure 5.2 General permeation patterns of micro- and mesoporous membranes.

zeolite membrane materials, other microporous membrane groups include pseudo-zeolitic materials containing tetrahedrally coordinated phosphorus, such as $AlPO_4$ [35] and $SAPO_4$ [36], and zeolite membranes containing Ti and V in their structure. These membranes, such as TS-1 and VS-1 [37], are potentially interesting from the point of view of catalytic applications. In all these special cases, the number of publications is very small when compared to that of MFI-type zeolite or LTA-, FAU-, and MOR-type zeolites. It seems that these new materials may broaden the scope of application of zeolite membranes and ZMRs.

Wide-pore zeolites, such as the recent ITQ-21, accessible through 6 circular 0.74-nm openings [38], and extra-large-pore zeolites, such as the phosphate-based VPI-5 or the more stable silicas UTD-1 and CIT-5 [6] with 0.8- to 1.2-nm openings comprising more than 12-membered rings, have potential interest as membranes, even if such materials have not hitherto been used in membrane preparation. One possible application would be to meet the demand posed by fuels and petrochemicals coming from crude oil in the near future. Other zeolite-like innovative microporous materials that could offer interesting properties as membranes are MOF that seem to be promising candidates for metal-organic heterogeneous catalysis and perhaps to separate and purify hydrogen, since these MOF-type solids have extremely high methane [34] and hydrogen [39] adsorption capacities.

Finally, coming back to amorphous silicate and aluminosilicate mesoporous materials such as MCM-41, MCM-48, and others, these materials would give access to far larger pore sizes (2–30 nm) able to process bulky molecules and macromolecules. Unfortunately, these materials lack thermal, hydrothermal, and mechanical stabilities, mainly due to their relatively thin amorphous silica walls. Many modifications have been carried out to improve the stability of these mesoporous materials. Perhaps the most studied concern the synthesis of zeolites with primary and secondary structures: for instance, from the partial transformation of MCM-41 into ZSM-5 [40]. In any event, the successful preparation of MCM-48 membranes has already been reported [41, 42].

1.3. Commercial aspects

Ceramic membranes (usually being the supports for zeolite membranes) are 10–100 times more expensive than the equivalent polymeric membranes [43]. This cost differential can be tolerated only in applications (e.g., high temperature) in which polymeric membranes completely fail to work. Figure 5.3 compares cost estimations for zeolite membranes given by several authors over recent years. These costs are compared to that of polymeric membranes (around 80 €/m^2, used for CO_2 sequestration, a gas application [44]). The 200 €/m^2 given by Maloncy *et al.* [45] is only a fitting value for the purpose of an economical evaluation of a zeolite membrane-based heptane hydroisomerization process.

Many scale-up aspects are common to zeolite and inorganic membranes in general [8]: (1) module reliability under extreme temperature cycling; (2) low permeabilities; (3) availability with larger areas and preparation with satisfactory reproducibility; and (4) durability under elevated pressures and temperatures, stability in a steam-containing atmosphere, and membrane regeneration. However, there is a large-scale pervaporation plant using tubular zeolite A membranes producing 600 liter/h of solvents

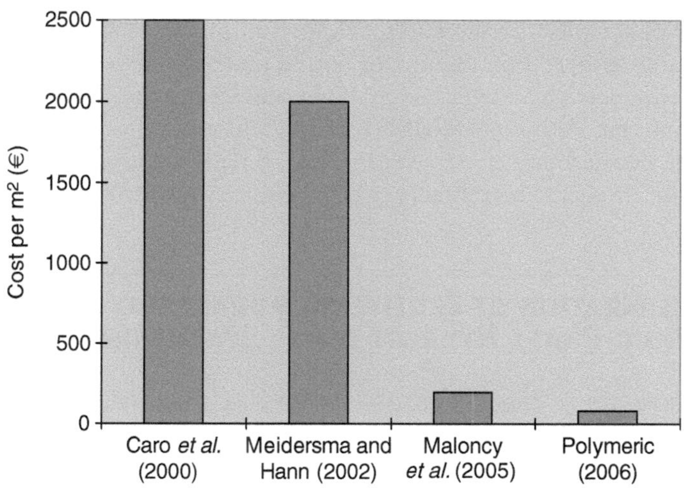

Figure 5.3 Estimative cost of zeolite membranes. Data extracted from Refs. [18, 44–46].

(methanol, ethanol, *i*-propanol, etc.) at less than 0.2 wt % of water [29]. This plant is equipped with 16 modules, each of which consists of 125 zeolite A tubular membranes with ~0.0275 m^2 of permeation area, in total 55 m^2.

As pointed out above, the preparation of zeolite membranes and films has undergone an impressive development during the last decade. This development has stimulated a great deal of interest in the possible uses at a relatively large scale of zeolite layers in separation applications (either as membranes or as adsorbents) and in reaction processes, although the current high price of zeolite membranes and the other inherent limitations mentioned above would suggest that the use of zeolite membranes would be more appropriate in small-scale and microscale applications [47]. There are clear advantages to small-scale zeolite membranes: (1) excellent temperature control capable of preventing hot-spot formation, (2) higher selectivity (e.g., by minimizing homogeneous reactions) and conversion (by working at higher temperatures and pressures, often in the explosive region compared to conventional reactors), (3) intrinsic safety, (4) easier scale-up and process integration in a highly compact way (process intensification), (5) synergy between reaction and permeation at a microscopic level, (6) higher probability of obtaining a defect-free interface, since this probability should increase for smaller membrane areas, (7) micromembranes could also be synthesized as single-crystal units, and (8) microscale membranes are amenable to preparation methods that use high-throughput synthesis procedures. Advantages (1) to (4) are general for microsystems [48] and appear when the surface to volume ratio (A/V) of the considered system decreases, while the others are more specific to zeolite membranes. Figure 5.4 shows a scheme of this scale evolution for applications related to sensors, microreactors, microseparators, micromembrane reactors, coated catalysts, hollow zeolite capsules, etc.

Finally, an elegant attempt at a better economical approach to zeolite membranes concerns the development of so-called MMMs, which has received much attention for gas separation [49]. MMMs have an organic phase which is a polymer and an inorganic phase which is a zeolite (see Fig. 5.1, Chapter 4 of this volume). This kind of membrane provides a promising way of getting the best of polymers (processability) and the best of a molecular sieve (high separation performance) for gas separation. MMMs are able to overcome the upper bound trade off curve between gas permeability and permselectivity established by Roberson [50] in the case of many industrial mixtures of interest. Although MMMs lead to an increase in selectivity, there is also some loss in permeability as shown in (Fig. 5.5). However, their lower prices compared with inorganic membranes make them very attractive.

2. Preparation of Zeolite Membranes by *In Situ* Liquid-Phase Hydrothermal Synthesis

The formation of a zeolite membrane requires the development of a continuous, defect-free, nearly bidimensional layer of zeolite crystals, so that only transport through the zeolite pores takes place. The usual procedure for this consists in

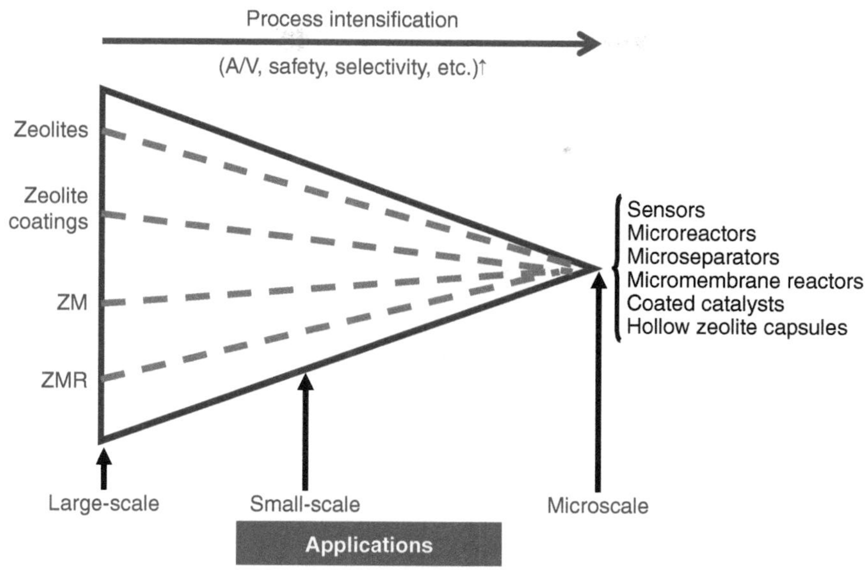

Figure 5.4 Scale evolution of zeolite membranes.

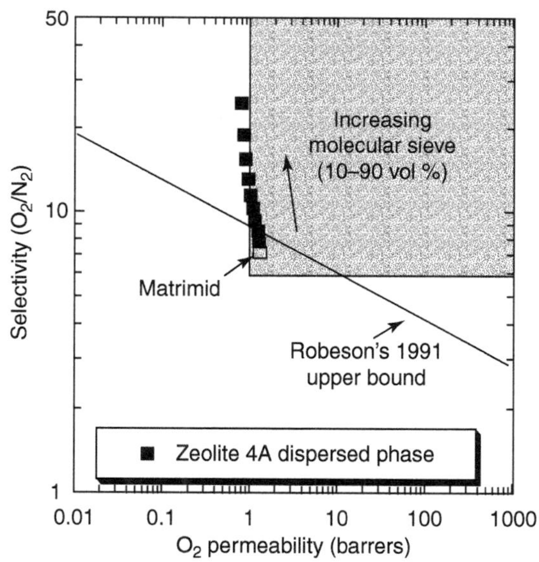

Figure 5.5 Predicted zeolite 4A—Matrimid membrane performance from Maxwell's equation. Reprinted with permission from [51], Copyright (2000) American Chemical Society.

depositing zeolite crystals onto a previously existing porous support, which confers the necessary mechanical strength and allows the development of more extensive structures (Fig. 5.6).

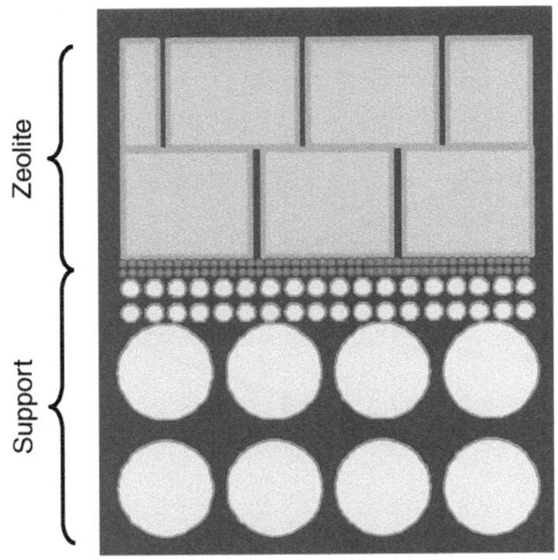

Figure 5.6 Scheme of a zeolite membrane.

2.1. Previous aspects

The typical supports for the preparation of zeolite membranes are porous inorganic tubes or plates made of alumina and stainless steel, although monoliths (also as wheels or rotors), stainless steel grids, wire gauze packings, glass fibers, ceramic and metal nonporous plates, glass and steel beads, etc. have been used as supports with different purposes [8].

In catalytic processes, monoliths exhibit high rates and selectivities, a high degree of operational flexibility, low energy consumption, and reasonable commercial costs; also, from an engineering point of view, the easy scale-up and high safety potential are appealing [52]. Other advantages of monoliths include their low pressure drop, good tolerance to plugging by dust, and, in the case of zeolites, with the fact that they can be prepared binderless. Monolithic structures such as ZSM-5 prepared on cordierite by the common liquid-phase hydrothermal synthesis [53] or by solid state *in situ* crystallization [54] are conceived mainly with the idea of automotive applications.

In terms of adsorption or membrane separation processes, zeolitic monoliths have found very interesting and fruitful applications as rotatory adsorbers for use in dehumidifiers and desiccant cooling processes [55] or VOC (volatile organic compound) treatment systems [56]. Alumina-coated, silicon carbide monolith supports have also been employed as a means of producing B-ZSM-5 membranes with larger surface area per volume for use in separation processes. With such membranes, these authors have reported *n*-butane/isobutane and H_2/isobutane separation selectivities of 35 and 77, respectively [57]. Also, silicalite membranes produced on stainless steel grids (Fig. 5.7) have performed well in the separation of *n*-butane/isobutane mixtures, with separation factors as high as 53 at 63 °C [58].

Before the actual process of preparing the zeolite membrane, several pretreatments can be carried out on the support to improve control over the characteristics of the membranes prepared by hydrothermal synthesis. These include (1) seeding the

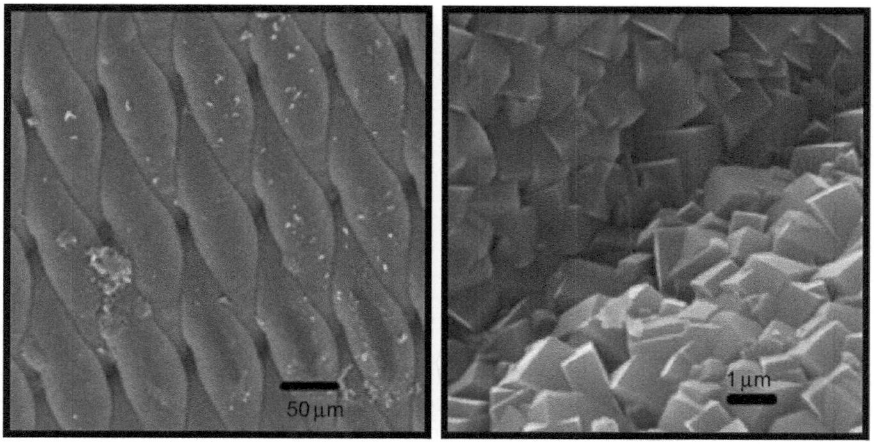

Figure 5.7 A silicalite membrane prepared on a stainless steel grid with nominal aperture of 5 μm. Reprinted with permission from [58]. Copyright (2005) American Chemical Society.

support with crystals of the zeolite to be synthesized, to control nucleation (this will be explained later), (2) using diffusion barriers to limit the penetration of the precursor species into the support pores [59], (3) masking the support with wax also to avoid penetration of the synthesis into the support pores and to obtain a very thin membrane [60], and (4) oxidation of stainless steel supports or deposition [61] of iron oxide by impregnation with iron precursors [62]. The seeding process can be made by simply rubbing [63], by filtration [64], by means of an ultrasonic bath to help to disperse a zeolite suspension of commercial or previously prepared crystals [65], by dip-coating using colloidal suspensions [27], by laser ablation [66], or by chemical bonding [31].

2.2. The method

Zeolite membranes are commonly prepared by *in situ* hydrothermal synthesis on porous supports in which the porous support is immersed into the zeolite precursor gel, and the membrane is synthesized under autogenous pressure in an autoclave. The autoclave is usually placed in an oven at 80–230 °C, depending on the zeolite to be crystallized, for several hours or even days. The gel usually contains water, silicon, or alumina sources, sodium hydroxide, and an organic structure-directing agent (SDA). At the end of the synthesis procedure (1) the zeolite can be preferentially deposited inside the porous structure of the support, or (2) most of the zeolite material exists as a thin layer on top of the porous support. Both types of membranes often exhibit a different separation behavior [67, 68]. After synthesis the membrane is subjected to characterization [mainly XRD (X-ray diffraction) and permeation measurements] and, depending on the results obtained, the synthesis is repeated until the desired result is achieved. Note that, in many instances, when an SDA is used, a defect-free membrane should be nonpermeable at the end of the synthesis because of the presence of this SDA inside the zeolitic pores of the membrane. The membrane is activated for permeation after the removal of the template by calcination. As examples of zeolitic membranes prepared by the *in situ* method,

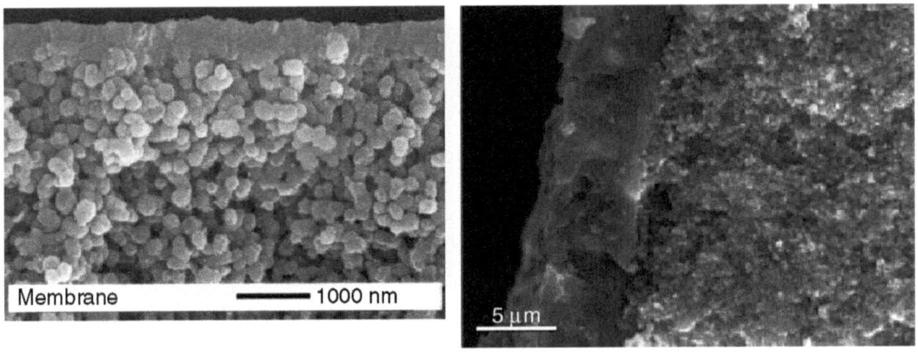

Figure 5.8 Examples (cross sections) of zeolitic membranes: silicalite (left Reprinted from Microporous and Mesoporous Materials, [60], Copyright (2002), with permission from Elsevier.) and titanosilicate umbite (right Reprinted with permission from [69] Copyright (2006) American Chemical Society).

Fig. 5.8 shows a silicalite membrane (left) prepared on top of an alumina support previously pretreated with wax to produce an ultrathin membrane 0.5 μm thick [60], and a titanosilicate umbite (right) membrane 5 μm thick prepared on a titania tubular support [69]. Both membranes perform efficiently in the separation of *p*/*o*-xylene and H_2/N_2 mixtures, respectively.

Even if one-synthesis membranes are preferred to simplify the preparation procedure of a zeolite membrane, the experimental number of synthesis cycles to be used depends on the particular zeolite and on the preparation conditions. Thus, for instance, Vroon *et al.* [70] used two consecutive hydrothermal treatments with different synthesis temperatures (98–186 °C) to prepare zeolite MFI-type membranes on α-alumina supports. They concluded that their membranes prepared in a single hydrothermal synthesis had defects because of a lack of connectivity between the individual particles, that is, one synthesis was not enough to achieve sufficient crystal intergrowth. On the other hand, zeolite membranes obtained by three or more hydrothermal treatments became too thick and probably cracked during the removal of the template. Nevertheless, sometimes more than one synthesis is needed from the point of view of the separation performance [71].

Vilaseca *et al.* [72] have studied the growth of silicalite membranes on nonporous alumina supports by *in situ* liquid-phase hydrothermal synthesis. Using AFM (atomic force microscopy) and SEM (scanning electron microscopy) images, they monitored the nucleus and crystal size evolution as a function of synthesis time (Fig. 5.9). At short synthesis times, nuclei are evident on both support particles and emerging zeolite crystals. After 2 h of synthesis the nuclei disappeared from the silicalite crystal surfaces, while these crystals continued to grow reaching a size of about 5 μm after 8 h. This clearly suggests that when the concentration of reactants in the synthesis gel drops below a certain level the heterogeneous nucleation stops and only crystal growth can take place, even when using dissolved nuclei as feed. One important result of the observations is that nuclei coexist with crystals for a long time during synthesis. Figure 5.10 depicts the mechanism proposal of the authors based on data from Fig. 5.9 for the formation of MFI-type zeolite films onto alumina supports.

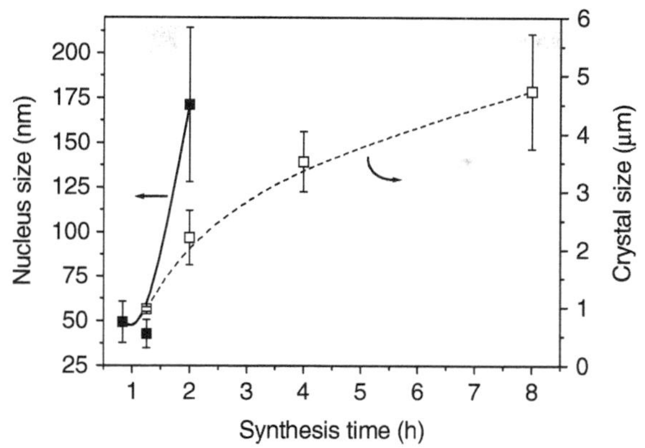

Figure 5.9 Nucleus (full squares) and crystal (hollow squares) sizes as a function of synthesis time. Reprinted from Microporous and Mesoporous Materials, [72], Copyright (2004), with permission from Elsevier.

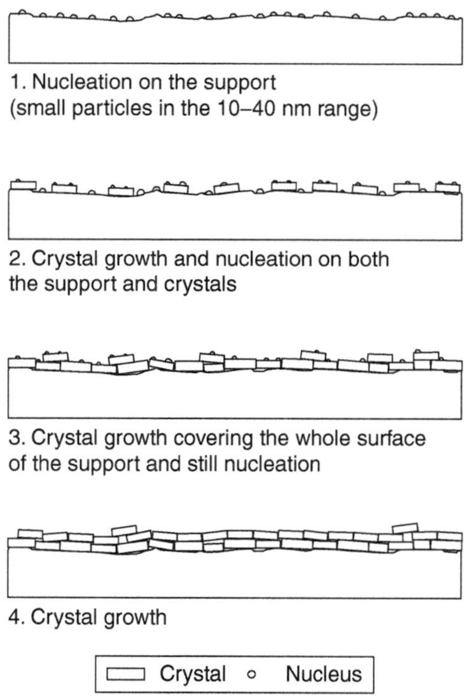

Figure 5.10 Mechanism for the formation of an MFI-type zeolite film. Reprinted from Microporous and Mesoporous Materials, [72], Copyright (2004), with permission from Elsevier.

3. Preparation of Zeolite Membranes by Secondary (Seeded) Growth

Because of the insufficient understanding of crystalline nucleation and growth phenomena, the success of *in situ* methods in yielding uniformly oriented MFI-type zeolite (the type most studied as a membrane) is limited. The most successful approach to controlling membrane formation, including crystal orientation, involves segregation of the processes of crystal nucleation and growth. Unlike the direct or *in situ* synthesis procedures just discussed, the first step in this method includes only nucleation and initial crystal growth, which are usually carried out as a homogeneous synthesis (in the absence of the porous support), yielding colloidal zeolite crystals. These can now be used as seeds, deposited on the support and brought into contact with a solution containing the necessary nutrients for growth [7, 27, 31]. Because the concentration needed for secondary growth is lower than that required for nucleation, further nucleation is strongly decreased and almost all of the crystal growth takes place over the existing crystal seeds. By controlling the composition and concentration of the secondary growth solution, the crystallization of undesired zeolite phases and the dissolution of the support can be avoided, and the rate and direction of crystal growth can be controlled.

This situation has been studied in depth for silicalite (MFI-type structure), which has 0.53 × 0.56 nm straight channels along [010] direction and 0.51 × 0.55 nm zigzag channels along [98] (Fig. 5.11). There are no channels in the MFI-type structure along [001] direction. All the silicalite-oriented membranes claimed before 2003 [31] were c-oriented membranes, a typical example being the tubular membrane shown in (Fig. 5.12). In this case, it was said that the crystals grow preferentially from the seeded support toward the bulk of the solution (where the nutrients are), giving rise to a columnar, well-intergrown layer that can be appreciated in the top view of Fig. 5.12 [68]. The preferential crystallographic orientation of a film is usually confirmed by XRD analysis.

However, from the point of view of transport, c-oriented membranes (with no channels perpendicular to the support surface) would be at a disadvantage with respect to b-, a-, or even randomly oriented membranes. In all these three cases, the possibility of having zeolitic channels perpendicular to the support is higher. Recently, Tsapatsis and coworkers reported the formation of uniformly b- [31] and a- [7] out-of-plane oriented by secondary growth of, respectively, b- and a-oriented seed layers. This was achieved by the appropriate use of SDAs that act as crystal shape modifiers to enable the

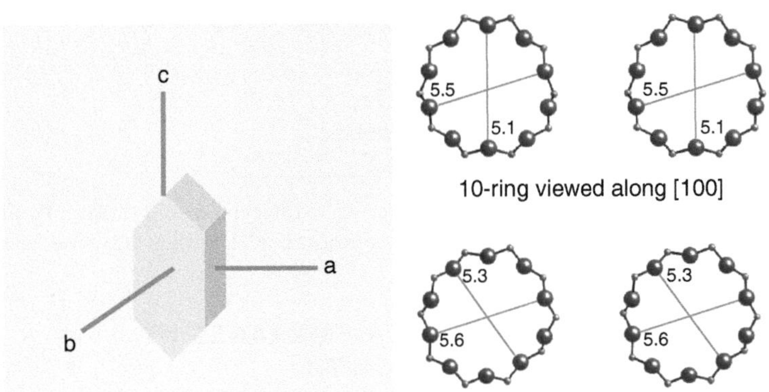

Figure 5.11 Typical crystal shape and channels [19] of the MFI-type structure.

Figure 5.12 The MFI-type zeolite films with c-out-of-plane orientations left, top view; right, cross section. Reprinted from Microporous and Mesoporous Materials, [68], Copyright (2003), with permission from Elsevier.

necessary orientation of the seed layer and the appropriate balance between in-plane and out-of-plane growth rates during secondary growth. As shown in Fig. 5.13, after coating a mesoporous silica layer on the polished side of an α-alumina porous disk by slip coating, the SDA (tetrapropylammonium hydroxide, TPA) used for seed formation (with the typical coffin shape) in the procedure leading to b-oriented films was used for secondary growth of a-oriented membranes. The SDA (trimer-TPA) used for secondary growth in the procedure leading to b-oriented films was applied to the synthesis of the MFI-type zeolite seed crystals (leaf-shaped) needed as seeds for the a-oriented membrane. Figure 5.14 shows the results that this method can produce. In fact, b-oriented membranes give a superior performance in terms of xylene isomer separation [31].

Besides MFI-type zeolites, other oriented zeolite structures have been prepared, among these are zeolite L [27], zeolite A [74], and zeolite UTD-1 [66], ETS-10 [65], and mordenite [75].

4. Preparation of Membranes by the Dry Gel Method

Despite the success of the two preceding methods of preparing zeolite membranes, they have two main disadvantages: first, some crystals may nucleate and grow in the bulk of the synthesis gel, and consequently be incorporated into the growing zeolite layer without the desired order, giving rise to additional defects in the final membrane. Second, a considerable excess of water and other reactants (such as silicon and aluminum sources and SDAs) is employed, leading to a more expensive synthesis. Obviously, the secondary (seeded) growth method, using seed crystals synthesized in the absence of the support, produces crystal growth under conditions that hinder further nucleation, reducing the influence of the homogeneous nucleation in the final characteristics of the zeolite membrane.

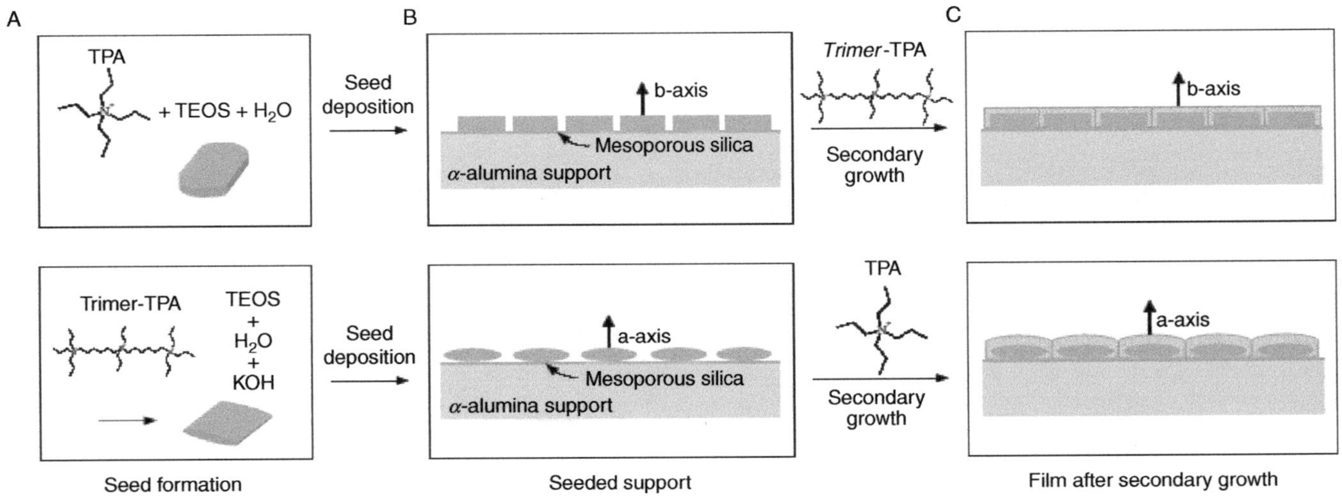

Figure 5.13 Preparation of the MFI-type zeolite films with b- and a-out-of-plane orientations [7]: A) seed formation, B) support seeding, and C) secondary growth.

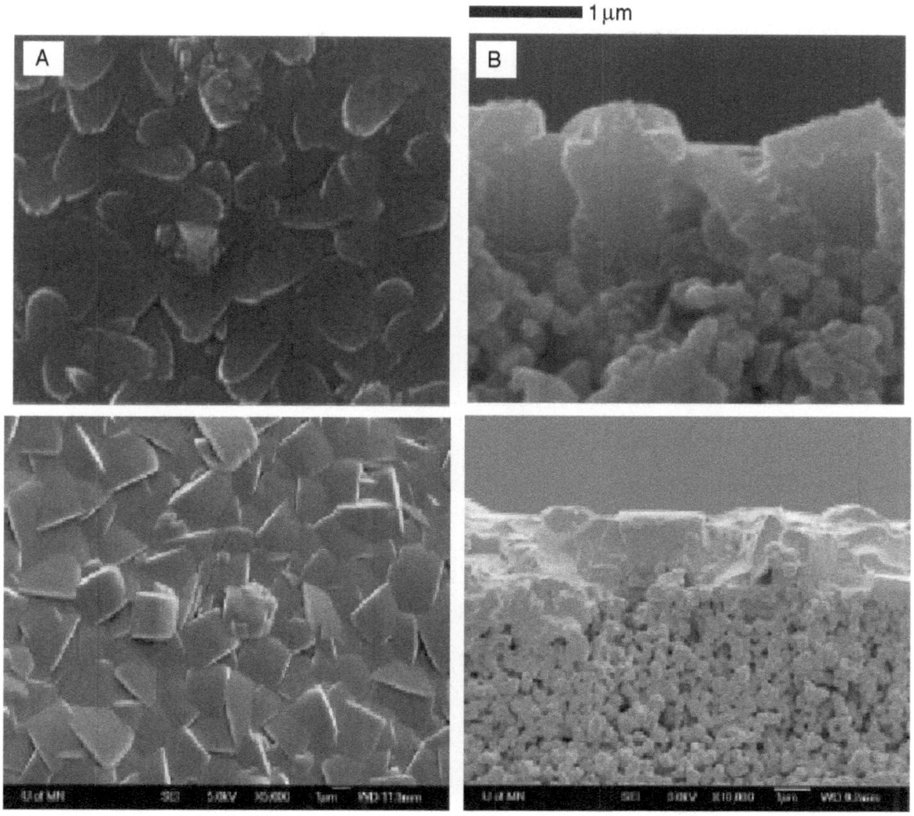

Figure 5.14 Top views (left) and cross sections (right) of the MFI-type zeolite films: top, b-oriented [73]; bottom, a-oriented [7].

An alternative approach to avoid the homogeneous nucleation and growth of crystals that could impair the quality of the zeolite membrane is to deposit a layer of dry aluminosilicate gel on the support and then transform this gel into zeolite in the presence of vapors. This has the additional advantage of minimizing the waste reactants. Xu *et al.* [76] studied the conversion of a dry aluminosilicate gel into an MFI-type zeolite by putting the gel in contact with water and amine vapors. Matsukata and coworkers [77, 78], using a similar procedure, prepared ZSM-5, ferrierite, mordenite, and analcime flat membranes. Similarly, zeolite membranes have also been prepared by vapor-phase regrowth of colloidal MFI-type zeolite particles deposited on disks [79], or by liquid-phase hydrothermal treatment of a dry gel barrier previously incorporated onto an alumina tubular support [80]. With this general approach, two routes can be distinguished [81]: (1) the VPT method (when the organic SDAs are not included in the dry parent gel), and (2) steam assisted crystallization (SAC), where the dry gel already contains the templating agents and only steam is supplied from the vapor phase. The dry gel method in its VPT version is schematized in (Fig. 5.15). The second strategy was followed to prepare, on a symmetric stainless steel tube, the silicalite membrane shown in Fig. 5.16 [82]. Although this method produces membranes able to separate an *n*-butane/isobutane

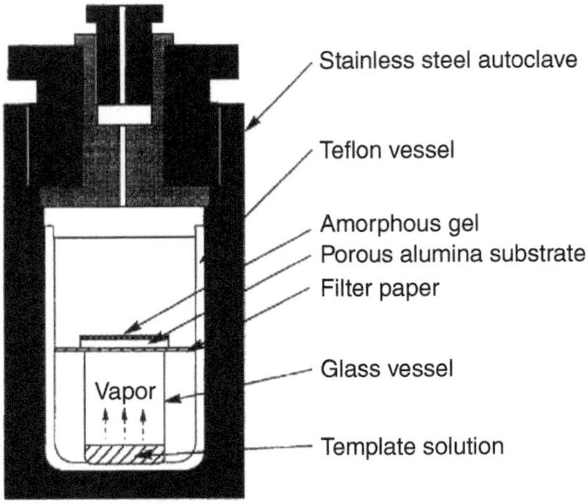

Figure 5.15 Scheme of the autoclave used for synthesizing zeolite membranes by the VPT method. Reprinted from Microporous Materials, [83], Copyright (1997), with permission from Elsevier.

Figure 5.16 A silicalite membrane prepared by SAC left, top view; right, cross section. Reprinted from Microporous and Mesoporous Materials, [82], Copyright (2001), with permission from Elsevier.

mixture with high selectivity like other similar membranes, its main drawbacks are the duration of the synthesis (many days) and the need for repeating several times the cycle of gel deposition-SAC.

5. Special Issues

There are some issues relating to zeolite membranes that require special attention. These include (1) the influence of the support, (2) calcination of the membrane, (3) possible posttreatments of the membrane, and (4) two-layered and zoned zeolite membranes.

5.1. Influence of the support

The presence of the support introduces a number of factors that hinder the reproducibility of the zeolite membrane synthesis: (1) The mechanism of nucleation changes because the surface of the support provides nucleation sites that are not present in homogeneous synthesis. (2) The support itself may be dissolved in the synthesis gel, leading to ZSM-5 rather than silicalite membranes. (3) Because the support can selectively restrict the diffusion of the gel components, synthesis inside the support pores may take place with a composition different from that of the bulk liquid.

An interesting example that illustrates these three aspects is the preparation of MFI-type zeolite membranes via the so-called pore-plugging synthesis based on zeolite crystallization within the pores of a host support [84]. This method produces a nanocomposite zeolite membrane structure on an α-alumina asymmetric tubular support, as shown in Fig. 5.17.

Transmission electron microscopy (TEM) has been used to investigate the material qualitatively described in Fig. 5.17 (right). In the smaller magnification micrograph in (Fig. 5.18), the support (α-alumina) corresponds to the darker areas. A crystalline material (see the larger magnification micrograph on the right of Fig. 5.18) fills up completely the pores generated by the alumina particles. This material shows regular crystallographic intervals and the fast Fourier transform method has revealed the MFI-type zeolite structure (Fig. 5.18, bottom right and top left). EDX (energy dispersive X-ray) elemental analyses of the zeolite crystal carried out on different spots using a probe size of about 10 nm indicate an average Si/Al ratio of about 10 all over the MFI-type crystals filling the pores, that is even if the precursor gel does not contain any aluminum source, a ZSM-5 membrane was produced instead of a silicalite

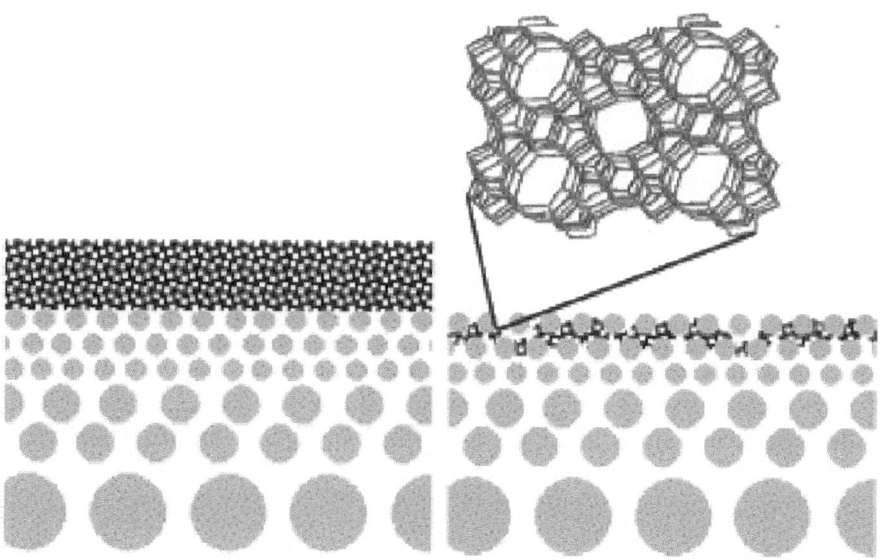

Figure 5.17 Schematic comparison between film (left) and nanocomposite zeolite membrane (right). Reprinted from Journal of Membrane Science, [84], Copyright (2006), with permission from Elsevier.

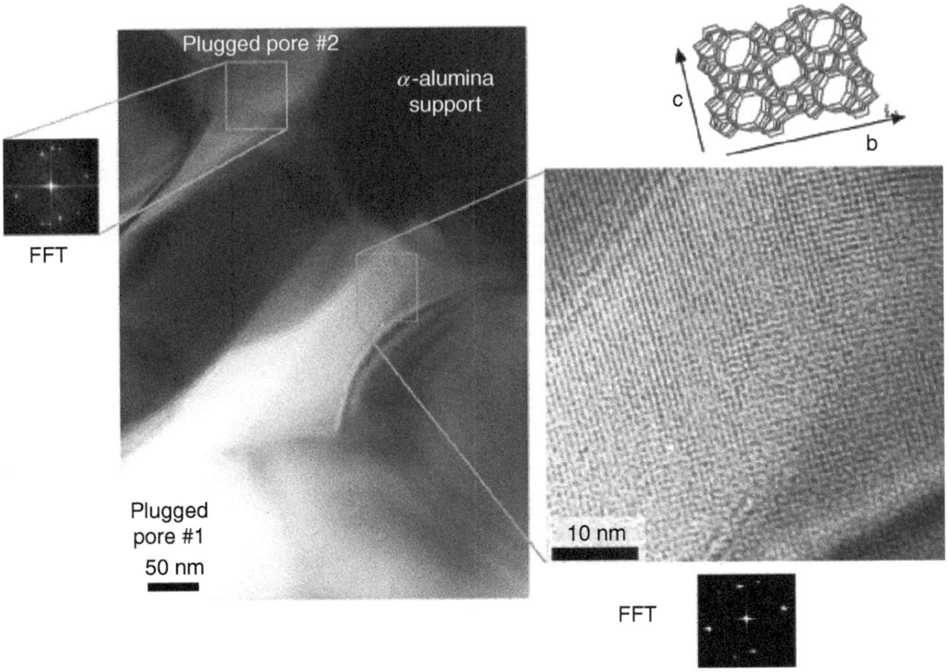

Figure 5.18 Transmission electron micrograph of the zeolite/alumina composite membrane. Reprinted from Journal of Membrane Science, [84], Copyright (2006), with permission from Elsevier.

one. From the point of view of the permeation properties, similar MFI-type membranes exhibit separation selectivities of some mixtures (*n*-butane/isobutane) at higher temperatures than usual [68].

Figure 5.19 is an example of a situation in which the support, of symmetric α-alumina, would not be an effective barrier for the diffusion of the reactants during the hydrothermal synthesis of an ETS-10 membrane [65]. In any case, EDS (X-ray spectrometry) analysis of the membrane cross sections indicates that ETS-10 membrane forms on the surface rather than inside the support pores, which may be due to the viscosity of the synthesis gel that hinders the penetration of the Si and Ti sources into the support pores. On the other hand, since this ETS-10 membrane was prepared under moderately alkaline conditions, a low degree of aluminum leaching from the support may be expected and, thus, the membranes are likely to consist mostly of ETS-10, although the coexistence of ETAS-10 (aluminum-containing ETS-10) with ETS-10 on the membrane is consistent with (Fig. 5.19). Also, it can be seen there that the membrane surface is rich in Si and Ti and these elements are homogeneously distributed in this region, while the bright regions in the support area (~10- to 20-μm deep) result from small ETS-10 particles (probably seeds or crystals already grown from seeds).

Finally, aluminum may be transferred from the support into the zeolite membrane in two paths [85], as shown in Fig. 5.20: (1) Aluminum can diffuse in the solid state to the zeolite layer, within a certain range (around 2.5–5 μm in thickness),

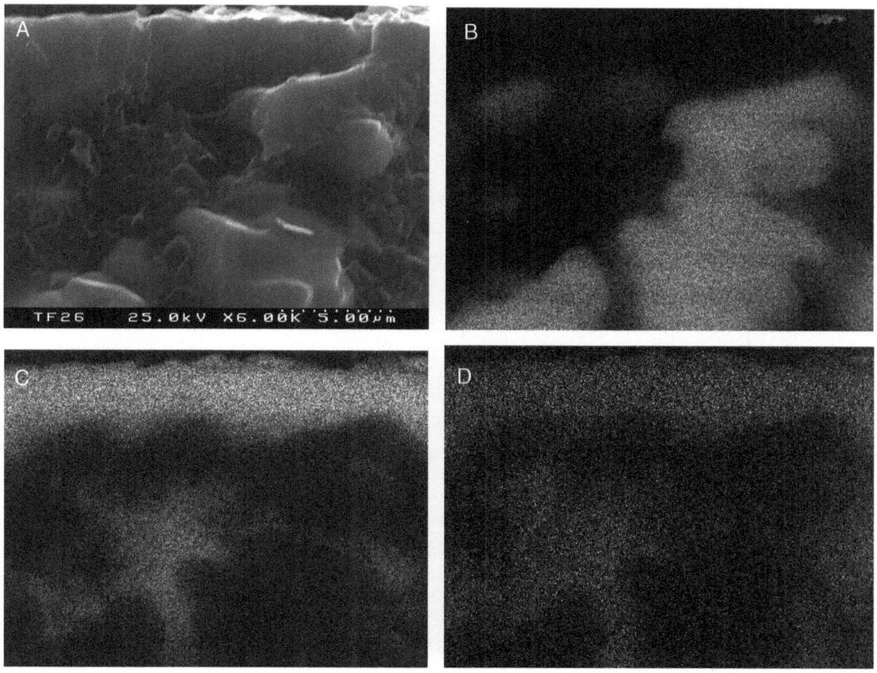

Figure 5.19 X-ray map images of the cross section of an ETS-10 membrane: (A) SEM image of the test area; (B) aluminum; (C) silicon; and (D) titanium element distribution. Bright = high concentration. Reprinted from Microporous and Mesoporous Materials, [65], Copyright (2004), with permission from Elsevier.

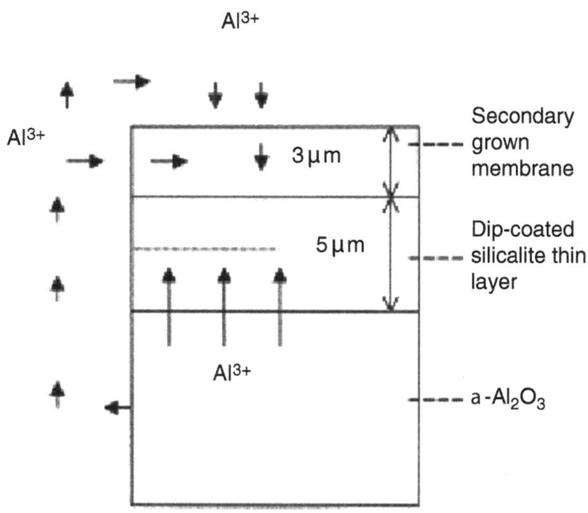

Figure 5.20 Influence of aluminum from alumina support on membrane composition. Reprinted from Microporous and Mesoporous Materials, [85], Copyright (2001), with permission from Elsevier.

during the calcination process. (2) As explained above, aluminum can also be transferred from the alumina support to the zeolite layer in the form of dissolved Al^{3+} via the synthesis solution during the hydrothermal growth process.

5.2. Calcination

Zeolite membranes prepared using an organic SDA need an activation stage. A calcination (at around 400–500 °C for MFI-type zeolite membranes) process usually decomposes and/or oxidizes the organic template releasing the microporosity of the zeolitic membrane. Dong *et al.* [86] have studied the template removal-associated microstructural evolution of MFI-type zeolite membranes prepared on ceramic supports. The MFI-type zeolite crystal framework shrinks during template removal at 350–500 °C (Fig. 5.21). In contrast, after template removal, the MFI-type zeolite framework expands while the substrate shrinks on cooling. As a consequence, high-quality MFI-type zeolite membranes can be obtained with a suitable calcination temperature program (i.e., controlling heating and cooling rates and using annealing periods at intermediate temperatures). Also, removal of the template creates or enlarges intercrystalline gaps (microporous non-zeolitic pores) due to a decrease in zeolite crystallite size after removal of the template which could affect permselectivity of the membrane.

The removal of the organic SDA molecules from small-pore zeolite requires higher temperatures (700–950 °C) [87, 88] than usual (500–600 °C), which in the case of a zeolite membrane could provoke the formation of cracks [86, 89]. Consequently, alternatives to calcination should be envisaged not only to prepare better and more reproducible existing zeolite membranes but also to afford the synthesis of small-pore zeolite membranes, useful for applications (i.e., purification of H_2 streams) that

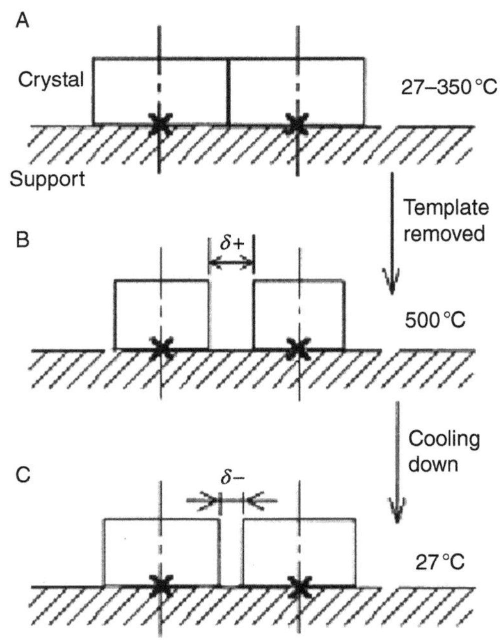

Figure 5.21 Schematic diagram showing template-removal-associated microstructural development of a zeolite film. Reprinted from Microporous and Mesoporous Materials, [86], Copyright (2000), with permission from Elsevier: A) heating at 27-350 °C, B) template removal at 500 °C, and C) cooling down.

are still asking for new membrane materials. A serious alternative to calcination is ozonization which, due to the active radical species formed by ozone decomposition [90], has proved to be more efficient in carefully removing the organic SDA molecules than the conventional treatment of calcination. Another interesting alternative, not yet applied to zeolite membranes, is the so-called room temperature detemplation of zeolite (beta) through H_2O_2-mediated oxidation using traces of Fe^{3+} [91].

5.3. Posttreatments

After the preparation of the zeolite membrane by any of the methods described above and given, in some cases, the presence of a certain number of defects (produced either by lack of connectivity between crystals or during the activation stage), the quality of a given zeolite membrane can be improved by a number of postsynthetic treatments: ion exchange [92], chemical vapor deposition (CVD) [93], coking treatments [94], or basic treatments [95].

A postsynthetic coking treatment was accomplished by impregnating ZSM-5 membranes with liquid 1,3,5-triisopropylbenzene (TIPB) at room temperature and then calcining in air at 500 °C [94]. TIPB coking selectively eliminates microdefects while the intracrystalline pore space of the zeolite is not affected (the size of TIPB molecules, 0.84 nm, is beyond that of the MFI-type zeolite pores). The elimination of non-zeolitic pores results in a large increase of *n*-butane/isobutane permselectivity (45 vs 320 at 180 °C) accompanied by a strong reduction of *n*-butane flux as expected (since the initial quality of the membranes in terms of permselectivity was considerably high). This suggests that the defects were in series as well as in parallel with the zeolite pathways, as represented in Fig. 5.22.

Another effective posttreatment to treat hydrothermally the membranes at 180 °C under moderately alkaline conditions was revealed as very efficient for mordenite membranes [95]. Dissolution of amorphous and mordenitic materials rich in Si and further recrystallization of the zeolite deposits formed could explain how the pervaporation performance of the mordenite membranes was improved leading to membranes with water/ethanol separation factors higher than 200 together with water fluxes over 1 kg/h m^2. In Fig. 5.23, top views of two mordenite membranes (before and after posttreatments) show randomly oriented prismatic crystals: cleaner and with less sharp edges in the case of the posttreated sample.

5.4. Zoned or two-layered zeolite membranes

Another interesting possibility would be to combine different zeolites (with catalytic and separation activity, for instance) on the same membrane, a subject that has scarcely been investigated. Salomon *et al.* [96] synthesized a mordenite/ZSM-5/chabazite membrane but the synthesis procedure used did not allow any type of control of the relative amounts and distribution of the different zeolites composing the intergrown layer. Li *et al.* [97] prepared, what they defined as, zoned MFI-type zeolite films by a two-step crystallization procedure; that is, films assembled by crystals propagating from the support (nonporous silicon and quartz substrates) to the top surface of the film with varying aluminum content along the length of the

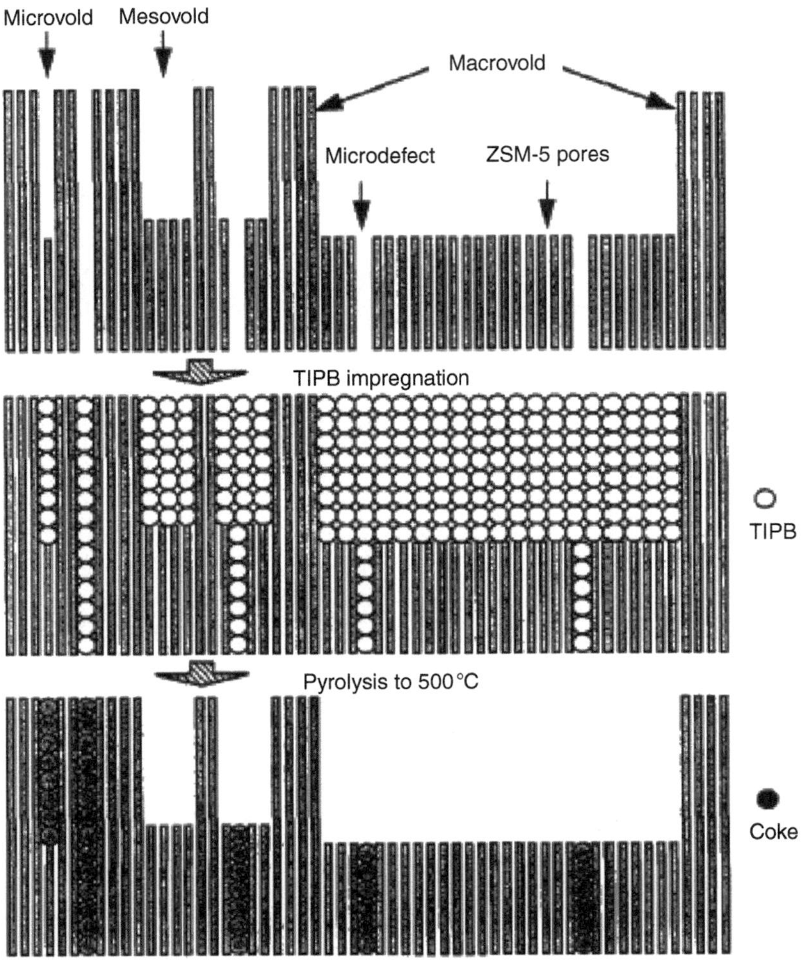

Figure 5.22 Postsynthetic TIPB coking process. Reprinted from Journal of Membrane Science, [94], Copyright (1997), with permission from Elsevier.

crystals, as shown in Fig. 5.24 (this is an example since, among other possibilities, it is possible to prepare the silicalite on the support and the ZSM-5 on the silicalite). Another strategy for preparing these zoned MFI-type zeolite films consists in coating the support with a well-defined precursor ZSM-5 film which is subsequently treated with hydrochloric acid and then hydrothermally autoclaved in a silicalite-1 synthesis solution [98].

Zhang *et al.* [99] dealt with the preparation of a two-layered zeolite A-silicalite membrane on porous alumina tubes, while de la Iglesia *et al.* [100] tried to combine the catalytic activity of H-ZSM-5 with the water pervaporation selectivity of mordenite membranes, that is, to produce true bifunctional two-layered H-ZSM-5-mordenite zeolite membranes by a two-step hydrothermal synthesis procedure. In this case, it was very important to be sure that the necessary thermal treatment stage, to remove the TPAOH (tetrapropylammonium hydroxide) template from the

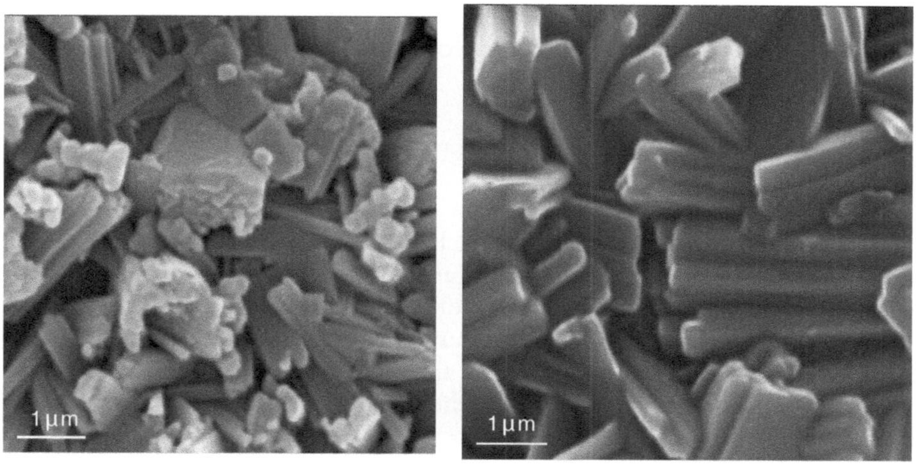

Figure 5.23 Top views of mordenite membranes: left, as synthesized; right, after basic posttreatment. Reprinted from Journal of Membrane Science, [95], Copyright (2006), with permission from Elsevier.

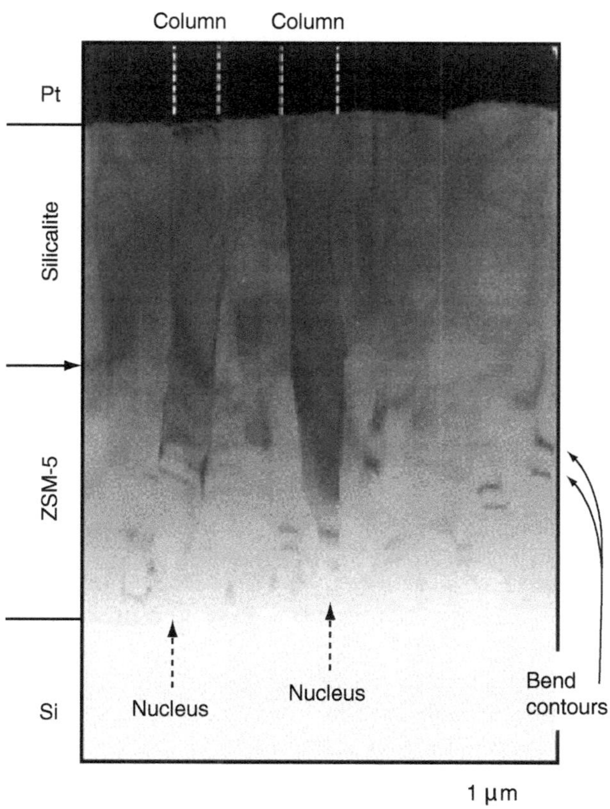

Figure 5.24 Cross-sectional TEM image of a zoned MFI-zeolite film. Reprinted from Microporous and Mesoporous Materials, [97], Copyright (2002), with permission from Elsevier.

H-ZSM-5 pores, did not alter the hydrophilic character of these two-layered H-ZSM-5-mordenite membranes, and consequently the calcination was carried out under a steam/air atmosphere.

6. Characterization

Many of the techniques used to characterize zeolite membranes are also commonly employed in powder characterization. Among the most widely used techniques, XRD is an indispensable tool to identify the zeotype formed in the synthesis and to evaluate the type of impurities that are present. XRD analysis also determines whether the zeolite crystals in the continuous zeolitic layer are oriented or not. SEM can be used to analyze the shape and size of crystals, and their distribution on the support. It can also provide a measurement of membrane thickness and a first impression of the existence of intercrystalline defects (large cracks). Surface analyzing techniques such as EPMA (electron-probe microanalysis) or XPS (X-ray photoelectron spectroscopy) can be used to measure the metal (usually, Si and Al) concentration profiles across the membrane radius. TEM can identify the existence of zeolite-support nanocomposites, while nuclear magnetic resonance (NMR) can give information concerning the oligomer size during the maturation time of the precursor gel [84]. In the case of thick (up to 25 μm) polycrystalline MFI-type membranes, fluorescence confocal optical microscopy using a fluorochrome molecule as dye allows the identification of features corresponding to grain boundaries [89].

Adsorption measurements can help to explain the permeation mechanism observed in membranes evaluated for separating a given mixture. For instance, adsorption measurements of components commonly used in pervaporation (water and alcohols) can help to define what the hydrophobicity is in relative terms: preferential adsorption of water over organic compounds. Also, a zeolite that adsorbs more water than another zeolite would still be considered more hydrophobic of the two if its organic/water adsorption ratio is higher [101]. These two considerations are summarized in Fig. 5.25.

Specific permeation measurements, either of single gases or of multicomponent mixtures, provide useful information on the effective pore structure of the membrane and on the existence of intercrystalline defects. A number of single-gas permeation experiments using molecules with different kinetic diameters can be used to gauge the effective pore size in defect-free membranes, as is depicted in (Fig. 5.26) for several zeolitic membranes [102]. The measurement of single-gas permeances of different gases and hydrocarbons as a function of the temperature, using a temperature-programmed permeation system [103], can give an idea of the permeation mechanism through a zeolitic membrane in terms of the combination of adsorption and diffusion phenomena. As an example, in (Fig. 5.27) the H_2, methane, ethane, propane, and *n*-butane single-gas permeances are plotted as a function of temperature for a tubular ZSM-5 membrane. The combination of activated diffusion and adsorption explains the maxima observed for ethane, propane, and *n*-butane permeances (already qualitatively represented in Fig. 5.2 at the beginning

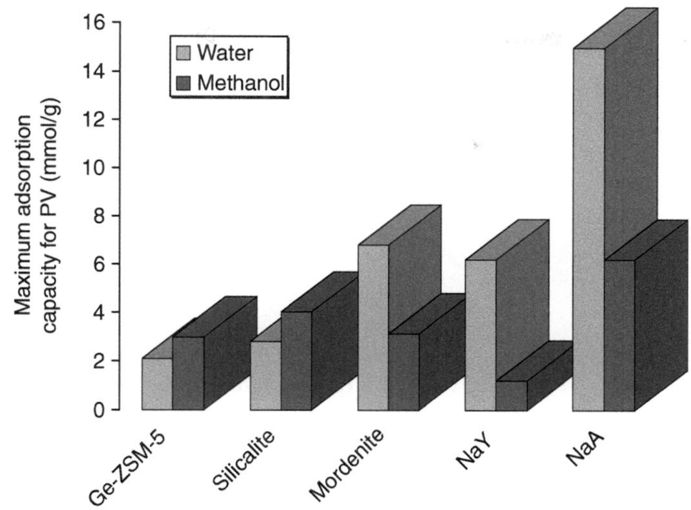

Figure 5.25 Maximum adsorption capacity for pervaporation (PV) as a function of the zeolite. Data adapted from Bowen *et al.* [101].

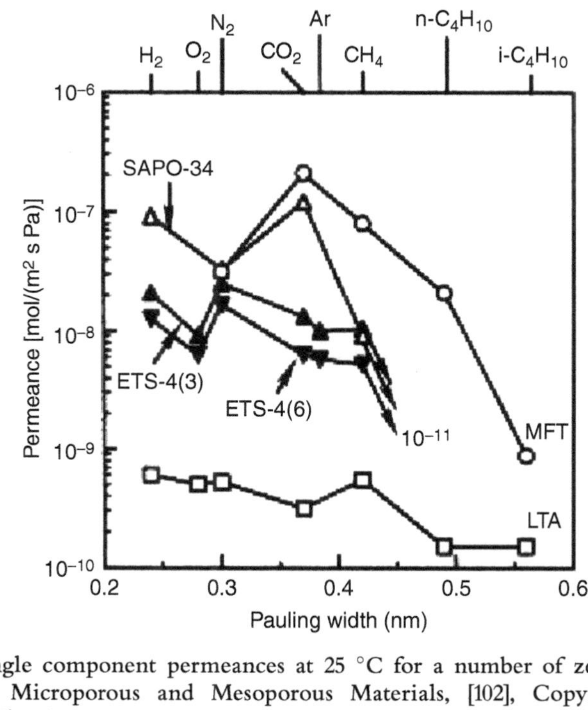

Figure 5.26 Single component permeances at 25 °C for a number of zeolitic membranes. Reprinted from Microporous and Mesoporous Materials, [102], Copyright (2001), with permission from Elsevier.

of the chapter). As the temperature increases the mobility of adsorbed species is enhanced, up to a point where the decline in occupancy prevails, which gives rise to a decrease in permeance. Eventually, the permeance is controlled by activated transport through micropores, increasing again with temperature. The sequence in the maxima

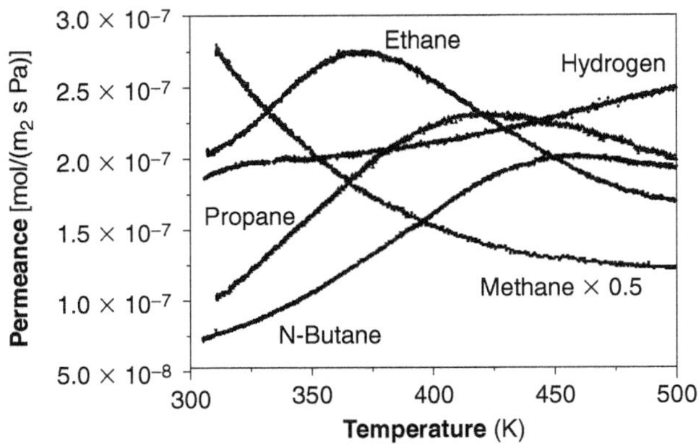

Figure 5.27 Hydrocarbon and H_2 single-gas permeances as a function of temperature for a ZSM-5 membrane [103].

follows the same order as the equilibrium amount adsorbed in silicalite [104]: methane ethane propane *n*-butane. H_2, a permanent gas, and CH_4, the smallest hydrocarbon, with relatively weak adsorption on the MFI-type zeolite, do not show maxima in the range of temperature tested.

Furthermore, selective blocking of membrane pores combined with permeation measurements can be used to evaluate defects. Thus, van de Graaf *et al.* [105] and Lin *et al.* [106] measured, respectively, krypton and N_2 or SF_6 permeation on membranes calcined at increasingly higher temperatures. Membranes calcined at low temperatures still had their zeolitic pores blocked by the template, and these authors attributed all of the permeation observed below 327–337 °C to permeation through microdefects.

7. Applications of Zeolite Membranes

Although it is quite clear that zeolite films have found applications in many different fields, membrane reactors and zeolite-based sensors are by far the main applications investigated up to now.

7.1. Separation of mixtures

The permeation of mixtures is a complex phenomenon and, in general, the behavior that is experimentally observed for mixtures cannot be predicted solely from the permeance of the individual components [107]. It is often the case that in a binary mixture, the component that permeates faster as a single gas (in general the more weakly adsorbed component) is the one giving the lower permeance in the binary mixture [67].

Four groups of separations using zeolite membranes can be distinguished [67]: (1) separation of mixtures of nonadsorbing compounds (e.g., H_2/CH_4, O_2/N_2, H_2/N_2), (2) separation of mixtures of adsorbing organic compounds (e.g., *n*-butane/isobutene, *n*-hexane/2,2-dimethylbutane), (3) separation of permanent gas from adsorbing compounds (e.g., CO_2/N_2, methanol/H_2, alcohol/O_2, *n*-butane/H_2), and (4) separation of water or polar molecules from organic compounds (e.g., alcohols and other organic compounds and water, usually by pervaporation). Although nonspecific mechanisms such as Knudsen diffusion and capillary condensation may also play a role, the majority of the separations reported with zeolite membranes can be explained in terms of surface diffusion and shape selectivity/molecular sieving. Surface diffusion in zeolites is related to specific adsorption of a given permeating compound on a given zeolitic membrane. The specificity of the adsorption process can be used to produce high selectivities of separations (2–4), where, depending on the operation conditions, capillary condensation can have some influence. Shape selectivity and molecular sieving become the dominant mechanisms in small-pore, defect-free zeolitic membranes when the adsorption strength of the potential sorbates on the zeolite pores decreases sufficiently (e.g., above the temperature value at which adsorption effects cease to be significant). This is of interest for the separation of mixtures of nonadsorbing permanent gases, provided that there are significant differences in their molecular sizes. A few separation examples using zeolite membranes are given below.

The recovery of hydrocarbons from natural gas is desirable for a number of reasons [62]: (1) the price of hydrocarbons is considerably higher than that of methane, (2) condensates from higher-molecular-weight hydrocarbons produce liquid slugs and give rise to partial dissolution/softening of plastic pipes and meters, and (3) the presence of hydrocarbons decreases the methane number below the tolerable value, causing detonation and engine-arrest problems in gas combustion and cogeneration engines. A membrane-based process could be an interesting alternative for the separation of higher hydrocarbons. As shown in Fig. 5.28, silicalite membranes are able to separate hydrocarbons from methane with a good selectivity, even at the low concentrations existing in natural gas [62]. The separation is governed by competitive adsorption of the different hydrocarbons. When the amount of *n*-butane present in a natural gas feed is increased it can be seen that the selectivity of *n*-butane with respect to all the other gases present in the mixture increases.

Another interesting example to study with zeolitic membranes is the separation of H_2 from its mixtures with gases [108]. An affordable, H_2-permselective membrane could be a definitive advantage in the implementation of challenging processes, among them are purification of hydrogen-containing streams and H_2 proton-exchange membrane (PEM) fuel cells [109]. However, in most membrane-separation processes water has a negative influence on membrane performance. Research in recent years has produced zeolitic membranes that can selectively separate water from its mixtures with permanent gases. In this case, due to the preferential adsorption and/or capillary condensation, water permeates selectively [110], and as a consequence the gas permeance is strongly decreased. A microporous titanosilicate with the structure of umbite (having pores below 0.3 nm) was prepared as a tubular membrane and tested for H_2/N_2 separation [69], even in the presence of moisture [111]. Figure 5.29 shows the H_2 permeance and the H_2/N_2 separation selectivity as a function of temperature. The permeance of H_2 is higher than that of N_2 in the 30–150 °C temperature range. As expected considering

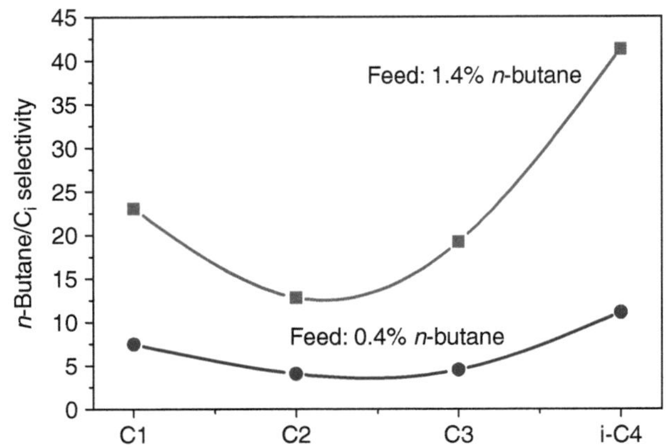

Figure 5.28 *n*-Butane/C_i separation selectivity for the separation of *n*-butane from natural gas through a silicalite membrane at room temperature. Data adapted from Arruebo *et al.* [62].

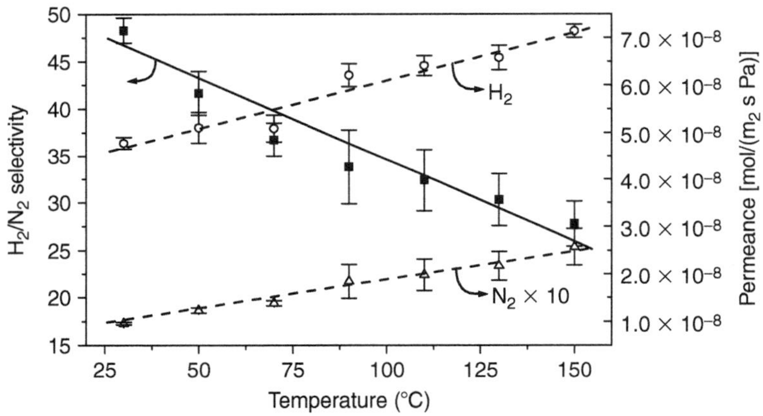

Figure 5.29 Separation of the H_2/N_2 mixture through a titanosilicate umbite membrane as a function of temperature. Reprinted with permission from [69], Copyright (2006) American Chemical Society.

the microporous character of these membranes, both permeances are activated, that is, they increase with increasing temperature. The H_2/N_2 selectivity decreases with temperature because the N_2 permeance increases faster than that of H_2. This different temperature activation may be explained assuming that H_2 and N_2 permeate through pores of different size: N_2 does not enter the umbite channel; it permeates only through intercrystalline defect pores. These defects must be of a size similar to that of the N_2 molecule otherwise such a high separation selectivity and activated transport would not be observed for the permeating molecules, even for N_2 whose transport for these membranes is allowed only through defects.

A final example is the separation of a permanent gas (O_2)/organic compound (cyclohexane) mixture through an MCM-48 mesoporous membrane. As stated in

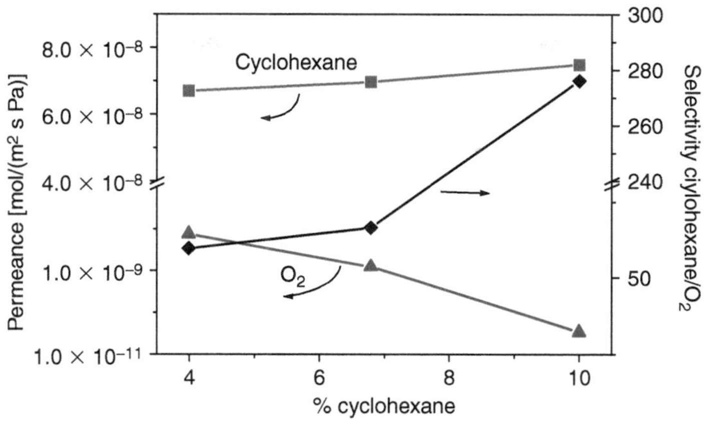

Figure 5.30 Separation of the cyclohexane/O_2 mixture through an MCM-48 mixture as a function of cyclohexane concentration in the feed at room temperature. Reprinted from Journal of Membrane Science, [42], Copyright (2006), with permission from Elsevier.

the introduction, MCM-48, being a nanostructured material with regular pores around 2–3 nm, can also be considered as a zeolitic material. In the results shown in Fig. 5.30, the application of Kelvin's equation indicated that at the higher partial pressures used ($P/P_0 = 0.41$ and 0.62) the Kelvin diameter for the mixture is higher than the mean pore diameter of the MCM-48 material (3.0–3.2 nm) obtained from N_2 adsorption analysis. This would mean that the capillary condensation of cyclohexane in the pores of the MCM-48 membrane is responsible for the high cyclohexane/O_2 selectivity. However, there is also some specific interaction of the organic molecule with the MCM-48, which de la Iglesia *et al.* [42] inferred from the fact that these membranes separated cyclohexane better than benzene and benzene better than *n*-hexane.

7.2. Zeolite membrane reactors

ZMRs belong to the area of inorganic membrane reactors. A membrane reactor is used to increase the yield and selectivity of a reaction by separating some components involved in it. The most common configuration of membrane reactors consists of a membrane prepared over a flat or tubular support. In this case, depending on the catalytic activity of the membrane, membrane reactors can be classified as inert membrane reactors, where the catalyst is located out of the membrane structure and the membrane is used exclusively as a separator, or as catalytic membrane reactors, with a catalytically active membrane. A third kind of reactor could be included in this first classification of ZMR, namely the particle level membrane reactor, in which there is a membrane coating on each catalyst particle. Figure 5.31 summarizes the classification of inorganic membranes following Coronas and Santamaría [112] and McLeary *et al.* [113].

Zeolite membranes, due to their selective separation ability, are highly suitable for use as membrane reactors. The application of ZMRs depends on the role that the membrane plays (Fig. 5.32). One of the most extensive applications is the removal of

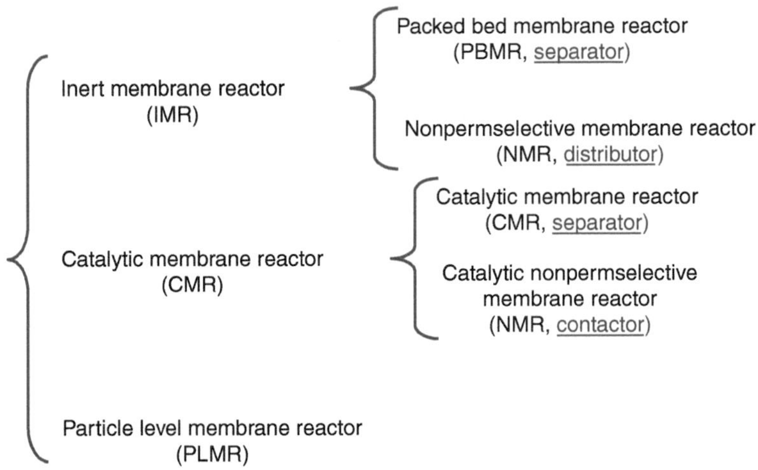

Figure 5.31 Classification of inorganic membrane reactors.

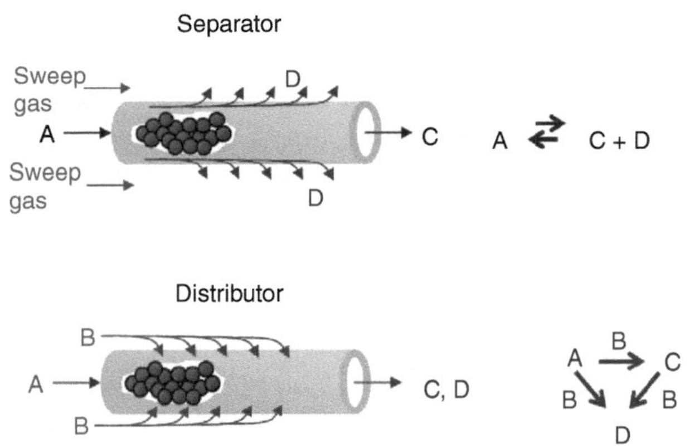

Figure 5.32 Examples of application of zeolite membrane reactors.

reaction products. This strategy is normally carried out in equilibrium-limited reactions to enhance the conversion by means of equilibrium displacement. Interesting examples of this application are the elimination of H_2 in dehydrogenation reactions [26, 114–116]) or the removal of water in esterification reactions [117]. Furthermore, ZMRs could also separate reaction products that inhibit catalyst activity, thus increasing the reaction rate.

Besides these applications, the removal of products is also used to increase selectivity in chain reactions when an intermediate product is the desired one. The separation of this product can avoid the formation of undesirable products. In this context, we can mention the work of Piera *et al.* [118], who used an inert

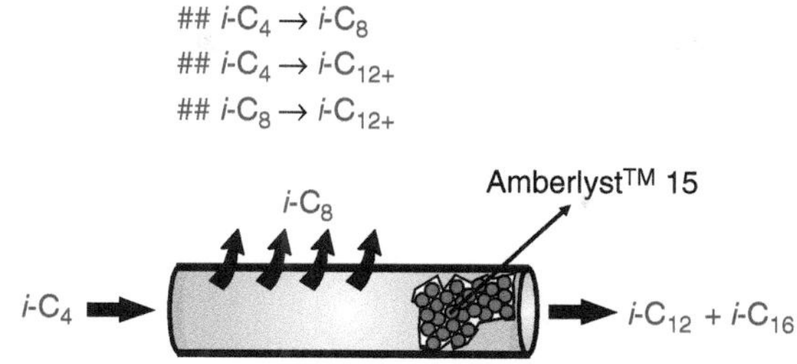

Figure 5.33 Scheme of the dimerization of *i*-butene in a zeolite membrane reactor.

silicalite membrane reactor with an acid resin (Amberlyst(15) as the catalyst for the liquid-phase dimerization of *i*-butene. As is shown in Fig. 5.33, a silicalite membrane was used for the selective removal of *i*-octene (the desired product) from the reaction environment, thus reducing the formation of unwanted C_{12} and C_{16} hydrocarbons. A comparison of the performance of the ZMR with that of a fixed bed reactor (FBR) with the same charge of catalyst demonstrated that the ZMR reached a conversion of *i*-butene near to 95% with a selectivity to *i*-octene of 40% while FBR had a similar conversion but with a selectivity 20% points lower. A second example of intermediate removal is related to xylene isomerization [119–121], where MFI-type membranes can have a key role due to their ability to selectively transport *p*-xylene over *m*- and *o*-xylene.

Another important use of ZMRs is their role as reactant distributors [122, 123] through the membrane. It can be very useful to maintain a low concentration of reactants to avoid side reactions (usually deep oxidation reactions) which are favored by high reactant (oxygen) concentrations.

Catalytically active ZMRs can also be used as contactors of reactants. In this case, the membrane keeps the reactants segregated on either side. This is useful for confining the reaction to a finite zone inside the porous structure, avoiding the slip of reactants from one side to the other. Further, the reactants arrive at the reaction zone in a stoichiometric ratio, which helps to reduce undesirable side reactions. This concept was applied to the catalytic combustion of *n*-hexane present at low concentrations in air with catalytically active Pt-exchanged ZSM-5 membranes, achieving a nearly complete *n*-hexane combustion at 210 °C [124].

All the application examples mentioned above correspond to traditional configurations of ZMR with inert or catalytic membranes. A new concept of ZMR is the particle level membrane reactor [15]. In this case, homogeneous silicalite layers with a thickness of 40 μm were synthesized covering spherical Pt/TiO_2 particles with a diameter of 0.5 mm (Fig. 5.34, left). The hydrogenation of a mixture of linear 1-hexene (1-Hex) and branched 3,3-dimethylbut-1-ene (3,3-DMB) was performed in an FBR filled with silicalite-coated Pt/TiO_2 particles. As seen in Fig. 5.34 (right), the composite silicalite/Pt/TiO_2 catalyst showed a 1-Hex/3,3-DMB selectivity of 12, while Pt/TiO_2 particles have an imperceptible selectivity under the same operation conditions (50 °C). This behavior is due to the selective permeation of 1-Hex into the Pt/TiO_2 particles through the silicalite layer. Besides, deactivation of the catalyst was also reduced, probably by protection against poisoning due to impurities present in the feed.

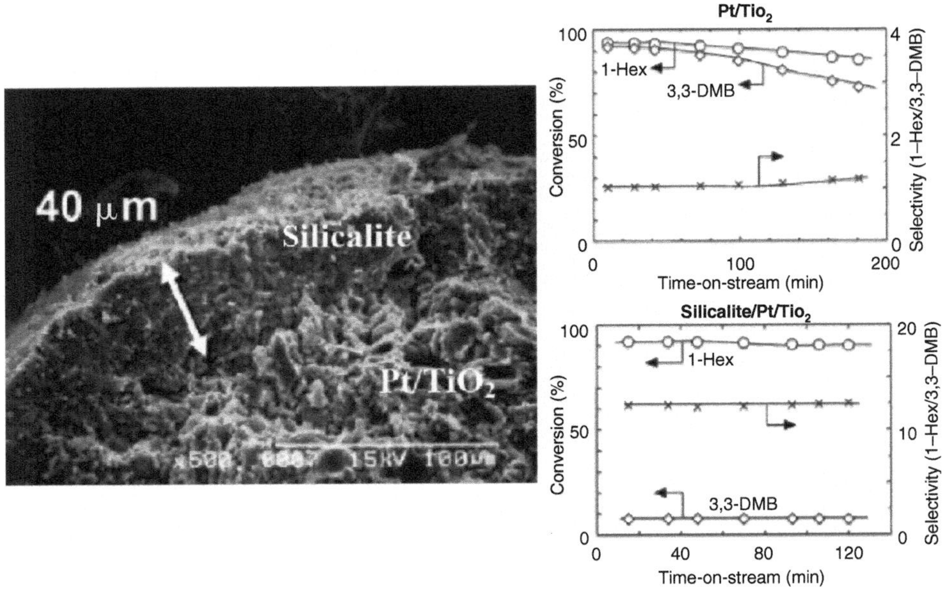

Figure 5.34 Left, SEM image of a silicalite/Pt/TiO_2 particle; right, reaction results [125].

7.3. Zeolitic microreactors

Microreactors include a three-dimensional structure with inner dimensions between 10 and 100 mm. The main characteristic of microstructured reactors is their high surface area/volume ratio, with values between 10,000 and 50,000 m^2/m^3, while those of traditional reactors are about 100 m^2/m^3, reaching values of 1000 m^2/m^3 only in rare cases. Microreactors also have a high heat-exchange coefficient, thus allowing reactions in virtual isotherm conditions with exactly defined residence times. Further, the development of hot spots is suppressed, avoiding undesirable side reactions. Moreover, mass transport is also considerably improved in microreactors because of their small dimensions, which implies that diffusion times are very short and the influence of mass transport on the speed of reaction can be considerably reduced.

Process parameters are more easily controlled in reactions that take place in small volumes. As a result, the main applications of microreactors are for strongly exothermic or explosive reactions as well as reactions involving toxic substances or high operating pressures. Indeed, these sorts of reactions can be carried out more safely than in traditional reactors [125]. Another fact to take into account is the easy scale-up of microreactors by replication, which allows a process to be transferred to the pilot and production scale in a short time.

Zeolites were introduced in the field of microreactors first as catalysts. Metal-exchanged zeolites became an alternative to the formerly used noble metal coating due to their microporous structure which provides a homogeneous distribution of catalytic active sites all over the reactor, with a lower cost than noble metal.

Also, they can act as molecular sieves, allowing a desired molecule to enter the active sites and reaction while other molecules are left out of the catalyst. As an example, zeolitic microreactors can consist of two plates with several channels constructed in them. When the plates are joined together, they form a device with an inlet and an outlet connected by means of the microchannels. These plates can be made of a nonporous material (ceramic [48], stainless steel [126], or silicon wafer [127]) or of a porous stainless steel [128]. Figure 5.35 shows one of these ceramic plates. In the case of nonporous supports, the zeolite layer can be deposited by one of the procedures discussed above inside the channels and act as a catalyst [126]. When the support is porous, the catalyst is placed over the channels and the zeolite layer is on the other side of the plate and has an exclusively separative function [9].

An example of an application of these zeolitic microsystems is the use of a microreactor coated with a Pt-ZSM-5 film for the selective oxidation of CO in the presence of H_2 [129]. Figure 5.36 (top) shows one of the microchannels previously described completely covered with MFI-type zeolite. This reaction is a means of purification of H_2 streams used for fuel cells in which a concentration of CO higher than 10 ppm can poison their catalytic electrodes. The Pt-ZSM-5 microreactor reached a total conversion of CO at 180 °C together with 25% selectivity (Fig. 5.36, bottom). This result demonstrates that zeolitic microreactors constitute an appropriate tool for the purification of these H_2 streams and are a feasible alternative to traditional reactors.

7.4. Zeolite-based sensors

A chemical sensor can be defined as a device that measures the concentration of a fixed chemical compound and transforms this magnitude into an electrical signal. This signal, previously calibrated, allows the detection of the presence of this compound in a very short time. Because of their specificity and molecular size

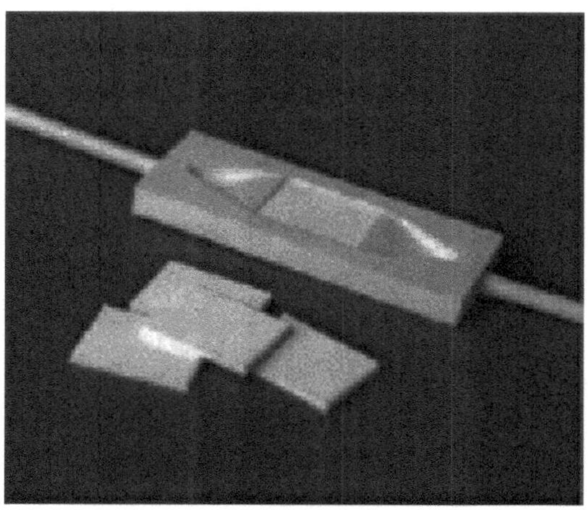

Figure 5.35 A nonporous ceramic microreactor. Reprinted from Catalysis Today, [48], Copyright (2004), with permission from Elsevier.

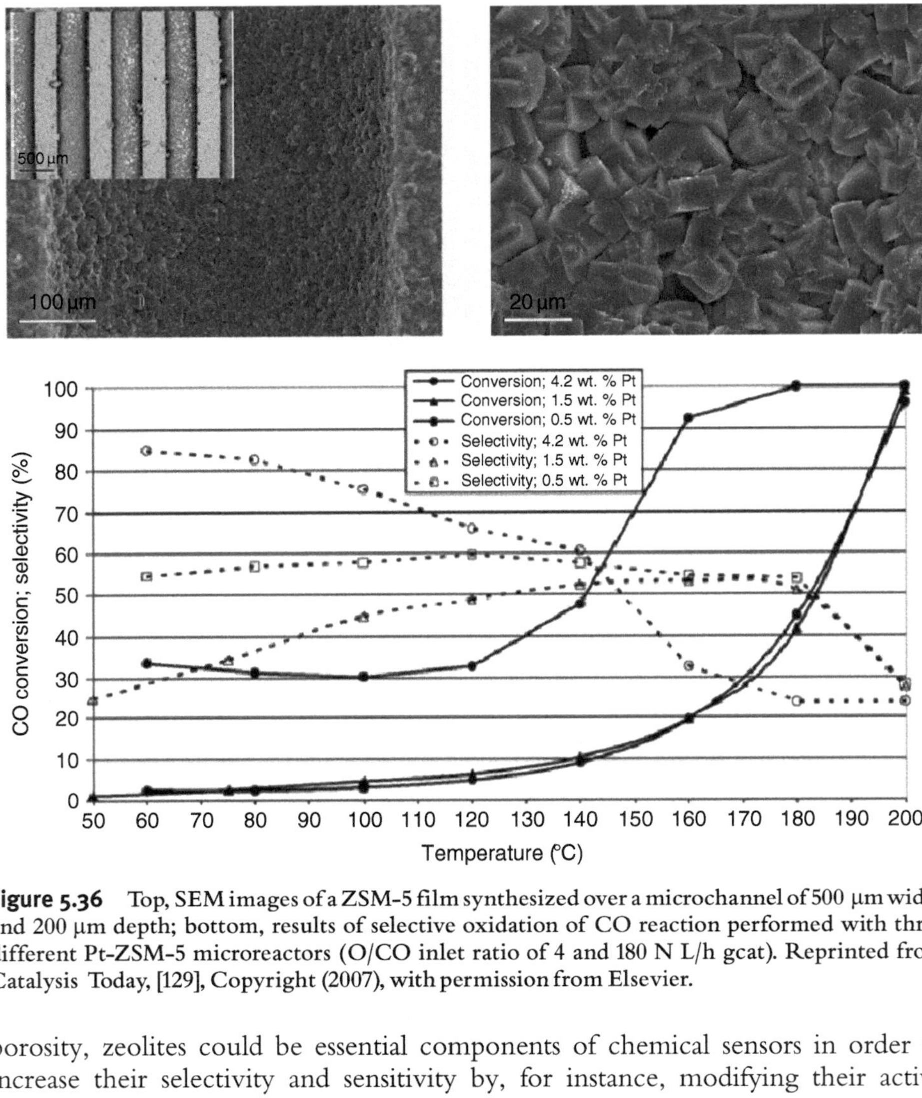

Figure 5.36 Top, SEM images of a ZSM-5 film synthesized over a microchannel of 500 μm width and 200 μm depth; bottom, results of selective oxidation of CO reaction performed with three different Pt-ZSM-5 microreactors (O/CO inlet ratio of 4 and 180 N L/h gcat). Reprinted from Catalysis Today, [129], Copyright (2007), with permission from Elsevier.

porosity, zeolites could be essential components of chemical sensors in order to increase their selectivity and sensitivity by, for instance, modifying their active surfaces.

One of the first devices to be improved by zeolites was the QCM. In this case, silicalite crystals were coupled to the surface of a QCM for sensing ethanol [24]. The regular micropores of the zeolitic material were found to effectively control molecular access to the device, thus increasing its sensitivity and selectivity. Also, QCM sensors can be modified by other zeolites such as $AlPO_4$-18 and zeolite A for gas sensing applications [130]. Figure 5.37 shows the response to increasing concentrations of water in the presence of a high concentration of propane (18%) for the $AlPO_4$-18-modified sensor compared to the silicalite one. The $AlPO_4$-18-modified sensor, because of the hydrophilicity of this material, responded quickly to each of the changes in water concentration in spite of the large concentration of propane. This is in strong contrast to the response of the silicalite-coated sensor: the

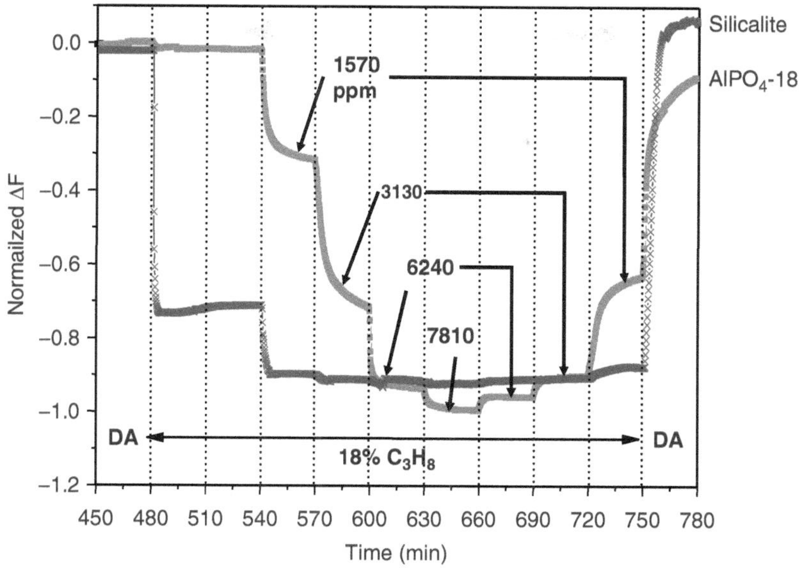

Figure 5.37 Frequency response in (A) AlPO$_4$-18 and (B) silicalite-coated sensors on introduction of 18% propane in air with increasing concentrations of water. Reprinted from Sensor. Actuat. B-Chem., [130], Copyright (2006), with permission from Elsevier.

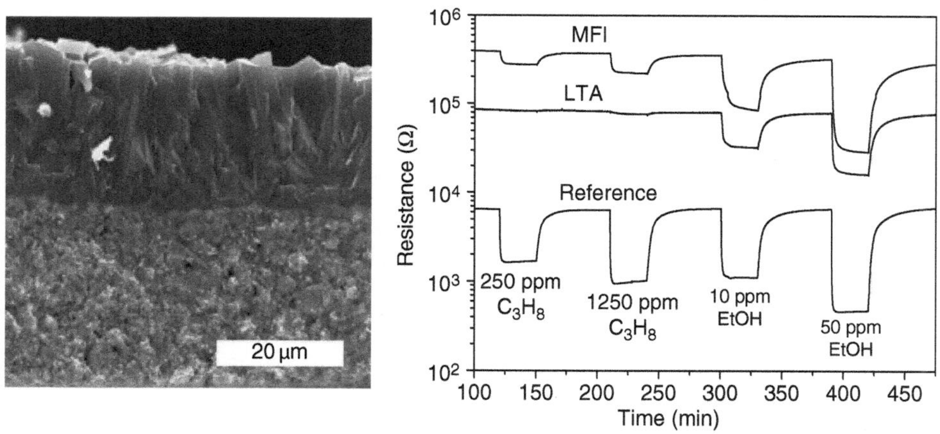

Figure 5.38 Left, SEM image of an MFI-type modified Pd-doped SnO_2 gas sensor; right, example of performance at 350 °C for methane and ethanol. Reprinted from Catalysis Today [10], Copyright (2003), with permission from Elsevier.

hydrophobicity of silicalite gave this sensor a sharp initial drop of frequency on the introduction of propane, while it was nearly insensitive to the changes in water concentration.

Another type of sensor is cantilever-based. Here, zeolite crystals attached to the sensor end make them sensitive because of the change in their vibration frequency occurring when some component is adsorbed in the zeolite [131]. Moreover, the fact that the adsorption of certain molecules modifies the dielectric constant of zeolites has been applied to produce zeolite-coated interdigitated capacitors for sensing gases, and butane at concentrations down to 10 ppm have been detected with a PtNaY-covered capacitor which showed no response to CO and H_2 [132].

Several authors have used beds of zeolite as adsorbent barriers to eliminate interfering molecules and thus increase the selectivity of reactive semiconductor-based sensors. Alternatively, the surface of a semiconductor (Pd-doped SnO_2) gas sensor can be coated with a zeolite film to improve its selectivity to different species [10]. To this end, either the MFI-type (as shown in Fig. 5.38, left) or the LTA-type have been grown onto the external SnO_2 sensing surface, acting as barriers for interfering species. As can be seen in Fig. 5.38 (right), the reference sensor (without zeolite) is sensitive either to methane or to ethanol. Nevertheless, the presence of the zeolite film strongly reduces and even suppresses (in the case of the LTA-type zeolite film) the sensor response to methane while maintaining its sensitivity to ethanol [10].

REFERENCES

[1] Corma, A., Rey, F., Rius, J., Sabater, M. J., and Valencia, S. (2004). Supramolecular self-assembled molecules as organic directing agent for synthesis of zeolites. *Nature* **431,** 287–290.

[2] Kuznicki, S. M., Bell, V. A., Nair, S., Hillhouse, H. W., Jacubinas, R. M., Braunbarth, C. M., Toby, B. H., and Tsapatsis, M. (2001). A titanosilicate molecular sieve with adjustable pores for size-selective adsorption of molecules. *Nature* **412,** 720–724.

[3] Stankiewicz, A. (2003). Reactive separations for process intensification: An industrial perspective. *Chem. Eng. Proc.* **42,** 137–144.

[4] Anastas, P. T., Kirchhoff, M. M., and Williamson, T. C. (2001). Catalysis as a foundational pillar of green chemistry. *Appl. Catal. A-Gen.* **221,** 3–13.

[5] Jeong, H. K., Krych, W., Ramanan, H., Nair, S., Marand, E., and Tsapatsis, M. (2004). Fabrication of polymer/selective-flake nanocomposite membranes and their use in gas separation. *Chem. Mater.* **16,** 3838–3845.

[6] Davis, M. E. (2002). Ordered porous materials for emerging applications. *Nature* **417,** 813–821.

[7] Choi, J., Ghosh, S., Lai, Z., and Tsapatsis, M. (2006). Uniformly a-oriented MFI zeolite films by secondary growth. *Angew. Chem. Int. Ed.* **45,** 1154–1158.

[8] Coronas, J., and Santamaría, J. (2004). State of the art in zeolite membrane reactors. *Top. Catal.* **29,** 29–44.

[9] Lai, S. M., Ng, C. P., Martin-Aranda, R., and Yeung, K. L. (2003). Knoevenagel condensation reaction in zeolite membrane microreactor. *Micropor. Mesopor. Mat.* **66,** 239–252.

[10] Vilaseca, M., Coronas, J., Cirera, A., Cornet, A., Morante, J. R., and Santamaría, J. (2003). Use of zeolite films to improve the selectivity of reactive gas sensors. *Catal. Today* **82,** 179–185.

[11] Mintova, S., Mo, S. Y., and Bein, T. (2001). Humidity sensing with ultrathin LTA-type molecular sieve films grown on piezoelectric devices. *Chem. Mater.* **13,** 901–905.

[12] Moos, R., Müller, R., Plog, C., Knezevic, A., Leye, H., Irion, E., Braum, T., Marquardt, K. J., and Binder, K. (2002). Selective ammonia exhaust gas sensor for automotive applications. *Sensor. Actuat. B-Chem.* **83,** 181–189.

[13] Cheng, X. L., Wang, Z. B., and Yan, Y. S. (2001). Corrosion-resistant zeolite coatings by *in situ* crystallization. *Electrochem. Solid St.* **4,** B23–B26.

[14] McDonnell, A. M. P., Beving, D., Wang, A. J., Chen, W., and Yan, Y. S. (2005). Hydrophilic and antimicrobial zeolite coatings for gravity-independent water separation. *Adv. Funct. Mater.* **15,** 336–340.

[15] Nishiyama, N., Ichioka, K., Park, D. H., Egashira, Y., Ueyama, K., Gora, L., Zhu, W., Kapteijn, F., and Moulijn, J. A. (2004). Reactant-selective hydrogenation over composite silicalite-1-coated Pt/tio2 particles. *Ind. Eng. Chem. Res.* **43,** 1211–1215.

[16] Valtchev, V., and Mintova, S. (2001). Layer-by-layer preparation of zeolite coatings of nano-sized crystals. *Micropor. Mesopor. Mat.* **43,** 41–49.

[17] Tavolaro, A., and Drioli, E. (1999). Zeolite membranes. *Adv. Mater.* **11,** 975–996.

[18] Caro, J., Noack, M., Kölsch, P., and Schäfer, R. (2000). Zeolite membranes—state of their development and perspective. *Micropor. Mesopor. Mat.* **38,** 3–24.

[19] Baerlocher, Ch., and McCusker, L. B, Database of zeolite structures. http://www.iza-structure.org/databases/.

[20] Kulprathipanja, S. (1986). US Patent 6,726,744.

[21] Suzuki, H. (1987). Oct. 13, US Patent 4,699,892.

[22] Ishikawa, A., Chiang, T. H., and Toda, F. (1989). Separation of water alcohol mixtures by permeation through a zeolite membrane on porous-glass. *J. Chem. Soc. - Chem. Commun.* **12,** 764–765.

[23] Haag, W. O., Tsikoyiannis, J. G., (1991). May. 28, US Patent 5,019,263.

[24] Yan, Y. G., and Bein, T. (1992). Molecular-sieve sensors for selective ethanol detection. *Chem. Mater.* **4,** 975–977.

[25] Matsukata, M., Nishiyama, N., and Ueyama, K. (1994). Preparation of a thin zeolitic membrane. *Stud. Surf. Sci. Catal.* **84,** 1183–1190.

[26] Casanave, D., Giroir-Fendler, A., Sánchez, J., Loutaty, R., and Dalmon, J. A. (1995). Control of transport-properties with a microporous membrane reactor to enhance yields in dehydrogenation reactions. *Catal. Today* **25,** 309–314.

[27] Lovallo, M. C., Tsapatsis, M., and Okubo, T. (1996). Preparation of an asymmetric zeolite L film. *Chem. Mater.* **8,** 1579–1583.

[28] Wan, Y. S. S., Chau, J. L. H., Gavriilidis, A., and Yeung, K. L. (2001). Design and fabrication of zeolite-based microreactors and membrane microseparators. *Micropor. Mesopor. Mater.* **42,** 157–175.

[29] Morigami, Y., Kondo, M., Abe, J., Kita, J. H., and Okamoto, K. (2001). The first large-scale pervaporation plant using tubular-type module with zeolite NaA membrane. *Sep. Purif. Technol.* **25,** 251–260.

[30] Bouizi, Y., Rouleau, L., and Valtchev, V. P. (2006). Bi-phase MOR/MFI-type zeolite core-shell composite. *Micropor. Mesopor. Mat.* **91,** 70–77.

[31] Lai, Z., Bonilla, G., Díaz, I., Nery, J. G., Sujaoti, K., Amat, M. A., Kokkoli, E., Terasaki, O., Thompson, R. W., Tsapatsis, M., and Vlachos, D. G. (2003). Microstructural optimization of a zeolite membrane for organic vapor separation. *Science* **300,** 456–460.

[32] Rocha, J., and Anderson, M. W. (2000). Microporous titanosilicates and other novel mixed octahedral-tetrahedral framework oxides. *Eur. J. Inorg. Chem.* **5,** 801–818.

[33] Cundy, C. S., and Cox, P. A. (2003). The hydrothermal synthesis of zeolites: History and development from the earliest days to the present time. *Chem. Rev.* **103,** 663–701.

[34] Schüth, F., and Schmidt, W. (2002). Microporous and mesoporous materials. *Adv. Mater.* **14,** 629–638.

[35] Vilaseca, M., Mintova, S., Valtchev, V., Metzger, T. H., and Bein, T. (2003). Synthesis of colloidal $AlPO_4$–18 crystals and their use for supported film growth. *J. Mater. Chem.* **13,** 1526–1528.

[36] Poshusta, J. C., Noble, R. D., and Falconer, J. L. (2001). Characterization of SAPO-34 membranes by water adsorption. *J. Memb. Sci.* **186,** 25–40.

[37] Au, L. T. Y., Chau, J. L., Tellez, C., and Yeung, K. L. (2000). The role of surface chemistry in zeolite membrane formation. *J. Memb. Sci.* **164,** 257–275.

[38] Corma, A., Díaz-Cabañas, M. J., Martínez-Triguero, J., Rey, F., and Rius, J. (2002). A large-cavity zeolite with wide pore windows and potential as an oil refining catalyst. *Nature* **418,** 514–517.

[39] Rosi, N., Eckert, J., Eddaoudi, M., Vodak, D. T., Kim, J., O'Keeffe, M., and Yaghi, O. M. (2002). Systematic design of pore size and functionality in isoreticular MOFs and their application in methane storage. *Science* **295,** 469–472.

[40] Verhoef, M. J., Kooyman, P. J., van der Waal, J. C., Rigutto, M. S., Peters, J. A., and van Bekkum, H. (2001). Partial transformation of MCM-41 material into zeolites: Formation of nanosized MFI type crystallites. *Chem. Mater.* **13,** 683–687.
[41] Nishiyama, N., Park, D. H., Koide, A., Egashira, Y., and Ueyama, K. (2001). A mesoporous silica (MCM-48) membrane: Preparation and characterization. *J. Memb. Sci.* **182,** 235–244.
[42] de la Iglesia, O., Pedernera, M., Mallada, R., Lin, Z., Rocha, J., Coronas, J., and Santamaría, J. (2006). Synthesis and characterization of MCM-48 tubular membranes. *J. Memb. Sci.* **280,** 867–875.
[43] Baker, R. W. (2002). Future directions of membrane gas separation technology. *Ind. Eng. Chem. Res.* **41,** 1393–1411.
[44] Ho, M. T., Allinson, G., and Wiley, D. E. (2006). Comparison of CO_2 separation options for geo-sequestration: Are membranes competitive. *Desalination* **192,** 288–295.
[45] Maloncy, M. L., Maschmeyer, T. h., and Jansen, J. C. (2005). Technical and economical evaluation of a zeolite membrane based heptane hydroisomerization process. *Chem. Eng. J.* **106,** 187–195.
[46] Meindersma, G. W., and de Haan, A. B. (2002). Economical feasibility of zeolite membranes for industrial scale separations of aromatic hydrocarbons. *Desalination* **149,** 29–34.
[47] Coronas, J., and Santamaría, J. (2004). The use of zeolite films in micro-scale applications. *Chem. Eng. Sci.* **59,** 4879–4885.
[48] Kolb, G., and Hessel, V. (2004). Micro-structured reactors for gas phase reactions. *Chem. Eng. J.* **98,** 1–38.
[49] Mahajan, R., De Vu, Q., and Koros, W. J. (2002). Mixed matrix membrane materials: An answer to the challenges faced by membrane based gas separations today. *J. Chinese Inst. Chem. Eng.* **33,** 77–86.
[50] Roberson, L. M. (1991). Correlation of separation factor versus permeability for polymeric membranes. *J. Memb. Sci.* **62,** 165–185.
[51] Mahajan, R., and Koros, W. J. (2000). Factors controlling successful formation of mixed-matrix gas separations materials. *Ind. Eng. Chem. Res.* **39,** 2692–2696.
[52] Kapteijn, F., Nijhuis, T. A., Heiszwolf, J. J., and Moulijn, J. A. (2001). New non-traditional multiphase catalytic reactors based on monolithic structures. *Catal. Today* **66,** 133–144.
[53] Ulla, M. A., Mallada, R., Coronas, J., Gutiérrez, L., Miró, E., and Santamaría, J. (2003). Synthesis and characterization of ZSM-5 coatings onto cordierite honeycomb supports. *Appl. Catal. A-Gen.* **253,** 257–269.
[54] Madhusoodana, C. D., Das, R. N., Kameshima, Y., Yasumori, A., and Okada, K. (2001). Characterization and adsorption behaviour of ZSM-5 zeolite film on cordierite honeycombs prepared by a novel *in situ* crystallization method. *J. Porous Mat.* **8,** 265–271.
[55] Kodama, A., Hieayama, T., Goto, M., Hirose, T., and Critoph, R. E. (2001). The use of psychrometric charts for the optimisation of a thermal swing desiccant wheel. *Appl. Therm. Eng.* **21,** 1657–1674.
[56] Shiraishi, F., Yamaguchi, S., and Ohbuchi, Y. (2003). A rapid treatment of formaldehyde in a highly tight room using a photocatalytic reactor combined with a continuous adsorption and desorption apparatus. *Chem. Eng. Sci.* **58,** 929–934.
[57] Kalipcilar, H., Gade, S. K., Falconer, J. L., and Noble, R. D. (2002). Synthesis and separation properties of B-ZSM-5 zeolite membranes on monolith supports. *J. Memb. Sci.* **210,** 113–127.
[58] López, F., Bernal, M. P., Mallada, R., Coronas, J., and Santamaría, J. (2005). Preparation of silicalite membranes on stainless steel grid supports. *Ind. Eng. Chem.* **44,** 7627–7632.
[59] Yan, Y., Davis, M. E., and Gavalas, G. R. (1997). Use of diffusion barriers in the preparation of supported zeolite ZSM-5 membranes. *J. Memb. Sci.* **126,** 53–65.
[60] Hedlund, J., Sterte, J., Anthonis, M., Bons, A. J., Carstensen, B., Corcoran, N., Cox, D., Deckman, H., Gijnst, W. D., de Moor, P. P., Lai, F., McHenry, J., *et al.* (2002). High-flux MFI membranes. *Micropor. Mesopor. Mat.* **52,** 179–189.
[61] Clet, G., Hansen, J. C., and van Bekkum, H. (1999). Factors controlling the coating of a metal with zeolite Y. *Stud. Surf. Sci. Catal.* **125,** 85–91.
[62] Arruebo, M., Coronas, J., Menéndez, M., and Santamaría, J. (2001). Separation of hydrocarbons from natural gas using silicalite membranes. *Sep. Purif. Technol.* **25,** 275–286.

[63] Casado, L., Mallada, R., Téllez, C., Coronas, J., Menéndez, M., and Santamaría, J. (2003). Preparation, characterization and pervaporation performance of mordenite membranes. *J. Memb. Sci.* **216,** 135–147.
[64] Pera-Titus, M., Llorens, J., Cunill, F., Mallada, R., and Santamaría, J. (2005). Preparation of zeolite NaA membranes on the inner side of tubular supports by means of a controlled seeding technique. *Catal. Today* **104,** 281–287.
[65] Lin, Z., Rocha, J., Navajas, A., Téllez, C., Coronas, J., and Santamaría, J. (2004). Synthesis and characterization of titanosilicate ETS-10 membranes. *Micropor. Mesopor. Mat.* **67,** 79–86.
[66] Balkus, K. J., Muñoz, T., and Gimon-Kinsel, M. E. (1998). Preparation of zeolite UTD-1 films by pulsed laser ablation: Evidence for oriented crystal growth. *Chem. Mater.* **10,** 464–466.
[67] Coronas, J., and Santamaría, J. (1999). Separations using zeolite membranes. *Separ. Purif. Method* **28,** 127–177.
[68] Bernal, M. P., Coronas, J., Menéndez, M., and Santamaría, J. (2003). On the effect of morphological features on the properties of MFI zeolite membranes. *Micropor. Mesopor. Mat.* **60,** 99–110.
[69] Sebastián, V., Lin, Z., Rocha, J., Téllez, C., Santamaría, J., and Coronas, J. (2006). Synthesis, characterization and permeation properties of Sn- and Ti-umbite membranes. *Chem. Mater.* **18,** 2472–2479.
[70] Vroon, Z. A. E. P., Keizer, K., Burggraaf, A. J., and Verweij, H. (1998). Preparation and characterization of thin zeolite MFI membranes on porous supports. *J. Mem. Sci.* **144,** 65–76.
[71] Bernal, M. P., Coronas, J., Menéndez, M., and Santamaría, J. (2004). Separation of CO_2/N_2 mixtures using MFI-type zeolite membranes. *AIChE J.* **50,** 127–135.
[72] Vilaseca, M., Mateo, E., Palacio, L., Prádanos, P., Hernández, A., Coronas, J., and Santamaría, J. (2004). AFM characterization of the growth of MFI-type zeolite films on alumina substrates. *Micropor. Mesopor. Mat.* **71,** 33–37.
[73] Lai, Z. P., Tsapatsis, M., and Nicolich, J. R. (2004). Siliceous ZSM-5 membranes by secondary growth of b-oriented seed layers. *Adv. Funct. Mater.* **14,** 716–729.
[74] Hedlund, J., Schoeman, B., and Sterte, J. (1997). Ultrathin oriented zeolite LTA films. *Chem. Commun.* **13,** 1193–1194.
[75] Li, G., Kikuchi, E., and Matsukata, M. (2003). The control of phase and orientation in zeolite membranes by the secondary growth method. *Micropor. Mesopor. Mat.* **62,** 211–220.
[76] Xu, W., Dong, J., Li, J., Li, W., and Wu, F. (1990). A novel method for the preparation of zeolite ZSM-5. *J. Chem. Soc. - Chem. Commun.* **10,** 755–756.
[77] Matsukata, M., Nishiyama, N., and Ueyama, K. (1994). Zeolitic membrane synthesized on a porous alumina support. *J. Chem. Soc. - Chem. Commun.* **3,** 339–340.
[78] Matsukata, M., and Kikuchi, E. (1997). Zeolitic membranes: Synthesis, properties, and prospects. *Bull. Chem. Soc. Jpn.* **70,** 2341–2356.
[79] Tsay, C. S., and Chiang, A. S. T. (2000). Supported zeolite membrane by vapor-phase regrowth. *AIChE J.* **46,** 616–625.
[80] Zhao, H., Jin, T., Kuraoka, K., Yazawa, T. (2000). A novel method for the synthesis of ZSM-5 zeolite membranes on a porous alumina tube: The role of a dry-gel barrier in pores, Chem. Commun. 1621–1622.
[81] Matsukata, M., Ogura, M., Osaki, T., Rao, P. R. H. P., Nomura, M., and Kikuchi, E. (1999). Conversion of dry gel to microporous crystals in gas phase. *Top. Catal.* **9,** 77–92.
[82] Alfaro, S., Arruebo, M., Coronas, J., Menéndez and, M., and Santamaría, J. (2001). Preparation of MFI type tubular membranes by steam-assisted crystallization. *Micropor. Mesopor. Mat.* **50,** 195–200.
[83] Kikuchi, E., Yamashita, K., Hiromoto, S., Ueyama, K., and Matsukata, M. (1997). Synthesis of a zeolitic thin layer by a vapor-phase transport method: Appearance of a preferential orientation of MFI zeolite. *Micropor. Mat.* **11,** 107–116.
[84] Miachon, S., Landrivon, E., Aouine, M., Sun, Y., Kumakiri, I., Li, Y., Prokopová, O. P., Guilhaume, N., Giroir-Fendler, A., Mozzanega, H., and Dalmon, J. A. (2006). Nanocomposite MFI-alumina membranes via pore-plugging synthesis: Preparation and morphological characterisation. *J. Memb. Sci.* **281,** 228–238.
[85] Pan, M., and Lin, Y. S. (2001). Template-free secondary growth synthesis of MFI type zeolite membranes. *Micropor. Mesopor. Mat.* **43,** 319–327.

[86] Dong, J., Lin, Y. S., Michael Hu, Z. C., Peascoe, R. A., and Payzant, E. A. (2000). Template-removal-associated microstructural development of porous-ceramic-supported MFI zeolite membranes. *Micropor. Mesopor. Mat.* **34,** 241–253.

[87] Villaescusa, L. A., Barrett, P. A., and Camblor, M. A. (1998). Calcination of octadecasil: Fluoride removal and symmetry of the pure SiO_2 host. *Chem. Mater.* **10,** 3966–3973.

[88] Tomita, T., Nakayama, K., and Sakai, H. (2004). Gas separation characteristics of DDR type zeolite membrane. *Micropor. Mesopor. Mat.* **68,** 71–75.

[89] Bonilla, G., Tsapatsis, M., Vlachos, D. G., and Xomeritakis, G. (2001). Fluorescence confocal optical microscopy imaging of the grain boundary structure of zeolite MFI membranes made by secondary (seeded) growth. *J. Memb. Sci.* **182,** 103–109.

[90] Heng, S., Lau, P. P. S., Yeung, K. L., Djafer, M., and Schrotter, J. C. (2004). Low-temperature ozone treatment for organic template removal from zeolite membrane. *J. Memb. Sci.* **243,** 69–78.

[91] Melian-Cabrera, I., Dapteijn, F., and Moulijn, J. A. (2005). Room temperature detemplation of zeolites through H_2O_2-mediated oxidation. *Chem. Commun.* **21,** 2744–2746.

[92] Kusakabe, K., Kuroda, T., and Morooka, S. (1998). Separation of carbon dioxide from nitrogen using ion-exchanged faujasite-type zeolite membranes formed on porous support tubes. *J. Memb. Sci.* **148,** 13–23.

[93] Nomura, M., Yamaguchi, T., and Nakao, S. (1997). Silicalite membranes modified by counter-diffusion CVD technique. *Ind. Eng. Chem. Res.* **36,** 4217–4223.

[94] Yan, Y., Davis, M. E., and Gavalas, G. R. (1997). Preparation of highly selective zeolite ZSM-5 membranes by a post-synthetic coking treatment. *J. Memb. Sci.* **123,** 95–103.

[95] Navajas, A., Mallada, R., Téllez, C., Coronas, J., Menéndez, M., and Santamaría, J. (2006). The use of post-synthetic treatments to improve the pervaporation performance of mordenite membranes. *J. Memb. Sci.* **270,** 32–41.

[96] Salomón, M. A., Coronas, J., Menéndez, M., and Santamaría, J. (1998). Synthesis of a mordenite/ZSM-5/chabazite hydrophilic membrane on a tubular support. Application to the separation of a water-propanol mixture. *Chem. Commun.* **1,** 125–126.

[97] Li, Q., Hedlund, J., Sterte, J., Creaser, D., and Bons, A. J. (2002). Synthesis and characterization of zoned MFI films by seeded growth. *Micropor. Mesopor. Mater.* **56,** 291–302.

[98] Öhrman, O., and Hedlund, J. (2006). Synthesis and catalytic evaluation of zoned MFI films. *Micropor. Mesopor. Mat.* **91,** 312–320.

[99] Zhang, X., Liu, H., and Yeung, K. L. (2005). Novel two-layered zeolite NaA-silicalite-1 membranes. *J. Phys. Chem. Solid.* **66,** 1034–1038.

[100] de la Iglesia, O., Irusta, S., Mallada, R., Menéndez, M., Coronas, J., and Santamaría, J. (2006). Preparation and characterization of two-layered mordenite-ZSM-5 bi-functional membranes. *Micropor. Mesopor. Mat.* **93,** 318–324.

[101] Bowen, T. C., Noble, R. D., and Falconer, J. L. (2004). Fundamentals and applications of pervaporation through zeolite membranes. *J. Memb. Sci.* **245,** 1–33.

[102] Guan, G., Kusakabe, K., and Morooka, S. (2001). Synthesis and permeation properties of ion-exchanged ETS-4 tubular membranes. *Micropor. Mesopor. Mat.* **50,** 109–120.

[103] Piera, E., Bernal, M. P., Salomón, M. A., Coronas, J., Menéndez, M., and Santamaría, J. (1999). Preparation and permeation properties of different zeolite tubular membranes. *Stud. Surf. Sci. Catal.* **125,** 189–196.

[104] Bakker, W. J. W., Kapteijn, F., Poppe, J., and Moulijn, J. A. (1996). Permeation characteristics of a metal-supported silicalite-1 zeolite membrane. *J. Memb. Sci.* **117,** 57–78.

[105] van de Graaf, J. M., van der Bijl, E., Stol, A., Kapteijn, F., and Moulijn, J. A. (1998). Effect of operating conditions and membrane quality on the separation performance of composite silicalite-1 membranes. *Ind. Eng. Chem. Res.* **37,** 4071–4083.

[106] Lin, X., Noble, R. D., and Falconer, J. L. (1998). Parallel pathways for transport in ZSM-5 zeolite membranes. *Chem. Mater.* **10,** 3716–3723.

[107] Coronas, J., Falconer, J. L., and Noble, R. D. (1997). Preparation, characterization and permeation properties of tubular ZSM-5 composite membranes. *AIChE J.* **43,** 1797–1812.

[108] Prabhu, A. K., and Oyama, S. T. (2000). Highly hydrogen selective ceramic membranes: Application to the transformation of greenhouse gases. *J. Memb. Sci.* **176,** 233–248.

[109] Farrauto, R. J. (2005). Introduction to solid polymer membrane fuel cells and reforming natural gas for production of hydrogen. *Appl. Catal. B-Environ.* **56,** 3–7.

[110] Aoki, K., Kusakabe, K., and Morooka, S. (2000). Separation of gases with an A-type zeolite membrane. *Ind. Eng. Chem. Res.* **39,** 2245–2251.
[111] Sebastián, V., Lin, Z., Rocha, J., Téllez, C., Santamaría, J., and Coronas, J. (2005). A new titanosilicate umbite membrane for the separation of H_2. *Chem. Commun.* **24,** 3036–3037.
[112] Coronas, J., and Santamaría, J. (1999). Catalytic reactors based on porous ceramic membranes. *Catal. Today* **51,** 377–389.
[113] McLeary, E. E., Jansen, J. C., and Kapteijn, F. (2006). Zeolite based films, membranes and membrane reactors: Progress and prospects. *Micropor. Mesopor. Mat.* **96,** 198–220.
[114] Ciavarella, D., Casanave, D., Moueddeb, H., Miachon, S., Fiaty, K., and Dalmon, J. A. (2001). Isobutane dehydrogenation in a membrane reactor. Influence of the operating conditions on the performance. *Catal. Today* **67,** 177–184.
[115] van Dyk, S., Miachon, S., Lorenzen, L., Torres, M., Fiaty, K., and Dalmon, J. A. (2003). Comparison of microporous MFI and dense Pd membrane performances in an extractor-type CMR. *Catal. Today* **82,** 167–177.
[116] Jeong, B.H, Sotowa, K. I., and Kusakabe, K. (2003). Catalytic dehydrogenation of cyclohexane in a FAU-type zeolite membrane reactor. *J. Memb. Sci.* **224,** 151–158.
[117] Bernal, M. P., Coronas, J., Menéndez, M., and Santamaría, J. (2002). Coupling of reaction and separation at the microscopic level: Esterification processes in a H-ZSM-5 membrane reactor. *Chem. Eng. Sci.* **57,** 1557–1562.
[118] Piera, E., Téllez, C., Coronas, J., Menéndez, M., and Santamaría, J. (2001). Use of zeolite membrane reactors for selectivity enhancement: Application to the liquid-phase oligomerization of i-butene. *Catal. Today* **67,** 127–138.
[119] van Dyk, L., Lorenzen, L., Miachon, S., and Dalmon, J. A. (2005). Xylene isomerization in an extractor type catalytic membrane reactor. *Catal. Today* **104,** 274–280.
[120] Tarditi, A. M., Irusta, S., and Lombardo, E. A. (2006). Xylene isomerization in a membrane reactor—part I: The synthesis of MFI membranes for the p-xylene separation. *Chem. Eng. J.* **122,** 167–174.
[121] Deshayes, A. L., Miro, E. E., and Horowitz, G. I. (2006). Xylene isomerization in a membrane reactor—part II. Simulation of an industrial reactor. *Chem. Eng. J.* **122,** 149–157.
[122] Pantzidis, A., Dalmon, J. A., and Mirodatos, C. (1995). Oxidative dehydrogenation of propane on catalytic membrane reactors. *Catal. Today* **25,** 403–408.
[123] Mota, S., Miachon, S., Volta, J. C., and Dalmon, J. A. (2001). Membrane reactor for selective oxidation of butane to maleic anhydride. *Catal Today* **67,** 169–176.
[124] Aguado, S., Coronas, J., and Santamaría, J. (2005). Use of zeolite membrane reactors for the combustion of VOCs present in air at low concentrations. *Chem. Eng. Res. Des.* **83,** 295–301.
[125] Veser, G. (2001). Experimental and theoretical investigation of H_2 oxidation in a high-temperature catalytic microreactor. *Chem. Eng. Sci.* **56,** 1265–1273.
[126] Rebrov, E. V., Seijer, G. B. F., Calis, H. P. A., de Croon, M. H. J. M., van den Bleek, C. M., and Schouten, J. C. (2001). The preparation of highly ordered single layer ZSM-5 coating on prefabricated stainless steel microchannels. *Appl. Catal. A-Gen.* **206,** 125–143.
[127] Wan, Y. S. S., Chau, J. L. H., Yeung, K. L., and Gavriilidis, A. (2004). 1-Pentene epoxidation in catalytic microfabricated reactors. *J. Catal.* **223,** 241–249.
[128] Wan, Y. S. S., Yeung, K. L., and Gavriilidis, A. (2005). TS-1 oxidation of aniline to azoxibenzene in a microstructured reactor. *Appl. Catal. A-Gen.* **281,** 285–293.
[129] Hessel, V., Kolb, G., Löwe, H., Nikolaidis, G., Zapf, R., Ziogas, A., Schouten, J. C., Delsman, E. R., de Croon, M. H. J., Santamaría, J., de la Iglesia, O., and Mallada, R. (2006). A review and an overview of own work on fuel processing for fuel cells. *Oil Gas European Magazine* **2,** 1–11.
[130] Vilaseca, M., Yagüe, C., Coronas, J., and Santamaría, J. (2006). Development of QCM sensors modified by $AlPO_4$–18 films. *Sensor. Actuat. B-Chem.* **117,** 143–150.
[131] Berger, R., Gerber, C., Lang, H. P., and Gimzewski, J. K. (1997). Micromechanics. A toolbox for femtoscale science: "Towards a laboratory on a tip". *Microelectron. Eng.* **35,** 373–379.
[132] Plog, C., Maunz, W., Kurzweil, P., Obermeier, E., and Sceibe, C. (1995). Impedance of zeolite-based gas sensors. *Sensor. Actuat. B-Chem.* **24–25,** 653–656.

CHAPTER SIX

Industrial Applications of Porous Ceramic Membranes (Pressure-Driven Processes)

Susana Luque, Daniel Gómez, *and* José R. Álvarez

Contents

1. Introduction: Pressure-Driven Membrane Processes

Pressure-driven membrane processes are among the most mature membrane technologies. They are used for liquid separations and are generally classified into four categories (Fig. 6.1). These are, in ascending order of size of solutes that can be separated, reverse osmosis (RO), nanofiltration (NF), ultrafiltration (UF), and microfiltration (MF):

- Microfiltration (0.05–10 μm)
- Ultrafiltration (0.1–0.001 μm)
- Nanofiltration (0.005–0.0005 μm)
- Reverse osmosis (<0.0005 μm, no "real" pores)

Department of Chemical and Environmental Engineering, University of Oviedo, 33071 Oviedo, Spain

Membrane Science and Technology, Volume 13
ISSN 0927-5193, DOI: 10.1016/S0927-5193(07)13006-0

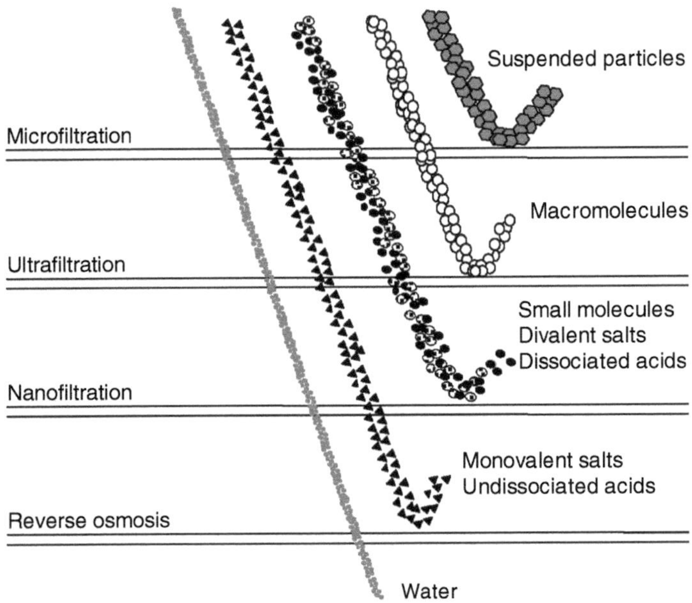

Figure 6.1 Pressure-driven membrane processes.

As pores (and porosity) become smaller, the resistance of the membranes to mass transfer increases and the applied pressure (the driving force) has to be increased. In all of the above pressure-driven processes, the membrane often separates a small fraction of the liquid feed (most often, water or an aqueous solution) into the *permeate* or filtrate, the stream which passes through the membrane. The remaining fraction of feed, called *retentate* or *concentrate*, contains the solutes rejected by the membrane.

Pressure-driven membrane processes at present have high industrial impact, with a market constantly growing. They are used in a wide range of separation processes and are considered among the *Best Available Technologies* (BAT), in the European Union environmental recommendations, because they present several advantages with respect to other separation processes:

- higher selectivity and fewer undesirable by-products, since no additives are required
- lower energy consumption, since they involve no phase change
- often simpler than alternative processes
- allow continuous operation
- carried out under mild conditions
- membrane properties can be tailored for the specific separation
- easily combined with other separation processes (in the so-called hybrid processes)
- lower cost
- linear scale-up

Depending upon the desired separation, membranes can be used either to (1) *remove contaminants*, that is from process or waste streams in order to allow the water to be recycled to the process or discharged; or to (2) *concentrate or recover valuable materials* while the rest is allowed to pass through the membrane. Both tasks are of great importance since they can be used to fulfill two objectives of many industrial processes: *water reuse, and recovery of raw materials or (by)products.*

MF is used for the removal of particles in the range of 0.05–10 μm and uses porous membranes with pore sizes in the same range. Particles are retained based on their dimensions (shape and size), that is, following a sieving mechanism. This is the membrane process which most closely resembles conventional coarse filtration.

Most of the MF membranes are made from polymeric materials, but inorganic materials are also widely used: aluminum (Al) oxide, zirconium (Zr) oxide, titanium (Ti) oxide, carbon, silica carbide, and various metals (steel, tin, nickel, etc.).

MF is typically used to retain suspended solids and can be used either in the clarification (e.g., of fruit juices, wine, and beer) or sterilization stages (e.g., in the cold sterilization of beverages and pharmaceuticals), or as a concentration step (i.e., in cell harvesting). MF is also used as pretreatment step for NF and RO, that is, for the production of potable water from surface water. Other applications include ultrapure water production in the semiconductor industry, metals recovery as colloidal oxides or hydroxides, wastewater and surface water treatment, and separation of oil–water emulsions.

In UF relatively large molecules in the range of 2–100 nm (proteins, polymers, colloids, and emulsions) are retained. Small molecules such as salts and low molecular weight organic solvents and solutes can freely pass through the membrane. The separation is based on a sieving mechanism, similar to MF.

As in MF, UF membranes can be made either of polymers or of inorganic materials (mainly Al and Ti oxides).

UF has many applications as a concentration/purification step in the dairy (in milk and whey processing, and cheese making) and food industries (i.e., in the production of potato starch, and protein recovery and hydrolysis). It is also used in metal and surface industries (for the treatment of oil–water emulsions and in electropaint recovery), textile (i.e., for indigo recovery), and automotive (electropaint recovery) sectors, and in water and wastewater treatment, to name a few.

NF is a membrane process in a size range between UF and RO, used to retain low molecular weight organic molecules (200–1000 Da) and multivalent ions, such as sulfates. Compared to RO, relatively low pressure is applied (in the range of 3–20 bars) whereas high fluxes are obtained. Separation in NF membranes can be performed based on particle size, particle charge, or a combination of both.

At present, most NF membranes are polymeric composites; however, there is an increasing number of NF ceramic membranes in the market with unique separation capabilities.

NF processes are mainly applied in the removal of small molecules such as proteins, enzymes, antibiotics, dyes, herbicides, pesticides, and insecticides in water softening (removal of calcium salts), solvent recovery, and wastewater treatment.

RO is a process which allows the retention of low molecular weight molecules such as salts. Retention of organic solvents is generally not that high because they tend to dissolve in the membrane.

Most common RO membrane materials used are cellulose triacetate, aromatic polyamides, and polybenzimidazoles. At present, there are no commercial RO membranes made of ceramic materials.

The main applications of RO are found in the desalination of brackish and seawater; the production of ultrapure water (electronic industry); concentration of food juice, sugars, and milk; and in the treatment of wastewater.

2. Porous Ceramic Membranes used in Pressure-Driven Filtration

The most common ceramic membranes are made of Al, Si, Ti, or Zr oxides. Other materials include nonoxides (carbides, borides, nitrides, and silicides) and composites, which are made of combinations of oxides and nonoxides. Al, Zr, and Ti oxides are suitable for food and pharmaceutical applications as they meet the US Food and Drug Administration (FDA) requirements detailed in 21 Code of Federal Regulations (Good manufacturing practices). They normally have an asymmetrical structure composed of at least two, mostly three, different porosity levels (Fig. 6.2). The active layer lies in the inside surface of the filtration channels and is in direct contact with the fluid to be treated. It provides selectivity and separation efficiency. Before applying the active, microporous top layer, a mesoporous intermediate layer is often applied in order to reduce the surface roughness. The macroporous support is the most porous part of the membrane and provides mechanical strength while minimizing mass transfer resistance.

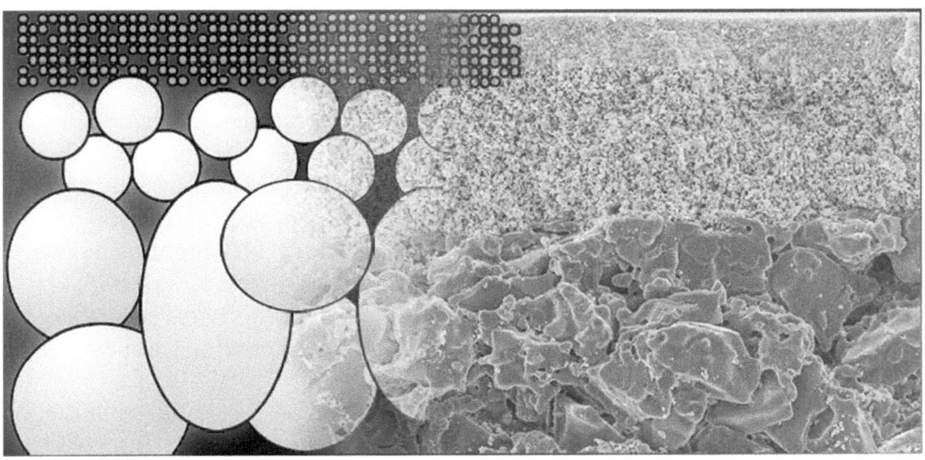

Figure 6.2 Schematic drawing combined with an electron microscopy image showing a cross section of a ceramic membrane (including the support). The multilayer construction is clearly visible. The size of the particles determines pore size (courtesy of Membraflow GEA Westfalia).

The performance of the ceramic membrane-based systems depends on the separation and permeation properties of the membrane as well as its mechanical integrity. These properties depend on the selective top layer and on the support system on which the active separation layer is coated. Therefore, pore size, porosity, surface roughness, and mechanical properties are all important parameters.

The ceramic membranes are often formed into a multichannel element (typically, up to 39 channels, though modules with over 60 channels are also marketed), also called a monolith, which provides higher membrane packing density (filtration area/unit volume). This reduces the system footprint and provides capital and operating cost savings. Ceramic membranes are available from several manufacturers in different shapes, mainly round and hexagonal, and with various channel diameters and shapes (Table 6.1 and Fig. 6.3). The newest production facility in Europe (ITN Halberstadt) has a capacity of 100,000 m^2 of membrane per year [1]. Ceramic membrane monoliths have external diameters which vary typically from 10 to 41 mm (some as large as 52 mm are available in the market place), while the hydraulic diameters are in the range from 2 to 6 mm. They are commercially available in 200–1200 mm length. Some of the channel shapes have evolved from the traditional (round) design to noncircular ones (i.e., star, ovoids, or the so-called intermingled rings), which provide more surface area for filtration compared to conventional round section channels. This, in turn, implies that less pumping power is required to achieve equivalent flux rates through the filters. In addition, noncircular channel shapes create a natural turbulence at the surface of the membrane which provides a scouring effect to help keep the filter surface clean and can increase operational efficiency by up to 50% compared with circular channels.

These elements are grouped together in housings (Fig. 6.4), available in stainless steel, coated steel, other alloys, and plastics as polyvinylidene. Housings can accommodate from 1 (starting at 0.01 m^2 membrane area) up to 99 elements. The latter is, at present, the largest housing commercially available in which 49.5 m^2 can be accommodated resulting in a very compact design, as given by the ratio of the filtration surface to the volume of the housing. For some applications, particularly where a high flow rate is required, up to 300 elements have been installed in one housing, reducing the unit cost (Fig. 6.5).

For applications in the food and pharmaceutical industry, special housings are available in order to comply with cleanability and sterilization capability requirements. The ceramic membrane elements have sealing gaskets attached at each end, which can be as simple as o-rings or designed seals that extend over the element. A typical industrial installation will have several of these modules arranged in series and/or parallel configuration.

Ceramic membranes are operated in the cross-flow filtration mode, a continuous process in which the feed stream flows parallel (tangential) to the membrane filtration surface and generates the two outgoing streams. This mode has the benefit of maintaining a high filtration rate for membrane filters compared with the direct flow filtration mode of conventional filters.

The separation is driven by the pressure difference across the membrane, the transmembrane pressure. The parallel flow of the feed stream, which creates turbulence,

Table 6.1 Commercial ceramic membrane manufacturers for pressure-driven applications

Company	Marketed ceramic membranes	Trade name(s)	Contact info
atech innovations gmbh	Tubular and Monolithic TiO_2 and Al_2O_3 ceramic membranes for MF and UF	ATTAXX®[a]	atech innovations gmbh, Innovationspark Wiesenbusch, Am Wiesenbusch 26, D-45966 Gladbeck, Germany. Tel.: +49 20 43 94 34 0; Fax: +49 20 43 94 34 34; e-mail: info@atech-innovations.com; http://www.atech-innovations.com
CEPAration	Al_2O_3 hollow fibers for MF, UF, and NF	CeraCep™	Hyflux CEPAration BV, P.O. Box 2128, 5700 DA Helmond, The Netherlands. Tel.: +31 492 565 330; Fax: +31 492 590 010; e-mail: info@ceparation.com; http://www.ceparation.com
CeraMem	Honeycomb-like SiC and TiO_2-coated membranes for MF, UF, and NF		CeraMem Corporation, 12 Clematis Avenue, Waltham, Massachusetts 02453. Tel.: +1 781 899 4495; Fax: +1 781 899 6478; http://www.ceramem.com
ECO Ceramics B.V.	Al_2O_3 ceramic membranes for MF	Velterop®; Sephi-Matic®	ECO Ceramics B.V., Staalstraat 10, 1951 JP Velsen—Noord, The Netherlands. Tel.: +31 251 229212; Fax: +31 251 229 615; e-mail: Info@ECOCeramics.nl; http://www.ecoceramics.nl
Fairey Industrial Ceramics	Monolithic Al_2O_3 ceramic membranes for MF (star-shaped channels)	Star-Sep™ FICL	Fairey Filtration Systems Limited, Normacot Road, Longton, Stoke-on-Trent, ST3 1PA, UK. Tel: +44(0)1782 377550; Fax: +44 (0) 1782 377599; e-mail: afrost@faireyfiltrationsystems.com; http://www.faireyfiltrationsystems.com

Hilliard Corp.	Honeycomb-like ceramic membranes for MF and UF	Hilco	The Hilliard Corporation, 100 West Fourth Street, Elmira, New York 14902. Tel.: +1 607 733 7121; Fax: +1 607 737 1108; e-mail: hilliard@hilliardcorp.com; http://www.hilliardcorp.com
inopor	Tubular and Monolithic Al_2O_3, TiO_2, ZrO_2, and SiO_2 ceramic membranes for MF, UF, and NF	inopor® nano; inopor® ultra; inopor® micro	inopor GmbH, Michael Faraday Str. 1, D-07629 Hermsdorf, Germany. Tel.: +49 3 66 01 6 39 02; Fax: +49 3 66 01 6 39 21; e-mail: info@inopor.com, contact@inopor.com; http://www.inopor.com
Kubota	Tubular ceramic membranes, used in external MBRs		Kubota Corporation Membrane System Department, 1 3, Nihombashi Muromachi 3 Chome, Chuo ku Tokyo 103 8310, Japan. Tel.: +81 3 3245 3665; Fax: +81 3 32453407; http://www.kubota.co.jp
Media and Process Technology	Hollow fiber MF, UF, and NF ceramic membranes made of Al_2O_3, packaged in bundles	MPTech	Media and Process Technology Inc., 1155 William Pitt Way, Pittsburgh, Pennsylvania 15238. Tel.: +1 412 826 3721; Fax.: +1 412 826 3720; e-mail: info@mediaandprocess.com; http://www.mediaandprocess.com
Membraflow	Monolithic ceramic membranes for MF and UF with a		Westfalia Separator Membraflow GmbH, Robert-Bosch-Straβe 99, 73457 Essingen, Germany. Tel.: +49 7365 9614 0; Fax: +49 7365 9614 40;

(*continued*)

Table 6.1 (*Continued*)

Company	Marketed ceramic membranes	Trade name(s)	Contact info
	pure α-Al_2O_3 support		e-mail: membraflow@gea-westfalia.de; http://www.westfalia-separator.com
MemPro	Tubular and Monolithic TiO_2, Al_2O_3, and ZrO_2 ceramic membranes on alumina support for MF and UF	ATTAXX®[a]	MemPro Ceramics Corporation, P.O. Box 3806, Copper Mountain, Colorado 80443. Tel.: +1 303 224 9999; Fax: +1 303 224 0709; e-mail: sales@attaxx.com; http://www.attaxx.com, http://www.memproceramics.com
NGK Insulators, Ltd.	Honeycomb-like ceramic membrane		NGK Filtech, Ltd., Head Office, 2791, Hagizono, Chigasaki, Kanagawa 253 0071, Japan. Tel.: +81 467 85 8555; Fax: +81 467 85 6749; http://www.ngk.co.jp/english/products/engineering/environmental/index.html
Novasep	Tubular and Monolithic TiO_2, Al_2O_3, and ZrO_2 ceramic membranes on alumina or carbon support for MF, UF, and NF. Compact, honeycomb-like	Kerasep™, Carbosep®	Novasep Process, 5, chemin du Pilon, Saint Maurice de Beynost, 01708 MIRIBEL Cedex, France. Tel.: +33 4 72 01 27 27; Fax: + 33 4 78 55 38 33; e-mail: membranes@novasep.com; http://www.novasep.com/Technologies/Membranes.asp
Pall Corp.	Tubular and Monolithic Al_2O_3, Al_2O_3/	Membralox®, Schumasiv™, AccuSep™	Pall Corporation, 25 Harbor Park Drive, Port Washington, New York 11050. Tel.: +1 516 484 3600 (1 866 905

	TiO_2, and ZrO_2/ stainless steel ceramic membranes for MF and UF		7255); Fax: +1 516 625 3610; e-mail: foodandbeverage@pall.com; http://www.pall.com
Spintek	Flat rotating TiO_2, Al_2O_3, and ZrO_2 ceramic membrane for UF and MF	Speedy™	SpinTek Filtration, 10851 Portal Drive, Los Alamitos, California 90720. Tel.: +1 714 236 9190; Fax: +1 714 236 9196; http://www.spintek.com
TAMI Industries	Tubular and Monolithic pure TiO_2 ceramic membranes for MF, UF, and NF	INSIDE CéRAM, Filtanium, Isoflux, INSIDE DisRAM	TAMI Industries, Z.A. Les Laurons, 26110 Nyons, France. Tel.: +33 475 264 769; Fax: +33 475 264 787; e-mail: info@tami-industries.com; http://www.tami-industries.com
Uniceram	Monotubular and monolithic Al_2O_3 MF membranes		Uniceram Advanced Materials, 3883 Rogers Bridge Road, Suite 304B, Duluth, Georgia 30097. Tel.: +1 678 957 0896/678 957 0893/973 941 7427; Fax: +1 678 475 1823; e-mail: info@uniceramusa.com; http://www.uniceramusa.com

[a] atech innovations gmbh and MemPro are in a global partnership for ceramic membranes.
Data includes membrane types, trade names, and contact information. System suppliers and companies using ceramic membranes from other manufacturers are not included.

Figure 6.3 Examples of commercial ceramic membranes. (Photographies courtesy of TAMI Industries, Inopor, Pall Membralox, and Fairey Industrial Ceramics.)

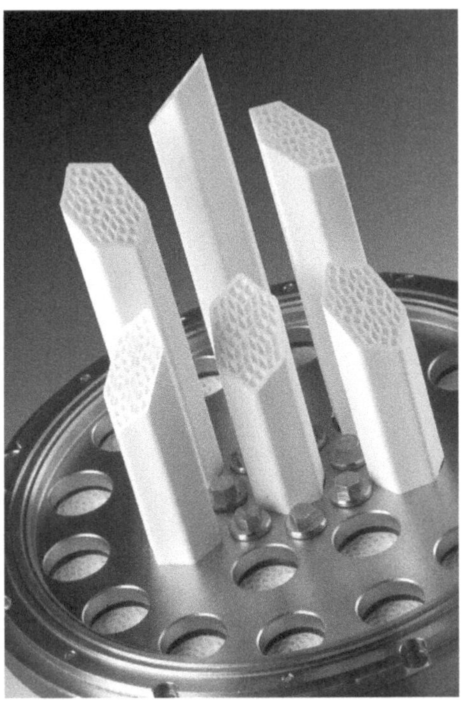

Figure 6.4 A stainless steel housing for ceramic membranes holding 23 modules. (Photography courtesy of Pall Membralox.)

Figure 6.5 One of the largest ceramic membrane housings, with over 250 modules. (Photography courtesy of Fairey Industrial Ceramics.)

continually sweeps away particles and other materials that would otherwise build up on the membrane surface.

Many industrial applications of membranes demand process stability, high availability, and low requirements for preliminary treatment, and minimum need for support and maintenance. For such demanding applications, in the chemical and pharmaceutical industries or in water and wastewater processing, *ceramic membranes* become the preferred option because of their extremely high chemical and physical (thermally and mechanically) stability, their outstanding separation characteristics, and their long working life. The most commonly encountered advantages of ceramic membranes, when compared to other membrane types used in pressure-driven membrane processes, include the following:

- long and reliable working lifetime
- resistance to high temperatures (up to 300 °C, though that also depends on the gaskets) across the entire pH range
- designed for pressures up to 30 bars (they normally can handle up to 90 bars)
- excellent chemical stability (i.e., organic solvents, oxidants, and hydrocarbons, including hydrofluoric and phosphoric acids when pure titania is used)
- corrosion and abrasion resistant
- high fluxes
- bacteria resistant and, frequently, bioinert
- compatibility with highly viscous fluids
- enhanced ease of cleaning and (chemical or steam) sterilization (back flushing is also possible)

These properties make them suitable for many applications where polymeric and other inorganic membranes cannot be used. Several membrane pore sizes are currently available covering the MF-UF-NF ranges (from 5 μm down to a few hundred daltons). Additionally, ceramic membranes are ideal for in-place chemical cleaning at high temperatures while using caustic, chlorine, hydrogen peroxide, ozone, and strong inorganic acids, and/or by using steam sterilization. Due to the unique properties of ceramic membranes, they can be cleaned online by means of reversing the flow through the membrane (backpulsing). By doing so, the filter cake is pushed away from the membrane surface, maintaining the filtration performance over a longer time and reducing downtimes for cleaning.

Permeability rates of most ceramic membranes are considerably higher than the rates of currently available polymeric membranes. Good retention of organics as well as low fouling tendency are confirmed in most applications. However, in contrast, salt rejection decreases strongly as electrolyte concentration rises, that is, during the treatment of pickling bath solutions. Therefore, the use of ceramic NF membranes in such media is not always a viable option. On the other hand, the rejection (in terms of both average cutoff and size distribution) given by ceramic membranes is not as sensitive to temperature as that of the polymeric materials, which could be an additional advantage in some applications.

The main disadvantages are their high weight and the considerable production costs of ceramic components. However, they typically have a long service life, up to 15 years, in applications which are not feasible for polymeric membranes. Moreover, in the event of membrane loss, they can be recoated therein decreasing element (membrane) replacement costs.

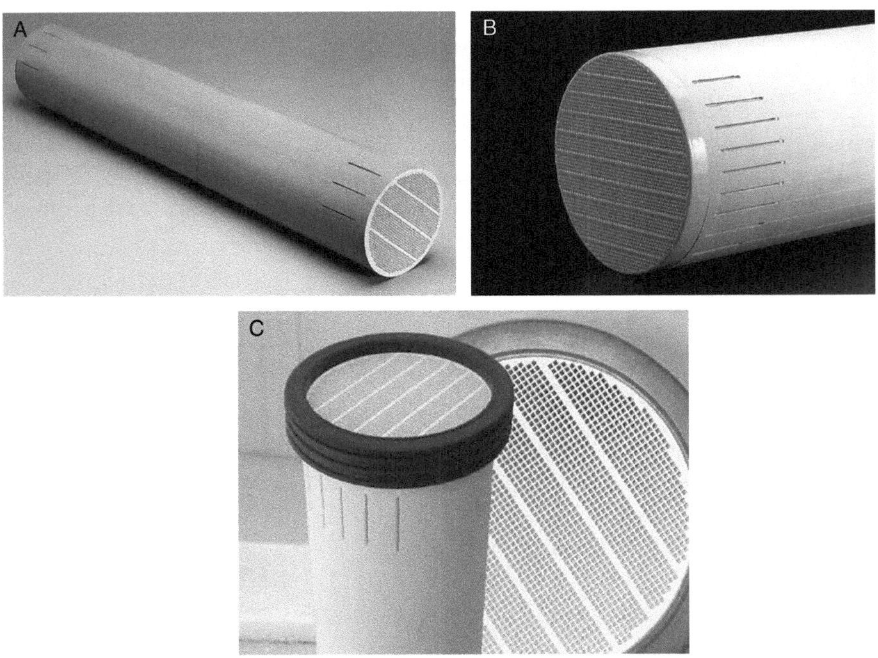

Figure 6.6 Honeycomb-like ceramic membranes. (A) Production membrane element with 10.7 m^2 membrane area. (Photography courtesy of CeraMem.) (B) NGK ceramic membrane element (180 mm diameter, 1000 mm length, and 15 m^2 membrane surface area). (Photography courtesy of NGK Insulators, Ltd.) (C) Hilco® ceramic membrane, with 0.2, 0.01, and 0.005 μm pore sizes. (Photography courtesy of Hilliard Corp.)

Another approaches used in the production of ceramic membranes include the honeycomb-like module for MF and UF (Fig. 6.6; [2]). They are abrasion-resistant chemically inert ceramic membranes applied to cordierite or silicon carbide large-diameter monolith supports and feature chemical and mechanical durability. They are very compact (over 760 m^2 membrane area/m^3) and they are reported to have costs competitive with many polymeric membranes. However, most industrial applications of this design are, at present, in gas separation. A similar MF membrane (0.1 μm) was developed in 1996 by NGK (Nagoya, Japan) for water purification systems (sold to small-scale purification facilities for public water systems), with the objective of elimination of turbidity and impurities such as colloids and bacteria contained in raw water (Fig. 6.6B). Other manufacturers producing these elements for UF are MF includes Hilliard (Elmira, NY; Fig. 6.6C). A similar concept has recently being launched by Novasep (Kerasep^{TM} Kompact) [3].

Ceramic hollow fibers also aim to decrease production costs and increase packing density (Fig. 6.7; [4]). Other ceramic monolith designs, such as the Dahlia configuration launched by TAMI Industries, are also marketed as a ceramic hollow fiber given its very low hydraulic diameter. Such system offers the additional advantage of not needing a potting material.

Table 6.2 provides some historical data for ceramic membranes used in the United States in 1992–2005, as well as Freedonia forecasts through 2007–2015.

3. Industrial Applications of Ceramic Membranes

Initially, ceramic membranes were developed for uranium enrichment and were also used in wastewater treatment. Over the past years, successful solutions and possible applications covered all industries and ceramic membranes are increasingly being used in industries such as biotechnology and pharmaceutical, dairy, food and beverage, as well as chemical and petrochemical, microelectronics, metal finishing,

Figure 6.7 Ceramic hollow fibers, single tubes, and Dahlia-designed tubes. (Photographies courtesy of CEPAration and TAMI Industries, respectively.)

Table 6.2 Membrane applications for monolithic ceramics in the United States (million dollars)

Item	1992	1997	2002	2007	2012
Gross domestic product (bil $)	6319	8318	10,460	13,500	17,400
$ ceramics/mil $GDP	0.95	2.89	5.26	7.41	9.20
Monolithic ceramic membranes demand	6	24	55	100	160
Pollution control membranes	3	13	30	55	90
Process membranes	3	11	25	45	70
Chemical processing products	neg	3	6	10	15
Other process membranes	3	8	19	35	55
Monolithic ceramics demand % membranes	0.2	0.4	0.8	0.9	1.1
Item	**1995**	**2000**	**2005**	**2010**	**2015**
Ceramic membranes (mill • m^2)	0.33	0.82	1.9	3.6	6.2
$\$/m^2$	39.9	45.2	50.9	55.4	61.0

Source: The Freedonia Group, Inc.

and power generation. Each industry has specific needs and opportunities. Some current industrial applications of ceramic membranes are listed in Table 6.3. Table 6.4 also shows applications reported by a membrane manufacturer for the honeycomb-type membrane.

Besides the applications that are industry-specific, a number of common approaches are being used in different industries. Two examples are oil/water separation and the recovery of cleaning chemicals [5–7]. Many examples can be found in food industry, metal fabrication and allied industries, automotive industry, chemical manufacturing industry, and oil refinery, including petroleum oils from tanker washers, spills, drilling, various processing steps, etc. Ceramic membranes are successfully being used in the treatment of wastewater, for example, for deoiling wastewater or compressor condensate, cleaning of washing water from metal or textile cleaning, pretreatment of landfill sewage, and bilge water treatment. The benefits of ceramic membranes are also employed in recycling technology, for example, for degreasing cleaning baths, paint, coating and enamel, or petrochemicals recycling. The wastewater typically contains emulsified oils that are difficult to separate with conventional treatment technologies such as coalescers and oil skimmers. Furthermore, many polymeric membranes are unsuitable due to their limited stability in aggressive chemical environments such as highly contaminated oily wastewater (e.g., lube oils, petroleum fractions). Long-term flux stability has been achieved on commercial scale installations equipped with backpulsing devices (300 liter/m^2 h was reported for the treatment of produced water containing 10–100 ppm oils and suspended solids using 0.8 mm Membralox® membranes over several months) [8].

Cleaning chemicals recovery enables the reuse of caustic- and acid-based cleaning chemicals, which often requires the use of ceramic membranes. The recovery of chemicals is site specific, but typically 95% of the active chemical solution is recovered with rejections of up to 75% of Kjeldahl nitrogen, up to 97% of oil and grease, and up to 100% of suspended solids. The payback depends on a number of

Table 6.3 Commercial-scale applications of ceramic membranes (pressure-driven processes)

Chemical/Petrochemical industry
Product seperation and cleaning
Concentration of polymer suspensions, latex, and metal hydroxide solutions
Catalysts recovery
Brine recycling
Recovery of dyes and pigments
Desalination of products (i.e., optical brighteners, surfactants, and dyestuffs)
Cleaning, recovery, and recycling of organic solvents (solvent NF)
Fuel oil and ethanol processing
Textiles/Pulp and Paper industry
Lignosulfonate fractionation
COD reduction of bleach plant effluents
Whitewater recovery from paper machines
Paper coating recovery
Recovery of rinse water from primary and secondary wool scouring effluent
Color removal
Hot, high quality water recovery for recycling
Dye bath recovery and reuse
Spin bath finish and size recovery
Water jet loom motive water recovery
Pharmaceutical, Cosmetic, and Biotechnology
Concentration, fractionation, isolation of pharmaceutical active agents (antibiotics, enzymes, proteins, amino acids, and vitamins), and plant extracts and sterilization
Separation, concentration, and dewatering of biomass and algae
Blood plasma filtering
Processing of alginates and other excipients
Disposal of fat emulsions
Desalination
Fermentation broth clarification, separation of yeast, concentration of bacterial suspensions
Endotoxin removal
Aqueous stream (by-)product recovery
Vaccine production
Nutraceuticals
High purity process water (pyrogen and bacteria-free)
CIP
Automotive/Mechanical/Mineral/Metal Industry/Surface Engineering
Recycling and disposal of degreasing and rinsing bathes
Treatment of oil/water emulsions
Recovery of heavy metals
Cleaning of wastewater from grinding processes
Treatment of wastewater form glass and glass fiber production
Pretreatment cleaner
Washing machine tanks
Oily wastewater
Food, Dairy, and Beverages
Clarification and stabilization of wine, beer, cider, soy sauce, and vinegar
Concentration of juice
Removal or microorganisms from milk and whey

(continued)

- Separation and fractionation of milk and whey ingredients
- Desalination of whey
- Purification of spirits
- Separation of soybean protein
- Vegetable oil processing
- Caramel color concentration and purification
- Dewatering of products
- Purification of drinking water
- Sugars and starches processing
- Recovery of product from beer and cider tank bottoms
- C.I.P. solution renovation
- Recovery of water from waste stream

Water, Recycling, and Environment

- RO pretreatment
- Wastewaters and potable water treatment
- COD, BOD, and suspended solids reduction by membrane bioreactor
- Removal of suspended solids
- Washing wastewater (i.e., laundry) and CIP chemical recycling
- Emulsified oil removal, edible oils, waste oil treatment
- Removal of pharmaceuticals and pesticides
- Removal of microorganisms
- Removal of heavy metals
- Gray water treatment
- Recycling of water form swimming pools
- Purification of the drain of sewage plants
- Sewer overflow treatment
- Methane digester water recovery
- Upgrading vacuum residual oil
- Landfill leachate processing
- Radioactive wastewater treatment
- Effluent recycling in zero-waste plant

NF, nanofiltration; COD, chemical oxygen demand; BOD, biochemical oxygen demand; RO, reverse osmosis; CIP, cleaned-in-place.

factors: spent cleaning solution characteristics, cost of new chemicals, percentage of recovery, number of times the permeate can be reused, and utility and waste disposal costs. However a payback of less than 1 or 2 years is typical [9].

Another generally applied use of ceramic MF membranes is in the pretreatment of different water sources prior to RO. In such application, ceramic membranes offer reliable performance under significantly varying feed conditions, and as result of that, RO membrane lifetime can be doubled or tripled [9].

The strategic areas in which membrane technology can be applied, either in standalone mode or combined with other separation/reaction technologies, are depicted in Fig. 6.8. Some of these applications are discussed below in more detail.

Cross-flow membrane technology offers a cost-effective solution to wastewater problems and is becoming widely used throughout the process industries. The technology, which can achieve a reduction in the volume of wastewater by a factor of typically 5–10, allows potential savings on effluent charges as well as on producing a clean water stream that can often be reused as wash water or cooling water within the facility.

In some cases the concentrated wastewater can be reused in the production process giving a true "zero discharge" treatment. Membrane technology is particularly effective in treating waste streams associated with pulp and paper, chemical, food and beverage, textile processing, and dye house waters.

Table 6.4 Commercial-scale applications and successful pilot trials of the CeraMem honeycomb-type membrane

Sanitary applications		Industrial applications	
MF	UF	MF	UF
Corn starch hydrolyzate clarification	Gelatin clarification	Secondary effluent disinfection (dead-end operation)	Shipboard bilge water deoiling
Beer clarification	Soy protein hydrolyzate clarification	Membrane bioreactor industrial effluents	Produced water deoiling
Apple juice clarification	Wheat starch hydrolyzate clarification	Evaporator condensate pretreatment for RO	Desalter bottoms deoiling
Cranberry juice clarification	Skim milk concentration	Fuel ethanol stillage clarification	Industrial laundry wastewater treatment
Wine clarification	Whole milk concentration	Hydrogenated oil catalyst fines removal	Vehicle wash water recycle
Whey protein clarification (defatting)			Glycerin purification
Cane and beet sugar process streams clarification			Ethylene glycol purification
			Deasphalting vacuum residual oil
			Waste lubricating oil purification
			Shipboard bilge water deoiling

Source: CeraMem Corp.
MF, microfiltration; UF, ultrafiltration; RO, reverse osmosis.

3.1. Chemical and petrochemical industries

The chemical industry is very diverse and uses all of the major process technologies. In many chemical process applications there is the need to (1) recover and reuse chemicals, and/or (2) treat highly variable wastewaters containing, that is, detergents, polymers, and organic solvents. However, the demand for higher product

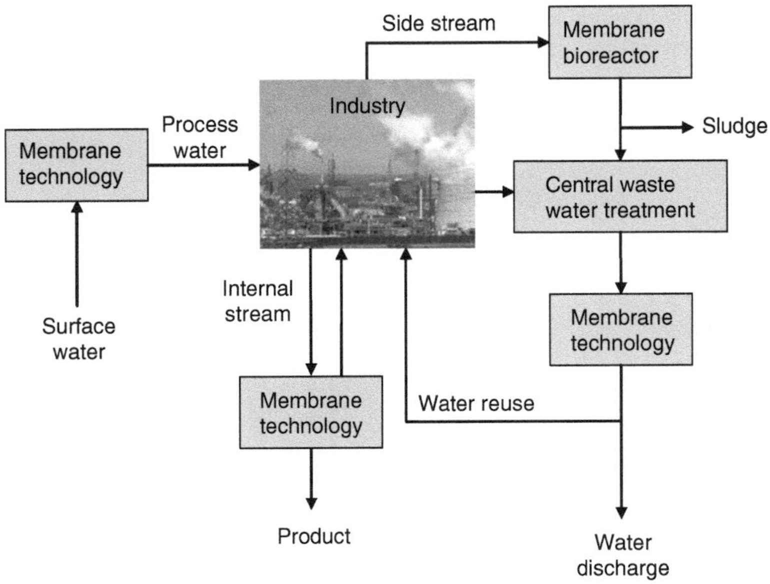

Figure 6.8 Potential industrial application strategies for membrane technology (adapted from a similar scheme byTriqua).

quality, shorter payback times, and reduced operating costs is increasing. As product separation becomes more complex, the use of membrane technology is increasing. This is particularly the case in the fine chemical industry where the membranes are used in a variety of applications, for example, in the concentration and desalting of dyestuffs and optical brighteners. In such scenario, the use of ceramic membranes, which can operate in more extreme conditions such as high temperature (above 200 °C), high/low pH, and are more resistant to solvents, can offer significant advantages over more conventional technologies [10].

Some current key applications of ceramic membranes in the chemical industry are the following:

- *Pigment, hydroxide, and catalyst recovery*. The recovered product is recycled back into the process stream, while the chemical oxygen demand (COD) of the resulting wastewater is reduced significantly.
- *Surfactant, optical brightener, and dyestuff desalting and concentrating*. Desalting is the process that removes inorganic matter during the manufacture of both powder and liquid dyestuffs. The dye is rejected by the membrane, while the salts and water pass through it, thus concentrating and desalting the dye simultaneously. Plant capacity can be increased using NF prior to drying in the production of powder dyes. In the manufacture of liquid dyes, the solubility of the dye is enhanced by almost complete desalting which helps to improve the stability of the product.
- *Silica sol and latex concentration*. Membrane technology provides a cost-effective solution for concentrating latex white water up to 15% solids. Latex can then be recovered and the treated water can be reused.

- *Solvent recovery (cleaning, recovery, and recycling).* While the aqueous applications of membranes are still dominating the pressure-driven membrane processes market, new solvent-resistant NF membranes reveal new opportunities in the chemical industry [11].
- *Oil/water separation.* For the disposal of mineral oil on surface water and sewage systems, most governments have special requirements on the quality of the wastewater. By means of a water–oil separator, a large amount of oil can be removed from the wastewater. However, chemically stabilized oil/water solutions should be managed in an appropriate manner. These solutions can be purified by means of membrane filtration in order to lead to dischargeable water. Typical results from waste stream to permeate are: oil/grease 136 ppm to 1 ppm, Pb 0.79 to 0.09, Ni 0.15 to 0.01, Cu 1.49 to 0.36, and Zn 5.9 to 0.37 [12]. Most applications often include the use of ceramic membranes as an intermediate filter or prefilter for more expensive RO membranes, reduction of oily bilge wastes, reduction of process wastes, or reduction of wash water wastes.
- *RO pretreatment.* Ceramic membranes can be continuously cleaned-in-place (CIP) by backpulsing, which minimizes the need for off-line cleaning. Thus, ceramic microfilters reduce RO replacement costs.
- *Chlorine-alkali plants* based on amalgam process produce sodium hydroxide (50% by weight) which is loaded with mercury downstream of the amalgam decomposer. This mercury, together with other solid impurities like graphite, can be removed by filtration using ceramic membranes. With 1–10 ppm Hg as inlet concentration, outlet concentration $<$10 ppb can be accomplished, partly with precoat or multistage filtration. The outlet concentration depends on the operating temperature (usually 60–75 °C) and the filtration velocity.

  In the new chlorine-alkali plants that use the ion membrane process, ceramic membranes can also be used for brine filtration. This allows the removal of suspended solids [magnesium hydroxide, calcium carbonate, iron (III) hydroxide, barium sulfate, and, in special cases, mercury sulfide] and, thus, helps protecting valuable plant components (i.e., membrane in the electrolytic cell and ion exchange unit).
- *Recovery of high flash spent solvents.* Several industries are moving from conventional solvents (toluene, xylene, acetone, etc.) to high flash solvents including dibasic and other esters, glycol ethers, terpenes, etc. Regulations worldwide are enforcing this change, since these high flash solvents are safer and environmentally friendlier. For instance, ~65–70% of the cleaning solvents used in the screen-printing industry, representing over $50 million per year in solvent sales, are high flash solvents. However, little recycling is practiced since distillation and filtration are not economically feasible.

  These spent solvents can be recovered (to near virgin quality) and reused by ceramic membranes. Virgin solvent requirement reductions of over 80% via this recycling of spent material can be achieved. The solvent maintains its cleaning efficiency on a long-term basis over a number of reclaim/reuse cycles. Moreover, the disposed waste volume is less than 10% of the original volume [13].

- *Refinery effluent treatment*. The current trend in industrial wastewater management focuses both on pollution prevention by source reduction/clean technologies and on closed water systems in which wastewater recycling plays a major role. Even if total recycling may not be required in all cases, it represents an alternative for industries with high water consumption, when either stringent discharge limits are imposed or limited fresh water resources exist. Recycling of refinery effluents for cooling water systems is one of the options that can be cost-effective in meeting or supplementing the plant water requirements. An analysis of the possibility of recycling industrial effluents as makeup for cooling towers has to consider a multitude of factors such as: effluent quality and quantity, makeup water quality requirements, evaluation of different treatment schemes, and cost analysis.
- *Spent oil recycling*. The oil re-refining process in general consists of two steps, a deashing step to remove particulate matter and a decolorization polishing step. Ceramic membranes have been used for deashing (removal of particulate matter) of spent car motor oils and synthetic oils [13]. Following deashing, spent oil can be sold as low ash high-quality burner fuel or further processed by decolorization and returned to the original or secondary markets (i.e., to lubricant packagers/blenders for reuse).

Industrial laundries generate highly contaminated wastewater, which can be processed on-site by ceramic membranes with the additional benefit of savings in fresh water and detergents (Fig. 6.9). Usually ceramic membranes are part of a two-stage installation comprising MF and NF or RO in which washing water and detergents are recycled. The MF skid removes first all emulsified and suspended contaminants. The clear filtrate still contains detergent elements that can be reused for prewashing and main washing processes. NF treats part of the MF permeate for reuse in rinsing cycles the filtrate in order to remove any remaining detergents and part of salts. The overall rate of recycling is $\sim$80% with 100 m^3/day filtrate [14].

A special and intensely researched niche for ceramic membranes is *integrated catalytic chemical reactor/membrane separator*.

Extraordinarily dirty RO operations such as hot stream/water recovery in certain petrochemical and refining operations may also dictate the use of ceramic membranes as pretreatment. An example of that can be found in the production of isopropyl alcohol/ethyl alcohol by chemical synthesis in which propylene is cracked with ethylene, using phosphoric acid as catalyst. Soda lye is added to the mixture generated during this process. As a result, sodium phosphate is formed. This is followed by distillation to retrieve the alcohol. The mixture remaining in the bottom of the distillation column, the effluent, is mainly composed of sodium phosphate, water, long-chain hydrocarbons produced by the cracking process, and alcohol residues. This effluent was conventionally treated by means of a gravity separator and disposed. Due to stringent regulations and the constantly rising costs for the disposal, a recycling system was implemented, consisting of a UF stage using multi-channel ceramic membranes (able to treat directly effluent mixture at a temperature of 80–90 °C), two RO units with spiral-wound modules, and one ion exchanger. UF removes the oil-containing constituents and particles from the effluent, and

Figure 6.9 Ceramic MF system for industrial laundry water and chemicals recycle. (Photography courtesy of Siva and TAMI Industries.)

concentrates them. The oils separated in the retentate are fed to a decanter and separated by a skimmer. Then, they are recycled to the in-house combustion plant for energy generation. A sludge with the particulate matter is extracted from the bottom of the decanter. The most important requirement to be fulfilled by the recycling system was economic efficiency combined with short payback periods because the up-front and the operating costs had to be balanced by savings on the fee levied for the discharge of effluents [14].

3.2. Metal, mechanical, and automotive industries

Many applications can be found in these industries, including cutting oil/lubricant filtration, recycling motor oils and glycols, electroplating solution recovery, etc. A number of small and medium size systems (ranging from 0.1 to 30 m^2 filter surface) are used on a wide range of feed stocks: solvents, degreasers, thermal heating oils, PVC dispersions, paints, brines, etc. One such application is the recycling of process water containing radioactive contamination (NORMs) within a Refurbishment Facility at Imerys Minerals Ltd., St. Austell in Cornwall (Fairey Industrial Ceramics) [15].

In the metal and surface processing industries, ceramic membranes are used in the following applications:

- Oil removal for recycling of degreasing, chemical, and rinsing bathes in metal machining and pickling
- Treatment of hot/cold rolling emulsified oily wastewater from metal processing
- Recovery of heavy metals
- Cleaning of wastewater from grinding processes, glass and glass fiber production

In the automotive and mechanical industries, main applications include the following:

- Pretreatment cleaner
- Washing machine tanks
- Oily wastewater

Other applications include the reclamation of nanopowders and the filtration of acid/alkali-containing liquids. More details on same examples of the above applications follow.

3.2.1. Recovery of cleaning alkaline solutions (and removal of oils)

In many of these industries, alkaline cleaning solutions are used at different times during manufacturing to remove oils, coolants, and other metalworking fluids, that is, prior to electroplating. The contaminated cleaner is pumped through a bag filter and into a settling compartment where free oils and sludges are removed. The solution is then pumped through the membrane at a high flow rate. Ceramic MF or UF membranes are used to remove oils and recover the alkaline cleaning chemicals. The contaminants are separated from the cleaner solution and concentrated. Alkaline cleaner recovery averages 95–98%, while contaminant removal efficiency ranges from 84–97% [9]. The purified cleaner solution flows back to the cleaner line. Moreover, there is a negligible oil drag-out into plating bath, finishing line, or downstream processes, together with consistently cleaner parts. Ceramic membranes tolerate nearly all chemicals except for phosphoric and hydrofluoric acids, and are able to work at temperatures up to 100 °C. This process provides an environmental benefit by extending the bath life of the alkaline cleaner, thereby reducing the amount of liquid and solid wastes produced by the cleaning operation without removing the cleaning constituents of the bath. This process has no hazardous waste disposal or costs due to membrane replacement, and the membrane life expectancy is over 10 years. As a direct result of the application of ceramic membranes, the overall volume of concentrated waste generated from alkaline cleaning can be reduced over 65% and the weight of total solids in the waste products can be reduced by almost 60%. The resulting process has typically payback periods between 1 and 2 years [16].

3.2.2. Further processing of membrane concentrates from oil/water emulsion filtration

High shear rotary ceramic membrane systems, such as the one manufactured by SpinTek [17], can be used to further treat the concentrate from a UF of an emulsified milling effluent resulting from a mixture of coolants and lubricants diluted with wastewater. The effluent, which initially contained about 0.5% oil/grease (O/G) is first concentrated by UF up to 5% at a volume reduction factor near 30. This stream is then concentrated to 50% oil using the ceramic membrane rotary system and then transferred to storage tanks equipped with skimmers. At high oil contents, phase separation can occur without chemical addition, and the oil skimmed from the tanks can be disposed of by an oil recycler.

In this type of rotary systems, high surface velocities are used (about four times typical cross-flow filtration values). As the concentrate thickens with treatment time, a conventional UF system is not able to maintain high velocities because of the difficulty in pumping viscous material at large flow rates. Since rotary systems do not rely on pumping to produce the required membrane surface velocities, they are able to treat extremely concentrated wastes.

3.3. Textile, pulp and paper, and tannery industries

The textile industry is very water intensive. Water is used for cleaning the raw material and for many flushing steps during production. Wastewater has then to be cleaned from fat, oil, color, and other chemicals, which are used during the several production steps. The cleaning process depends on the kind of wastewater (not every plant applies the same production process) and also on the amount of used water. Moreover, not all plants use the same chemicals, especially companies with a special (environmental) standard try to keep water cleaned in all steps of production.

It is quite difficult to define a general quality standard for textile water reuse because of the different requirements of each fiber (silk, cotton, polyester, etc.), of the textile process (scouring, desizing, dyeing, washing, etc.), and because of the different qualities required for the final fabric. This results in wastewaters with a wide range of pH (3–13), temperature (20–90 °C), and composition (sometimes with the presence of whiteners and bleaching agents as hydrogen peroxide [11]).

Nevertheless, membrane filtration is a preferable option compared to other wastewater treatment techniques because of the constant quality of effluent, which is partly or almost completely softened and free of color and surfactants. This technology allows cost-effective recycling solutions to the ever-growing environmental and legal pressures on the industry to cut down on the use of large amounts of quality water. Moreover, membrane technology has a reduced impact on factory space by having minimal footprint. Main applications of ceramic membranes in these industries are the following:

- Recovery of rinse water from wool scouring effluent. The recovered water is suitable for use in the process without any further treatment.
- Color removal. Depending on the membrane selected and the application, almost complete removal of color is achieved and almost all the soluble and insoluble components, including heavy metals, are also removed.
- Hot, high-quality water recovery (even above 90%) for recycling, achieving savings on energy, water charges, and a significant reduction in effluent disposal costs. This has the additional benefit of offering a secure water supply at a consistent quality and temperature for reuse.
- Dye bath recovery and reuse.

- Spin bath finish and size recovery.
- Water jet loom motive water recovery.

Most of the aforementioned applications address effluent issues, but also involve a reduction in operational costs, such as those associated with softening, heating, cooling, and disposal. In some applications, ceramic membranes are used in the front stage on a filtration plant in order to remove, that is, particles and long-chain organic molecules at high temperature (95 °C). The permeate is then fed to an NF or RO stage, which removes almost all the dissolved dyes and salts from the water. The purified process wastewater can be reintroduced into all areas of textile finishing without negatively affecting the end product. One process of this kind has been installed at Van Clewe GmbH textile finisher plant in Dingden, Germany. The prefiltration step in the recovery of wastewater uses ceramic membranes, which have been in continuous, near to failure-free, operation with practically no personnel expenditure for more than 5 years [18].

Treatment of the wastewaters directly by NF using ceramic membranes has also been reported [11] for decolorizing wastewaters from textile finishing. A 25 m^2 skid equipped with inopor NF membranes (Fig. 6.10) has been in operation from 2002. The plant treats the washing water directly after dyeing, in a feed and bleed mode with volume reduction factors between 10 and 20. Clear permeate, which can be directly reused is obtained at a flow rate of 5 m^3/h at a transmembrane pressure of 15 bars. The concentrate is decolorized by cathodic reduction.

Membrane technology can also be used to concentrate the chromium and salts from the tannery operation for reuse. This not only cost-effectively solves disposal problems, but it also reduces manufacturing costs for a more profitable waste operation.

Figure 6.10 Ceramic NF system for decolorizing wastewater from textile finishing. (Photography courtesy of inopor.)

As in the textile sector, the forest products industry produces large volumes of effluent and uses high volumes of water as part of the manufacture of pulp and paper. Because of the increased environmental awareness and stringent legislation, the paper and cellulose industry is forced to reduce water consumption. Normally, the wastewater from a paper plant is biologically treated; however, even though the quality of the effluent may be good enough for disposal, it is often not high enough for reuse as process water. Examples of membrane applications include the reduction of COD from pulp bleaching effluent using UF or NF (COD removal efficiencies are on the order of 60–80% for UF, and between 85 and 95 for NF, depending on process and operating conditions), fractionation of lignosulfonates, and the recovery and reuse of white waters and coating chemicals from paper production [19].

Like in other fields of application, ceramic membranes are not only able to produce a purified water stream, but also to recover valuable materials which can be returned to the process. An example of that in paper manufacture is the treatment of coating solutions. To be printed, the paper must receive a coating from a suspension or coating slip that contains latex, PVA, titania, dye, etc. This coating is applied in the form of an aqueous solution, and the wastewater from this processing step, since it still contains part of the coating chemicals, must be treated before their disposal. Using ceramic membranes, it is possible to split this waste stream in (1) a permeate that is clear water (with some traces of soluble products) that can be recycled or disposed in the sewage plant, and (2) a retentate containing all suspended maters that can be recycled. As a result, new coating slip is produced out of the retentate, and there is a reduction in water consumption and in disposal costs in the sewage plant.

3.4. Biotechnology, cosmetic, and pharmaceutical industries

Membranes have also been used in the biotechnology and pharmaceutical industry for over 30 years, primarily for the separation/filtration of bulk pharmaceuticals, product recovery, and concentration and desalting of low molecular weight active products, including antibiotics and other target molecules. This technology is vital for microbial cell separation and concentration, antibiotic broth clarification, enzyme separation, endotoxin removal, and amino acid concentration. Pharmaceutical plants are built to a high specification including internal and external polishing of pipe work in order to meet the high pharmaceutical standards required by the customers.

Ceramic membranes are used for *fermentation broth clarification* at numerous installations worldwide, competing with other technologies such as vacuum filtration, precoat filtration, centrifugation, and other membrane processes using polymeric membranes. A number of applications can be found in the recovery of enzymes or antibiotics (Figs. 6.11 and 6.12) from whole and preclarified fermentation broths. Benefits include more effective processes, with higher product yields, eliminating the need of precoat filter aids. Moreover, the system is easily cleaned (CIP) and steam sterilized. Better process control, and frequently improvements of the downstream UF purification step are also reported advantages [20].

Figure 6.11 Fermentation broth clarification plants. [Photographies courtesy of PCI Membranes (A), Siva and TAMI Industries (B).]

Figure 6.12 Steam sterilizable unit for vaccine production. (Photography courtesy of TAMI Industries.)

Typical operating conditions include feed temperatures of 30–60 °C, high crossflow velocities (4–8 m/s), and transmembrane pressures up to 6 bars, with typical values in the range of 1.5–3 bars. Feed pressures up to 10 bars may be required for processing large volumes with multiple modules arranged in series and parallel [10].

Figure 6.11A shows an eight stages-in-series ceramic membrane UF broth clarification plant, designed to process up to 20,000 liter/h of fermentation broth. This particular system is one of the largest plants of its kind. Designed to run in "carousel" mode, this is an automatic computer-controlled system which offers continuous broth clarification with minimal production downtime required for membrane cleaning [21]. Figure 6.11B shows a 554 m^2 ceramic membrane UF unit also used for broth clarification.

Ceramic membranes are also used to purify viscous solutions of biopolymers. The products are used in consumer medical products as well as more complex surgical procedures. They are also often used to remove pyrogens to levels that meet water for injection (WFI) criteria, that is, in process water, with installed capacities of 6500 liter/h [22].

Membrane technology is also ideal for the variable composition of cosmetics processing wastewater, producing consistent, high-quality effluent with minimal operating labor and without the use of chemicals. UF can remove all suspended materials and provide biological and CODs reduction.

3.5. Food and beverages

3.5.1. Juices and Wine

Clarification of wine and natural fruit juices such as apple, cranberry, and grape is one of the most successful and widely practiced industrial applications of ceramic membranes (Fig. 6.13). This process safely removes suspended solids in wine and juice filtration applications, providing exceptional clarity without stripping color. Within a single step, ceramic membranes are able to replace the conventional clarification process scheme (including gelatin and/or bentonite addition, holding/decanting, diatomaceous earth, cake filtration, and polishing) producing superior clarity juice with higher yields [10]. Ceramic UF can be applied to enzymatically treated juice or to raw juice (in that case, backpulsing and frequent cleaning are

Figure 6.13 (A) Clarification of apple juice. (Photography courtesy of Unipektin AG, CH-Zürich, IBMEM, and atech innovations gmbh.) (B) Wine clarification skid. (Photography courtesy of Siva and TAMI Industries.)

needed, or even a combination of both, to reduce system downtime). Another process alternative includes MF membranes in combination with absorption to remove polyphenols and proteins. The latter has been applied, that is, to citrus juices, which often contain etheric oil, capable to destroy polyester, typically used as support in polymeric membranes [23].

3.5.2. Beer production

Beer clarification, in order to eliminate yeast and colloidal particles responsible for haze (chill and permanent hazes), is typically carried out using conventional dead-end filtration with filter-aids such as diatomaceous earth, which retains solid particles (yeast cells, macrocolloids, and suspended matter) and solutes responsible for haze. In addition, this operation should also ensure the biological stability of the beer. This operation should comply with the haze specification of a lager beer in order to produce a clear bright beer [European Brewery Convention (EBC) norms]. The variety of compounds (chemical diversity, large size range) to be retained makes this operation one of the most difficult to control and, therefore, MF is an alternative [24–26].

Ceramic membranes are also used to recover valuable materials from beer and cider tank bottoms. This process can recover more than 90% of the alcohol thereby increasing the yield and reducing effluents. The dilute yeast slurry, after clarification, can account for up to 3% of the total raw beer batch volume and 90% of this can be reclaimed as usable product. Another application of ceramic membranes in beer manufacturing is the production and concentration of the biocatalyst, which involves the combination of MF (1.4 μm) for broth prefiltration and further concentration of yeast by a 50-kDa UF membrane. The process, developed by IBMEM is in operation at a brewery in Landsberg, using technologies from biotech (fermtec) and membrane (TAMI) companies [1].

3.5.3. Sugars and starch

Ceramic membranes can be used in several different stages in the filtration of sugar cane juice, both in the raw and refine sugar production (Fig. 6.14). A typical example is the use of UF or MF as a pretreatment of the clarified juice (7–14 Brix) and/or the preevaporated juice (20–25 Brix) prior to ion exchange or chromatographic separations Typical operating conditions include feed temperatures of 90–100 °C, a high cross-flow velocity (4–7 m/s) and transmembrane pressures up to 5 bars [10]. The need to purchase, use, and dispose of the conventionally used filter aids is, thus, eliminated.

Figure 6.14A shows a full-scale plant used in the clarification of cane sugar juice, with a capacity of 280 m^3/h of clarified juice, using a technology developed by Applexion (nowadays Novasep) [27]. This clarification process yields a substantial increase in product quality, reducing color bodies by 40%, ashes by 55%, and turbidity by 70%. The rate of sugar crystallization is also increased by 20%. A similar approach is used to remove pulp, some proteins, fats, and starch using a ceramic MF skid of 66 m^2 in a sugar factory in Africa. The filtrate is free of pulp (NTU reduction from over 1000 down to 0.5 units), has a dark color, and approximate 15 Brix. The sugar juice (at 70–80 °C) is filtered batch by batch (to a concentration factor of 10–12) with an hourly output of 7.5 m^3. In order to minimize product losses and the

Figure 6.14 (A) The first commercial scale cross-flow MF unit for sugarcane juice at the Cane Sugar Mill of HC&S in Puunene, Hawaii (in operation in 1994). (B) Ceramic membrane MF in the processing of raw sugar melt. (C) Unit for glucose syrup clarification. (Photgraphies courtesy of Novasep.)

resulting discharge, diafiltration in then employed. The final concentrate is used as animal feed [23].

Figure 6.14B shows another full-scale plant used in processing of raw sugar melt, with a capacity of 200 t/day of raw sugar. In this processing step, ceramic MF membranes are replacing the steps of affination, and phosphatation or carbonation and the subsequent clarification or filtration [28].

With a similar approach, ceramic MF membranes are used to remove unliquified starch, polysaccharides, proteinaceous matter, and other impurities from the saccharification tank liquor. The process has been successfully applied to sweeteners derived from various starch sources—corn, wheat, tapioca, potatoes, or cassava. The process eliminates the use of diatomaceous earth in rotary vacuum filters while at the same time producing a superior quality product.

3.5.4. Dairy Products

3.5.4.1. Bacteria and spores reduction Demand for products with low bacteria and spore counts is increasing. Examples of these are extended shelf life (ESL) milk, pretreatment of cheese milk to prevent gas formation during the cheese aging process, production of skim milk and whole milk powders with very low bacteria and spore counts, and better microbiological quality of whey protein concentrates (WPC). Plants equipped with ceramic membranes can give at least a 4 log (99.99%) reduction in bacteria and spores depending on the loading in the feed.

3.5.4.2. Fractionation of casein and whey proteins in skim milk In cheese making processes, *Clostridium* and other spore formers can survive normal pasteurization at 80 °C. Such temperature can actually activate the spores which then thrive in the ripening storage environment. These bacteria produce butyric acid and hydrogen gas and can destroy large quantities of cheese by late blowing. This can be controlled by nitrate addition, but this practice is criticized and even banned in some countries.

On the other hand, separation of casein from whey proteins in skim milk and its concentration gives an opportunity to fortify cheese milk without increasing the whey protein content. This has a positive influence on increasing the capacity of cheese making equipment without affecting the quality of cheese. As an added benefit, whey proteins produced by such a process have some unique functional properties since they have not come in contact with cheese starter, rennet, or cheese additives.

Ceramic MF membranes make these two processes feasible, with superior performance to that of polymeric membranes. This has been accomplished by using systems with uniform transmembrane pressure.

In conventional MF, as the feed streams flows down the tubular element, there is a natural hydrodynamic pressure drop from the inlet to the outlet of the flow channel, which results in a decreasing transmembrane pressure. The uneven permeate flux distribution along the length of the flow channel can be very significant due to the resulting concentration polarization and fouling phenomena, particularly at the higher pressure inlet. This alters solutes transmission through the membrane, and also with time because of the uneven performance.

A number of commercial approaches (Allcross® from TetraPak [29], Isoflux® [30] from TAMI Industries in systems designed and engineering by GEA filtration, Membralox GP from Pall Membralox [31]) have focused in developing a consistent transmembrane pressure system. This is typically accomplished by the use of a variable thickness active membrane layer, or a variable resistance in the membrane support layer (a decreasing porosity along the module length), in order to produce a uniform permeate flux along the entire length of the element [20]. This optimizes the MF of complex, transmission sensitive feed streams.

Applications of this uniform transmembrane pressure approach are mostly in the dairy industry, particularly for the removal of heat-resistant bacteria and spores from all types of dairy products, without altering the protein content or other properties.

MF with ceramic membranes using uniform transmembrane pressure efficiently removes spores and bacteria from cheese milk and guarantees production of high-quality cheese without nitrate addition. The MF module separates the skim milk into two parts—the main stream passing through the membrane (permeate) which includes serum, proteins, minerals, salt, and small fat globules, while a small fraction (between 0.5% and 5% of the initial feed) containing at least 99.5% of the bacteria (retentate), including spores. Other pasteurization-resistant bacteria than those spore formers which can cause problems for cheese makers, for example, strains of *Enterococci* are also effectively removed by MF. This bacteria-rich retentate can be heated at a higher temperature, 120–130 °C for 2–3 s during the pasteurization process, which is an effective treatment for spore formers, before being remixed with the main milk stream or be continuously bled off. The ceramic membrane plants also feature efficient CIP cleaning and disinfection which eliminates the risk of recontamination.

More applications of this technology can be found in the reduction of fat in whey protein isolates (WPI) to increase clarity while still retaining important subfractions and maintaining biological activity; and the fractionation of casein and whey proteins from skim milk for use as protein supplements and to add nutritional value to sports drinks. Other food applications include the clarification of raw fruit and vegetable juices; the clarification and concentration of dextrose and maltodextrins; and the concentration of vegetable and plant proteins.

Additional applications of ceramic MF are, that is, in the recovery of fat cream from cheese cooker water or salt whey drippings (Fig. 6.15). Both side streams in cheese manufacture allow an increase in process yield. The permeate from the MF can be directly recycled as process water or further processed by RO or NF to recover nonfat solids [32].

3.5.5. Whey protein concentrates and whey protein isolates

In whey processing, the clarified raw whey is separated into whey cream and whey. The whey cream is removed and can be used for whey butter or for cheese production. The separated whey is then pasteurized and cooled to 50 °C. The limitation of fat removal from whey by mechanical separation results in high fat content in WPC. This high fat level limits the maximum protein content in the final WPC powder, usually 80–84% depending on the feed quality. WPI require reduction of fat content in the final product to values below 0.5%. This can be

Figure 6.15 Ceramic membrane unit for cheese side stream processing. (Photography courtesy of GEA Filtration.)

accomplished by reducing the fat content further than using mechanical separation only by MF with either ceramic or spiral wound membranes [7].

The separated and pasteurized whey is then concentrated by UF to raise the total solids and protein content. Diafiltration, a procedure in which water is added to the feed as filtration proceeds in order to wash out low molecular weight components such as lactose and minerals, can be used to raise the concentration of protein relative to total solids. The whey retentate is then evaporated and dried to make WPC products.

Another processing scheme starts with UF of the cheese whey to recover the proteins. From the UF retentate, it is possible to further purify the proteins. This is carried out by a combination of MF and UF. The UF retentate is concentrated by a factor of 10 by MF. Ninety percent of the feed passes through the microfilter membrane as permeate with nearly all the fat, denatured proteins, bacteria, and spores remaining in the retentate. The MF permeate is then sent to the second UF plant for final concentration and then dried to produce a WPI.

3.5.6. Sanitary conditions

In the food and beverage industry, hot caustic solutions (pH 12–13.5) are used to clean a variety of equipment such as tanks, mixers, evaporators, and crystallizers. This is necessary to maintain proper sanitary conditions between batches. CIP, which requires caustic and acid cleaning solutions, is widely used to maintain sanitary standards. The spent caustic solutions contain suspended solids (pulp, low molecular weight proteins, organic polymers, fats, and soil), dissolved matter (sugars, polysaccharides, and oligosaccharides), and additives (such as wetting agents and surfactants). Ceramic membranes are able to retain the suspended solids (over 99%) and allow permeation of the additives (Fig. 6.16). Typical operating conditions include feed temperatures of 50–70 °C, a high cross-flow velocity (4–7 m/s), and transmembrane pressures up to 5 bars [10, 33].

4. Ceramic Membrane Applications in Water and Wastewater Treatment

In the end of the 1980s, together with increased environmental awareness within society and industry, "zero discharge" changed from a technical description of 100% wastewater recycling to a "goal." The principle of "zero discharge" is recycling of all industrial wastewater. This means that wastewater will be treated and used again in the process. Because of the water reuse, wastewater will not be released on the sewer system or surface water.

4.1. Water management in industry

The concept of energy management has been adopted by many industries in their quest for improved efficiency; however, many have yet to grasp the opportunities offered by another important area: water management. With charges for fresh water,

Figure 6.16 A caustic recovery system using ceramic modules, at the Southern Gardens Citrus processing plant in Florida. (Photography courtesy of GEA Filtration and Membralox.)

sewage, and trade effluent continuing to rise, cost savings as high as 50% are achievable.

An estimated 1.4 billion people will face water scarcity by 2025, however there are few industries that do not use large volumes of water, much of it taken from public water supplies. During this time of growing pressures on national and global water resources, effective water management can not only enable companies to reduce operational costs but also help them establish their "green" credentials and fulfill their corporate and social responsibilities.

Efficiency measures may deliver savings in water use as high as 90% for industrial users, depending on the processes involved. Installing automatic shut-offs will reduce water use by an average of 15%, while introducing closed loop recycling of water will achieve a 90% savings. Other measures could include reusing wash water (50% savings) and cleaning in place (60%). Commercial users can reduce their use of water by an average of 40% through making low cost improvements to toilet cisterns, urinals, showers, and other facilities.

It is important to recognize that good water management does not just translate as reducing the amount of mains water used. The application of recycling, that is the reuse of water for the same or a similar duty, normally relies on suitable process technology for water purification. Membrane technology can help address water shortages worldwide by treating waste streams and regenerating water for reuse.

Effective management also stretches to the indirect expenses associated with water. The energy consumed in heating, cooling, pumping, storing, and treating water as well as essential maintenance such as replacing corroded pipe work and leaks presents significant additional costs. Heating or treating water can push up overall costs considerably. The overheads for chlorinating water are estimated at up to 3€ a cubic meter, softening or demineralising water at 6€ a cubic meter, and producing steam up to 24€ a ton. By reducing the volumes of treated or heated water used, total savings are much greater than the cost of the water alone.

The basics of effective water management, much like energy efficiency, often equate with common sense. Good housekeeping measures that start with turning off water-using equipment when not in use can ease consumption. Technologies such as rainwater harvesting systems that collect rainwater and use it for suitable purposes do involve more investment, however with an increasing population, such investment will be necessary.

Of course, the concept of water management extends to the water companies themselves as climate change impacts on the water industry. Obvious hazards such as flood, drought, health risks, and damage to assets are combined with greater uncertainty in weather patterns that affect land use, water usage, water storage, consumer demand (for both water and energy), and, last but not least, ecosystems. As climate change models are only capable of simulating change at large scales, building this level of uncertainty into long-term strategic planning presents additional challenges.

Industry strategies for adapting to climate change include sustainable drainage systems (SUDS), the operation of dams and reservoirs, water conservation, use of reclaimed wastewater, using pricing and market mechanisms to increase efficiency of water use, and improved ability to shift water within and between sectors to accommodate for shortages.

The reuse of process and wastewater becomes more and more interesting because:

- The cost of tap water is constantly increasing while concessions for the extraction of ground water are rarely granted anymore. It is expected that drinking water will soon double in price.
- Many industrial processes require heated or cooled process water. By reusing process water, the energy requirement demands are lowered, and therefore savings can be made on energy expenses.
- Stricter demands are being made on the quality of the water flowing through processes. When preparing process water, certain elements are removed to increase the performance of the system, while other substances are added to improve the water quality and promote the effectiveness of the process. When this water is reused, it is free of unwanted elements and already contains those elements which are needed by the process, therefore lowering costs.
- The cost of draining off wastewater has risen in the past years and are expected to rise even more. Some companies have already placed pretreatment units for treating wastewater. In many cases, posttreatment of the water is also possible, therefore making it suitable for reuse in the process. The effluent can also undergo a less effective treatment and can be used as cooling or cleaning water.

Influences such as increasingly stringent regulations, shortages of acceptable quality water, higher costs, and greater public awareness have created a growing concern for water and wastewater treatment and reuse. Membrane separation technologies, with their clear benefits, are becoming more widely utilized to accomplish these goals.

Ceramic membranes have found applications in water and wastewater treatment as RO pretreatment, in the removal of suspended solids, microorganisms, heavy metals, radioactive substances, and pharmaceuticals and pesticides. They can also be used for oil/water separation, edible oils, and waste oil treatment. In membrane bioreactors (MBRs) they can reduce COD, biochemical oxygen demand (BOD), and suspended solids. As part of a zero-liquid discharge approaches, they allow recycling of washing wastewater, CIP chemicals, and effluents. They can also be used in the purification of the drain of sewage plants, landfill leachate processing, and in upgrading vacuum residual oil.

4.2. Secondary and tertiary wastewater treatment

Conventional water treatment may include chemical addition (aluminum sulfate, lime, and polymers), coagulation, flocculation, sedimentation, filtration, and disinfection, usually with chlorine. Indeed all these conventional systems, based on physical and chemical principles, cannot give an absolute guarantee in terms of separation efficiency, while MF and UF can provide a complete removal of most bacteria and all protozoan cysts of concern as long as the membrane and associated system components are intact and operating correctly. In the case of UF, virus removal is also achieved [34].

Membrane systems (basically MF, UF, and NF) have been used in the last years as substitutes for secondary settling basins in conventional activated sludge treatment plants, thus eliminating decantation problems and making it possible to work with high biomass concentrations in the biological reactors: the advantage is a reduction in the reactor volume while ensuring the same treatment efficiency. Additional space economy is achieved through the replacement of a settling basin by a filtration unit. Immersed membranes in the biological settling basins have also been used to produce a clarified effluent by means of intermittent suction. Another application of membrane technology in the field of wastewater treatment is focused on the tertiary treatment in order to obtain a high-quality final effluent that can be reused for different purposes. Basically, these technologies aim at a reduction of suspended matter; organic pollution abatement; reduction of potential toxic compounds associated with the solid fraction; and wastewater disinfection, free from the risk of organochloride buildup associated with chlorine treatment. Nevertheless, even though technical feasibility is very well documented, their implementation is sometimes limited because of the high investment and operational costs involved.

The large-scale (180 mm in diameter) honeycomb-like ceramic membrane developed by NGK in 2001 was selected by the Tokyo Waterworks Department and resulted in the largest ceramic membrane system ever used in Japan, purifying 3400 m^3 of water per day. Another MF system for secondary effluent disinfection using similar membranes is shown in Fig. 6.17.

Figure 6.17 42-Module MF system (secondary effluent disinfection) at Berlin Water (40 m^3/h) with honeycomb CeraMem membranes. (Photography courtesy of CeraMem Corp.)

Low cost (hollow fiber) ceramic membrane technology has also been proven as a cost-effective single-step treatment option for small community drinking water treatment [13], able to significantly reduce turbidity, bacteria, virus, and trihalomethane precursors.

4.3. Membrane bioreactors

Biological treatment technologies have been utilized in wastewater reclamation for over a century. Out of the many different processes employed, the activated sludge system has proven to be the most popular. The implementation of membranes within the treatment sequence of a water pollution control facility was initially limited to tertiary treatment and polishing. UF, MF, or RO units were utilized in areas where discharge requirement were very stringent or direct reuse of the effluent was desired. High capital and operational costs as well as inadequate knowledge on membrane application in waste treatment were predominant factors in limiting the domain of this technology. However, with the emergence of less expensive and more effective membrane modules and the implementation of ever-tightening water discharge standards, membrane systems regained interest. Membrane modules have evolved from being utilized solely in tertiary wastewater treatment to being integrated into secondary wastewater treatment.

These systems are now most commonly referred to as MBRs (Membrane Bioreactors). There are several advantages associated with the MBR which make it a valuable alternative over other treatment techniques. First of all, the retention of all suspended matter and most soluble compounds within the bioreactor leads to excellent effluent quality capable of meeting stringent discharge requirements and opening the door to direct water reuse. The possibility of retaining all bacteria and viruses results in a sterile effluent, eliminating extensive disinfection and the corresponding hazards related to disinfection by-products.

Since suspended solids are not lost in the clarification step, total separation and control of the solid retention time (SRT) and hydraulic retention time (HRT) are possible enabling optimum control of the microbial population and flexibility in operation. The absence of a clarifier, which also acts as a natural selector for settling organisms, enables sensitive, slow-growing species (nitrifying bacteria, bacteria capable of degrading complex compounds) to develop and persist in the system even under short SRTs. The membrane not only retains all biomass but also prevents the escape of exocellular enzymes and soluble oxidants creating a more active biological mixture capable of degrading a wider range of carbon sources. MBRs eliminate process difficulties and problems associated with settling, which is usually the most troublesome part of wastewater treatment.

The potential for operating the MBR at very high SRTs without having the obstacle of settling allows high biomass concentrations in the bioreactor [35]. Consequently, higher strength wastewater can be treated and lower biomass yields are realized. This also results in more compact systems than conventional processes significantly reducing plant footprint making it desirable for water recycling applications. High molecular weight soluble compounds, which are not readily biodegradable in conventional systems, are retained in the MBR. Thus, their residence time is prolonged and the possibility of oxidation is improved. The system is also able to handle fluctuations in nutrient concentrations due to extensive biological acclimation and retention of decaying biomass. The disadvantages associated with the MBR are mainly cost related. High capital costs due to expensive membrane units and high energy costs due to the need for a pressure gradient have characterized the system. Concentration polarization and other membrane fouling problems can lead to frequent cleaning of the membranes, which stop operation and require clean water and chemicals. Another drawback can be problematic waste-activated sludge disposal. Since the MBR retains all suspended solids and most soluble organic matter, waste-activated sludge may exhibit poor filterability and settleability properties. Additionally, when operated at high SRTs, inorganic compounds accumulating in the bioreactor can reach concentration levels that can be harmful to the microbial population or membrane structure.

System configurations: MBRs are composed of two primary parts, the biological unit responsible for the biodegradation of the waste compounds and the membrane module for the physical separation of the treated water from mixed liquor. MBR systems can be classified into two major groups according to their configuration.

The first group, commonly known as the integrated MBR, involves outer-skin membranes that are internal to the bioreactor. The driving force across the membrane is achieved by pressurizing the bioreactor or creating negative pressure on the permeate side. Cleaning of the membrane is achieved through frequent permeate backpulsing and occasional chemical backwashing. A diffuser is usually placed directly beneath the membrane module to facilitate scouring of the filtration surface. Aeration and mixing are also achieved by the same unit. Anoxic or anaerobic compartments can be incorporated to enable simultaneous biological nutrient removal. This internal MBR system uses typically polymeric hollow fiber or flat sheet membranes.

The second configuration is the recycle (external) MBR, which involves the recirculation of the mixed liquor through a membrane module that is outside the bioreactor. Both inner-skin and outer-skin membranes can be used in this application. The driving force is the pressure created by high cross-flow velocity along the membrane surface. The emergence of less expensive and more resilient polymeric membranes along with lower pressure requirements and higher permeate fluxes have accelerated the worldwide commercial use of submerged MBRs. Several types and configurations of membranes have been used for MBR applications. These include tubular, plate and frame, rotary disk, hollow fiber, organic (polyethylene, polyethersulfone, polysulfone, polyolefin, etc.), metallic, and inorganic (ceramic) MF and UF membranes. The pore size of membranes used ranged from 0.01 to 0.4 μm. The fluxes obtained ranged from 0.05 to 10 m/d (m^3/m^2 d), strongly depending on the configuration and membrane material. Typical values for inner-skin membranes are reported as 0.5–2.0 m/d and for outer-skin membranes as 0.2–0.6 m/d at 20 °C. The applied transmembrane pressure ranges from 20 to 500 kPa for inner-skin membranes and from −10 to −80 kPa for outer-skin membranes.

Although a number of studies have been performed with MBRs involving ceramic membranes few references can be found with full-scale processes, as reviewed by [36].

The Suez-Lyonnaise-des-Eaux/Infilco-Degremont consortium marketed in-series MBRs with tubular ceramic membranes in an external loop for an anaerobic reactor. Aerobic and anaerobic MBR processes (BRM®) for industrial and municipal wastewater were first developed by the Centre International de Recherche sur l'Eau et l'Environnement (CIRSEE) in 1993. The first BRM was installed at a cosmetic factory in northern France in 1993. It has a design capacity of 6600 liter/h and 1200 kg COD/day [36].

Kubota Inc. patented submerged-type ceramic membrane process (BIOREM® and FILCERA®), for treating domestic wastewaters, and today, eight wastewater treatment plants incorporating their MBRs were operating in Japan in 1998. These two are basically the same, but the difference lies in their application. One system (BIOREM) is intended for treating wastewaters, while the other (FILCERA) is intended for water purification. Both consist of a tank-submerged-type membrane module, which brings together multiple external pressure tubular-type ceramic membrane elements of an asymmetric two-layer structure. Typical results include more than 95% BOD, COD, and TSS (total suspended solids) removals. The first system was developed in 1988 years ago and was running using the original ceramic membrane for over 10 years [36].

CeraMem has also patented (US Pat 6,767,455 and US Pat 6,921,483) a novel airlift MBR process that uses its standard membranes.

ACKNOWLEDGMENTS

Authors are grateful to the following people and companies, who were so kind as to review this chapter, provide useful comments and corrections, and give most of the pictures shown. They are listed below in alphabetical order of their company name: Peter Bolduan from atech innovations gmbh; Bruce Bishop from CeraMem; Andrew Frost from Fairey Filtration Systems Limited; Pierre Strauch from IBMEM; Ingolf Voigt from inopor GmbH; John Finley from MemPro Ceramics; David Dubbin and Bob Keefe from Niro GEA Filtration; Rishi Sondhi from Pall Corporation; and Dominique Girard from TAMI Industries and The Freedonia Group, Inc.

REFERENCES

[1] Strauch, P. (2006). Membrane Filtration with Ceramic Membranes, Frankfurt: Achema, Frankfurt.

[2] CeraMem Technology Brief December (2005). Available online at http://www.ceramem.com/techbrief.pdf

[3] Kerasep™ Kompact Technical Bulletin (2005), Novasep, NVS-10-2.3.P1.v1 10/05 (2005).

[4] Hyflux CEPAration Technical Bulletin. CeraCep™ (2006). Long lasting Ceramic Hollow Fibre Membranes for Extreme Conditions, SIP 0406-V1.

[5] Cheryan, M., and Rajagopalan, N. (1998). Membrane processing of oily streams. Wastewater treatment and waste reduction. *J. Membr. Sci.* **151,** 13–28.

[6] Bhave, R. R. (1991). Inorganic Membranes Synthesis, Characteristics and Applications. New York: Van Nostrand Reinhold, New York.

[7] Cheryan, M. (1998). Ultrafiltration and Microfiltration Handbook Lancaster, PA: Technomic, Lancaster, PA.

[8] Sondhi, R., and Bhave, R. (2001). Role of backpulsing in fouling minimization in crossflow filtration with ceramic membranes. *J. Membr. Sci* **186,** 41–52.

[9] Finley, J. (2005). Ceramic membranes: A robust filtration alternative. *Filtr. Sep.* **42**(11), 34–37.

[10] Sondhi, R., Bhave, R., and Jung, L. (2003). Applications and benefits of ceramic membranes. *Membr. Technol.* **11,** 5–8.

[11] Voigt, I. (2006). Nanofiltration with Ceramic Membranes, Ceramic World – New Ceramic Spheres 19. *In* "Ceramic Forum International -Yearbook" pp. 63–68. Göller Verlag.

[12] Hilliard Hilco Division Technical Bulletin: Ceramic Membrane Crossflow Liquid Filtration System CMS-3. (2005). THC-500-10/05.

[13] Ciora, R. J., Jr., and Liu, P. K. T. (2003). Ceramic membranes for environmental related applications. *Fluid/Part. Sep. J.* **15**(1), 51–60.

[14] Bolduan, P., and Latz, M. (2003). Filtration by Means of Ceramic Membranes—Practical Examples from the Chemical and Food Industries Germany: atech innovations gmbh, Germany.

[15] Read, D., Rabey, B., Black, S., Glasser, F. P., Grigg, C., and Street, A. (2004). Implementation of a strategy for managing radioactive scale in the China Clay industry. *Miner. Eng.* **17**(2), 293–304.

[16] Novasep commercial information. (2006). Available online at http://www.novasep.com/Download/index.htm

[17] Speedy™ by SpinTek, Rotating Ceramic Ultrafilter Technical Bulletin (2007). Available online at http://www.spintek.com/pdfs/speedy.pdf

[18] Van Clewe, A., Gehrlein, H., and Bottger, D. (2004). Production integrated environmental conservation. *Pakistan Textile J.* freely downloadable at http://www.ptj.com.pk/Web%202004/05-2004/benninger.html
[19] Nyström, M., Nuortila-Jokinen, J. M. K., and Mänttäri, M. J. (2005). Nanofiltration in the pulp and paper industry. *In* "Nanofiltration, Principles and Applications" (A. I. Schäfer, A. G. Fane, and T. D. White, eds.), Elsevier, Oxford, UK.
[20] Keefe, R. J., and Dubbin, D. M. (2005). Benefits using microfiltration for enzyme cell separation. *Chem. Eng. freely downloadable at* http://www.geafiltration.com/filtration_library/specifying_microfilt-ration_systems.pdf
[21] PCI Membranes, Commercial information™ (2006). Available online at http://www.pcimem.com/
[22] Kerasep™ Technical Bulletin (2006) Novasep, NVS-67-2.3.P1.v1 10/06.
[23] Bolduan, P., and Latz, M. (2000). Ceramic membranes and their application in food and beverage processing. *Filtr. Sep.* **37**(3), 36–38.
[24] Noordman, T. R., Berghuis, O. A. E., Mol, M. N. M., Peet, C. J., Muller, J. L. M., Broens, L., and Van Hoof, S. (1999). Membrane filtration for bright beer, an alternative to Kieselguhr filtration. Proceedings of the 27th EBC Congress Cannes, France, pp. 815–822.
[25] Van Hoof, S. C. J. M., Noordman, T., Berghuis, O., Mol, M., Peet, C., and Broens, L. (2000). Membrane filtration for bright beer, an alternative to Kieselguhr. *MBAA Tech. Q.* **37**(2), 273–276.
[26] Fillaudeau, L., and Carrère, H. (2002). Yeast cells, beer composition and mean pore diameter impacts on fouling and retention during cross-flow filtration of beer with ceramic membranes. *J. Memb. Sci.* **196,** 39–57.
[27] Lancrenon, X. (2003). Membrane cross flow filtration for the production of refined sugar at the mill and in the refinery. *Int. Sugar J.* **105**(1257), 390–394. Hamilton Island, Australia.
[28] Kwok, R. J. (1996). Production of super VLC raw sugar in Hawaii. Experience with the new NAP ultrafiltration/softening process. *Int. Sugar J.* **98**(1173), 490–496.
[29] Rojo, U. (1999). Applications of membrane techniques in the dairy industries. *In* "Membrane Applications in the Food & Dairy Industry" (J. Coca and S. Luque, eds.), pp. 231–253. Universidad de Oviedo, Oviedo (Spain), 84-669-0176-1.
[30] Isoflux® Tubular ceramic membranes Technical bulletin. (2006). REV. 15/09/06.
[31] Pall Exekia Membralox GP Technical bulletin. (2006). Available online at http://www.pall.com/pdf/
[32] Niro Filtration Product Bulletin. Cheesemaking By-product Recovery Process. (2005). *Niro GEA Filtration* available online http://www.niroinc.com/html/filtration/membrane-filtration_division_lit.htm
[33] Bhave, R., Jung, G., and Sondhi, R. (2001). Potential savings through ceramic membrane caustic reclaim. *Food Process.* **62**(12), 49–52. Available online at http://www.pall.com/pdf/foo_pdf_CausticRecovery-2001.pdf
[34] Bottino, A., Capannelli, G., Del Borghi, A., and Colombino, M. (2001). Water treatment for drinking purpose: Ceramic microfiltration application. *Desalination* **141,** 75–79.
[35] Adham, S., Gagliardo, P., Boulos, L., Oppenheimer, J., and Trussell, R. (2001). Feasibility of the membrane bioreactor process for water reclamation. *Water Sci. Tech.* **43**(10), 203–209.
[36] Cicek, N. (2003). A review of membrane bioreactors and their potential application in the treatment of agricultural wastewater. *Can. Biosystems Eng.* **45,** 6.37–6.49.

CHAPTER SEVEN

Pervaporation and Gas Separation Using Microporous Membranes

Clara Casado Coterillo,* Ana María Urtiaga Mendia,† *and* Inmaculada Ortiz Uribe†

Contents

1. Introduction

A membrane is a physical barrier allowing selective transport of mass species. Membranes consist of three large groups: biological, organic, (polymeric) and inorganic membranes. Inorganic membranes can be further classified into ceramic and metallic membranes.

The synthesis and characterization of ceramic membranes has been intensely studied due to their potential applications in the domains of gas separation [5–6, 21, 104-105], pervaporation [18–19], reverse osmosis [156] and nanofiltration [62, 68, 152], and the development of membrane reactors using catalytically active or passive inorganic membranes [33, 39–41, 74, 124, 150, 157–158]. Disk and single-tube are the most common geometries of the microporous inorganic membranes, although some researchers [36, 127–129] reported synthesis of microporous membranes in the hollow fiber system, but this has already been presented in Chapter 3, and flat sheet

* European Membrane Institute Twente, University of Twente, 7500AE Enschede, The Netherlands
† Department of Chemical Engineering, Universidad de Cantabria, 39005 Santander (Cantabria), Spain

Membrane Science and Technology, Volume 13
ISSN 0927-5193, DOI: 10.1016/S0927-5193(07)13007-2

geometries ([23], for instance). The inorganic membranes are usually tubular membranes.

Dense metal or oxide tubes are examples of symmetric membranes, which consist of a single layer of a structured material. Single-walled symmetric systems have a considerable thickness for mechanical strength. In order to obtain larger fluxes, the thickness of the separation layers must be reduced as much as possible. This is achieved by using asymmetric membranes which consist of a support with large pores on top of which are layers with gradually decreasing thickness and pore size ([24]; Fig. 7.1).

According to the membrane pore size (or the size of the molecules that can move through the membranes), both organic and inorganic membranes can be divided into microfiltration (MF), ultrafiltration (UF), nanofiltration (NF), reverse osmosis (RO), and gas separation membranes. According to the definition of IUPAC in terms of the material pore size, [105] the MF, UF, and NF membranes correspond, respectively, to macroporous (pore diameter $d_p > 50$ nm), mesoporous (2 nm $< d_p < 50$ nm), and microporous ($0 < d_p < 2$ nm) membranes.

In the previous decades, the research and development of inorganic membranes has been focused on decreasing the pore size, thus broadening their application to gas separation and pervaporation [5, 170]. Microporous inorganic membranes are made as thin films on a multilayer porous support and have a network of connected micropores of 0.5–1 nm diameters. The transport of small molecules in this network is driven by a chemical potential gradient. The separation performance of a membrane is determined by its resistance to the mass transport of a certain species, and its selectivity for this particular species with respect to others. In general, the membrane must have a high flux for the species of interest combined with low fluxes

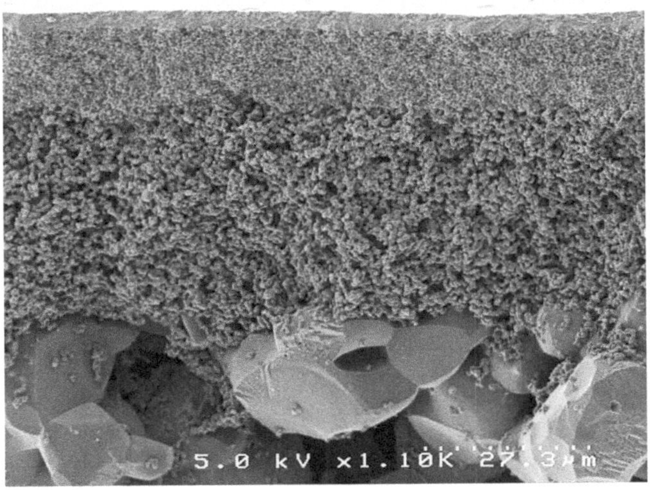

Figure 7.1 Field emission scanning electron microscopic (FE-SEM) picture of the cross section of a commercial alumina tubular support (PALL Exekia, Bazet, France). The average pore size of the support, of the two intermediate layers, and the top γ-Al_2O_3 layer are 10 μm, 0.8 μm, 0.2, μm and 5 nm, respectively.

for other species. Several transport mechanisms can occur in porous media [14, 82]; and in the membrane systems studied here, the dominant transport mechanisms can differ from layer to layer, as a function of the pore diameter [24]. This is discussed in chapter 2 in this book. Viscous (Poiseuille) flow and molecular diffusion are nonselective. Nevertheless, they play an important role in the macroporous substrate(s) supporting the separation layer and can seriously affect the total flow resistance of the membrane system when the microporous membranes contain some mesoporous defects where permeation also takes place. Mesoporous separation layers or supports are frequently in the transient regime between Knudsen diffusion (flow) and molecular diffusion, with large effects on the separation factor (selectivity). In other words, when the microporous membranes are not defect-free, the transport takes place through mesopores as well as micropores, thus more than one mechanism applies.

Normally, transport mechanism in porous membranes is characterized by comparison to Knudsen diffusion or the separation based on different molecular weights of the components of the feed mixture through the membrane.

The properties of gas flow in porous media depend on the ratio of the number of molecule–molecule collisions to that of the molecule–wall collisions. The Knudsen number K_n is a characteristic parameter defined by $K_n = \lambda / d_{\mathrm{p}}$ with λ being the average free path length of the gas molecules and d_{p} the characteristic pore diameter (sometimes the hydraulic pore radius is taken).

Molecular sieving [47, 122, 130, 181] or configurational or micropore diffusion [131, 175] takes place in microporous membranes, where the pore size could discriminate between two molecules due to their shape or size. The membrane acts as a sieve which only allows the permeation of the molecules that can diffuse through the pore, due to their smaller size, and excludes the bigger molecules.

Another mechanism occurring in microporous membranes is capillary condensation or the partial condensation of some components of a gas mixture in the pores with [182] the exclusion of others and subsequent transport of the condensed molecules through the pores.

Surface diffusion consists of the selective adsorption of the more strongly adsorbed components of a mixture onto the pore surface, followed by diffusion of the adsorbed membranes across the pore surface.

The models predicting transport mechanisms through microporous membranes for gas separation applications are basically the same as the ones applied to describe the pervaporation performance. In the case of pervaporation, the coverages on the feed side are close to saturation, making competitive adsorption and molecule interactions more important in multicomponent diffusion [14, 124].

Microporous inorganic membranes are generally composed of a thin film of amorphous silica or other metallic oxide, carbon, or zeolites on a porous support. The selection of the actual membrane composition and supporting structure depends on the final application. The success of this designed choice depends critically on fundamental studies of membrane transport, strength, and operational stability.

This chapter addresses the application of microporous silica and zeolite membranes on gas separation and pervaporation and ends with a review of the transport models available so far.

2. Types of Microporous Membranes

Structurally, microporous inorganic membranes include crystalline (mainly zeolites) and amorphous (mainly metallic oxides or carbon) materials. The membranes are usually prepared as thin films on porous inorganic supports that provide the mechanical strength. The thickness of the microporous film varies from a few tens of nanometers to a few micrometers. In most cases, the microporous membrane film is coated on the surface of the UF (or mesoporous) inorganic membrane layer.

2.1. Zeolite membranes

Zeolites are microporous crystalline aluminosilicates which have uniform, molecular-sized pores. Zeolites can be used as membrane material to separate gaseous or liquid mixtures, as these membranes constitute a special type of microporous ceramic membranes in which the pore size can be modulated up to molecular size [18, 123]. The molecular sieve-type MFI with a pore size of 0.54 nm can separate, for example, mixtures of methanol and methyl tert butyl ether (MTBE) (molecular diameters 0.39 and 0.63 nm, respectively) by molecular sieving. Large pore membranes of the type FAU [133–134] with a pore size of 0.74 nm are typically used for separations based on energetic interactions which influence mixture adsorption equilibrium and mixture diffusion coefficients, for example, pervaporation. Therefore, transport through ideal molecular sieve membranes takes place exclusively through the zeolite pores, but real membranes are usually influenced by non-zeolite pores of low selectivity such as intercrystalline defects or pinholes.

Zeolite membranes are usually prepared via *in situ* hydrothermal synthesis in the liquid phase on ceramic supports [18, 40].

One important feature of zeolite membranes is polarity. The polarity of the zeolite is an important parameter determining its separation properties, and it can be tuned by the Si/Al ratio. Lower Si/Al indicates more polar, hydrophilic membrane [18, 85, 124]. This has been largely studied for its use on pervaporation of different mixtures. In Table 7.5, zeolite membranes are ordered according to increasing Si/Al ratio. The low Si content membranes are very hydrophilic. However, these are not very stable in acid media or at low pH, thus some authors have searched for a means of modifying the surface by increasing the Si/Al ratio, as on ZSM-5 membranes [97] or even silicalite-1 [3, 17, 113–114, 136] to increase stability without losing hydrophilic character. Nevertheless, these membranes provide a lower flux than zeolite A membranes. Although FAU membranes (type X and type Y) [92, 99–100, 133–134] are water selective, their separation factor is somehow inferior to that of zeolite A membranes. This phenomenon might be attributed to the higher Si/Al value and decreasing hydrophilicity of the membrane.

Only a few studies used this property to improve the stability of hydrophilic zeolite membranes for pervaporation at low pH [43, 100, 113].

2.2. Silica-based membranes

Since the beginning of preliminary studies on ceramic membranes, silica has been one of the preferred materials to prepare top layers on mesoporous supports because of its high porosity due to its amorphous structure in a wide temperature range, as well as the possibility of being sintered in dry atmosphere at high temperatures around 400–700 °C, which results in an easier fabrication of microporous silica membranes [10–11, 86, 89].

Several procedures have been proposed for the fabrication of silica membranes: sol-gel method, chemical vapor deposition (CVD) method, surfactant-templated method, explained in detail in Chapter 2. Silica membranes can be produced by CVD or sol-gel. Chemical vapor infiltration (CVI) membranes are produced from the reaction of a gaseous silica precursor, such as tetraethylorthosilicate (TEOS), with the reaction taking place inside the pores of a macro- or mesoporous support [122] or by counterdiffusion CVD using two reactants on either side of the substrate [124].

In the past decade, sol-gel-derived microporous silica membranes have been studied extensively [93, 154–155, 160–161] for preparing ceramic membranes of pore size lower than 1 nm. Sol-gel is most referred to in the literature due to its flexibility and soft condition procedures of fabrication of microporous ceramic membranes from different kinds of inorganic precursors [5, 24–25, 105]. Sol-gel consists of the hydrolysis and condensation of organometallic precursors [130], allowing modification of the chemical composition of the membrane active layer, by mixing various oxides to improve its characteristics [6–10, 30, 156, 178–179]. Other improvements in the synthesis of supported silica membranes have resulted in very thin defect-free layers [5, 22, 27, 80–81, 95].

Sol-gel synthesis processes are divided into three main procedures: the route of the polymeric gel and slip casting [12], the route of the colloidal sol and hot coating [5–11, 80–81, 152–157, 178–182], and the surfactant-templated method [78, 130, 176]. Due to the small pore size, these silica membranes show high fluxes for small molecules like H_2, He, CO_2, and O_2 and high selectivities for these molecules with respect to larger gas molecules, such as SF_6 and various hydrocarbons. Silica membrane preparation has been extensively explained in chapter 2 in this book.

3. Applications

Applications of microporous membranes are molecular separations of gas/liquid mixtures. The molecular sizes of the molecules whose separation is attempted by microporous membranes are mostly less than 1 nm. Therefore, porous membranes are required to have pores of the order of 1 nm or less in order to offer high selectivity. Although it is still doubtful whether it is possible to fabricate a porous membrane whose pore size is controlled between 0.292 (H_2) and 0.368 (N_2) [5], many efforts have been focused on the potential of microporous inorganic membranes for gas separation and pervaporation, as they are discussed below.

3.1. Gas separation

Transport data of membranes can be expressed, according to the IUPAC convention, as "permeance" or pressure normalized flux, that is, the transport flux per unit transmembrane driving force (kmol/m^2 s kPa). Data are given as permeability, that is, thickness normalized permeance (mol m/m^2 s Pa).

It is generally acknowledged that the driving force for transport through microporous ceramic membranes is the chemical potential gradient [13, 18, 20, 90–91]. From experimental data, a preexponential factor for permeance and an apparent activation energy may be obtained. Burggraaf [24–25] mentioned an alternative, semiempirical expression

$$J_t = aP_{\mathrm{d}}^{b}\Delta P \tag{7.1}$$

where the exponent b of the dimensionless pressure P_{d} ($=P/P_{\mathrm{ref}}$) is a measure for the extent of the viscous flow ($b = 1$ for viscous flow and $b = 0$ for Knudsen diffusion). The reference pressure P_{ref} is chosen as a typical or average pressure for the range of applications concerned. This is one way of analyzing the contribution of these transport mechanisms, whose separate description was outlined above.

The selectivity of a membrane, α, is defined as

$$\alpha = \frac{C_i^{\mathrm{P}}/C_j^{\mathrm{P}}}{C_i^{\mathrm{F}}/C_j^{\mathrm{F}}} \tag{7.2}$$

where C^{P} and C^{F} are the mass concentrations of component i or j in the permeate or feed mixture, respectively [24–25].

Table 7.1 summarizes microporous inorganic membranes and gas separation applications reported by several research groups in the last decade. Some microporous membranes have also shown good liquid/vapor properties.

Nowadays, many papers can be found in the literature regarding inorganic membranes and most of them are focused on gas separation applications (Table 7.1). Lately, two trends can be distinguished from the rest: hydrogen purification and carbon dioxide removal. In the first case, due to the fact that 80% of the energy consumed in the world comes from scarce resources of fossil fuels, many papers are either dedicated to H_2 separation [11, 60, 80–81, 94–95] on the light of H_2 as an energy source that produces only water as by-product [146]. In the second, many efforts have been directed to the potential of membrane technology for CO_2 separation [20] due the need to remove CO_2 from flue gas streams or capture it ["CO_2 capture project" (http://www.co2captureproject.com)]. The first one requires a high selectivity and a separation mainly based on molecular sieving [80–81]. The second one belongs to the cases where higher selectivity than expected could be attained by the interaction of permeating molecules and atoms or firmly adsorbed molecules on the pore surface. CVI-silica membranes show high permselectivities toward H_2; values as high as 3000 have been measured for H_2/N_2. However, their resistance for mass transport is high as well

Table 7.1 Overview of microporous inorganic membranes for gas permeation and separation, including synthesis procedure

Membrane material	Synthesis method	Permeation/ separation data available (gas phase)	References
SiO_2	Polymeric sol-gel	He, H_2, CO_2, CH_4, C_3H_6	[161]
SiO_2	Colloidal/polymeric sol-gel	He, Ar, N_2, CH_4, C_2H_4, C_2H_6, C_3H_6, C_3H_8	[5]
SiO_2	Polymeric sol-gel	He, H_2, CO_2, N_2, CH_4, C_3H_6, *i*-C_4H_{10}	[87]
SiO_2	Surfactant templated sol-gel	He, CO_2, N_2, CH_4, SF_6	[130]
SiO_2	Polymeric sol-gel	H_2, CO_2, O_2, N_2, CH_4	[171]
SiO_2	Polymeric sol-gel	H_2-CH_4	[91]
SiO_2	Polymeric/ templated sol-gel	H_2, N_2, CH_4, CO, CO_2	[152]
SiO_2	Colloidal sol-gel	H_2, CO_2, N_2, CH_4, C_2H_6, C_3H_8	[10]
SiO_2	Polymeric sol-gel	H_2, C_3H_8	[134]
SiO_2	Polymeric sol-gel	He, H_2, CO_2, O_2, CH_4, N_2	[47]
SiO_2	Polymeric sol-gel	H_2, N_2	[98]
SiO_2	Polymeric, colloidal sol-gel	CO_2, N_2	[181]
SiO_2	CVD[a]	CO_2, N_2	[42]
SiO_2	Thermal decomposition of TEOS[b]	H_2, N_2, CO, CO_2	[23]
SiC	CVD/CVI	He, N_2	[38]
SiO_2-ZrO_2	Colloidal sol-gel	He, H_2, CO_2, N_2, O_2, Vapor/liquid phase	[153]
ZrO_2	Polymeric sol-gel	H_2	[67]
TiO_2	Polymeric sol-gel	CO_2, N_2	[2]
SiO_2-ZSM-5	Hydrothermal	CO_2/He/H_2/O_2/ SF_6/CO	[17]
MFI	Hydrothermal	H_2, propane	[79]
ZSM-5	Hydrothermal	CO_2, N_2	[140]
H-ZSM-5	Alkali-free hydrothermal	N_2, CO_2, CH_4, butane	[59]

(continued)

Table 7.1 (*Continued*)

Membrane material	Synthesis method	Permeation/ separation data available (gas phase)	References
Silicalite-1	Hydrothermal	CH_4, C_2H_6, C_3H_8, n-C_4H_{10}, i-C_4H_{10}, n-C_6H_{12}, ...	[87]
Zeolite T	Hydrothermal	He, H_2, CO_2, O_2, N_2, CH_4, C_3H_8	[43]
Zeolite NaA and NaY	Hydrothermal	n-C_4H_{10}, i-C_4H_{10}	[97]
Zeolite LTA	Microwave	H_2, N_2, C_3H_8, n-C_4H_{10}	[101]
Zeolite NaA	Hydrothermal	H_2, N_2, SF_6	[183]
Aluminophosphate		H_2, C_3H_8	[141]

[a] CVD, chemical vapor deposition.
[b] TEOS, tetraethoxysilane, common inorganic precursor of silica-based microporous membranes.

($F_{H_2} = 2 - 4 \times 10^{-8}$mol/m^2 s Pa at 200 °C) due to the presence of nearly dense silica plugs inside the pores of the supporting structure [14].

Regarding CO_2 applications, Table 7.2 collects the inorganic membranes reported to separate CO_2 from N_2 at different temperatures. This is one of the popular applications nowadays as the effect of CO_2 on the green house effect is generally acknowledged and these gases are both present in larger quantities in the flue gases from power generation plants [1, 20]. Now, this is a separation widely studied at low temperature and industrially implemented. The interest in inorganic membranes arises from the potential of performing this separation at high temperatures, thus avoiding the high energy costs of cooling down the effluent and heating up again for reuse. As shown in Table 7.2, Gu *et al.* [66] managed to separate CO_2 from N_2 using FAU membranes, but the separation factor at 200 °C was not higher than 4.26. At high temperature, the transport mechanism that allows this separation is not molecular sieving due to the same order of magnitude of the kinetic diameters of CO_2 and N_2, but is based on the polarity and affinity of CO_2 on the material of the active separation layer of the membrane. Thus for N_2 the transport through the inorganic membrane might still be described by Knudsen as diffusion or molecular sieving, but for CO_2 the transport is governed by the adsorption on the membrane surface. Thus, Table 7.2 collects data on the transport mechanisms as studied by several research groups and their values of permeability and selectivity for both gases. Since the mechanism in most commercially available membranes is Knudsen diffusion, which gives a very low selectivity for CO_2/N_2, many researchers focused on modifying the surface of the commercial membranes to improve their separation performance. Thus, for instance, in Table 7.2, Abidi *et al.* [2] claimed that grafting the alumina or TiO_2 membranes with fluorinated triethoxysilane increased the CO_2 selectivity. These adsorption differences at high

Table 7.2 Data on CO_2/N_2 separation using ceramic membranes and transport mechanism governing the separation

Membrane material	Temperature (K)	ΔP (atm)	Permeance (mol/m^2 s Pa) CO_2	Permeance (mol/m^2 s Pa) N_2	S.F. CO_2/N_2	Transport mechanism	References
γ-Alumina (PALL)	293	0.73	9×10^{-15}	2×10^{-15}	4.50	Knudsen	[2]
		2.22	9×10^{-15}	2.8×10^{-15}	3.21		
TiO_2—fluorinated TEOS grafted		0.73	2×10^{-13}	4×10^{-14}	5.00	Solution diffusion	
		2.22	2.4×10^{-13}	1.4×10^{-13}	1.71		
Silica ZSM-5	298	1.09	5.9×10^{-7}	2×10^{-7}	2.95	Surface diffusion	[17]
Zeolite FAU	473	0.86	6×10^{-8}	2×10^{-8}	4.60	Surface diffusivity	[67]
Zeolite FAU/NaY	313	1	2×10^{-6}	1×10^{-7}	20	Selective adsorption + surface diffusion	[92]
Zeolite ZSM-5/silica-modified surface	298	0.4	3.6×10^{-8}	6.63×10^{-10}	54	CO_2 (adsorption/surface diffusion + MP), N_2 (micropore diffusion)	[140]
Si-NH_2	298		3×10^{-7}		80	Molecular sieving, selective adsorption	[177]

temperature and high pressure were also used by Li *et al.* [96] to attempt the separation of CO_2 and CH_4, which is also the cause of energy inefficiency and corrosion problems in natural gas. These differences among transport mechanisms have been recently used by other groups in order to obtain enriched oxygen by perovskite membranes [51]. However, selectivity is seen to decrease at higher temperature, where the adsorption of CO_2 or O_2 diminishes. Table 7.2 only collects the data for CO_2/N_2 separation for clarity purposes, however many efforts have also been focused on the separation of CO_2 from CH_4, also important in flue gas applications [96].

Other researchers chose to rely on the molecular sieving mechanism, but since the molecular size of CO_2 and N_2 is so similar, the pore size distribution has to be carefully tuned. Xomeritakis *et al.* [176–177] attempted this by coating several layers on the alumina support: first a surfactant-templated silica layer that covers the defects or pinholes of the support, and then successive polymeric silica were spin coated in order to reduce the pore size of the membrane to achieve the desired CO_2/N_2 separation.

Molecular sieving is also the common mechanism dominating the transport in the systems focused on H_2 removal from gas mixtures. In Table 7.3 several membranes reported for H_2 separation from various gas mixtures are presented. Zeolite membranes attempting H_2 separation from other gases [98] should be carefully tuned in order to avoid intercrystalline pores through which the small H_2 molecule could pass. High separation factors can be obtained using zeolite membranes: such as ZSM-5 [98]. Other materials such as aluminophosphates have been recently investigated [141] where the calcination temperature of the membrane is most important factor controlling the performance process, as the layer is grown *in situ* on the support. These researchers reported aluminophosphate microporous membranes where mesopores were excluded, thus favoring the preferential separation of H_2 from propane. Zeolite-MFI membranes have also been reported for the separation of difficult mixtures like *n/i*-C_4H_{10}, where one of the components of the mixture is supposed to block the other one and thus improve the separation factor of the membrane [118].

These membranes have only been reported to operate at low temperatures. For high temperature applications, silica-based membranes are most commonly found in literature, for H_2 separation. Sol-gel processes allow fine-tuning of the microporous layer with careful consideration of the coating solution concentration and temperature, coating procedure, drying, and sintering [86, 94, 167]. Successful silica and silica-zirconia membranes for H_2 separation in steam reforming reactors at 500 °C have been reported [80–81]. Recently, metal-doped silica membranes have also developed with high flux and separation factor and high stability in water vapor [81]. Some of their results are also included in Table 7.3.

As seen in Table 7.3, sol-gel membranes provide in general higher fluxes than zeolite membranes, a reason why may be the possibility of forming a very thin uniform layer without defects on a smooth porous surface of the mesoporous support of 3–5 nm pore size [105].

Table 7.3 Data on H_2 separation from gas mixtures using microporous inorganic membranes [107]

Membrane material	Temperature (K)	ΔP (atm)	Mixture	S.F.	Permeance (mol/m² s Pa)	References
SiO_2	473	3	H_2/C_3H_6	156	26×10^{-7}	[160]
	543	3	H_2/N_2	185	0.5×10^{-7}	[86]
	473	3	H_2/CH_4	30	20×10^{-7}	[94]
	423	1	H_2/N_2	76	10^{-8}	[167]
	298			844	10^{-8}	
	773	1	H_2/CO_2	210	5.8×10^{-7}	[153]
SiO_2/stainless steel	564	4	H_2/N_2	19	394×10^{-7}	[23]
			H_2/CO_2	35	495×10^{-7}	
Ni-doped SiO_2	773	0.90 (vapor)	H_2/N_2	400	46×10^{-7}	[80]
Zeolite LTA	293	0.20	H_2/C_3H_8	9.17	1.71×10^{-7}	[101]
Zeolite ZSM-5	293	1	$H_2/n\text{-}C_4H_{10}$	129	24×10^{-7}	[97]
			$H_2/i\text{-}C4H_{10}$	1548		
$Al_{100}P_{60}O_z$	303	2	H_2/C_3H_8	15.5	1.9×10^{-7}	[141]

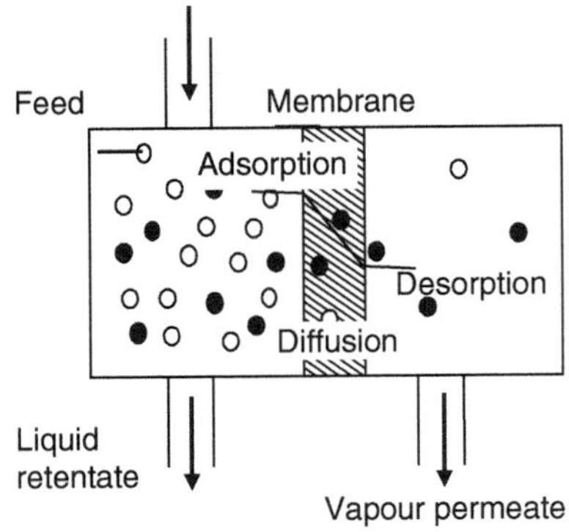

Figure 7.2 Schema of the pervaporation process.

3.2. Pervaporation

Pervaporation is a membrane separation process where the components of a liquid mixture (feed) are separated by placing the mixture in contact with one side of a membrane and the permeated product (permeate) is removed as a low-pressure vapor from the other side [13, 18, 48, 52, 55, 76, 83]. Figure 7.2 shows the principle of pervaporation. The liquid feed mixture to be separated flows along one side of the membrane while the various feed components are permeating in and through the membrane at different rates. Therefore, the liquid retentate leaving the unit on the same side of the membrane as the feed enters is depleted in the components permeating preferentially. Consequently, the vapor permeate is enriched in the preferentially permeating component.

The use of pervaporation as an alternative or complement to conventional processes, such as distillation for difficult separations, has been suggested for a number of cases [4, 56–57, 77, 108–109, 111–112, 159]. Although dehydration has been the first and most widely studied or implemented application, increasing attention has been directed toward the separation of organic compounds from aqueous mixtures [66] and organic–organic separations [35, 53, 63–65, 106–107, 143, 148], using polymeric membranes.

The performance of a pervaporation membrane is characterized by three main features: membrane productivity, selectivity, and stability.

The *productivity* of a membrane is given by the permeate flux, that is, the amount of the permeating component passing through the unit membrane area per unit time, commonly measured as $kg/m^2\,h$.

The permeate total flux is calculated by the following expression:

$$J = \frac{m}{A\Delta t} \tag{7.3}$$

where m is the collected permeate weight, kg; A, the membrane effective area; and Δt, the time interval during which the permeate simple was collected.

Water flux, therefore, is given by the following expression:

$$J_{\text{water}} = JC_{\text{water,p}}(\text{wt\%}) \tag{7.4}$$

The *selectivity* is quantified by parameter α, defined as in gas separation according to Eq. (7.2) [24]. Several authors call this selectivity [61, 166, 175] and many others, separation factor [4, 21, 41, 43, 50, 56–58, 69, 73, 97, 109, 117].

Regarding pervaporation, the physical meaning of α is not so clear as in gas separation, as the components of the feed mixture do not permeate independently, but the flux of one affects the flux of the other; sometimes the coupling of fluxes is very strong and one component can end up blocking the other components in the mixture [133, 183] and the flux coupling, the concentration-dependent diffusion coefficient, together with the equilibrium on the membrane surface should be considered.

The stability of the membrane consists of its ability to maintain the same permeability and selectivity for long periods of time. Given the extensive time requirements of stability tests, there are but few reports [10, 27] published related to the characterization of pervaporation membranes in transient conditions.

The best-known examples of microporous inorganic membranes for dehydration of organic compounds are NaA-type zeolite membranes [32, 85, 117, 126, 163], but amorphous silica [30, 91, 166–167], mixed oxides like silica with zirconia and titania [8–10, 162, 180], and X- and Y-type zeolites [97, 126] have also been investigated.

In Table 7.4, some pervaporation literature data for silica and mixed metal oxide membranes, such as silica-zirconia, prepared by the different routes of sol-gel procedure, regarding the dehydration of organics, are summarized.

Following Table 7.4, an overview of zeolite membranes for pervaporation is presented in Table 7.5. These are ordered according to increasing Si/Al ratio. Besides these data, relevant process and membrane parameters are also given, such as feed composition, selectivity, flux, as well as some intrinsic membrane properties such as top layer thickness and average pore size.

Sol-gel enables obtaining separation layers much thinner than in the case of zeolite membranes shown in Table 7.5, that is, between 8 and 4000 nm, which in general gives elevated permeate fluxes for the sol-gel-derived membranes, such as 9 kg/m^2 h at 75 °C on the separation of water/tetrahydrofuran [6].

Only [29] reported a silica membrane with a separation factor value typical of zeolite membranes, for the separation of water/butanol at 95 °C, and a flux of 4 kg/m^2 h.

On the other hand, zeolite membranes, because of their intrinsic characteristics, such as a narrow pore size distribution, showed higher separation factors in the dehydration of alcohols [126, 139], and hydrophobic zeolite membranes, such as silicalite-1, have been successfully used to remove organic components from water [109], despite the smaller size of the water molecule with respect to organic molecules. Flux and selectivity data are also collected in Table 7.5. These data describe the functional characterization of zeolite membranes on pervaporation.

For zeolite membranes, the larger the Si/Al ratio, the lower is the water flux through the membrane, which turns in general more hydrophobic. The separation factors are therefore one to three orders lower in hydrophobic than in dehydration processes.

The transport mechanism in pervaporation is a combination of adsorption and diffusion. This applies both to sol-gel [166–167] and zeolite membranes [139].

Table 7.4 Microporous inorganic membranes for pervaporation and synthesis method employed for their preparation

Membrane	Sol-gel method	Active layer thickness (nm)	Pore size diameter (nm)	A/B separation	T (°C)	A in feed (wt %)	Flux (kg/m^2 h)	Separation factor (–)	References
Flat SiO_2/γ–	Sol-gel, *in situ* TEOS hydrolysis	8	0.55	Water/Acetone	25	5[a]	0.04	1200	[28]
				Water/Isopropanol			0.1	1000	
Al_2O_3		2000–4000	5	Water/Tetrahydrofuran	25	5	0.05	9860	[89]
					60		0.1	1275	
Tubular SiO_2	Sol-gel, polymeric route from TMOS and *templates*	NA[b]	0.6	Ethanol/Cyclohexane	20	10	NA	40 wt % A in the permeate	[45]
		300	0.9	Water/Ethanol	25	10	1	17	[35]
SiO_2 [ECN, Sulzer]	Sol-gel, polymeric route	150–200	0.4	Water/Acetone	50	10	0.75	33	[166]
				Water/Isopropanol	80	4.5	1.86	1150	
				Water/Isopropanol	70	7	2.8	690	[167]
				Water/Butanol					
		100–200	0.4	Water/Isopropanol	80	10	2.8	800	[144–145]
		100–200	0.4	Water/Isopropanol	70	8.2	1.9	53	[56–57]
				Water/Butanol	60	10	3.5	144	
		60–150	0.4	Water/Butanol	95	5	4	10,000	[27]
SiO_2 (Sulzer)	Sol-gel, polymeric route	150–200	0.4	Water/Isopropanol	75	10	0.8	714	[162]
(Pervatech)		100					4.8	480	

SiO_2 (Pervatech)	Sol-gel, polymeric or colloidal route	100	0.3–0.55	Water/Methanol	60	15	1.2	12	[50]
				Water/DMF			0.6	60	
				Water/Dioxane			2.5	125	
	(not disclosed)	100	0.3–0.55	Water/Isopropanol	70	5	1	100	[44]
Flat SiO_2	Sol-gel, polymeric or colloidal route	NA	NA	Methanol/MTBE	35	9	0.41	19	[61]
SiO_2 hollow fiber	Sol-gel, polymeric route and *dip coating*	20–60	≈0.36	Water/Butanol	80	5	2–3	500–1200	[128–129]
				Water/DMF	98		0.7–0.36	17–25	
SiO_2-ZrO_2	Sol-gel, colloidal sol route	<1000	0.5–0.6	Water/Isopropanol	n.b.p.[c]	10	9	2000	[7–9]
				Water/ Tetrahydrofuran			7	2000	
		<500	0.5–0.6	Water/Isopropanol	n.b.p.[c]	10	9	500	[8]
		<1000	0.5–0.6	Water/Acetone	55	10	2	800	[179]
				Water/Isopropanol	75		4	1800	
				Water/Dioxane	75		6	4000	
SiO_2-ZrO_2	Sol-gel, polymeric gel route	2000	2–2.5	Water/Isopropanol	80	10	0.86	300	[137]
				Water/2-butanol		5	0.61	360	

(continued)

Table 7.4 (*Continued*)

Membrane	Sol-gel method	Active layer thickness (nm)	Pore size diameter (nm)	A/B separation	T (°C)	A in feed (wt %)	Flux (kg/m^2 h)	Separation factor (–)	References
SiO_2-TiO_2	Sol-gel, colloidal sol route			Water/Isopropanol	80	10	0.78	210	
				Water/2-butanol		5	1.16	190	
SiO_2-ZrO_2	Sol-gel, colloidal gel route	500	≤0.5	Water/Isopropanol	75	10	3.65	1534	[162]

[a] Concentration expressed as vol %.
[b] NA, Not available in the cited reference.
[c] n.b.p., Normal boiling point.

Table 7.5 Zeolite pervaporation membranes and functional characteristics at 10 wt % preferentially permeating component (A)

Zeolite layer	Si/Al	Active layer thickness (μm)	Pore size (nm)	A/B Separation	A in feed (wt %)	T (°C)	Flux (kg/m² h)	Separation factor (–)
Zeolite NaA (LTA) [57]	1	5–10	0.42	Water/*tert*-butanol	10	60	7	16000
Zeolite NaA [126]	1	30		Water/Methanol	10	75	0.57	2100
				Water/Ethanol		75	2.15	10000
				Water/IPA		75	1.76	10000
				Water/Acetone		50	0.91	5600
				Water/Dioxane		60	1.87	>9000
				Water/DMF		60	0.95	>9000
Zeolite NaA [139]	1	30		Water/IPA	0	60	2.00	NA
					100		0.21	
Zeolite NaA [19]	1	7		Water/Ethanol	10	Room temperature	0.1×10^{-2a}	3040
Zeolite NaA [74]	1	NA		Water/Ethanol	10	50	0.12	20000
				Water/IPA		70	0.14	5000
Zeolite NaA [117]	1	10		Water/Ethanol	5	50	0.396	4800

(continued)

Table 7.5 *(Continued)*

Zeolite layer	Si/Al	Active layer thickness (μm)	Pore size (nm)	A/B Separation	A in feed (wt %)	T (°C)	Flux ($kg/m^2 h$)	Separation factor (–)
Zeolite NaA [163]	1	5–10		Water/Acetone	3	50	0.31	50
				Water/Tetrahydrofuran	7	55	0.98	1240
Zeolite X (FAU) [98]	1.4	25–50	0.74	1,3-propanediol/glycerol/water	Water with 100 g Org/litre	35	2.6	>200
					5 g Org/litre		3.6	42
Zeolite Y (FAU) [97]	1.5–3	20–35	0.74	Water/THF	6.7	60	2.4	230
Zeolite T [150]	3.6	NA	0.36 × 0.51	Water/Ethanol	10	75	0.8	830
							0.60	4400
Zeolite T [43]	10	10–20		Water/Methanol	10	50	0.16	37
				Water/Ethanol		75	1.10	900
				Water/IPA		75	2.20	8900
				Water/Acetone		50	0.45	880
				Water/DMF		60	0.45	2900
Mordenite (MOR) [32]	5.1	40–70	0.65 × 0.7	Water/Ethanol	10	90	0.023	80
			0.26 × 0.57			122	0.08	40
MOR [100]	10.2	< 1		Water/IPA	10	75	0.658	4832
				Water/Acetic acid	50	80	0.614	299

MOR [100]	1.5–3	20–35		Water/THF	6.7	60	2.4	230
ZSM-5 [97]	100	20–35	0.51 × 0.55	Acetone/Water	5	30	0.29	160
			0.56 × 0.53			50	0.68	330
ZSM-5 [99]	600	3–5		Water/IPA	10	30	0.15	NA
ZSM-5 [54]	600	NA		C_6 isomers		27	12[a]	10
						77	49[a]	24
Silicalite-1 [109]	>1300	NA	0.51 × 0.55	Methanol/Water	16.5	32	1.3	14.3
			0.56 × 0.53					
				Acetone/Water	0.8	30	0.2	255
Silicalite-1 [113]	Al = 0	32		Water/Acetone	20 kmol/m^3	45	0.06	NA
				Water/Ethanol				
				Water/Acetic acid				

[a] Flux expressed as mol/m^2 h.

Diffusion favors water permeation because the organic components to be separated are in general larger than water molecules. The hydrophobic zeolites overcome this problem as a result of the preferential adsorption of organic compounds. Some organic mixtures also exhibit differences in adsorption and diffusion [46, 53, 102] thus allowing the separation.

To sum up, sol-gel membranes give in general higher fluxes and lower separation factors than zeolite membranes, which provide higher separation factors because of their intrinsic characteristics, such as the narrow pore size distribution, allowing discriminating the large molecules in favor of the small ones in the mixture. Recently, several groups have developed methylated-silica [27] membranes or silica doped with stable metal oxides as zirconia or titania [7–11, 31, 137, 178] providing high flux and selectivity of the same order of magnitude as zeolite membranes.

4. Modeling of Mass Transport Through Microporous Membranes

For a proper understanding and prediction of separation processes in membranes, an accurate description of mass transport phenomena occurring in multicomponent mixtures is needed. The separation performance of a membrane is determined by its resistance to mass transport of a certain species and its selectivity for this particular species with respect to others. In general, the membrane must have high flux for the species of interest combined with low fluxes for the other species. It is also generally not sufficient to only take into account the microporous inorganic material, or the interactions of molecules with the mesoporous intermediate layer and the macroporous support, as well as with the other mobile species. Theoretical descriptions of multicomponent mass transport are abundant in the literature, but are still the subject of much confusion [172].

The limited definition of state-of-the-art membranes justifies the use of ideal gas thermodynamics for gases and Langmuir thermodynamics for surface adsorption, for example. Surface diffusion has been reported as the main mass transfer mechanism governing the pervaporation performance of zeolite A membranes [139]. Multicomponent permeation through the microporous inorganic membranes has been limited to pervaporation [97, 101–102] and the permeation of gas mixtures containing more than two components are limited to microporous membranes, although practical applications of the ceramic membranes most likely involve multicomponent systems [14].

Microporous inorganic membranes reviewed here include two groups of membranes with distinguishable macroscopic and mesoporous structures: amorphous and crystalline membranes. The former, mainly amorphous silica-based membranes prepared by sol-gel procedures, are macroscopically uniform across the microporous film. The latter group, zeolite membranes, have a structure defined by packing small microporous crystallites with an intercrystalline boundary (or grain boundary or intercrystalline region). Although a strict theoretical treatment through the second group of membranes is more complex, requiring a consideration of diffusion and adsorption in both the intracrystalline and intercrystalline regions, previous studies on the theories of transport through microporous membranes tended to assume a macroscopically homogeneous microporous film, thus treating the polycrystalline film as a single crystal film [105] in order to simplify the calculations. Recently, a different approach [151] has been proposed to describe mass transfer through pervaporation zeolite A membranes by treating intercrystalline and intracrystalline zeolite pores indifferently, as in amorphous sol-gel membranes [22].

Description of the transport through microporous amorphous silica and zeolite membranes has been attempted for meso- and macroporous membranes based on the concepts of Knudsen diffusion (gases only), viscous flow, or Maxwell–Stefan (MS) multicomponent transport (Dusty Gas Model for gases). For gas permeation through microporous membranes, these mechanisms do not apply and the transport is dominated by microporous (configurational) diffusion [93, 182]. However, if a microporous membrane contains defects or pinholes larger than 2 nm, the overall permeance and separation properties are determined by both microporous and macro/mesoporous transport mechanisms.

For single gases there are several transport mechanisms available, depending on the pore diameter distribution and/or the temperature–pressure combination, the contribution of each of these mechanisms to the total transport must be quantified.

In gas mixtures, the permeation and selectivity of components is only identical to that of single gases under special conditions (high temperature and low pressure). This difference is of importance in the transition region between molecular diffusion (Poiseuille flow) and Knudsen diffusion and in that of Knudsen and configurational diffusion. In multicomponent gas mixtures, general descriptions make use of MS equations and, for example, the extended Dusty Gas Model. For binary gases these more complicated models converge to Fickian-type equations and relatively easy-to-obtain solutions for permeation (and thus for ideal) separation factors [14]. There are several works studying mass transfer of gases through amorphous silica [93, 182] and zeolite [21] membranes.

The transport descriptions as mentioned are near equilibrium approaches that incorporate semiempirical expressions for the chemical potential of the permeating component. Some aspects of pervaporation separations are similar to those of gas-phase separations using zeolite membranes, but feed-side coverage is close to saturation during pervaporation, making competitive adsorption and molecule interactions more important in multicomponent diffusion [124].

A phenomenological approach considers that the permeation of a gas or liquid through a microporous membrane should consider five consecutive steps [65]: (1) mass transfer (diffusion) of a component from the bulk of the feed to the membrane surface; (2) partition of the component between the feed and the membrane, that is, mass transfer across the surface in which the molecules change from a fluid state to a state within the porous material; (3) transport (diffusion) of the component across the microporous membrane; (4) desorption of the component on the permeate side of the membrane; and (5) transfer of the component from the membrane surface to the permeate bulk. The surface reaction (sorption) and diffusion steps mentioned above play an important role in transport through microporous membranes. Their relative importance depends on membrane material characteristics such as pore size, sorption strength, and fluid properties such as molecular size, shape, and concentration.

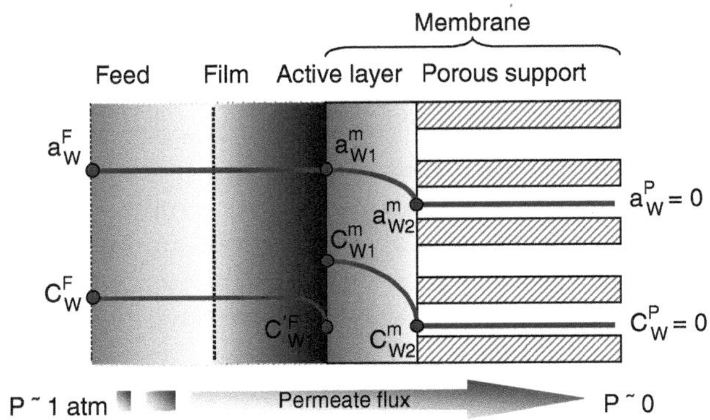

Figure 7.3 Transport through composite pervaporation membranes. Simplified solution–diffusion model [127].

Nevertheless, a simplified approach accounting for steps (2–4) is usually undertaken [127]; a more simplified model that accounts only for the membrane mass transport, that is, steps (2–4), has been widely used [70–71]. This simplified approach for pervaporation is drawn in the schema shown in Fig. 7.3.

The primary theory used for the description of pervaporation performance through polymeric membranes is the solution–diffusion model [52]. This has been recently applied to inorganic membranes for dehydration of organic solvents [18, 22, 127, 137, 151], considering that the molecules in the feed are adsorbed on the membrane surface, and then diffuse along the pores according to the chemical potential gradient between both sides of the membrane.

Membrane transport modeling for pervaporation of dilute organic–water mixtures have been studied for polymeric membranes, including the Fickian approach (both the classic and the thermodynamic), the thermodynamics of irreversible processes and the MS equation. Compared to Fick's laws, the lack of simplicity of the last two transport models is the main drawback for the development of simple flux equations incorporating adequate design parameters. The classic Fick's law is typically referred to in the literature simply as Fick's law and can be considered as a simplified form of the thermodynamic Fick's law, but that simplification is only applicable to binary systems exhibiting thermodynamic ideality.

The thermodynamic Fick's law considers the flux of a species being proportional to the gradient of its chemical potential across the membrane, instead of the gradient of its concentration [115]. The chemical potential is related to the solute activity, which can be determined by a suitable thermodynamic method.

The thermodynamics of irreversible processes have been used to describe coupled diffusion of species and, consequently, coupled fluxes through pervaporation membranes. This requires a lot of experimental data. The MS model [22, 88, 139, 168] combines the solution–diffusion model and the influence of coupled diffusion based on the thermodynamics of irreversible processes. It considers that the driving force, the gradient of the chemical potential, and a friction force that acts on every component moving through a medium are balanced with the friction opposed by the medium. This approach has been generally recognized as accurately describing mass transfer across pervaporation membranes because it takes into account the thermodynamic nonidealities and the influence of external force fields. Due to the complex nature of MS equations, however, many authors continue to use simplified models based on the thermodynamic Fick's law and the solution–diffusion model.

The generalized MS theory, proposed by [88] and van den Broeke *et al.* [21], provides an accurate description of multicomponent transport in microporous materials, as the Dusty Gas model is used for meso- and macroporous materials, and has the general form

$$-\frac{1}{RT}\frac{d\mu_i}{RT} = \frac{x_j}{Đ_{ij}}\left(v_i - v_j\right) + \frac{x_M}{Đ_{iM}}\left(v_i - v_M\right) \tag{7.5}$$

where R is the gas constant (J/mol K), T the temperature (K), μ_i the chemical potential gradient of species i, z is the direction of mass transfer, x_j the molar fraction of j in the adsorbed phase (membrane), $Đ_{ij}$ MS micropore diffusivity between i and j (m^2/s), $Đ_{iM}$ MS micropore diffusivity in the membrane (m^2/s), and v_i and v_j the velocities of i and j in the membrane (m/s), respectively.

Equation (7.5) may be simplified by including the concentration inside the membrane (unknown) x_M into the membrane diffusion coefficient, $Đ_{iM}$:

$$-\frac{1}{RT}\frac{Đ\mu_i}{RT} = \frac{x_j}{Đ_{ij}}\left(v_i - v_j\right) + \frac{1}{Đ_{iM}}v_i \tag{7.6}$$

Assuming that the components are transferred as individual species via activated or surface diffusion in the gas phase [93], the driving force can be expressed as a function of the activity or partial pressure, instead

$$\mu_i = \mu_i^0 + \ln p_i \tag{7.7}$$

where μ_i^0 and p_i are the standard chemical potential and the partial pressure of i. The linear velocity v_i may be expressed as the ratio J_i/c_i, thus Eq. (7.7) is converted to

$$-\frac{1}{p_i}\frac{dp_i}{RT} = \frac{x_j}{Đ_{ij}}\left(\frac{J_i}{c_i} - \frac{J_j}{c_j}\right) + \frac{1}{Đ_{iM}}\frac{J_i}{c_i} \tag{7.8}$$

Verkerk *et al.* [168] even reduced the flux equation of the main permeating component, which on pervaporation is generally water, to

$$J_w = -\frac{Đ_{SW}Đ_{wM}}{Đ_{SW} + x_{aS}Đ_{wM}}\frac{c_w}{p_w}\frac{dp_w}{dz} \tag{7.9}$$

Taking into account the work of Wolf and Schlündler [174] on the adsorption of water/isopropanol mixtures onto silica gel and silica-coated ceramics, which revealed linear adsorption isotherms of water onto supported silica, Verkerk *et al.* [168] concluded that preferential transport is not influenced by adsorption. Thus, they considered only diffusive transport, where the simplest case is that where $Đ'_{iM}$ is not dependent on concentration, in order to integrate [Eq. (7.9)] over the entire membrane active layer thickness. This led to a linear relationship between the flux and the partial pressure difference:

$$J_w = -\frac{Đ_{SW}Đ_{wM}}{Đ_{SW} + x_s Đ_{wM}} \frac{A_w}{L} (p_w^* - p_w^0) \tag{7.10}$$

This approach is similar to the well-known model proposed by Wijmans and Baker [173] for the description of pervaporation through polymeric membranes. This model has also been successfully applied to the description of solvent dehydration and organic–organic separations using ultrathin silica membranes by Elshof *et al.* [50] and Sommer and Melin [144–145] and for the transport of pure components by Bettens *et al.* [15]. From experimental data, a preexponential factor for permeance and an apparent activation energy are obtained. The advantage of this modeling approach is its simplicity compared to models based on the MS or Onsager formulations for mass transport. For the latter two approaches, detailed information on the material properties and the adsorption behavior of the components in the material must be known, as clearly outlined by Benes [14] and Verweij [169–170].

One assumption underlying the use of partial pressure differences as the driving force is that the adsorption of the components from the bulk feed into the membrane can be described by Henry adsorption, as shown by Elshof *et al.* [50]. Bettens *et al.* [16] found that this also applies for pure component pervaporation through a silica membrane. They excluded any contribution of convective transport since their experimental data did not show any correlation with the viscosities. Bowen *et al.* [18] assumed saturated monolayer coverage at the feed side of a zeolite membrane during pervaporation, that is, limiting case of gas separation as mentioned before. These authors further stated that the application of modeling as proposed by Wijmans and Baker [173] is accurate if the (Fickian) diffusion coefficients are independent of concentration and the isotherms are linear.

When the temperature of the gas is such that adsorption on pore walls is important, experimental results show that the preceding laws for gaseous flow are no longer valid. This mechanism is rather complicated and has been treated in detail by other authors [178–179].

For relatively low surface concentrations, the surface flux J_s for a single gas is generally described by the two-dimensional Fick's law:

$$J_s = -\frac{\varepsilon}{\tau} D_s \frac{dC_s}{dz} \tag{7.11}$$

where C_s is the surface concentration (mol/m^2). Expressed in terms of directly measurable parameters, this gives

$$J_s = -\rho(1-\varepsilon)\left(\frac{\varepsilon}{\tau}\right)D_s\frac{dq}{dz} \tag{7.12}$$

where $c_s = q\rho(1-\varepsilon)$ and $\rho(1-\varepsilon)$ is the density of the porous material. To demonstrate the influence of the pore size on the magnitude of the surface flow, it is considered that

$$q = \theta_s S_w c_{sat} \tag{7.13}$$

and

$$S_w = 2\frac{\varepsilon}{\rho\bar{r}} \tag{7.14}$$

where θ_s is the occupancy, defined as the mole fraction occupied by adsorption relative to a monolayer with sorption capacity c_{sat} in mol/m^2 and S_w is the surface area of the porous medium.

Substituting Eq. (7.14) in (7.12) one obtains:

$$J_s = -\frac{2\varepsilon^2}{\tau}\frac{D_s}{\bar{r}}c_{sat}\frac{d\theta}{dL} \tag{7.15}$$

This expression shows that J_s increases strongly with decreasing average pore size.

Assuming local adsorption equilibrium (adsorption processes are fast), Eq. (7.15) can be converted in terms of pressure instead of concentration using $dq/dL = dq/dP \cdot dP/dL$ and the expressions for the adsorption isotherm which relate q or θ_s to P, for example, for Henry's law:

$$q = bP \tag{7.16}$$

for Langmuir adsorption:

$$\theta_i = \frac{K_i P}{1 + K_i P} \tag{7.17}$$

with $c_{i,sat}$ the saturation concentration of i (mol/kg or mol/m^2) and $q = \theta \cdot c_{i,sat}$ (kg/mol) $= \theta \cdot c_{sat}$ (mol/m^2) $\cdot S_w$.

Although the selective features of the multilayered membrane system are mainly determined by the thin microporous layer, the porous supporting structure also influences the mass transport of molecules. Benes *et al.* [14], de Bruijn [22] and Zah *et al.* [183] studied the contribution of the support to the overall mass transport through the microporous amorphous silica membrane for gas separation and pervaporation, respectively. However, when the materials do not show as broad a pore size distribution as silica, this influence may be negligible. The active separation top layer is also considered the rate-limiting process when there is surface diffusion of adsorbed molecules along the micropores, in the case of zeolite membranes [139, 151].

According to the adsorption–diffusion model and considering negligible variation of temperature and pressure within the pervaporation separation process [115], the mass transport flux can be expressed as

$$J_{i,\mathrm{mass}} = -D_{T,i}(C_i^m)C_i\left(\frac{d\ln a_i{}^m}{dz}\right) \tag{7.18}$$

where D_i, the thermodynamic diffusion coefficient is

$$D_{T,i} = m_i RT \tag{7.19}$$

When equilibrium at the surface of the membrane is considered

$$\mu_i^{m(z=0)} = \mu_i^f \tag{7.20}$$

and therefore

$$a_w^{m(z=0)} = a_w^F \tag{7.21}$$

The most extended correlation to describe the diffusive transport of one component through a homogeneous membrane is reported as [75, 119],

$$D_w = D_{w,0}\ \exp(\tau C_w^m) \tag{7.22}$$

where τ is considered as a plasticization coefficient in polymer membranes and related to adsorption equilibrium in porous inorganic membranes.

As it was mentioned earlier in this text, the adsorption of a component as a function of its partial pressure or activity fugacity in gas mixtures separation) is usually described by a Langmuir-type strongly nonlinear isotherm that shows linearity at low fugacities, that is, activities, and the approach of saturation at high fugacities for gases onto ceramic microporous membranes. However, as has been mentioned before [168, 174] the adsorption of water into silica has shown a linear isotherm in the range of operating conditions applicable in pervaporation. Thus, by introduction of Eq. (7.22) and (7.21) is converted to

$$J_i = \frac{\rho_i\ D_{i,0}}{\tau\ \delta}\left[\exp(\tau k_i a_i^F) - 1\right] \tag{7.23}$$

where δ is the thickness of the active separation membrane. Equation (7.23) can also be written as follows:

$$\ln\left[J_i + \frac{\rho_i D_{i,0}}{\delta\tau}\right] = ln\left(\frac{\rho_i D_{i,0}}{\delta\tau}\right) + (\tau k_i)a_i^F \tag{7.24}$$

or, for the usual values of the parameters $D_{i,0,\tau}$ and δ in pervaporation dense membranes, the term $\rho_i D_{i,0}/\delta\tau$ is negligible in comparison to the values of J_i, except for low concentrations in the membrane and thus Eq. (7.24) can be approached by

$$\ln J_w = ln\left(\frac{\rho_w D_{w,0}}{\delta\tau}\right) + (\tau k_w)a_w^F \tag{7.25}$$

Thus, Eq. (7.25)—which is a simplified expression obtained from the combination of the thermodynamic Fick's law and the solution–diffusion model—can be applied to the prediction of the pervaporation flux of a component contained in a liquid mixture, provided that the sorption isotherm onto the membrane has a linear shape.

Figure 7.4 demonstrates how from Eq. (7.26) the transport parameters predicting the water flux across a hydrophilic microporous silica membrane can be predicted in a simple way, using Eq. (7.26).

Drioli *et al.* [49] also expressed this monotonicity of the curves flux versus water concentration in the feed, as indicator of the interaction between the penetrants and the membrane results in coupled transport, and developed a similar semiempirical flux equation for predicting the pervaporation performance of dense and composite polymeric membranes based on three parameters much like the ones shown before. They treated them as deviation phenomenological parameters, expressing the deviation from ideal situation, that is, Fick's law, to take thermodynamic and coupling effects discussed into account. Titus *et al.* [151] developed an adsorption–diffusion model for the description of the dehydration of water–ethanol mixtures through zeolite pervaporation membranes, treating the zeolite layer as a whole "amorphous" layer with homogeneous properties based on three parameters with physical

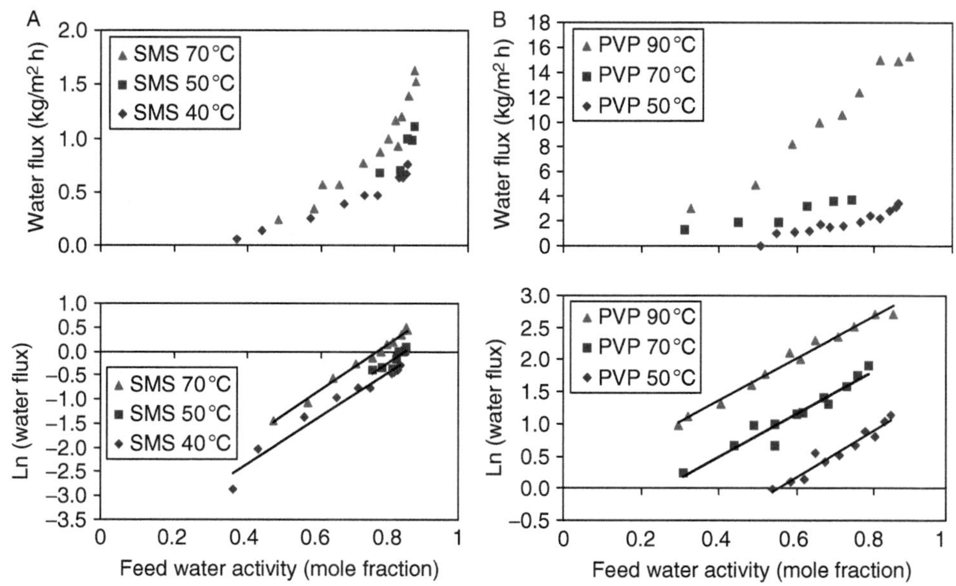

Figure 7.4 LnJw versus water activity in the feed solution for the Pervap SMS (Sulzer Chemtech)—water–ketones (A) and the Silica Pervatech—water–isopropanol system (B) [29–30].

meaning: surface diffusivity of water at zero coverage, water adsorption constant, and saturation vapor pressure of water. However, although they had observed a nonlinear dependence of water flux versus partial pressure, that is, water activity in the feed, they recognized this nonideality as due to the effect of the non-zeolite pores in the separation active layer.

The adsorption–diffusion model is thus one of the most widely used in recent years to describe mass transfer through amorphous silica membranes [22]. Titus *et al.* [151] also developed this model for zeolite membranes performance on the dehydration of ethanol by pervaporation. For this purpose, they consider the zeolite layer, which really consists of intercrystalline and intracrystalline pores, as a whole homogeneous layer where similar mechanisms apply.

5. Conclusions

Taking into account the specific transport properties of microporous inorganic membranes, they have been widely studied for decades, regarding their potentialities for molecular separations, where many factors and mechanisms are involved.

They consist of a thin layer on top of an asymmetric support that provides mechanical strength. The thin layer may be made of amorphous ceramic materials (silica, metallic oxides, or mixed oxides) or crystalline zeolites. However, the transport mechanisms of small molecules along the micropores are the same despite the material: adsorption on the membrane surface, diffusion across the membrane, and desorption at the downstream side. Due to the unavailability of specific transport mechanisms through microporous membranes, most researchers applied well-known transport mechanisms in porous media to explain what is happening inside the membrane. Due to the small size of the molecules whose separation is attempted with these membranes, the performance in pervaporation or gas separation seems to follow the same criteria. The driving force for the separation is the chemical potential gradient between both sides of the membrane.

For a better understanding of the phenomena taking place inside the membrane, modeling of the transport mechanisms is desirable. There are well-known models describing fluid transport in porous media, which have been applied to porous membranes in order to gain knowledge in the porous structure of the membrane. For an accurate description, however, very complicate differential equations and many empirically obtained parameters—not always feasible—are necessary. Thus, simplification is generally done, and semiempirical equations have been reported that describe the transport across microporous inorganic membranes with acceptable accuracy, each of them a step forward on the knowledge of the phenomena actually happening inside the membranes that may lead, eventually, to a practical scale-up and implementation of the inorganic membrane technology in the industrial processes.

REFERENCES

[1] Aaron, N. D., and Tsouris, C. (2005). Separation of CO_2 from flue gas: A review. *Sep. Sci. Technol.* **40,** 321–348.

[2] Abidi, N., Sivade, A., Bourret, D., Larbot, A., Boutevin, B., Guida-Pietrasanta, F., and Ratsimihaty, A. (2006). Surface modification of mesoporous membranes by fluoro-silane coupling reagent for CO_2 separation. *J. Membr. Sci.* **270,** 101–107.

[3] Algieri, C., Bernardo, P., Golemme, G., Barbieri, G., and Drioli, E. (2004). Permeation properties of a thin silicalite-1 (MFI) membrane. *J. Membr. Sci.* **222,** 181–190.

[4] Aptel, P., Challard, N., Cuny, J., and Néel, J. (1976). Application of the pervaporation process to separate azeotropic mixtures. *J. Membr. Sci.* **1,** 271–287.

[5] Asaeda, M. (2006). Microporous silica-based membranes. Present status and prospective. *In* "Proceedings of the Ninth International Conference on Inorganic Membranes, ICIM9," pp. 38–47. June 25-29. Lillehammer, Norway.

[6] Asaeda, M., Ishida, M., and Tasaka, Y. (2005). Pervaporation Characteristics of Silica-Zirconia Membranes for Separation of Aqueous Organic Mixtures. *Sep. Sci. Technol.* **40,** 239–254.

[7] Asaeda, M., Kanezashi, M., Yoshioka, T., and Tsuru, T. (2002). Gas permeation characteristics and stability of composite silica-metal oxide membranes. *In* "Membranes-Preparation, Properties and Applications" (V. N. Burganos, R. D. Noble, M. Asaeda, A. Ayral, and J. D. L. LeRoux, eds.), **752,** pp. 213–218. Materials Research Society, Warrendale, PA.

[8] Asaeda, M., and Tasaka, Y. (2002). Pervaporation and vapor permeation characteristics of a porous SiO_2-ZrO_2 membrane for separation of organic solvents/water solutions. *In* "Membranes-Preparation, Properties and Applications," **752,** pp. 251–256. Materials Research Society, Warrendale, PA.

[9] Asaeda, M., Yang, J., and Sakou, Y. (2002). Porous silica-zirconia (50%) membranes for pervaporation of isopropyl alcohol (IPA)/water mixtures. *J. Chem. Eng. Jpn.* **35,** 365–371.

[10] Asaeda, M., Sakou, Y., Yang, J., and Shimasaki, K. (2002). Stability and performance of porous silica-zirconia composite membranes for pervaporation of aqueous organic solutions. *J. Membr. Sci.* **209,** 163–175.

[11] Asaeda, M., and Yamasaki, S. (2001). Separation of inorganic/organic gas mixtures by porous silica membranes. *Sep. Purif. Technol.* **25,** 151–159.

[12] Ayral, A., Julbe, A., Roualdes, S., Rouessac, V., Durand, J., and Sala, B. (2006). Silica membranes: Basic principles. *Periodica Polytech. Ser. Chem. Eng.* **50,** 67–79.

[13] Baker, R. W. (2004). Membrane Technology and Applications. Wiley & Sons, Chichester, UK.

[14] Benes, N. E., Nijmeijer, A., and Verweij, H. (2000). Silica microporous membranes. *In* "Recent Advances in Gas Separations by Microporous Membranes, Membrane Science and Technologies Series 6" (N. Kannellopoulos, ed.), pp. 335–372. Elsevier, Amsterdam.

[15] Bettens, B., Degrève, J., van der Bruggen, B., and Vandecasteele, C. (2007). Transprot of Binary Mixtures in Pervaporation through a Microporous Silica Membrane. Shortcommings of Fickian Models. *Sep. Sci. Technol.* **42,** 1–23.

[16] Bhave, R. R. (1990). Inorganic Membranes, Synthesis, Characteristics and Applications. Van Nostrand Reinhold, New York.

[17] Bonhomme, F., Welk, M. E., and Nenoff, T. M. (2003). CO_2 selectivity and lifetimes of high silica ZSM-5 membranes. *Micropor. Mesopor. Mat.* **66,** 181–188.

[18] Bowen, T. C., Noble, R. D., and Falconer, J. L. (2004). Fundamentals and applications of pervaporation through zeolite membranes. *Review. J. Membr. Sci.* **245,** 1–33.

[19] Braunarth, C. M., Boudreau, L. C., and Tsapatsis, M. (2000). Synthesis of ETS-4/TiO_2 composite membranes and their pervaporation performance. *J. Membr. Sci.* **174,** 31–42.

[20] Bredesen, R., Jordal, K., and Bolland, O. (2004). High-temperature membranes in power generation with CO_2 capture. *Chem. Eng. Proc.* **43,** 1129–1158.

[21] Broeke, L. J. P., van den, Kapteijn, F., and Moulijn, J. A. (1999). Transport and separation properties of a silicalite-1 membrane-II. Variable separation factor. *Chem. Eng. Sci.* **54,** 259–269.

[22] De Bruijn, F. (2006). Pervaporation and Vapour Permeation of Methanol and MTBE through a Microporous Methylated Silica Membrane. Ph.D. thesis, Technical University of Delft, The Netherlands.

[23] Brunetti, A., Barbieri, G., Drioli, E., Lee, K. H., and Lee, D. W. (2007). WGS reaction in a membrane reactor using a porous stainless steel supported silica membrane. *Chem. Eng. Process.* **46,** 149–156.

[24] Burggraaf, A. J., and Cot, L. (1996). Fundamentals of inorganic membrane science technology. *In* "Membrane Science and Technology Series, **4**," pp. 1–681. Elsevier, Amsterdam.

[25] Burggraaf, A. J., and Keizer, K. (1991). Synthesis of inorganic membranes. *In* "Inorganic Membranes: Synthesis, Characteristics and Applications" (R. R. Bhave, ed.), pp. 10–63. Van Nostrand Reinhold, New York.

[26] Burshe, M. C., Netke, S. A., Sawant, S. B., Joshi, J. B., and Pangarkar, V. G. (1997). Pervaporative dehydration of organic liquids. Marcel Dekker, Bombay.

[27] Campaniello, J., Engelen, C. W. R., Haije, W. G., Pex, P. P. A. C., and Vente, J. (2004). Long term pervaporation performance of microporous methylated silica membranes. *Chem. Commun.* **7,** 833–834.

[28] Caro, J., Noack, M., and Kölsch, P. (1998). Chemical modification of ceramic membranes. *Micropor. Mesopor. Mat.* **22,** 321–332.

[29] Casado Coterillo, C. (2005). Comportamiento de membranas cerámicas de pervaporación en la deshidratación de disolventes orgánicos industriales. Ph.D. thesis, Spain: University of Cantabria, Santander, Spain.

[30] Casado, C., Gorri, D., Urtiaga, A., and Ortiz, I. (2005). Pervaporative dehydration of organic mixtures using a commercial silica membrane. Determination of kinetic parameters. *Sep. Purif. Technol.* **42,** 39–45.

[31] Casado, C., Julbe, A., Urtiaga, A. M., and Ortiz, I. (2004). SiO_2-ZrO_2 membranes for the dehydration of isopropanol by pervaporation. *In* "Proceedings of Euromembrane 2004." Hamburg, Germany.

[32] Casado, L., Mallada, R., Téllez, C., Coronas, J., Menéndez, M., and Santamaría, J. (2003). Preparation, characterization and pervaporation performance of mordenite membranes. *J. Membr. Sci.* **216,** 135–147.

[33] Casanave, D., Giroir-Fendler, A., Sanchez, J., Loutary, R., and Dalmon, J. A. (1995). Control of transport properties with a microporous membrane reactor to enhance yields in dehydrogenation reactions. *Catal. Today* **25,** 309–314.

[34] Chang, C. H., Gopalan, R., and Lin, Y. S. (1994). A comparative study on thermal and hydrothermal stability of alumina, titania and zirconia membranes. *J. Membr. Sci.* **91,** 27–45.

[35] Chentoufi, M., Dlimi, L., Frouji, A., Elmidaoui, A., Guizard, C., and Cot, L. (1998). Separation of organic mixtures using inorganic membranes in pervaporation. *Phys. Theor. Chem.* **1**(8), 503–508.

[36] Chowdhury, S. R., Peters, A. M., Blank, D. H. A., and ten Elshof, J. E. (2006). Influence of porous substrate on mesopore structure and water permeability of surfactant templated mesoporous silica membranes. *J. Membr. Sci.* **279,** 267–281.

[37] Chung, S. J., Park, J. H., Li, D., Ida, J. I., and Kumakiri, I. (2005). Dual-phase metal-carbonate membrane for high temperature carbon dioxide separation. *Ind. Eng. Chem. Res.* **44,** 7999–8006.

[38] Ciora, R. J., Fayyaz, B., Liu, P. K. T., Suwannethanond, V., Mallada, R., Sahimi, M., and Tsotsis, T. T. (2004). Preparation and reactive applications of nanoporous silicon carbide membranes. *Chem. Eng. Sci.* **59,** 4957–4965.

[39] Coronas, J., Noble, R. D., and Falconer, J. L. (1998). Separations of C-4 and C-6 isomers in ZSM-5 tubular membranes. *Ind. Eng. Chem. Res.* **37,** 166–176.

[40] Coronas, J., and Santamaría, J. (2004). State-of-the-art in zeolite membrane reactors. *Top. Catal.* **29,** 29–44.

[41] Coronas, J., and Santamaría, J. (1999). Separations using zeolite membranes. *Sep. Purif. Met.* **28,** 127–177.

[42] Cuffe, L., MacElroy, J. M. D., Tacke, M., Kozahok, M., and Mooney, D.a. (2006). The development of nanoporous membranes for separation of carbon dioxide at high temperatures. *J. Membr. Sci.* **272,** 6–10.

[43] Cui, Y., Kita, H., and Okamoto, K. I. (2004). Zeolite T. membrane: Preparation, characterization, pervaporation of water/organic liquid mixtures and acid stability. *J. Membr. Sci.* **236,** 17–27.
[44] Cuperus, F. P., and van Gemert, R. W. (2002). Dehydration using ceramic silica pervaporation membranes-the influence of hydrodynamic conditions. *Sep. Purif. Technol.* **27,** 225–229.
[45] Dabadie, T., Ayral, A., Guizard, C., Cot, L., Robert, J. C., and Poncelet, O. (1994). New microporous silica materials for membrane processes. *In* "Proceedings of the Third International Conference on Inorganic Membranes," pp. 411–419. Worcester, Massachusetts, USA.
[46] Dafinov, A., Garcia-Valls, R., and Font, J. (2002). Modification of ceramic membranes by alcohol adsorption. *J. Membr. Sci.* **196,** 69–77.
[47] Diniz da Costa, J. C., Lu, G. Q., Rudolph, V., and Lin, Y. S. (2002). Novel molecular sieve silica (MSS) membranes: Characterisation and permeation of single-step and two-step sol-gel membranes. *J. Membr. Sci.* **198,** 9–21.
[48] Drioli, E., and Romano, M. (2001). Progress and new perspectives on integrated membrane operations for sustainable industrial growth. *Ind. Eng. Chem. Res.* **40,** 1277–1300.
[49] Drioli, E., Zhang, S., and Basile, A. (1993). On the coupling effect in pervaporation. *J. Membr. Sci.* **81,** 43–55.
[50] Elshof, J. E., Rubio Abadal, C., Sekulic, J., Doy Chowdhury, S., and Blank, D. H. A. (2003). Transport mechanism of water and organic solvents through microporous silica in the pervaporation of binary liquids. *Micropor. Mesopor. Mat.* **65,** 197–208.
[51] Feldhoff, A., Martynczuk, J., and Wang, H. (2007). Advanced $Ba_{0.5}Sr_{0.5}Zn_{0.2}Fe_{0.8}O_{3-\delta}$ pervovskite-type ceramics as oxygen selective membranes: Evaluation of the synthetic procedures. *Prog. Solid. State Chem.* **35,** 339–353.
[52] Feng, X., and Hang, R. Y. M. (1997). Liquid separation by membrane pervaporation: A review. *Ind. Eng. Chem. Res.* **36,** 1048–1066.
[53] Figoli, A., Donato, L., Carnevale, R., Tundis, R., Statti, G. A., Menichini, F., and Drioli, E. (2006). Bergamot essential oil extraction by pervaporation. *Desal.* **193,** 160–165.
[54] Flanders, C. L., Tuan, V. A., Noble, R. D., and Falconer, J. L. (2000). Separation of C6 isomers by vapor permeation and pervaporation through ZSM-5 membranes. *J. Membr. Sci.* **176,** 43–53.
[55] Fleming, H. L., and Slater, C. S. (1992). Pervaporation. *In* "Membrane Handbook" (W. S. W. Ho and K. K. Sirkar, eds.), pp. 105–116. Chapman & Hall, New York.
[56] Gallego-Lizón, T., Ho, Y. S., and Freitas dos Santos, L. (2002). Comparative study of commercially available polymeric and microporous silica membranes for the dehydration of IPA/water mixtures by PV/VP. *Desal.* **149,** 3–8.
[57] Gallego-Lizón, T., Edwards, E., Lobiundo, G., and Freitas dos Santos, L. (2002). Dehydration of water/t-butanol mixtures by pervaporation: Comparative study of commercially available polymeric, microporous silica and zeolite membranes. *J. Membr. Sci.* **197,** 309.
[58] Gao, Z., Yue, Y., and Li, W. (1996). Application of zeolite-filled pervaporation membrane. *Zeolites* 70–74.
[59] Gardner, T. Q., Falconer, J. L., and Noble, R. D. (2002). Adsorption and diffusion properties of zeolite membranes by transient permeation. *Desal.* **149,** 435–440.
[60] Gavalas, G. R., Megris, C. E., and Nam, S. W. (1989). Deposition of H_2-permselective SiO_2 films. *Chem. Eng. Sci.* **44,** 1829.
[61] Van Gemert, R. W., and Cuperus, F. P. (1995). Newly developed ceramic membranes for dehydration and separation of organic mixtures by pervaporation. *J. Membr. Sci.* **105,** 287–291.
[62] Van Gestel, T., Vandecasteele, C., Boukenhoudt, A., Dotremont, C., Luyten, J., Van der Bruggen, B., and Maes, G. (2003). Corrosion properties of alumina and titania NF membranes. *J. Membr. Sci.* **214,** 21–29.
[63] Gómez, P., Ibáñez, R., Ortiz, I., and Grossmann, I. (2006). Optimum design of PV processes for dehydration of organic mixtures. *Desal.* **193,** 152–159.
[64] González González, B. (2000). Aplicación de la pervaporación a la separación de mezclas azeotrópicas alcohol/éter. Ph.D. thesis, University of Cantabria, Santander, Spain.
[65] González González, B., and Ortiz, I. (2000). Mathematical modelling of the pervaporative separation of methanol-methyltertbutyl ether mixtures. *Ind. Eng. Chem. Res.* **40,** 1720–1731.

[66] Gorri Cirella, E. D. (1999). Separación de compuestos organoclorados volátiles de disoluciones acuosas mediante pervaporación. Ph.D. thesis, University of Cantabria, Santander, Spain.
[67] Gu, Y., Kusakabe, K., and Morooka, S. (2001). Sulfuric acid-modified zirconia membrane for use in hydrogen separation. *Sep. Purif. Technol.* **24,** 489–495.
[68] Guizard, C., Ayral, A., and Julbe, A. (2002). Potentiality of organic solvents filtration with ceramic membranes. A comparison with polymer membranes. *Desal.* **147,** 275–280.
[69] He, W., Chan, W. H., and Ng, C. F. (2001). Water-alcohol separation by pervaporation through zeolite-modified poly(amidesulfomamide). *J. appl. Polym. Sci.* **82,** 1323–1329.
[70] Heintz, A., and Stephan, W. (1994). A generalized solution-diffusion model of the pervaporation process through composite membranes. Part I. Prediction of mixture solubilities in the dense active layer using the UNIQUAC model. *J. Membr. Sci.* **89,** 143–151.
[71] Heintz, A., and Stephan, W. (1994). A generalized solution-diffusion model of the pervaporation process through composite membranes. Part II. Concentration polarization, coupled diffusion and the influence of the porous support layer. *J. Membr. Sci.* **89,** 153–169.
[72] Van Hoof, V., Dotremont, C., and boukenhoudt, A. (2006). Performance of Mitsui NaA type zeolite membranes for the dehydration of organic solvents in comparison with commercial polymeric pervaporation membranes. *Sep. Purif. Technol.* **48,** 304–309.
[73] Huang, R. Y. M., Shao, P., Feng, X., and Anderson, W. A. (2002). Separation of ethylene glycol-water mixtures using sulfonated poly(ether ether ketone) pervaporation membranes: Membrane relaxation and separation performance analysis. *Ind. Eng. Chem. Res.* **41,** 2957–2965.
[74] Jafar, J. J., Budd, P. M., and Hughes, R. (2002). Enhancement of esterification reaction yield using zeolite A vapour permeation membrane. *J. Membr. Sci.* **199,** 117–123.
[75] Jiraratanon, R., Chanachai, A., and Huang, R. Y. M. (2002). Pervaporation dehydration of ethanol-water mixtures with chitosan/hydroxyethylcellulose (CS/HEC) composite membranes. II. Analysis of mass transport. *J. Membr. Sci.* **199,** 211–222.
[76] Jonquières, A., Clément, R., Lochon, P., Néel, J., Dresch, M., and Chrétien, B. (2002). Industrial state-of-the-art of pervaporation and vapour permeation in the western countries. *J. Membr. Sci.* **206,** 87–117.
[77] Jou, J. D., Yoshida, W., and Cohen, Y. (1999). A novel ceramic-supported polymer membrane for pervaporation of dilute volatile organic compounds. *J. Membr. Sci.* **162,** 269–284.
[78] Julbe, A., Farruseng, D., de Ménorval, L. C., and Guizard, C. (2003). Characterization of MFI/ αAl_2O_3 and V-MFI/ αAl_2O_3 composite membranes by 129Xe NMR. *Sep. Purif. Technol.* **32,** 168–173.
[79] Julbe, A., Guizard, C., Larbot, A., Mouchet, C., Vacassy, R., Metz, R., and Cot, L. (1994). Sol-gel processing of zirconia and titania membranes. *In* "Proc. ICIM3" (Y. H. Ma, ed.), pp. 17–25. Worcester, Massachussets, USA.
[80] Kanezashi, M., and Asaeda, M. (2006). Hydrogen permeation characteristics and stability of Ni-doped silica membranes in steam at high temperature. *J. Membr. Sci.* **271,** 86–93.
[81] Kanezashi, M., Fujita, T., and Asaeda, M. (2005). Nickel-doped silica membranes for separation of helium from organic gas mixtures. *Sep. Sci. Technol.* **40,** 225–238.
[82] Kärger, J., and Ruthven, D. M. (1991). Diffusion in Zeolites and Other Microporous Solids, pp. 1–605. Wiley & Sons, New York, USA.
[83] Karlssson, H. O. E., and Trägardh, G. (1993). Pervaporation of dilute organic-water mixtures. A literature review on modelling studies and applications to aroma compound recovery. *J. Membr. Sci.* **76,** 121–146.
[84] Kawamura, H., Yamaguchi, T., Nair, B. N., Nakagawa, K., and Nakao, S. I. (2005). Dual-ion conducting lithium zirconate-based membranes for high temperature CO_2 separation. *J. Chem. Eng. Jpn.* **38,** 322–328.
[85] Kita, H. (2003). Fabrication and application of zeolite membrane. (V. N. Burganos, R. D. Noble, M. Asaeda, A. Ayral, and J. D. LeRoux, eds.), **752,** p. 239.
[86] Kitao, S., and Asaeda, M. (1990). Separation of organic acid/water mixtures by thin porous silica membrane. *J. Chem. Eng. Jpn.* **23,** 367–370.
[87] Krishna, R., and Paschek, D. (2000). Separation of hydrocarbon mixtures using zeolite membranes: A modelling approach combining molecular simulations with the Maxwell-Stefan theory. *Sep. Purif. Technol.* **21,** 111–136.

[88] Krishna, R., and Wesselingh, J. A. (1997). The Maxwell-Stefan approach to mass transfer. *Chem. Eng. Sci.* **52,** 861–911.

[89] Kölsch, P., Sziládi, M., Noack, M., Caro, J., Kotsis, L., and Sieber, I. (2002). Ceramic membranes for water separation from organic solvents. *Chem. Eng. Technol.* **25,** 357–362.

[90] Koriabkina, A. O., de Jong, A. M., Hensen, E. J. M., and van Santen, R. A. (2005). Concentration and temperature dependence of the diffusivity of n-hexane in MFI-zeolites. *Micropor. Mesopor. Mat.* **77,** 119–129.

[91] Koukou, M. K., Papayannnakos, N., Markatos, N. C., Bracht, M., van Veen, H. M., and Roskam, A. (1999). Performance of ceramic membranes at elevated pressure and temperature. Effect of non-ideal flow conditions in a pilot scale membrane separator. *J. Membr. Sci.* **155,** 241–259.

[92] Kusakabe, K., Kuroda, T., and Morooka, S. (1998). Separation of carbon dioxide from nitrogen using ion-exchanged faujasite-type zeolite membranes formed on porous support tubes. *J. Membr. Sci.* **148,** 13–23.

[93] De Lange, R. S. A., Keizer, K., and Burggraaf, A. J. (1995). Analysis and theory of gas transport in microporous sol-gel derived ceramic membranes. *J. Membr. Sci.* **104,** 81–100.

[94] Lee, D., Zhang, L., Oyama, S. T., Niu, S., and Saraf, R. F. (2004). Synthesis, characterisation, and gas permeation properties of a hydrogen permeable silica membrane supported on porous alumina. *J. Membr. Sci.* **231,** 117–126.

[95] Lee, D. W., Lee, Y. G., Sea, B., Ihm, S. K., and Lee, K. H. (2004). Improvement in thermal stability of stainless steel supported silica membranes by the soaking-rolling method. *J. Membr. Sci.* **236,** 53–63.

[96] Legrand-Bucema, C., Malibert, C., and Bach, S. (2002). Elaboration and characterisation of thin films of TiO_2 prepared by sol-gel process. *Thin Solid Films* **418,** 79–84.

[97] Li, S., Tuan, V. A., Noble, R. D., and Falconer, J. L. (2001). Pervaporation of water/THF mixtures using zeolite membranes. *Ind. Eng. Chem. Res.* **40,** 4577–4585.

[98] Li, J., Nguyen, Q. T., Zhou, L. Z., Want, G., Long, Y. C., and Ping, Z. H. (2002). Preparation and properties of ZSM-5 zeolite membrane obtained by low-temperature chemical vapour deposition. *Desalination* **147,** 321–326.

[99] Li, S., Tuan, V. A., Falconer, J. L., and Noble, R. D. (2003). X-type zeolite membranes: Preparation, characterization and pervaporation performance. *Micropor. Mesopor. Mat.* **58,** 137–154.

[100] Li, G., Kikuchi, E., and Matsukata, M. (2003). Separation of water acetic acid mixtures by pervaporation using a thin mordenite membrane. *Sep. Purif. Technol.* **32,** 199–206.

[101] Li, Y., Chen, H., and Yang, W. (2006). Microwave synthesis of LTA zeolite membranes without seeding. *J. Membr. Sci.* **277,** 230–239.

[102] Li, Y., Zhu, G., Liu, J., and Yang, W. (2007). Pervaporation and vapour permeation dehydration of Fischer-Tropsch mixed-alcohols by LTA zeolite membranes. *J. Membr. Sci.* **297,** 10–15.

[103] Lin, S. H., and Wang, C. S. (2004). Recovery of isopropyl alcohol from water solvent of a semiconductor plant. *J. Hazardous Mat.* **106B,** 161–168.

[104] Lin, Y. S. (2001). Microporous and dense inorganic membranes: Current status and perspective. *Sep. Purif. Technol.* **25,** 39–55.

[105] Lin, Y. S., Kumakiri, I., Nair, B. N., and Alsyouri, H. (2002). Microporous inorganic membranes. *Sep. Purif. Met.* **31,** 229–379.

[106] Lipnizki, F., Hausmanns, S., and Field, R. W. (2004). Influence of impermeable components on the permeation of aqueous 1-propanol mixtures in hydrophobic pervaporation. *J. Membr. Sci.* **228,** 129–138.

[107] Lipnizki, f., Olsson, J., and Trägardh, G. (2002). Scale-up of pervaporation for the recovery of natural aroma compounds in the food industry. Part I: Simulation and performance. *J. Food Eng.* **54,** 183–205.

[108] Lipnizki, F., Field, R. W., and Ten, P. K. (1999). Pervaporation-based hybrid process: A review of process design, applications and economics. *J. Membr. Sci.* **153,** 183–210.

[109] Liu, Q., Noble, R. D., Falconer, J. L., and Funke, H. H. (1996). Organics/water separation by pervaporation with a zeolite membrane. *J. Membr. Sci.* **117,** 163–174.

[110] Livage, J., Babonneareau, F., Chatry, M., and Coury, L. (1997). Sol-gel synthesis and NMR characterization of ceramics. *Ceram. Intl.* **23,** 13–18.

[111] McCabe, W., Smith, J., and Harriot, P. (1993). Unit Operations of Chemical Engineering, 5th ed. McGraw-Hill, New York.

[112] Martin, N. (2003). *Reprint,* in: Sulzer Technical Rev.

[113] Masuda, T., Otani, S. H., Tsuji, T., Kitamura, M., and Mukai, S. R. (2003). Preparation of hydrophilic and acid-proof silicalite-1 zeolite membrane and its application to selective separation of water from water solutions of concentrated acetic acid by pervaporation. *Sep Purif. Technol.* **32,** 181–189.

[114] Matsuda, H., Yanagishita, H., Kitamoto, D., Baraya, K., Nekane, T., Takada, T., Idemoto, Y., Koura, N., and Sano, T. (2001). Preparation of silicalite pervaporation membrane with ethanol permselectivity by a 2-step hydrothermal synthesis. *Sep. Sci. and Technol.* **36,** 3305–3310.

[115] Matsui, S., and Paul, d.R. (2004). A simple model for pervaporative transport of binary mixtures through rubbery polymeric membranes. *J. Membr. Sci.* **235,** 25–30.

[116] McCool, B. A., and DeSisto, W. J. (2004). Synthesis and characterization of silica membranes prepared by pyridine-catalyzed atomic layer deposition. *Ind. Eng. Chem. Res.* **43,** 2478–2484.

[117] Morigami, Y., Masakazu, K., Abe, J., Kita, M., and Okamoto, K. (2001). The first large-scale pervaporation plant using tubular-type module with zeolite NaA membrane. *Sep. Purif. Technol.* **25,** 251–260.

[118] Motuzas, J., Heng, S., Ze Lau, P. P. S., Yeung, K. L., Beresnevicius, Z. J., and Julbe, A. (2007). Ultra-rapid production of MFI membranes by coupling microwave-assisted synthesis with either ozone or calcination method. *Micropor. Mesopor. Mat.* **99,** 197–205.

[119] Mulder, M. (1996). Basic Principles of Membrane Technology, Kluwer Academic Publishers Dordrecht.

[120] Nair, S., and Tsapatsis, H. (2003). Infrared reflectance measurements of zeolite film thickness, refractive index and other characteristics. *Macropor. Mesopor. Mat.* **58,** 81–89.

[121] Néel, J. (1997). Pervaporation. Tec & Doc Lavoisier, Paris.

[122] Nijmeijer, A., bladengroen, B. J., and Verweij, H. (1998). Low-temperature CVI modification of γ-alumina membranes. *Micropor. Mesopor. Mat.* **25,** 179–184.

[123] Noack, M., Kölsch, P., Schäfer, R., Toussaint, P., and Caro, J. (2002). Molecular sieve membranes for industrial application: Problems, progress, solutions. *Chem. Eng. Technol.* **25,** 221–230.

[124] Noble, R., Falconer, J. L., and Bowen, T. (2005). *In* "Proceedings of AIChE Annual Meeting, United States .

[125] Nomura, M., Ono, K., Goplakrishnan, S., Sugawara, T., and Nakao, S. (2005). Preparation of a stable silica membrane by a counter-diffusion chemical vapour deposition method. *J. Membr. Sci.* **251,** 151–158.

[126] Okamoto, K., Kita, M., Horii, K., Tanaka, K., and Kondo, M. (2001). Zeolite NaA membrane: Preparation, single-gas permeation, and pervaporation and vapor permeation of water/organic liquid mixtures. *Ind. Eng. Chem. Res.* **40,** 165–175.

[127] Ortiz, I., Gorri, D., Casado, C., and Urtiaga, A. (2005). Modelling of the pervaporative flux through hydrophilic membranes. *J. Chem. Technol. Biotechnol.* **80,** 397–405.

[128] Peters, T. A., Benes, N. E., and Keurentjes, J. T. F. (2005). Zeolite-coated ceramic pervaporation membranes, pervaporation-esterification coupling and reactor evaluation. *Ind. Eng. Chem. Res.* **44,** 9490–9496.

[129] Peters, T. A., Fontalvo, J., Vorstman, M. A. G., Benes, N. E., van Dam, R. A., Vroon, Z. A. E. P., van Soest-Vercammen, E. L. J., and Keurentjes, J. T. F. (2005). Hollow fibre microporous silica membranes for gas separation and pervaporation synthesis, performance and stability. *J. Membr. Sci.* **248,** 73–80.

[130] Raman, N. K., and Brinker, C. J. (1995). Organic "Template" approach to molecular sieving silica membranes. *J. Membr. Sci.* **105,** 273–279.

[131] Sah, A. (2006). Chemically modified ceramic membranes – Study of structural and transport properties, Ph.D. thesis, University of Twente, The Netherlands.

[132] Sakohara, S., Muramoto, F., Sakata, T., and Asaeda, M. (1990). Separation of acetone/water mixture by thin acrylamide gel membrane prepared in pores of thin ceramic membrane. *J. Chem. Eng. Jpn.* **23,** 40–45.
[133] Sato, K., Sugimoto, K., Nakane, T., and Matsukata, M. (2006). Dehydration properties of MFI-membrane for various kinds of solvents. *In* "Proceedings of the Ninth International Conference on Inorganic Membranes, ICIM9," pp. 333–336. Lillehammer, Norway.
[134] Sato, K., Sugimoto, K., Sekine, Y., Takada, M., Matsukata, M., and Nakane, T. (2007). Application of FAU-type zeolite membranes to vapour/gas separation under high pressure and high temperatures up to 50 MPa and 180°C. *Micropor. Mesopor. Mat.* **101,** 312–318.
[135] Schäfer, R., Noack, M., Kölsch, P., Thomas, A., Seidel-Morgenstern, A., and Caro, J. (2001). Development of a H_2-selective silica membrane for the catalytic dehydrogenation of propane. *Sep. Purif. Technol.* **23,** 39.
[136] Sebastián, V., Lin, Z., Rocha, J., Téllez, C., Santamaría, J., and Coronas, J. (2006). Synthesis, characterization and separation properties of Sn- and Ti-silicalite umbite membranes. *Chem. Mater.* **18,** 2472–2479.
[137] Sekulic, J., ten Elshof, J. E., and Blank, D. H. A. (2006). Microporous silica and doped silica membrane for alcohol dehydration by pervaporation. *J. Membr. Sci.* **254,** 267–274.
[138] Semenova, S. I., Ohya, H., and Soontarapa, K. (1997). Hydrophilic membranes for pervaporation. An analytical review. *Desal.* **110,** 251–286.
[139] Shah, D., Kissick, K., Ghorpade, A., Hannah, R., and Bhattacharyya, D. (2000). Pervaporation of alcohol-water and dimethylformamide-water mixtures using hydrophilic zeolite NaA membranes: Mechanism and experimental results. *J. Membr. Sci.* **179,** 185–205.
[140] Shin, D. W., Hyun, S. H., Cho, C.H., and Han, M. H. (2005). Synthesis and CO_2/N_2 gas permeation characteristics of ZSM-5 zeolite membranes. *Micropor. Mesopor. Mat.* **85,** 313–323.
[141] Sklari, S. D., and Zaspalis, V. T. (2007). A novel system of $Al_{100}P_{60}O_8$ microporous ceramic membrane for hydrogen separation from hydrogen/propane mixtures. *Micropor. Mesopor. Mat.* **99,** 176–180.
[142] Skoulidas, A. I., and Sholl, d.S. (2001). Direct tests of the darken approximation for molecular diffusion in zeolites using equilibrium molecular dynamics. *J. Phys. Chem.* **105,** 3151–3154.
[143] Smitha, B., Dhanuja, G., and Sridhar, S. (2006). Dehydration of 1,4-dioxane by pervaporation using modified blend membranes of chitosan and nylon 66. *Carbohydrate Polymers* **66**(4), 463–472.
[144] Sommer, S., and Melin, T. (2004). Design and optimization of hybrid separation processes for the dehydration of 2-propanol and other solvent. *Ind. Eng. Chem. Res.* **43,** 5248–4259.
[145] Sommer, S., and Melin, T. (2006). Influence of operation parameters on the separation of mixtures by pervaporation and vapour permeation with inorganic membranes. Part I: Dehydration of solvents. *Chem. Eng. Sci.* **60,** 4509–4523.
[146] Sushil, A., and Fernando, S. (2006). Hydrogen Membrane Separation Techniques. *Ind. Eng. Chem. Res.* **45,** 875–881.
[147] Suzuki, H. (1987). Composite membrane having a surface layer of an ultrathin film of cage-shaped zeolite and processes for production thereof. *US Patent* 4699892.
[148] Takacs, L., Vatai, G., and Korany, K. (2005). Production of alcohol free wine by pervaporation. *J. Food Eng.* **78,** 118–125.
[149] Takaba, H., Mizukami, K., Oumi, Y., and Kubo, M. (1999). Application of integrated computational chemistry system to the design of inorganic membranes. *Catal. Today* **50,** 651–660.
[150] Tanaka, K., Yoshikawa, R., Ying, C., Kita, H., and Okamoto, K. (2001). Application of zeolite membranes to esterification reactions. *Catal. Today* **67,** 121–125.
[151] Titus, M. P., Llorens, J., Tejero, J., and Cunill, F. (2006). Description of the pervaporation dehydration performance of A-type zeolite membranes: A modelling approach based on the Maxwell-Stefan theory. *Catal. Today* **118,** 73–84.
[152] Tsai, C. Y., Tam, S. Y., Lu, Y., and Brinker, C. J. (2000). Dual-layer asymmetric microporous silica membranes. *J. Membr. Sci.* **169,** 255–269.
[153] Tsuru, T., Wada, S-i., Izumi, S., and Asaeda, M. (1998). Silica-zirconia membranes for nanofiltration. *J. Membr. Sci.* **149,** 127–135.

[154] Tsuru, T., Hino, T., Yoshioka, T., and Asaeda, M. (2001). Permporometry characterization of microporous ceramic membranes. *J. Membr. Sci.* **186,** 257–265.

[155] Tsuru, T., Takata, Y., Kondo, H., Hirano, F., Yoshioka, T., and Asaeda, M. (2003). Characterization of sol-gel derived membranes and zeolite membranes by nanopermporometry. *Sep. Purif. Technol.* **32,** 23–27.

[156] Tsuru, T., Miyawaki, M., Yoshioka, T., and Asaeda, M. (2006). Reverse Osmosis of Nonaqueous Solutions through Porous Silica-Zirconia Membranes. *AIChE J.* **52,** 522–531.

[157] Tsuru, T., Shintani, H., Yoshioka, T., and Asaeda, M. (2006). A bimodal catalytic membrane having a hydrogen-permselective silica layer on a bimodal catalytic support: Preparation and application to the steam reforming of methane. *App. Cat. A: Gen.* **302,** 78–85.

[158] Tsuru, T., Kan-no, T., Yoshioka, T., and Asaeda, M. (2006). A photocatalytic membrane reactor for VOC decomposition using Pt-modified titanium oxide porous membranes. *J. Membr. Sci.* **280,** 156–162.

[159] Tuan, V. A., Li, S., Falconer, J. L., and Noble, R. D. (2002). Separating organics from water by pervaporation with isomorphously-substituted MFI zeolite membranes. *J. Membr. Sci.* **196,** 111–123.

[160] Uhlhorn, R. J. R., and Burggraaf, A. J. (1991). *In* "Inorganic Membranes: Synthesis, Characteristics and Applications" (R. R. Bhave, ed.), pp. 155–176. Van Nostrand Reinhold, New York.

[161] Uhlhorn, R. J. R., Veld, M. H. B. J. H. I., Keizer, K., and Burggraaf, A. J. (1989). High permselectivities of microporous silica-modified γ-alumina membranes. *J. Mat. Sci. Lett.* **8,** 1135–1138.

[162] Urtiaga, A., Casado, C., Asaeda, M., and Ortiz, I. (2006). Comparison of SiO_2-ZrO_2-50% and commercial SiO_2 membranes on the pervaporative dehydration of organic solvents. *Desal.* **193,** 97–102.

[163] Urtiaga, A., Gorri, E. D., Casado, C., and Ortiz, I. (2003). Pervaporative dehydration of industrial solvents using a zeolite NaA commercial membrane. *Sep. Purif. Technol.* **32,** 207–213.

[164] Urtiaga, A., Casado, C., Aragoza, C., and Ortiz, I. (2003). Dehydration of industrial ketonic effluents by pervaporation. Comparative behaviour of ceramic and polymeric membranes. *Sep. Purif. Technol.* **38,** 3473–3491.

[165] Urtiaga, A., and Ortiz, I. (2001). Water recovery from an acetone-water effluent by means of pervaporation. *In* "SeparationTechnology Topical Conference, AIChE Annual Meeting" **1,** pp. 400–406. Reno (Nevada, USA).

[166] Van Veen, H. M., van Delft, Y. C., Engelen, C. W. R., and Pex, P. P. A. C. (2001). Dewatering of organics by pervaporation with silica membranes. *Sep. Purif. Technol.* **22,** 361–366.

[167] Verkerk, A. W., van Male, P., Vorstman, M. A. G., and Keurentjes, J. T. F. (2001). Properties of high flux ceramic pervaporation membranes for dehydration of alcohol/water mixtures. *Sep. Purif. Technol.* **22,** 689–695.

[168] Verkerk, A. W., van Male, P., Vorstman, M. A. G., and Keurentjes, J. T. F. (2001). Description of dehydration performance of amorphous silica pervaporation membranes. *J. Membr. Sci.* **193,** 227–238.

[169] Verweij, H., Lin, Y. S., and Dong, J. (2006). Microporous silica and zeolite membranes for hydrogen purification. *Mat. Res. Soc. Bull.* **31,** 756.

[170] Verweij, H. (2003). Ceramic membranes: Morphology and transport. *J. Mat. Sci.* **38,** 4677–4695.

[171] De Vos, R. M., Maier, W. F., and Verweij, H. (1999). Hydrophobic silica membranes for gas separation. *J. Membr. Sci.* **158,** 277–288.

[172] Wesselingh, J. A., and Krishna, R. (2000). Mass Transfer for Multicomponent Mixtures. University Press, Delft.

[173] Wijmans, J. G., and Baker, R. W. (1995). The solution-diffusion model: A review. *J. Membr. Sci.* **107,** 1–21.

[174] Wolf, H. E., and Schlünder, E. U. (1999). Adsorption equilibrium of solvent mixtures on silica gel and silica gel coated membranes. *Chem. Eng. Proc.* **38,** 211–218.

[175] Xiao, J., and Wei, J. (1992). Diffusion mechanism of hydrocarbons in zeolites-I. Theory. *Chem. Eng. Sci.* **47,** 1123–1141.

[176] Xomeritakis, G., Nair, S., Braunbarth, C. M., Cornelius, C. J., Pardey, R., and Brinker, C. J. (2003). Organic-templated silica membranes. I. Gas and vapour transport properties.
[177] Xomeritakis, G., Liu, N. G., Chen, Z., Jiang, Y. B., Köhn, R., Johnson, P. E., Tsai, C. Y., Shah, P. B., Khalil, S., Singh, S., and Brinker, C. J. (2007). Anodic alumina supported dual-layer microporous silica membranes. *J. Membr. Sci.* **287,** 157–161.
[178] Yang, J. (2002). A Study on Pervaporation Separation of Aqueous Solutions of Organic Solvents With Microporous SiO_2-ZrO_2 Membranes. Ph.D. thesis, University of HiroshimaJapan.
[179] Yang, J., and Asaeda, M. (2003). Permeation mechanism of water through microporous silica-zirconia membranes for separation of aqueous solutions of organic solvents by pervaporation. *Sep. Purif. Technol.* **23,** 29–36.
[180] Yang, J., Yoshioka, T., Tsuru, T., and Asaeda, M. (2006). Pervaporation characteristics of aqueous-organic solutions with microporous SiO_2-ZrO_2 membranes. Experimental study on separation mechanism. *J. Membr. Sci.* **284,** 205–213.
[181] Yoshioka, T., Tsuru, T., and Asaeda, M. (2001). Experimental Studies of Permeation through Microporous Silica Membranes. *AIChE J.* **47,** 2055–2063.
[182] Yoshioka, T., Tsuru, M., and Asaeda, M. (2003). Condensable Vapor Permeation through Microporous Silica Membranes with Molecular Dynamics Simulation. *Sep. Purif. Technol.* **32,** 231–237.
[183] Zah, J., Krieg, H. M., and Breytenbach, J. C. (2007). Single gas permeation through compositionally different zeolite NaA membranes: Observations on the intercrystalline porosity in an unconventional, semicrystalline layer. *J. Membr. Sci.* **287,** 300–316.
[184] Zhong, J., Wei, S., Xia, Q., Li, H. B., and Yuan, C. Z. (2006). Molecular dynamics simulation of nitrogen/acetone mixture permeation through microporous ceramic membranes. *In* "Proceedings of the Ninth International Conference on Inorganic Membranes, ICIM9," pp. 543–546. Lillehammer, Norway.

Synthesis, Characterization, and Applications of Palladium Membranes

Angelo Basile,* Fausto Gallucci,* *and* Silvano Tosti†

Contents

* Institute on Membrane Technology, ITM-CNR, c/o University of Calabria, 87030 Rende (CS), Italy
† ENEA, Dipartimento Fusione Tecnologie e Presidio Nucleare, C.R. ENEA Frascati, Frascati, Roma 00044, Italy

Membrane Science and Technology, Volume 13
ISSN 0927-5193, DOI: 10.1016/S0927-5193(07)13008-4

1. Introduction

A membrane is a permeable phase, often in the form of a thin film, made of a variety of materials ranging from inorganic solids to different types of polymers.

The main role of the membrane film is to control the exchange of materials between the two adjacent fluid phases. A membrane is able to act as a selective barrier, which separates different species either by sieving or by controlling their relative rate of transport through itself. Transport processes across the membrane are the result of a driving force, which is generally associated with a gradient of concentration, pressure, temperature, electric potential, etc.

According to the IUPAC definition, a membrane reactor is a device that combines the separation properties of membranes with the typical characteristics of catalytic reaction steps in only one unit. In particular, the membrane does not only play the role as a separator but also as a part of the reactor itself. In other words, a membrane reactor is an engineering device that selectively removes a product from the reaction system, giving the possibility of achieving higher conversion than a traditional process under the same operation conditions.

2. Preparation of Palladium-Based Membranes

Most of the progress in the membrane separation and membrane reactor areas has happened in the last twenty years mainly owing to the development of new membrane materials able, for example, to resist at high temperature, mechanical strength, and so on. In particular, inorganic membranes offer several advantages over organic membranes, because of their stability at a relatively high temperature (>373 K), and good chemical and mechanical resistances. The synthesis of stable microporous or dense inorganic materials for the preparation of membranes surely was the key factor for increasing the application of membrane-based reactive separations in the catalysis field. The significant progress in the field of membrane reactors is reflected in the increasing number of publications that have grown exponentially over the last few years, as also indicated by McLeary *et al.* [1].

Porous membranes with an average pore diameter >50 nm are classified (IUPAC) as macroporous, and those with average pore diameters in the intermediate range between 2 and 50 nm as mesoporous; microporous membranes have average pore diameters <2 nm. Table 8.1 summarizes their main characteristics with reference to the hydrogen permeation (and separation).

Macroporous membranes show no selectivity toward the gas permeating through the pores. Mesoporous membranes suffer from poor membrane selectivity, since the permeation is mainly based on a Knudsen diffusion mechanism, and separation is affected only by molecular size. For each kind of porous membrane used in a membrane reactor, a big drawback is the permeation of part of the reactants from the reaction zone to the shell side (the so-called *reactant loss problem*) or, as an extreme case, the permeation of all reactants to the permeation side, where they are generally

Table 8.1 Different characteristics of porous and dense membranes

Membrane	ϕ_{pore} (nm)	Diffusion mechanism	$\alpha_{H_2/other\ gas}$	H_2 permeability	Reactant loss
Macroporous	>50	Poiseuille (viscous flow)	1	Very high	High
Mesoporous	2–50	Knudsen	α_{H_2/N_2}=3.74	High	Average
Microporous	<2	Activated process	High	Average	Low
Dense Pd	–	Fick	Infinite	Very low	–

lost to further conversion. If a sweep gas is fed in the permeation side, the back-permeation of the inert sweep gas from the permeate side to the reaction side is also important.

On the other hand, dense palladium membranes feature exclusive selectivity to hydrogen, meeting the criterion of high selectivity. The level of permeability is limited by the thickness of the membrane, which has to be designed taking into account the mechanical strength and durability of the membrane itself. A dense palladium membrane, of course, avoids the reactant loss problem. The use of commercial dense palladium membranes (wall thickness around 100–150 μm) involves two important drawbacks: (1) hydrogen fluxes are low and (2) these membranes are very much expensive because of the cost of palladium (last year the cost was ranging between 5 and 10 €/g).

To meet the dual challenge of selectivity and permeability, the idea is a compromise: to deposit thin layers of highly selective membrane material on supports having high permeability. In fact, porous membranes can be used as substrate onto which a very thin but continuous film of a selective metal, such as palladium or palladium-alloy, can be deposited. This results in a palladium-based composite membrane. Membranes of this type offer clear advantages over either thick-walled palladium-based ones in terms of higher hydrogen fluxes or porous materials from the standpoint of higher selectivities.

In the field of surface engineering techniques, the fabrication of thin films has evolved greatly in recent years. Among the various methods reported in the specialized literature, we can mention: physical vapor deposition, chemical vapor deposition, ion implementation, sol-gel, electrodeposition, laser deposition, and thermal pyrolysis. The choice of a specific method may depend on such factors as the substrate, the support, the nature of the application, the manufacturing facilities available on site, the required thickness, the cost, the surface area, the geometrical form, the purity, and so on. Unfortunately, no method has all the attributes, and every choice is necessarily a result of various compromises. Some of these methods are today largely used to deposit thin active metallic films on inorganic membranes. In particular, it should be considered that on the laboratory scale the principal requirements are the thickness, continuity, and imperviousness of the film.

In most cases, where the palladium layer is thicker than some micrometers, hydrogen flux is inversely proportional to the thickness. Therefore, the history of the development of supported metal membranes could also be seen as the challenge to obtain sufficient thickness defects-free of a metal layer without compromising the mechanical integrity.

In particular, in the last few years, much effort has been devoted to the development and establishment of potential routes to form a thin metal layer on a porous substrate. Several techniques are today available for reducing the thickness of the palladium-based film.

2.1. Dense palladium-based membranes

Palladium and palladium alloys are able to separate hydrogen selectively because of the high solubility and diffusivity of this gas in their lattice. The hydrogen transport mechanism through palladium membranes is due to the interaction of hydrogen atoms and palladium metal. Molecular hydrogen is dissociated and chemisorbed on one side of the membrane surface and dissolved in the palladium matrix. The atomic hydrogen then diffuses to the other side of the membrane surface. The driving force for the permeation becomes the difference in the square root of the hydrogen partial pressure on both sides of the membrane. The experimental evidence shows that only hydrogen can permeate through the dense film of palladium-based membranes, whereas no other gas molecules are allowed to transfer through it.

2.2. Rolled Pd and Pd–Ag thin wall permeator tubes

Pd and Pd–Ag thin wall tubes have recently been produced, for example, by a procedure developed at the ENEA (Frascati, Italy) laboratories [2–7]. Typically, these permeator tubes have a diameter of 10 mm, length 150–200 mm, and the wall thickness of 40–50 μm. According to this procedure, Pd or Pd–Ag tapes of width about 40 mm are cold rolled in order to obtain foils of thickness 50 μm. Then the thin foils are wrapped around an alumina bar, and their overlapped edges are joined by diffusion welding. This technique consists of a thermal treatment carried out, while the metal parts to be joined are mechanically pressed. In Fig. 8.1 the cross section of the welded zone of a permeator tube is shown, while in Fig. 8.2 the scheme of the press used for welding the composite membrane tubes is shown.

As a consequence of the high mobility of the Ag atoms in the solid phase of the Pd–Ag alloy, in this process the formation of the diffusion bonding was performed under moderate-operating conditions (temperature and time). In fact, the heat treatment performing the diffusion welding is carried out under vacuum or controlled atmosphere (i.e., inert gas) and consists of the following steps:

- heating up to a temperature of 720 °C at a ramp rate of 6 °C/min,
- heating up to a temperature of 1100 °C at a ramp rate of 12 °C/min,
- (holding) dwell at 1100 °C for 1 h,
- free cooling down to the room temperature (in about 3 h).

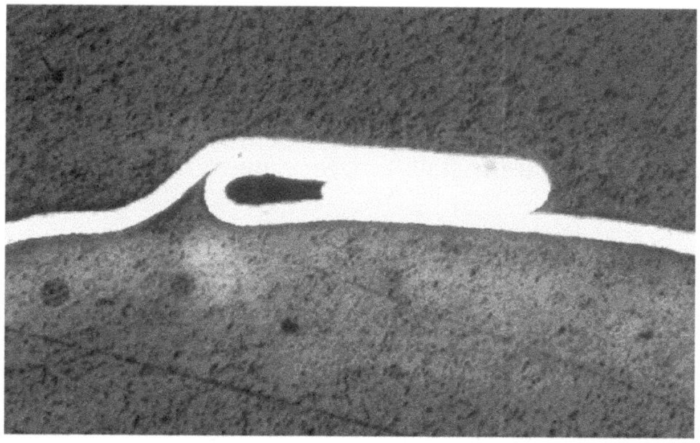

Figure 8.1 Cross section of the welded zone of a permeator tube.

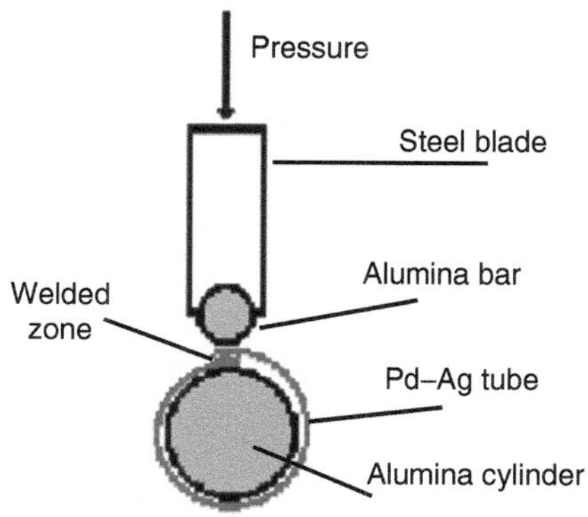

Figure 8.2 Scheme of the press used for welding the composite membrane tubes.

In long-term permeation tests these Pd–Ag permeator tubes have shown high hydrogen permeation fluxes, a good capability to produce ultrapure hydrogen and both good chemical and physical stabilities [197]. Today, they are used to build membrane reactors for carrying out dehydrogenation reactions [8, 9].

2.3. Palladium-based composite metal membranes

The major aim of the study concerning palladium alloy membranes is the reduction of membrane thickness. The flux will increase as the membrane gets thinner; therefore, the overall costs will decrease because of both the reduced operating and material costs. According to the applications where the ultrapure hydrogen production is required, commercial self-supporting dense palladium-based membranes have wall thicknesses >100 μm for keeping a sufficient mechanical strength and for assuring a defect-free surface and the complete hydrogen selectivity. However, these membranes are too thick to obtain a satisfactory hydrogen flux. Moreover, the low hydrogen permeance, thick palladium membranes are too expensive.

Indeed, the price of palladium does not permit the production of low-cost permeators for a wide use in medium-large-scale applications. For a practical use, it is necessary to reduce the thickness of the palladium layer. Composite metal membranes and composite porous membranes seem to be an advisable choice to do it.

Several kinds of composite metallic membrane tubes, based on palladium-based membranes, were studied, such as [5–10]:

- supported membranes, where metallic supports (grids, perforated foils) have been joined to Pd–Ag foils by using diffusion welding;
- laminated metal membranes, where thick non-noble metals have been covered with Pd–Ag foils and eventually cold-rolled.

2.4. Supported palladium-based membranes

In order to produce low-cost and permselective membranes for hydrogen separation, stainless steel grids and nickel perforated foils are also used to support Pd–Ag thin foils. The steel or nickel supports give the membrane the mechanical strength needed to withstand high-operating differential pressures, while the reduced thickness of the Pd–Ag foils provides low-mass transfer resistance and reduced costs. Furthermore, also in this case a diffusion-welding procedure assuring the production of reliable and durable joints between the palladium foils and the support is used [5]. An example is reported in Fig. 8.3, where a supported membrane tube obtained by supporting a Pd–Ag foil of thickness 42 μm on a nickel perforated sheet (thickness 210 μm, holes diameter 2.5 mm) is shown. The hydrogen permeability through this composite membrane is 1.60×10^{-9} and 1.8×10^{-9} mol/m/s/Pa$^{0.5}$, at 350 and 375 °C, respectively.

2.5. Laminated metal membranes

Hydrogen permeability through various metals has been extensively studied. Figure 8.4 is a graph showing the values extrapolated from the literature data [11].

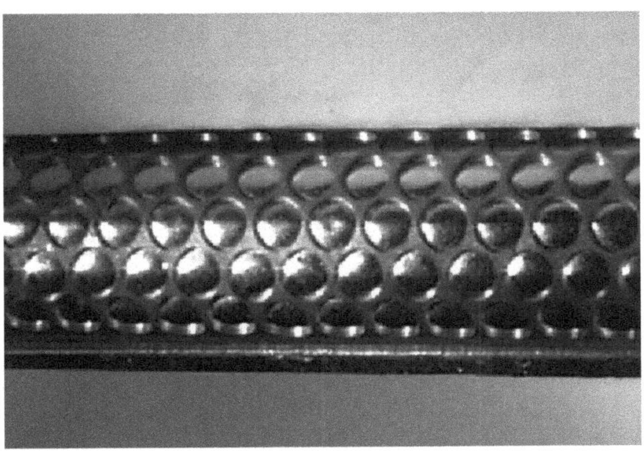

Figure 8.3 Photograph of a membrane tube obtained by supporting a Pd–Ag foil on a nickel perforated sheet [5].

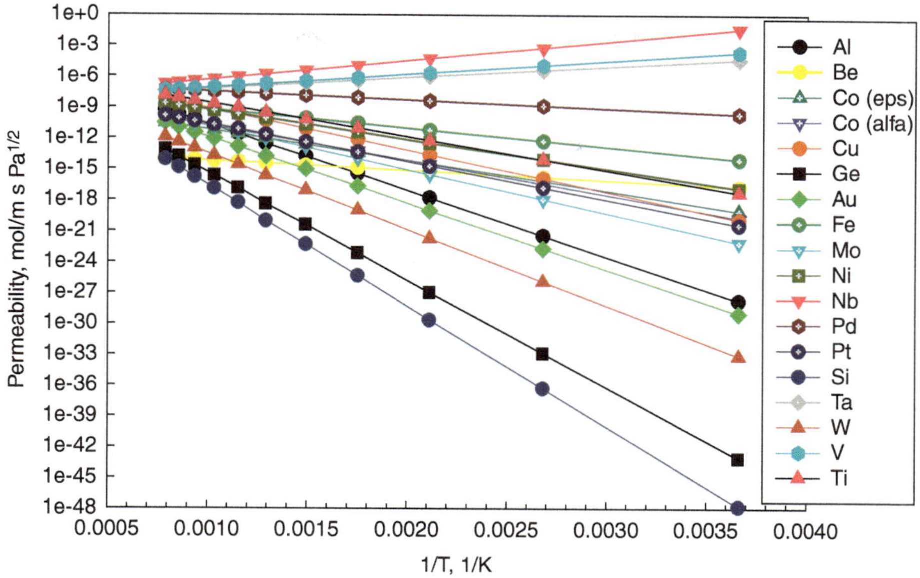

Figure 8.4 Hydrogen permeability through various metals.

Accordingly, several metals (including Nb, Ta, and V) have been predicted to have higher-hydrogen permeability values than palladium and palladium alloys, while other low-cost metals (Ni and Fe) present significant permeability values. Unfortunately, all these metals have a stronger surface resistance to hydrogen transport than Pd: these poor surface proprieties are associated with high reaction to gases under many operating conditions. Thin palladium coatings over these metals avoid the formation of oxides on the metallic surfaces resulting in a reduction of the hydrogen adsorption activation energy and consequently in an increase of the hydrogen permeation flux. In this case, the dense non-noble metal is the support structure for the thin Pd–Ag layers coated over its surfaces: a proper thickness of the non-noble metal assures the mechanical strength, the absence of defects and the complete hydrogen selectivity of the membrane. Samples of laminated membranes have been produced by covering non-noble metal sheets with Pd–Ag thin layers by the diffusion welding technique. After this step, the laminated metal obtained needs to be cold-rolled in order to reduce the total thickness of the composite membrane down to an optimum value for supporting the required characteristics of permeability, chemical and physical stabilities, and also reduced costs (i.e., to reduce the Pd–Ag layer). In Fig. 8.5, a metallographic picture of the cross section of an Ni-laminated membrane sheet, obtained according to the procedure described above, is shown before rolling: the total thickness of this composite membrane has been reduced down to 141 μm through cold-rolling. Hydrogen permeance values, in the range 250–400 °C, are shown in the Arrhenius plot reported in Fig. 8.6 [6].

Figure 8.5 Metallographic picture of the cross section of an Ni-laminated membrane sheet before rolling.

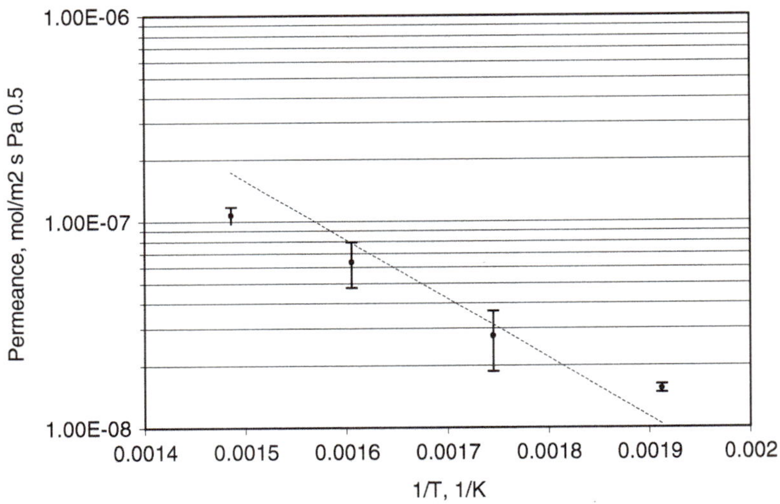

Figure 8.6 Arrhenius plot of the hydrogen permeance versus temperature through Ni-laminated membrane sheet (temperature range: 250–400 °C).

2.6. Other studies on metal membranes

Several recent studies use metal membranes other than Pd and Pd alloys. For example, Hara *et al.* [12] produced and tested $Zr_{36}Ni_{64}$ membranes of thickness 30 μm. These membranes exhibit a permeation rate of $> 1\text{cm}^3_{H_2}$ (STP)/cm^2/min. The V–15%Ni alloy was used by Nishimura *et al.* [13] and Zhang *et al.* [14] for preparing membranes covered by a thin Pd layer (0.1 μm): the permeation tests showed a slight decrease of the hydrogen permeability versus time. Ozaki *et al.* [15]

reported characterization tests on V–Ni–Al: at 623 K hydrogen permeability values of 6.29×10^{-8} $mol_{H_2}/m/s/Pa^{0.5}$ were measured, about twice that of the V–15%Ni membranes. Hashi *et al.* [16] investigated Ni–Ti–Nb alloys: the hydrogen permeability of Ni–Ti membranes increases when alloying with Nb. In particular, the $Ni_{30}Ti_{31}Nb_{39}$ alloy provides a hydrogen permeability of 1.9×10^{-8} $mol_{H_2}/m/s/Pa^{0.5}$.

A composite metal membrane consisting of Ta foil of thickness 13 μm coated with a 1 μm of palladium using a vapor deposition technique was studied by Peachy *et al.* [17] at Los Alamos National Laboratory. The authors tested the membrane at 300 and 365 °C observing hydrogen permeability lower than pure palladium membrane and demonstrating the presence of surface effects that involve a pressure dependence different than $p^{0.5}$.

Ti–Ni–Pd membranes, with low content of Pd (2, 5, and 9%), were studied by Tereschenko *et al.* [18]. They found that, even though the dense membranes produced have a lower permeability than the ones with high Pd content, their use in membrane reactors for the cyclohexane dehydrogenation allowed them to overcome the conversion of a traditional system up to 20%.

2.7. Palladium-based composite porous membranes

Palladium-based composite porous membranes consist of a thin dense layer of palladium (or a palladium alloy) deposited on a porous support. Tubes or disks of porous glass, porous ceramic, or porous stainless steel (PSS) can all be used as supports.

Around the 1990s, two types of porous membranes were mainly used as supports in the basic studies of supported metal membranes: the porous Vycor glass (silica gel) with symmetric structure and a mean pore size of 4 nm, supplied by Corning Glass Works; and a microfiltration porous ceramic such as porous α-alumina with asymmetric multilayer structure, for example Membralox, supplied by SCT/US Filter.

Porous Vycor glass (thickness of about 1 mm) has a symmetric structure and (compared with a metallic layer) gives larger resistance to gas permeation through its pores. Furthermore, the surface was so smooth that a few micrometers thick fabricated metal film was easily stripped off owing to the loss of an anchor effect, which supplies a mechanically assisting bond between the metal film and the porous support.

Another type of porous glass material made of SiO_2, Al_2O_3, and B_2O_3 is also used for supported palladium membranes. This support gives an excellent anchor effect and adherence. Nevertheless, sufficient thermal and mechanical durability was scarcely obtained as a metal support usable at high temperature for catalytic reactions (max operating $T = 823$ K) and in some cases the glass was broken at the sites sealed by a gas tight o-ring.

Thin Pd films on a ceramic support may in principle combine the selectivity characteristics of the Pd thin film to the robustness characteristics of the ceramic membrane. But this poses new questions regarding the ceramic–metallic adherence and interface.

In order to decrease the thickness of the metal layer, surface modification or coating of a supporting substrate is required. Further information will be necessary on kinetics, especially about the rate-determining step for hydrogen permeance, as well as adhesion between the metal layer and the support surface, relating to surface roughness, homogeneity, and chemistry.

Recently, PSS has also been considered as a valid support because of its mechanical durability, its thermal expansion coefficient close to that of palladium, and the ease of gas sealing. Unfortunately, PSS support forms an alloy with the palladium at relatively high temperatures, leading the degradation of hydrogen permeability. To prevent the alloy formation, a diffusion barrier of TiN with a thickness of about 0.1 μm was formed on the surface of a 316L PSS support by the sputtering technique [19]. Modification and pretreatment of a PSS support are still under investigation to reduce the thickness of a palladium film.

Also other materials, for example porous silica, tungsten, tantalum oxide, magnesia, and alumina, have been used as a diffusion barrier.

The upper temperature limits of the porous membranes depend on the material, the chemical atmosphere, and the pore characteristics such as porosity and pore diameter.

Macroporous α-alumina membranes have the superior upper temperature limit of 1573 K because α-alumina is the most thermodynamically stable form while, in the case of a mesoporous γ-alumina membrane the permeability is scarcely affected below 673 K, but is drastically changed at 773–873 K [20]. In Table 8.2 the upper temperature limits of some supporting materials are shown [20].

Many methods have been developed to prepare a thin palladium-coated composite membrane, some of which are briefly outlined in the following.

2.7.1. Physical deposition methods

Various deposition techniques consist of a physical deposition of the vapor: physical vapor deposition, spray pyrolysis, sputtering and magnetron sputtering, solvated metal atom deposition and electroplating. In the following each of them is briefly presented.

Table 8.2 Upper temperature limits of some supporting materials

	Upper temperature limit [K]	
Material	Oxidizing atmosphere	Reducing atmosphere
Alumina	973	973
Alumina/titania	673	673
Stainless steel	673	813
Zirconia/titania	623	623
α-alumina	1573	1573
γ-alumina	673	673

2.7.2. Physical vapor deposition

In a physical vapor deposition process the solid material to be deposited (generally Pd or Pd–Ag alloy) is firstly evaporated in a vacuum system (<1.3 mPa) by means of physical techniques (thermal evaporation, sputtering, etc.). Subsequently, there is a condensation and deposition as a thin film on a cooler substrate. The material of the heating element must have a fusion temperature considerably higher than the evaporating species, so it must not contaminate the deposition process. In fact, tungsten and molybdenum are commonly chosen for their high melting point (3422 and 2610 °C, respectively), while palladium is easily evaporated at 1553 °C and yields good deposit [21]. For example, Ilias and Govind [22] have used the thermal evaporation method to form thin films of such materials as Pd, Ag, and Au on porous supports, while Basile *et al.* [23] used this technique for depositing a palladium thickness of 0.5–1.0 μm on the outer surface of a tubular ceramic support. In the last case, in order to achieve a uniform coating, the tubular membrane was located on a rotating device in a vacuum chamber, where the vacuum level was (6–7) × 10^{-7} torr and a palladium wire with a diameter of 0.5 mm and parallel to the tubular membrane was heated ohmically to produce the vapor for film deposition.

Physical vapor deposition (PVD) is not a laborious technique, provides faster deposition rates, and allows for better control of the film thickness. The film can be obtained at relatively low temperatures, avoiding damaging the substrate. Depending on the shape of the support to be coated, rotation may be necessary to achieve uniform coatings. The achievement of a nonuniform film thickness is one of the most important disadvantages of this technique.

Resuming, PVD is a very versatile technique for the manufacturing of films of pure metals, alloys, or compounds in the thin to medium thickness range, easily up to 50 μm. Evaporation can be performed simply by resistive heating of palladium. Near or above the fusion temperature of the evaporant, the partial pressure of the vapor is high enough to ensure a significant mass transport, yielding a rate of deposition usually around 1 $g/m^2 \cdot s$ (or >1 μm/min). Thermally ejected atoms migrate with little obstruction in straight lines in all directions and condense on whatever target lies in their trajectory. Alternatively, heating with an electron gun, laser beam, or other sources has the advantage that there is no contact with any crucible and it is more efficient than conventional resistance heating.

2.7.3. Spray pyrolysis method

This method is very simple: a solution of metal salts is sprayed into a heated gas stream and is pyrolyzed. It can be used when a very high purity of hydrogen is not required owing to the relatively low H_2/other gases selectivity that can be obtained by using this technique. Probably it is due to the some pinholes still present at the end of the deposition. Only recently it has been applied for the first time to the formation of composite palladium membranes with the aim of hydrogen separation. The method has been successfully applied for the deposition of a Pd–Ag alloy membrane on the outer surface of a porous alumina hollow fiber by spray pyrolysis of $Pd(NO_3)_2$ and $AgNO_3$ mixed solution in a hydrogen–oxygen flame [22, 24]. The thickness of the resulting alloy membrane is 1.5–2.0 μm, the separation factor

H_2/N_2 24, and the hydrogen permeation coefficient is about 8×10^{-7} mol/m^2/s/Pa at 773 K. Compared with the other deposition techniques, the spray pyrolysis method shows a quite low separation factor, indicating that the technique needs some improvements, especially in producing a dense film. Studies are in progress for a better understanding of the methodology for controlling and reducing the thickness of the metallic layers.

2.7.4. Sputtering and magnetron sputtering

Sputtering is another evaporation technique used for PVD under vacuum, which may be classified as a cold extraction technique. In the simplest form, a sputtering system consists of a vacuum chamber containing a target (a plate of the material to be deposited) and the substrate (i.e., the membrane). A sputtering gas (e.g., argon) is introduced to provide the medium in which a glow discharge, or plasma, may be initiated and maintained. Afterwards, positive ions strike the target and remove target atoms and ions by momentum exchange. The condensation of these species over the support produces a thin film. To obtain a good quality film, three basic requirements are necessary: sufficient ionization near the substrate, high ion substrate current density (>1 mA/cm^2), and high deposition rates. The magnetic fields, generated by permanent magnets, increase the ionization giving higher deposition rates. The main disadvantage, in this case, is a strong decrease in the substrate ion current when increasing the substrate-to-target distance. The rates of deposition are generally lower (<0.1 μm/min) than the rates related to the thermal evaporation. Before sputtering the metal, the ion bombardment of the support involves an efficient cleaning of its surface, and as a consequence film adherence is better. This method has been used for the deposition of a thin palladium film on a porous substrate [25, 26]. For example, Basile *et al.* [23] used ultra high-purity argon as a sputtering gas. The film was deposited at 6.6×10^{-7} MPa partial pressure of argon; the distance between the target (Pd–Ad alloy) and the substrate (porous ceramic membrane) was 15.24 cm. The power supply setting coating was a ramp time: 0.025 kW/min. The deposition rate was 0.2 μm/min, giving an estimated thickness on the outer surface of the membrane of 10 μm.

Unfortunately, direct deposition of alloys is not easily controlled owing to the different partial pressures and evaporation rates from the melting of the components. Nevertheless, alloy films alternatively may be produced by simultaneous sequential evaporation from independent sources of pure metals. Composition may be controlled by the evaporation rates and a thermal treatment at relatively high temperature is usually necessary to homogenize the composition of a multilayer deposit (i.e., forming the alloy via a heat treatment).

2.7.5. Solvated metal atom deposition method (or *co-condensation* technique)

The introduction of the metal phase on the inner surface of a tubular membrane is easily performed using this technique.

In a typical experiment, palladium vapor obtained by the resistive heating of a *Sylvania alumina crucible* loaded with Pd shots (0.8 g) is co-condensed in 40 min with 1-hexene-mesitylene (80 ml, 1:1) in a typical glass reactor at liquid nitrogen temperature. At the end of the reaction, the flask is allowed to warm up to –40 °C and the resulting yellow-brown solution siphoned under argon and handled at low temperature, using the Schlenk tube technique. The amount of palladium in the isolated solution is determined by X-ray fluorescence (4.6 mg/ml). Palladium particles are deposited on the inner surface of a tubular membrane by filling the membrane tube (fitted with Teflon stoppers at the ends) with the above solution and heating up to room temperature. This technique was used, for example, by Basile *et al.* [23] and by Capannelli *et al.* [27] for depositing a palladium film on the inner surface of a commercial tubular ceramic membrane. Figure 8.7 shows the conceptual procedure of the co-condensation deposition technique.

2.7.6. Electroplating

In this deposition process, a substrate used as the cathode is coated with a metal or an alloy in a plating bath. This technique may be implemented with a basically simple equipment: a controllable DC source and a bath to maintain the electrolyte of the chosen composition at a constant temperature [28].

Palladium is easily deposited [29] in thin and ductile deposits. A good control is exercised on the composition of the bath, its temperature, and current density. The main investment in plating precious metals is in electrolyte. Hydrogen pumping, however, may be a problem, and it may be necessary to anneal the deposit under vacuum to ensure its ductility. Films of tubular forms are easily obtained using concentric electrodes.

The thickness of deposited film can easily be mastered by controlling electroplating time and current density, and values from a few microns to millimeters are also easily obtained.

Vice versa, alloy compositions are not so easy to control since the relative deposition of two metals simultaneously from the same solution depends on the ease of controlling chemical complexing in the bath.

A Japanese patent [30] describes a manufacturing method for porous-alloy-plated membranes by electroplating the surface of heat-resistant porous supports with

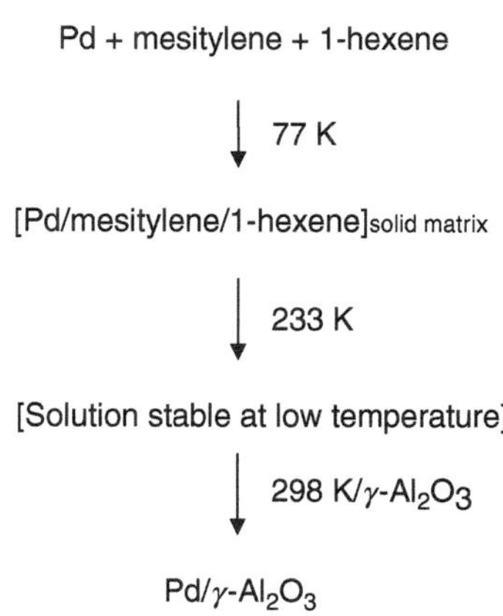

Figure 8.7 Schematic of the *co-condensation* deposition method.

average pore diameter between 20,000 and 30,000 Å, with Pd–Cu alloys. The alloy-coated membranes might contain 71–94 wt% Pd and 6–29 wt% Cu. Thus, a porous glass support (average pore diameter 3000 Å) was plated with a Pd–12wt% Cu alloy to obtain a membrane that was tested for hydrogen permeation at 214 ml/min at <300 °C.

Recently, Tong *et al.* [31] prepared a 3 μm pinhole-free membrane on a PSS tube. Their main problem was the lifetime of the membrane in the long term. This problem must be overcome before industrialization of this preparation method.

2.7.7. Reactive depositions

It is believed that in physical vapor deposition methods the film formation takes place on the top of the material used as a support, that is to say that it is difficult to control and produce uniformly thin metal films and good adhesion on the support materials. Thus, there is increasing attention to vapor phase processes such as reactive depositions producing good film quality and fast film growth.

2.7.8. Chemical vapor deposition

In this method, a chemical reaction involving a metal complex in the gas phase is performed at a controlled temperature. The metal produced by this reaction deposits as thin film by nucleation and growth on the substrate. Owing to their high volatility, such reactants as metal carbonyls, hydrides, halides, or organometallic compounds are usually utilized.

Deposition takes place on the hot substrate positioned in the Chemical vapor deposition (CVD) reactor. As in the case of PVD, the reaction temperature can be reached by resistive heating of the substrate or by other heating sources such as IR, RF, or laser beam and plasma.

This process was effectively used, especially, in the case of palladium deposition [21]: palladium films of 10–300 nm were prepared at low temperature (300–470 K) and inside pores of ceramic support (α-alumina tubes). By using this thechnique Uemia *et al.* deposited a dense palladium film [32].

CVD can successfully be used for the layers deposition consisting of different compounds. The support could also be rough or porous surface substrates.

Recently, Bobrov *et al.* [33] used the CVD method for preparing inorganic membranes with molybdenum selective layer. These membranes are considered a valuable substitution for the present noble metal-coated membranes for dehydrogenation process. The maximum selectivity to propylene for Mo–ceramic membranes was 63% (at $T = 853$ K, conversion to propane 17%), while the maximum selectivity to propylene for Mo–carbon membranes (Mo coated on carbon support) was 84% (at $T = 853$ K, conversion to propane 28%). Itoh *et al.* [34] prepared a thin (2–4 μm) palladium membrane using the CVD method under a forced flow carried out by heating under regulated vacuum pressure. The palladium membrane has a selectivity $H_2/N_2 > 5000$.

Generally, it is also possible to combine different methods of deposition. For example, Tong *et al.* [35] prepared a thin Pd membrane (thickness of 6 μm) with high hydrogen permeability and selectivity on a commercially macroporous stainless steel tube by a combined method of electroless plating and chemical vapor deposition: this membrane shows both high hydrogen permeability and selectivity.

2.7.9. Electrochemical vapor deposition

Essentially, electrochemical vapor deposition (EVD) is a variation of the CVD technique. In the EVD process, the porous substrate separates a mixture of chlorine vapors (ZCl_3, YCl_3, etc.) and an oxygen source (water, vapor, or oxygen). Initially, the reactants from both sides of the support interdiffuse into the pores and form solid oxides, as in the CVD process. When the pores are closed, oxygen ions are conducted across the solid oxide and the oxide film grows on the chlorine side. The EVD technique was mostly used by Han *et al.* [36, 37] for depositing Zr on inorganic substrates such as alumina. Composite membranes were prepared by using a laboratory-scale EVD reactor, realizing membranes with different thicknesses. The membranes prepared by this technique were effective for oxygen separation by gas mixtures.

2.7.10. Sol-gel technique

The sol-gel process is a technique adopted for the preparation of thin materials on which the morphological characteristics (e.g., thickness and porosity) must be accurately controlled. Composite membranes resulting from this process are usually microporous and mesoporous on which permeation of gases is mainly controlled by surface transport and/or the Knudsen flow mechanism. For example, Méndez-Vivar and Mendoza-Serna [38] used the sol-gel method technique for obtaining new microporous SiO_2–TiO_2 solids. In particular, polymeric Si–Ti sols prepared from titanium isopropoxide and tetraethoxysilane were used to deposit membranes on α-Al_2O_3 supports. The separation factors $\alpha(He/CO_2)$ and $\alpha(N_2/CO_2)$ exceed, at 773 K, the theoretical Knudsen values. Also Schafer *et al.* [39] used the sol-gel technique. They deposited SiO_2 layers on commercial alumina ceramic tubes with an asymmetric cross section. Their membranes show hydrogen fluxes $>20\ m^3$ (STP)/m^2 h bar and selectivities hydrogen/propane in the range 30–75 in the interval 450–550 °C. The composite membranes were also tested for carrying out the reaction of propane dehydrogenation, founding a propane yield in the membrane reactor higher than in a conventional fixed-bed reactor.

2.7.11. Electroless plating deposition

The principle of this technique is based on the controlled autocatalyzed reduction of a metastable metallic complexes salt on the target surface. In the case of palladium deposition, ammine complexes such as $Pd(NH_3)_4(NO_2)_2$, $Pd(NH_3)_4Br_2$, or $Pd(NH_3)_4Cl_2$ may be used to deposit films in the presence of a reducing agent, typically hydrazine or sodium hypophosphite. Usually, in order to have a uniform coating, the target must be cleaned and activated with palladium nuclei in an activation solution.

For some applications, this technique provides great advantages such as the uniformity of deposits on the surfaces of any geometry. Palladium and some of its alloys are among the few metals that can be deposited in this way. Figure 8.8 shows

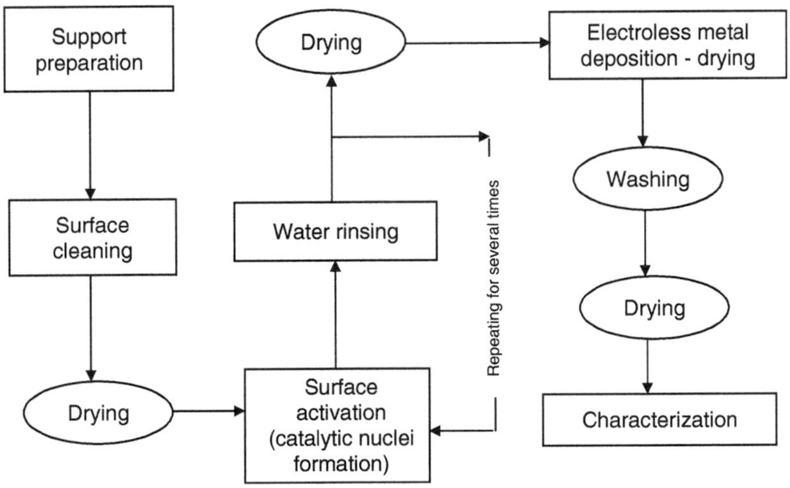

Figure 8.8 Schematic procedure of the electroless metal deposition technique.

the schematic procedure of the electroless metal deposition. However, the method is not free of problems. In particular, following the ref. [28], these are:

1. the thickness control is not easy;
2. a possible decomposition in the bath may induce costly "losses" of palladium;
3. the purity of the deposit is not guaranteed.

The electroless plating technique has been particularly promoted for the manufacture of palladium-based membranes by Japanese researchers [40]. For example, Uemiya *et al.* [41] prepared a composite membrane consisting of a thin palladium film coated over a microporous glass tube (300 nm pore size). The thickness of the palladium film was calculated to be 13 μm. The supported membrane prepared using this deposition technique eliminates the difficulties in fabrication techniques and poor mechanical properties of thick metallic commercial membranes (having thickness of 150 μm). This membrane, however, still had the problem of hydrogen embrittlement when exposed to hydrogen below 573 K. The hydrogen embrittlement temperature was reduced to 473 K by adding silver with palladium [42]. The permeability of this Pd–23 wt% Ag alloy film of thickness 5–8 μm is higher than the Pd membrane alone. Altinisik *et al.* [43] prepared and characterized a palladium-plated porous glass membrane using the electroless plating method. The selectivity ratio for H_2/N_2 was found to be about 7 at 200 °C.

Lin *et al.* [44] prepared palladium alloy membranes by sequential electroless plating steps on PSS, obtaining a homogeneous deposit over and inside the porous substrate. In other words, the deposited Pd–Ag alloy membrane retained the same composition over and through the porous substrate. The authors concluded that a combined mechanism of Knudsen diffusion and viscous flow was conducted in the porous membrane.

The co-deposition behavior of palladium and silver on PSS using the electroless deposition technique was studied by Shu *et al.* [45]. They found that simultaneous deposition was passivated by the preferential deposition of silver. An improved deposition procedure with effective Pd activation achieved co-deposition in separate phases. Also Tanaka *et al.* [46] used the electroless plating procedure for preparing a composite membrane by depositing simultaneously palladium and silver on the porous α-alumina tube. By optimizing the solution composition, the authors attained a uniform distribution of two metals with desired ratio up to 30 wt% Ag.

Recently, Volpe *et al.* [47] have proposed a new bath for the electroless deposition of palladium on anodic alumina (AA) membranes. The optimal conditions for the uniform deposition of palladium, with minimal damage to the AA membranes, were under conditions of pH 8.4 and plating times <30 min.

Summarizing the main advantages of membrane synthesis via electroless plating are uniform deposition on surfaces with complex geometries and large surface areas without pinholes and cracks, the hardness of the deposited film, good adhesion to the substrate, simple equipment, and ease of scale-up.

2.7.12. Molecular layering technique

One of the most promising methods of membrane modification on the atomic level is that of molecular layering (ML) developed, quite recently, in Russia [48–50]. The ML method is based on the chemisorptions of reagents on a solid substrate surface and consists of the irreversible interaction of low-molecular reagents and functional groups of a solid substrate surface under the conditions of continuous reagent feed and the subsequent removal of the formed gaseous products. The method was firstly proposed by the academician Aleksovskii in 1952 [51]. Recently and for the first time, Malygin *et al.* [52] have successfully applied the ML method to the preparation of a composite membrane catalyst. This membrane catalyst was used for the selective oxidation of methane to formaldehyde [53]. Also Pellin *et al.* [201] recently demonstrated the possibility to produce, by this technique coupled with atomic aluminium oxidation, ultra uniform inorganic catalytic membranes that exhibit novel and interesting catalytic properties owing to the possibility to control the pore wall composition and diameters.

2.7.13. Some general problems in composite membrane preparation

Each method, of course, has its advantages and its disadvantages. For example, both CVD and electroless plating techniques are time consuming, but they are able to coat a complex-shaped component with a uniform thickness layer. Unfortunately, unwanted compounds and impurities may be formed and incorporated in the Pd layer, reducing the flux of hydrogen through the film. Yet by the electroless plating method it is not easy to control the thickness of the film. Vice versa, an important advantage of electroless coating is that it is well suited to applications on available commercial tubular membranes. CVD is not an economic process because of the strict conditions required for the process. However, owing to its simplicity, relatively low costs and uniformity of the deposition, electroless plating seems to be the most popular and most successful procedure so far.

There are several problems with the palladium (composite or not) membranes. Two of them, in particular, are underlined by Armor [54]:

- The first is hydrogen embrittlement that is caused by the large distortion in the lattice structure due to the α–β phase transformation. When the cyclic stress is present by adsorption and desorption of hydrogen, the palladium becomes very brittle. This mechanical instability is somewhat decreased by making alloys with other metals such as Ag, Ru, Rh, or other rare earths.
- The sensitivity to poisoning in the presence of CO, H_2S, SO_2, sulfur, arsenic, chlorine, mercury, zinc, unsaturated hydrocarbons, and so on.

Other specific problems related to palladium-based membranes are reported in the following.

Thin Pd films on a ceramic support may in principle combine the selectivity characteristics of the Pd thin film to the robustness characteristics of the ceramic membrane. However, other problems regarding the ceramic–metallic adherence and interface arise. One way to avoid them and so to obtain a stable composite palladium membrane was suggested by Basile *et al.* [23], who considered the following three-step procedure:

1. depositing a thin layer of palladium on a ceramic support;
2. utilizing the composite membrane for carrying out a reaction;
3. depositing another layer of palladium; in case, continue with steps 2 and 3.

The conclusion of their study was that using the so-called procedure of "deposition and use" of the composite membrane, it is possible to achieve both a very selective and a stable composite palladium membrane.

As previously said, the Pd–Ag layer thickness is important in determining both the hydrogen permeability and the membrane costs. Especially in Pd–ceramic composite membranes, it plays a key role in terms of physical and chemical stability as well as of membrane selectivity. In a cited work [5], a theoretical evaluation of the interfacial metal/ceramic stresses rising in electroless and sputtered composite membranes has been carried out. In fact, in these composite membranes, as a consequence of the thermal and hydrogenation/dehydrogenation cycling at the interface of the Pd–Ag layer and the ceramic support, shear stresses take place owing to the differential metal elongation with respect to the ceramic one: it is demonstrated that these stresses are proportional to the metal layer thickness. In Fig. 8.9 the shear stress calculated at the interface metal/ceramic are reported versus the temperature as a function of the Pd–Ag layer thickness. As an example, in a first approximation, assuming as palladium–ceramic (electroless and sputtered) membranes adhesion values around 10 MPa, the rupture of the Pd–Ag layers of thickness about 15 and 8 μm can be foreseen at 500 and 600 K, respectively.

By increasing the film thickness, of course the pores of the ceramic support get progressively closed (high selectivity): however, the stresses at the interface ceramic/metal, due to thermal and hydrogenation cycles, can produce membrane failure (formation of cracks and peeling of the coating). On the other hand, ceramic porous supports having pores (i.e., diameter < 1 μm) can be made selective by depositing Pd–Ag layers of similar reduced size involving low shear stresses and then good durability: in this case, however, the mass transfer resistance through the ceramic becomes high (i.e., low permeability). This analysis demonstrates that highly permeable Pd–ceramic membranes can be produced either selective or durable.

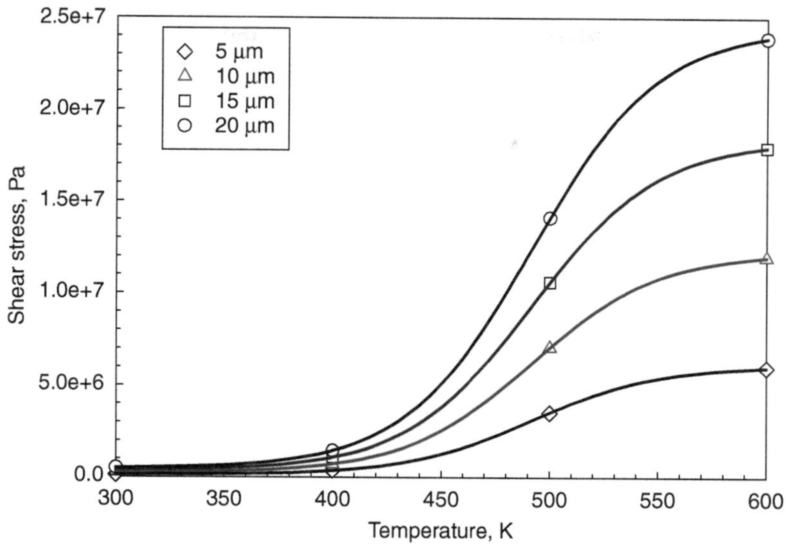

Figure 8.9 Shear stress calculated at interface metal/ceramic versus temperature at various thicknesses of the metal film.

Finally, in Table 8.3 is shown a comparison among some different palladium-based composite membranes prepared using various deposition techniques, as recently reported in the literature. Many parameters are reported in the table: membrane type and membrane thickness, temperature and pressure ranges of the permeation experiments, hydrogen flux, hydrogen permeance (referred to the Pd-based composite membrane), *ideal* separation factor, apparent activated energy, pre-exponential factor, preparation method of the metallic thin layer, and reference of the authors. Considering that the ability of a membrane to effect mixtures separation is determined only by two parameters (the permeability and separation factor), particular attention should be paid to the separation factor α_{H_2/N_2}. In effect, due to a lack of information in the papers considered, only the *ideal* separation factor (the so-called permselectivity or, shortly, selectivity), defined as the ratio between a target pure gas flux (generally hydrogen) to the flux of another pure gas (e.g., nitrogen) at the same operative conditions, is reported.

Generally, porous palladium-based membrane has high gas fluxes and low selectivities, while dense palladium-based membrane, vice versa, shows low hydrogen flux and high hydrogen selectivity.

In conclusion, the main goal in the palladium-based membrane preparation is the production of a dense, thin, stable, and durable layer. It seems very important to have a dense palladium alloy layer; in fact, it is well known that hydrogen permeates through palladium via a solution–diffusion transport mechanism [55] with an infinite value of the permselectivity H_2/other gas. In this case, it is possible to selectively remove only hydrogen from a gas mixture both in a permeator and in a membrane reactor.

Table 8.3 Permeation data of different palladium-based composite membranes reported in the literature

Membrane type	T [K]	ΔP (kPa)	Thickness [μm]	$Flux_{H_2}$ (mol/m^{2}s)	Pe_{H_2} mol/m^{2}s Pa)	α_{H_2/N_2}	Ea_{H_2} (kJ/mol)	P_o (mol m/m^{2}s Pa$^{0.5}$)	Preparation method	References
Pd–ZrO_2–PSS	773	100	10	0.083	8.3×10^{-7a}	–	7.1	0.6	ELP	[56]
Pd–αAl_2O_3	573	290	1	2.05	3.8×10^{-3a}	10^{-6}	–	–	MCVD	[57]
Pd–Al_2O_3	723	–	4.8	–	3×10^{-6}	60	–	–	ELP	[58]
Pd/Ag–Al_2O_3	–	142	10	0.142	1×10^{-6a}	1500	–	–	ELP	[59]
Pd/Ag–PSS	723	1000	29	–	–	8	–	–	ELP	[60]
Pd–PSS	593–773	–	20	–	–	–	12.54	56.41	ELP	[61]
Pd–Al_2O_3	573	30	2–4	0.1–0.2	6.66×10^{-6a}	5000	–	–	CVD	[34]
Pd–GCM	473	10	15	0.223	2.23×10^{-5a}	7	–	–	ELP	[43]
Pd–Al_2O_3	801	–	2–3	–	1.162×10^{-6}	<18	–	–	ELP	[62]
Pd/Ag–PSS	673–773	162	50	–	–	8	33.31	1.66×10^{-5}	LT	[63]
Melt-Spun Zr–Al–Co–Ni–Cu	673	390	44	0.0044	1.13×10^{-8a}	–	–	–	MST	[64]
Pd alloy/PNS	773	358	–	0.083	–	3.7	–	–	MST	[65]
Pd–Cu alloy	725	–	0.75	1.6	–	500(H_2/He	–	–	DST	[66]
Pd/Ni	673	20	2.5	0.31	1.15×10^{-5a}	–	–	–	DC SP	[67]
Pd/Ag alloy	620	400	5.5	–	10×10^{-7}	>4500	9.8	–	SELP	[68]
Pd/MPSS	773	100	6	0.302	3.02×10^{-6a}	–	16.7	–	ELP	[69]
Pd	473	51	95	0.0267	5.235×10^{-7a}	–	23.16	–	Oxidation	[70]
Pd_{84}–Cu_{16}/ZrO_2–PSS	753	250	5	0.6	5.265×10^{-4a}	8	14.5	–	ELP	[71]
Pd/PSS	773	100	11.7	–	7.91×10^{-4}	–	–	–	–	[72]
Pd_{90}–Ag_{10}–αAl_2O_3	473–616	80–250	20	0.14	1.25×10^{-6a}	30–178	–	–	ELP	[73]
Pd–CS	623–773	400	2	–	1.688×10^{-6}	1140–12900	12.19	144.6	ELP	[74]
Pd/Pd–Ag–PSS	673–773	100	2–3	0.3	3×10^{-6a}	–	25.7	0.17	ELP	[31]
Pd–Ag/αFe_2O_3/PSS	773	100–500	16–20	–	4.91×10^{-4}	3500–5000	–	–	SELP	[75]
Pd–Cu/αAl_2O_3	723	345	11	0.8	2.31×10^{-6a}	1150	–	–	ELP	[76]
Pd/TiO_2	773	45	0.3–0.4	0.283[a]	6.28×10^{-6a}	1140	–	–	ELP-UV	[77]
Pd–Ni/SS	623–823	68	0.8	0.138–1.315	2.02×10^{-6a} -1.93×10^{-5a}	300–4700	–	–	VED	[78]
Pd–Ru–In/SS	645	100	1.5	0.049	4.9×10^{-7a}	8	–	–	SD	[79]
Pd–Al_2O_3	673	100	5	0.155	1.55×10^{-6a}	100–200	–	–	ELP	[80]
Pd/PSS	793	150	10	0.175	1.166×10^{-6a}	–	–	8.862×10^{-4}	ELP	[81]

[a]Calculated from reference data.

3. Characterization of Palladium-Based Membranes

An important step before using a membrane in a separation or a reaction system is its characterization in terms of permeation as well as morphology: both Pd and Pd alloy composite membranes have been investigated by several authors. In the case of composite Pd-based membranes, the first characterization concerns the interface between the Pd film and the substrate because its physical properties are important in obtaining thermally stable metallic films.

3.1. Scanning electron microscopy

This characterization, widely used by all the authors preparing membranes, can be done using Scanning Electron Microscopy (SEM). Specifically, Uemiya *et al.* [82] used SEM to follow the structural change at every step (nontreated glass tube, deposition of palladium nuclei, and electroless plating) during the Pd–Ag membrane synthesis. Also Cheng *et al.* [83] used SEM to examine the microstructure of electroless deposited Pd on Vycor glass at different stage of the process concluding that the nucleation and growth took place simultaneously (Fig. 8.10). They also used SEM micrographs to measure the thickness of the film deposited.

Ma *et al.* [84] also used SEM to show the morphological change during the membrane synthesis (Fig. 8.11) and to measure the membrane thickness. The thickness measured with this technique gave an average value in good agreement

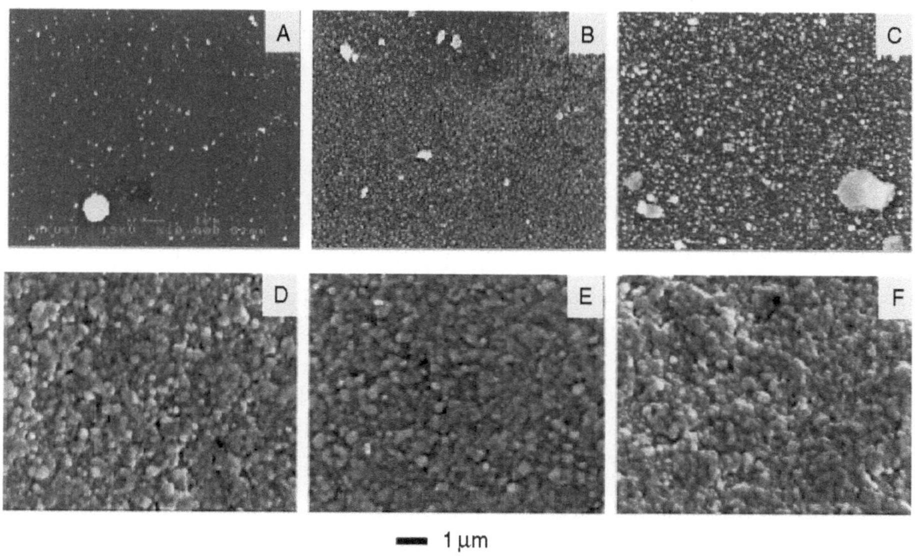

Figure 8.10 Evolution of film membrane microstructure during electroless plating of porous Vycor support. (A) Seeded surface ($t = 0$), (B) 10 min, (C) 20 min, (D) 30 min, (E) 40 min, and (F) 60 min. ($[Pd^{2+}] = 5$ mM, $[Ag^{+}] = 5$ mM, $[N_2H_4] = 10$ mM, $[NH_4OH] = 3$ M, and $T =$ 328 K) [83].

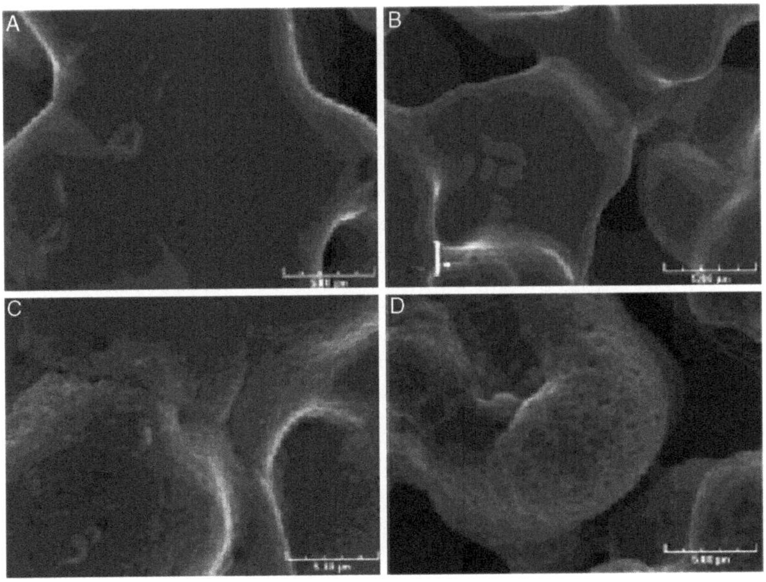

Figure 8.11 SEM micrographs of the supports after oxidation: (A) unoxidized, (B) oxidized at 400 °C, (C) oxidized at 600 °C, and (D) oxidized at 800 °C [87].

with the thickness determined with the gravimetric method, generally used for this determination [85]. Shu *et al.* [45] used SEM to observe the membrane surface morphology and to show the preferential deposition of Ag on the surface of the PSS substrate, leaving the pores uncovered. Basile *et al.* [23] also used SEM to determine the thickness of the ceramic support and of the palladium film coated on it. Souleimanova *et al.* [86] used SEM to study the pressure effect on the Pd morphology and reported the formation of finer Pd particles at high osmotic pressure.

All these studies show that SEM is an important technique in palladium membrane characterization. In fact, SEM pictures together with nitrogen permeability tests can show that the palladium membrane prepared is defect free [87].

Moreover, SEM pictures can also show the presence of boundaries or interfaces among the Pd particles deposited, suggesting that a heat treatment (above the so-called Tamman temperature) is necessary to eliminate these interparticle boundaries [88] (Fig. 8.12).

Finally, SEM can be used to study the membrane surface during the membrane preparation [89] in order to control the film deposition process.

3.2. X-ray diffraction analysis

Different papers extensively presented the theory of hydrogen diffusion through palladium. This process mainly involves the reversible dissociative chemisorption of hydrogen on both membrane surfaces, the dissolution of surface atomic hydrogen into the bulk, and the diffusion of atomic hydrogen through the bulk. As already

Figure 8.12 SEM photographs of (A) the stainless steel substrate and (B) the Pd/SS composite membrane [88].

mentioned for the Pd–H system, Pd often suffers embrittlement owing to $\alpha-\beta$ phase changes [90]. Adrover *et al.* [91] showed, by means of numerical simulations and analytical arguments, that the dynamics of propagation of the β phase are influenced by self-stresses in different ways in planar or tubular membranes. In each case these diffusion-induced stresses influence the transport equation and make the membrane brittle. The addition of Ag to Pd gives a high stabilizing effect. It has been found that alloys with more than 20% Ag show none of the Pd cycling effects; however, there is a rapid drop in hydrogen permeability for silver contents higher than 23% [42, 92].

For these reasons, an analysis of the alloy concentration is important before using the membrane for hydrogen permeation. Although different techniques are available for this analysis (i.e., electron probe microanalysis [92], etc.), the most used method is the X-ray diffraction analysis (XRD).

Specifically, Cheng *et al.* [83] used XRD to calculate the amounts of palladium and silver deposited with plating time by analyzing the bulk composition of the membrane produced, while Tebus *et al.* [93] used this technique to investigate palladium alloy properties degradation during long-term tritium exposure. XRD analysis was also used by Mundschau *et al.* [94] for determining the changes occurring during the deposition of palladium and silver alternating layers due to the interdiffusion progresses at temperatures of 400, 450, 500, and 550 °C.

An interesting derivation of the equation for concentration calculation from the XRD data was reported by Shu *et al.* [45]:

$$C_{\mathrm{Pd}} = \left[1 + \frac{I_{\mathrm{Ag}}}{I_{\mathrm{Pd}}}\left(\frac{V_{\mathrm{Ag}}}{V_{\mathrm{Pd}}}\right)^2 \times \frac{(1+\cos^2 2\theta_{\mathrm{Pd}})(\sin^2\theta_{\mathrm{Ag}}\cos\theta_{\mathrm{Ag}})}{(1+\cos^2 2\theta_{\mathrm{Ag}})(\sin^2\theta_{\mathrm{Pd}}\cos\theta_{\mathrm{Pd}})}\right]^{-1}$$

where C_{Pd} is the overall Pd content (atomic percentage) in the crystalline phase, I, V, and θ are the diffraction intensity, the volume of unit cell (m^3), and the Bragg angle, respectively, for the corresponding diffraction indices of both palladium and silver in the XRD pattern of the mixed composite membrane.

McCool *et al.* [95] used XRD to determine the thickness of the supported Pd–Ag films in the membrane produced. In fact, qualitatively, it is known that the intensity of the X-ray deflected from a support is attenuated by a thin film deposited on the support surface because the film absorbs the energy of the incident and deflected X-ray. But a thicker film gives a stronger intensity of the X-ray deflected from the film. Thus, the ratio of the intensity of the X-ray deflected by the deposited film (e.g., Pd–Ag (111) deflection peak) to that from the support (e.g., α-Al_2O_3 (104) deflected peak) increases monotonously with the thickness of the film (Fig. 8.13).

By combining the basic equation of the X-ray deflected from the support and the deposited film, respectively, with the exponential law of the X-ray absorption by the

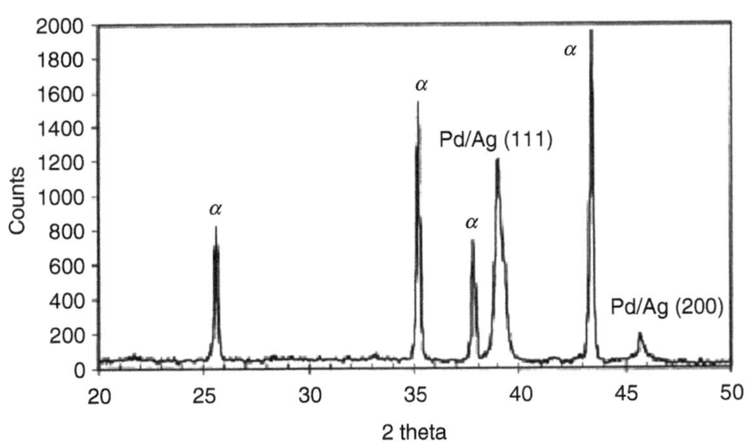

Figure 8.13 Typical X-ray diffraction pattern of Pd–Ag membranes prepared by sputter deposition.

film, the equation that correlates the X-ray intensity ratio to the thickness of the deposited film must be derived:

$$R = K\left\{\exp(\beta \times t) - \exp\left[\beta \times t\left(\frac{1-\sin(\Theta)_s}{\sin(\Theta)_f}\right)\right]\right\}$$

where R is the relative intensity of the membrane peak to the support peak, t the thickness of the membrane layer, Θ_s and Θ_f are the angle of reflection for the support and the membrane, respectively. K and β are constants that are independent of the thickness of the film. The values of K and β were obtained by SEM calibration of the three membranes with different Pd–Ag film thickness (Fig. 8.14). The calculation reported by these authors shows a relative error of 5% for the thickness determined.

In conclusion, this analysis is particularly interesting thanks to the possibility of calculating both the thickness and the composition by means of only one XRD measurement.

3.3. Auger electron spectroscopy analysis and energy dispersion spectrometry analysis

Other analyses are possible for the Pd-based membranes such as Auger Electron Spectroscopy (AES) for the surface composition [96, 97], mercury intrusion experiments for pore size distribution [72], and positron spectroscopy for the investigation of micro-defect behavior in the crystal lattice of the alloy [93]. For example, Yepes *et al.* [75] using the AES analysis found that the Auger profiles did not show the presence of Fe in the Pd–Ag film, confirming that the γ-Al_2O_3 layer effectively blocked the intermetallic diffusion.

Figure 8.14 SEM of a Pd–Ag membrane (Pd–Ag layer thickness is 525 nm).

Energy Dispersion Spectrometry (X-EDS) analysis was used, for example, by Basile *et al.* [23] in order to determine the distribution of the percentage of palladium along the cross section of a composite palladium membrane obtained using the *co-condensation* technique. In their analysis (reported in Fig. 8.15) they found that palladium is present only on the top layer of the inner surface of the supporting ceramic membrane (i.e., γ-Al_2O_3 structure) and does not penetrate the α-Al_2O_3 pores. This means that the pores of the γ-Al_2O_3 layer have been filled effectively without palladium dispersion.

3.4. Optical microscopy

Also optical stereomicroscopy has been used to show changes in the membrane surface morphology before and after permeation testing [98] (Fig. 8.16). The surface modifications (enlargement of the grain size) were observed after preliminary tests when higher hydrogen permeability values were also measured.

3.5. Gas permeation analysis

The next characterization is the gas permeation analysis to be done for both porous and dense Pd-based membranes. For dense membranes, it is also necessary to control the leaks before and after the thermal treatments and before and after the permeation tests. This can easily be operated by pressuring the membrane lumen with an inert gas. Usually, to verify the absence of defects, cracks and micro-holes, it is possible to close both the permeate and the retentate side of the module and measure the upstream and downstream pressures for a long time [98].

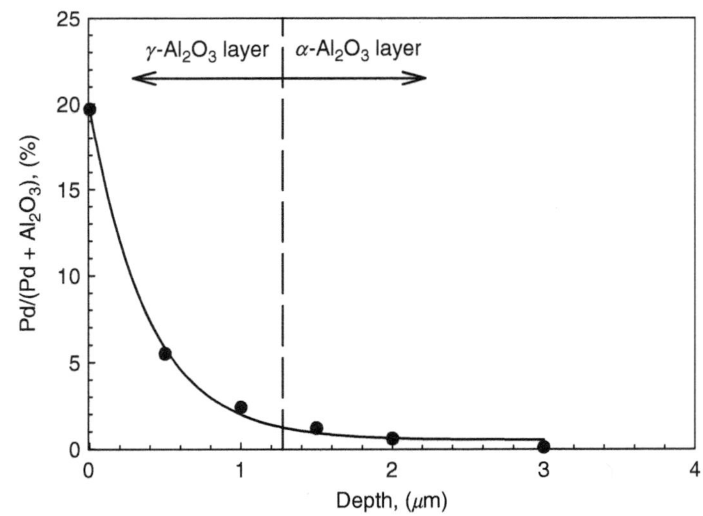

Figure 8.15 Palladium distribution along the cross section of a composite palladium membrane obtained using the co-condensation technique.

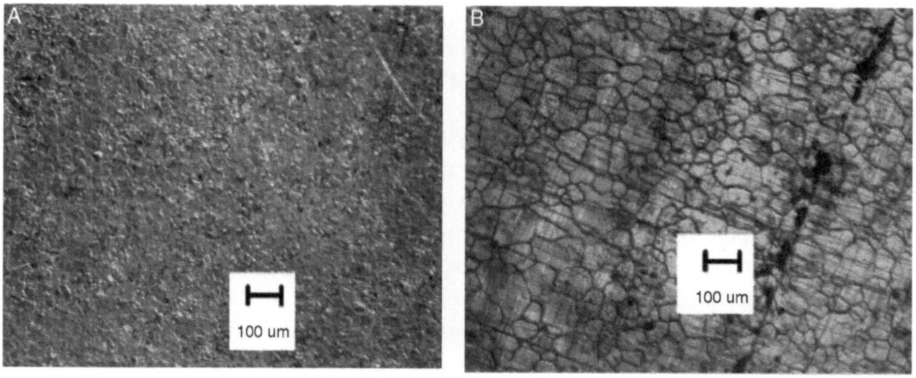

Figure 8.16 Pd–Ag membrane surface before testing (A) and after testing (B) [98].

In fact, the characterizations previously described in the last section are mainly devoted to analyzing the membrane surface and the membrane bulk, while the gas permeation test shows how the gas permeates through the membrane under different operative conditions. These data are very important to better understand what happens when membrane reactors are operating and also for both modeling the process and designing the membrane reactors and palladium-based membranes as a permeator itself.

Regarding the porous palladium membranes, it has to be noticed that, in principle, all components of a feed mixture can permeate through a porous membrane. Neglecting surface and thermo-diffusion effects, the Knudsen diffusion, the molecular diffusion and the viscous (or Poiseuille) fluxes are the main contributions to the overall flux [99, 100]. In contrast to the viscous flux, both the Knudsen diffusion and the molecular diffusion are selective. A limiting selectivity obtainable in porous systems is usually the Knudsen selectivity related to the molecular weights of the components [101]. To calculate the diffusive flux through the porous membranes the dusty gas model (DGM) gives better results for the process description [101]. The DGM is based on the combination of the Stefan–Maxwell diffusion equations and the characteristics of mass transfer in porous solids. In the case of linear pressure profile across the membrane, the DGM can be reduced to describe a single component diffusion as follows:

$$J_i = \frac{1}{RT}\left(\frac{Pe}{\eta}\frac{p_{\text{in}} + p_{\text{out}}}{2} + D^e_{K,i}\right)\frac{p_{\text{in}} - p_{\text{out}}}{t_{\text{Sup}}} \tag{8.1}$$

Where: η is the viscosity, Pe the permeability constant of the porous substrate which is related to the viscous flux, t_{Sup} the support thickness, and $D^e_{k,i}$ the effective Knudsen diffusivity.

To calculate the constants in the formula, permeation experiments for each i-species are necessary. By performing experiments at different pressures and by plotting $Ji/\Delta p$ versus the main pressure $[(p_{\text{in}} + p_{\text{out}})/2]$, both the effective Knudsen diffusivity $D^e_{K,i}$ and the permeability constant can be determined by using a linear regression. These constants can be used in the model for the gas permeation. It is clear that the permselectivity and the flux through the porous membrane strongly depend on the thickness and the quality of Pd-based layer deposited on the support.

3.5.1. The effect of pressure on the hydrogen flux

As is well known, palladium has considerable solubility for hydrogen and absorbs about 600 times its volume of hydrogen at room temperature [102]. However, a serious disadvantage of Pd is that it undergoes a phase transition between the α- and β-phases near room temperature when exposed to hydrogen [103]. This makes the Pd material brittle and eventually leads to micro-cracks in the metal. To avoid this problem, it is necessary to operate above both 298 °C and 2 MPa. In fact, as is shown in Fig. 8.17, where the equilibrium solubility isotherms of PdH_n for the bulk Pd at different temperatures are reported, in this region the β-phase is not present [104].

The transport mechanism through palladium membranes can be divided into two types:

- traditional solution–diffusion mechanism for dense metal membranes such as palladium-based membranes, and
- Knudsen diffusion mechanism or combined mechanism of Knudsen diffusion and surface diffusion for porous metal membranes such as stainless steel and microporous noble metals.

Specifically, concerning the solution–diffusion mechanism, the hydrogen permeation through a palladium layer is a complex process that consists of the adsorption and dissociation of hydrogen molecules, followed by a diffusion of hydrogen atoms through the metal lattice, the recombination of hydrogen atoms on the low pressure side and the desorption of molecular hydrogen according to the following 7-steps activated mechanism [55]:

1. adsorption of hydrogen molecules on the high hydrogen pressure side of the membrane surface;
2. dissociation of hydrogen molecules on the same surface;

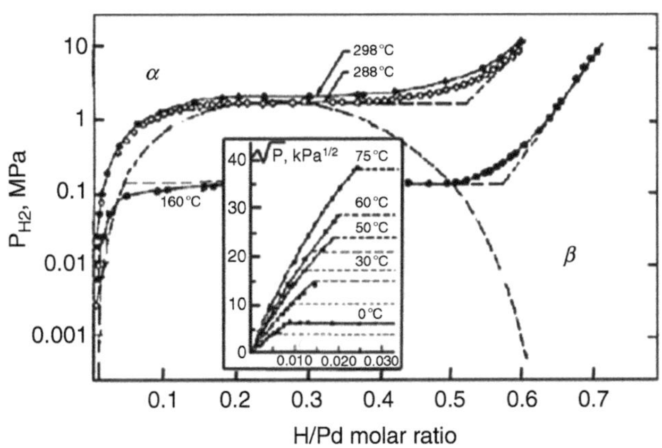

Figure 8.17 Palladium–hydrogen phase diagram; the low pressure region is also shown [119].

3. the chemisorbed molecule dissociates into atomic form H•, proton, and electron;
4. the dissociated hydrogen atoms H• dissolve into the lattice of the metal;
5. the proton and electron H• diffuse through the lattice from the high hydrogen pressure of the membrane surface side to the low hydrogen pressure of the membrane surface side;
6. on the low hydrogen pressure of the membrane surface side the proton and electron re-associate and recombine into hydrogen molecules H_2;
7. the hydrogen molecule desorbs from the low hydrogen pressure of the membrane surface side to the bulk.

In particular, compared with those of surface reactions, including hydrogen adsorption, dissociation, reassociation, and desorption at relatively high temperatures, for commercial available palladium-based membranes, step 5 (diffusion of hydrogen atoms through the lattice) is particularly slow. Under these conditions, the thermodynamic equilibrium between the hydrogen molecules in the gas phase and hydrogen atoms dissolved at the interface would be established.

The hydrogen permeation at constant temperature through a dense palladium membrane is described by the following general expression:

$$J_{H_2} = \frac{P_{H_2}(p^n_{H_2,ret} - p^n_{H_2,perm})}{t} \tag{8.2}$$

where J_{H_2} is the hydrogen flux through the membrane; P_{H_2} the hydrogen permeability; t the membrane thickness; p the hydrogen partial pressure either on the retentate (high hydrogen partial pressure) side or on the permeate (low hydrogen partial pressure) side; n the dependence factor of the hydrogen flux on the hydrogen partial pressure, generally in the range 0.5–1.

In the region of low pressures (see Fig. 8.17) the previous expression becomes:

$$J_{H_2,Sievert} = \frac{D_{H_2} \times K_S(p^{0.5}_{H_2,ret} - p^{0.5}_{H_2,perm})}{t} \tag{8.3}$$

K_S being Sievert's constant of Sievert's law:

$$C_{H_2} = K_S \times p^{0.5}_{H_2} \tag{8.4}$$

where C_{H_2} is the hydrogen concentration on the membrane surface.

To be observed the particular value of n (equal to 0.5) reported in this expression.

The product of the diffusion coefficient D_{H_2} and Sievert's constant K_S is the permeability coefficient $P_{H_2,Sievert}$:

$$P_{H_2,Sievert} = D_{H_2} \times K_S \tag{8.5}$$

Equations (8.2) and (8.3) describe the fact that under diffusion controlled conditions the hydrogen flux increases with $1/t$. Equation (8.3) is valid only for very low value of n (i.e., 0.5), which corresponds to the α phase at relatively low pressure, as is indicated in Fig. 8.17. At high pressures, hydrogen–hydrogen interactions within the palladium bulk are not negligible and Eq. (8.3) is no longer valid. For instance, at 160 °C, the hydrogen adsorption isotherm (hydrogen loading, H/Pd, as a function of $p^{0.5}$) is linear at low pressures, but starts to curve as the miscibility gap is approached.

The term $(P^{0.5}_{H_2,ret} - P^{0.5}_{H_2,perm})$ represents the driving force of the hydrogen permeation at relatively low pressures. However, the dependence factor of the hydrogen flux on the hydrogen partial pressure sometimes deviates from the value 0.5. In this way, the value of n is often used as an indicator for the rate-controlling step of the hydrogen permeation through the palladium (composite or not) membrane. In fact, if the diffusion of atomic hydrogen through the dense metal layer is the rate-limiting step, then the hydrogen flow is proportional to the difference of the square root of the hydrogen partial pressure on both sides of the membrane. In other words, Sievert's law is valid and $n = 0.5$.

Deviations from Sievert's law (i.e., $n > 0.5$) have been reported in the literature, and are generally expected for very thin membranes where the bulk diffusion of hydrogen through the palladium layer is faster than the surface dissociation/recombination reactions at either/or both sides of the membrane (see, e.g., [78, 105, 106]).

For thick membranes (<1 μm), deviations from Sievert's law can be caused by high hydrogen pressure values. From an experimental point of view, the hydrogen permeation through thick palladium foils was studied, for example, by Morreale *et al.* [107] at very high pressures, and n values were calculated based on permeation data from four pressure ranges (1.01–1.75, 1.01–7.75, 1.00–16.00, and 1.01–26.00 bara). They reported an increase in n values from 0.53 (the limit value) to 0.65 as the pressure range increased. From a theoretical point of view, based on a number of palladium membranes reported in the literature that presented different thicknesses and different n values, a hydrogen permeation model was proposed by Ward and Dao [108]. Their study was mainly directed to understand under what conditions hydrogen permeation was controlled by surface reactions or bulk diffusion. Their calculations show that, at temperatures above 400 °C, bulk diffusion was the rate-limiting step for membranes thicker than 1 μm if the hydrogen desorption step activation energy is equal to 12 kJ/mol. This result is in good agreement with several other works [79, 82, 108].

In contrast, for thinner membranes, where higher hydrogen fluxes can be achieved, the surface reaction processes gain importance and can become the rate-limiting steps, and the flux of hydrogen is proportional to the difference of the partial pressures:

$$J_{H_2} = \frac{P_{H_2,linear}\left(p_{H_2,ret} - p_{H_2,perm}\right)}{t} \tag{8.6}$$

$P_{H_2,linear}$ being the hydrogen permeability coefficient (and $n = 1$).

However, in practice there is an interplay of diffusion and surface control. Unfortunately, it is not yet clear at which palladium thickness surface processes actually begin to control the hydrogen permeance, and no detailed information is yet available in the literature on the influence of the microstructure of differently produced palladium layers on the hydrogen permeance and the hydrogen permselectivity.

Another important aspect to also be considered is the way the film is deposited on a porous support. For example, membranes prepared using CVD and electroless-plating techniques showed a complicated behavior for hydrogen permeation. In fact, the supported palladium membranes prepared by Uemiya *et al.* [118] and Yan *et al.* [79], though giving almost the same hydrogen permeance, exhibited a different dependence of the hydrogen flux on the hydrogen partial pressure: 0.5 for Uemiya *et al.* and 1.0 for Yan *et al.*

Uemiya *et al.* [118] concluded that hydrogen permeates through the CVD palladium membrane via the same mechanism as that observed for the membrane prepared using the electroless-plating technique.

Vice versa, Yan *et al.* [79] considered that a combination of the diffusion through palladium lattice and adsorption of hydrogen onto the surface of palladium was responsible for the transport through the membrane. In some cases, an increase in hydrogen permeance was also observed and attributed to surface morphology changes.

3.5.2. The effect of temperature on the hydrogen flux

The hydrogen transport through a dense palladium film is an activated process, and in many practical cases Sievert's law defines the pressure dependence of the hydrogen permeation. It is assumed that the hydrogen pressure exponent does not depend on temperature. Concerning the influence of temperature on the hydrogen permeability, the relationship between the hydrogen permeation rate and the temperature is expected to be described by an Arrhenius law:

$$P_{H_2} = P^{o}_{H_2} \exp\left(-\frac{E_a}{RT}\right) \tag{8.7}$$

where $P^{o}_{H_2}$, $E_{a,}$, R, and T are the pre-exponential factors, the apparent activation energy, the universal gas constant, and the absolute temperature, respectively.

Consequently, the hydrogen flux can be written in terms of the so-called Richardson's equation:

$$J_{H_2} = \frac{P^{o}_{H_2}[\exp(-E_a/RT)](p^{0.5}_{H_2,ret} - p^{0.5}_{H_2,perm})}{t} \tag{8.8}$$

In order to determine the two parameters $P^{o}_{H_2}$ and E_a (and, eventually, the constant n different from 0.5, if Sievert's law is not valid), permeation tests at different pressures and temperatures are necessary. Different authors estimated both the values of the pre-exponential factor and the apparent activation energy for palladium alloy dense membranes using an Arrhenius plot of the hydrogen permeance against the reciprocal temperature.

By referring to a dense Pd/Ag membrane, in Fig. 8.18, a typical example of (unpublished) experimental data related to the hydrogen flux versus the square root of the hydrogen partial pressure for pure hydrogen permeation at $T = 350\,^{\circ}\mathrm{C}$ is reported, while in Fig. 8.19 the Arrhenius plot for pure hydrogen permeation tests using a Pd–Ag membrane, $p_{\mathrm{lumen}} - p_{\mathrm{shell}} = 4.0$ bar, is shown.

Figures 8.18 and 8.19 show that both the Sievert and Arrhenius laws, respectively, are followed in the case of a Pd–Ag membrane.

Table 8.4 proposes a comparison of permeation parameters for Pd–Ag dense membranes, as reported in the recent specialized literature, while in Table 8.3 similar data for Pd-based composite membranes are also collected.

By summarizing, in Table 8.4 typical values for E_a and $P^{\mathrm{o}}_{\mathrm{H_2}}$ are in the range 12.5–48.5 kJ/mol and $3.8 \times 10^{-6} - 9.3 \times 10^{-5}$ mol m/s m^2 kPa$^{0.5}$, respectively; while in Table 8.3, E_a and $P^{\mathrm{o}}_{\mathrm{H_2}}$ are in the range 7.1–33.3 kJ/mol and 1.7×10^{-5}–1.4×10^{2} mol m/m^{2}s Pa$^{0.5}$, respectively.

3.5.3. The effect of contamination of the palladium surface on the hydrogen flux

Apart from the consideration reported above, the deviations from Sievert's law may also result from a surface contamination of mercury vapor, carbon from organic materials, and compounds containing chemical elements such as S, As, I, or Br. Membranes may show a permeation controlled by the rate of adsorption on heavily contaminated surfaces. In all these cases, the rate-limiting step is not bulk hydrogen diffusion: the hydrogen dissociative adsorption plays an important role.

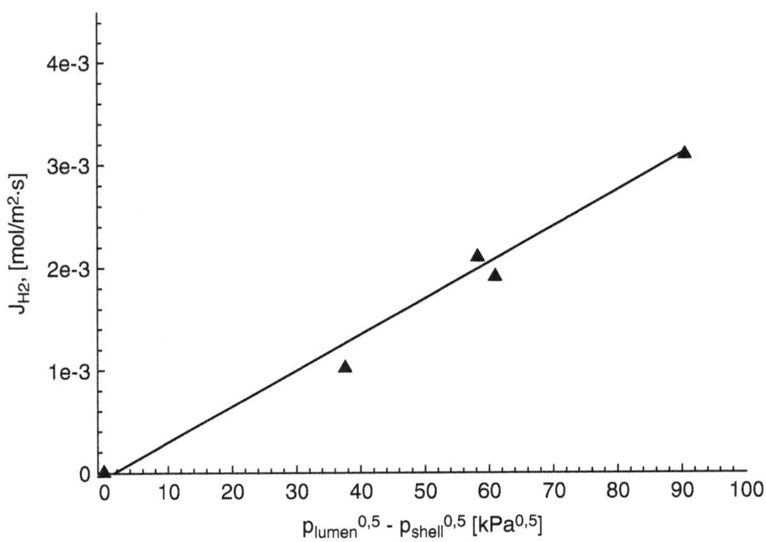

Figure 8.18 Sievert plot for pure hydrogen permeation using a dense 50 μm Pd–Ag membrane at $T = 350\,^{\circ}\mathrm{C}$.

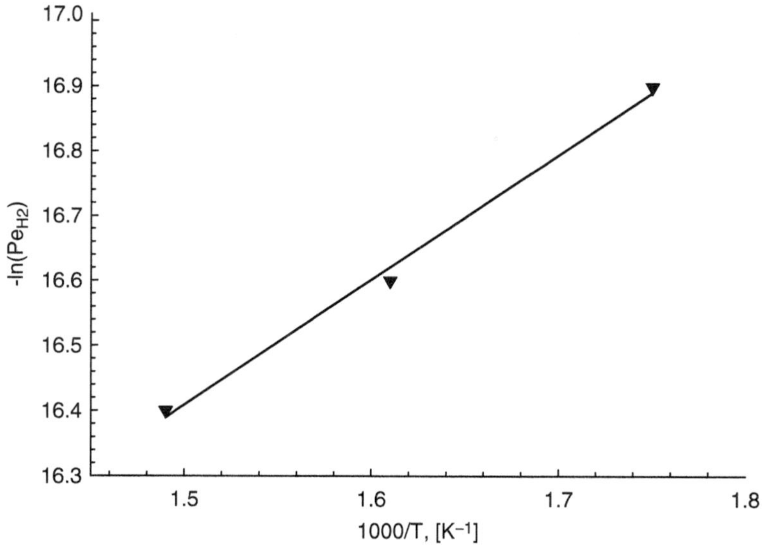

Figure 8.19 Arrhenius plot for pure hydrogen permeation tests using a dense 50 μm Pd–Ag membrane, $p_{lumen} - p_{shell} = 4.0$ bar.

Table 8.4 Apparent activation energy and pre-exponential factor for hydrogen permeation through Pd–Ag membranes from the literature

E_a (kJ/mol)	$P^O_{H_2}$ [10^{-5} mol m/(s m^2 kPa^{05})]	References
48.50	9.33	[109]
29.73	7.71	[110]
15.70	2.19	[111]
15.50	2.54	[112]
12.48	0.38	[113]
33.31	1.66	[114]

For thick films, deviations from Sievert's law can also be due to a decrease in the surface reactions rate after adsorption of contaminants, such as C, CO, CO_2, hydrocarbons, sulfur, or chlorine, on the palladium surface. In these cases, for Pd–Ag membranes an air oxidation at 300–450 °C for the removal of adsorbed contaminants is necessary [5, 115, 116].

In particular, it is well known that the presence of carbon monoxide decreases hydrogen permeation through a palladium membrane because of a blockage of their surface. The lower the temperature the stronger is the influence of this effect. Amandusson *et al.* [96] found a complete blockage of the palladium at 100 °C, and therefore, no permeation of hydrogen through the membrane at this temperature was observed. At 250 °C a decrease in the flux of only 10% was observed, while at temperatures above 350 °C no evidence for any influence of carbon monoxide was found at all. Many explanations were proposed for the observed deviation: contamination of membrane surface, dependence of the hydrogen diffusion coefficient on the dissolved hydrogen, hydrogen leakages through cracks or defects, and so on. However, a definite evidence has not been presented.

3.5.4. The effect of other metals

The superiority of palladium–silver alloy with respect to pure Pd, particularly at the silver content of 23 wt%, is due to both the hydrogen permeability and the hydrogen embrittlement suppression by the transformation of palladium hydride from the α to β phases. As can be seen in the palladium–hydrogen phase diagram (Fig. 8.17), this transformation cannot be avoided when pure palladium faces hydrogen below 298 °C. When the temperature is below this temperature and the pressure below 2 MPa, the β hydride may nucleate from the α phase that results in severe lattice strains so that the pure palladium membranes become brittle after a few cycles of $\alpha \Leftrightarrow \beta$ transitions. It is interesting to observe that these transitions do not correspond to a change of the lattice structure, but only to a lattice dilatation. For instance, the face-centered cubic (f.c.c.) lattice constant of palladium is 3.891 Å, the one for $PdH_{0.03}$ is 3.894 Å, and for $PdH_{0.6}$ it is 4.018 Å.

As can be seen in Table 8.5, where the diffusion coefficient and the parameters for determining the solubility in Pd–Ag alloys ($n \ll 1$, $T = 260$–640 °C at a specific pressure) are reported, it is clear that by adding silver to palladium there is an increase in hydrogen solubility [calculated using Eq. (3.9)], while the hydrogen diffusion coefficient significantly decreases.

Sievert's constant depends on the temperature as follows:

$$K_S = \left(\frac{1}{n_s}\right) \exp\left(\frac{\Delta H^o}{RT} - \frac{\Delta S^o}{R}\right) \tag{8.9}$$

where n_s is the geometrical-limiting H/Pd ratio; ΔH^O the heat of reaction and ΔS^O the entropy of hydrogen dissolution.

Thanks to the simultaneous increases in hydrogen solubility and decreases in hydrogen diffusion coefficient with the silver content, at a fixed temperature and

Table 8.5 Diffusion and solubility parameters in Pd–Ag alloys, $n \ll 1$, $T = 260$–640 °C, (Shu *et al.* [119])

Ag, atom (%)	$D_o \cdot 10^7$ (m²/s)	E_D (kJ/mol)	ΔH^o (kJ/mol)	ΔS^o (J/mol·K)
0	2.94	22.0	8.4	48.7
10	2.69	22.2	11.6	49.4
20	2.33	22.3	14.1	50.2
30	2.21	24.7	15.3	49.2
40	1.50	26.4	16.8	51.0
50	1.82	33.5	14.8	50.4

pressure the hydrogen flux through a Pd–Ag membrane plotted versus the silver content reaches a maximum value. In Fig. 8.20 the hydrogen flux versus the weight alloy content of different alloy elements is shown.

Specifically, at 350 °C and 2.2 MPa, this maximum value is reached for 23% Ag alloy and corresponds to a permeability value 1.7 times higher than the hydrogen permeability related to pure palladium.

In the same figure the hydrogen flux through other palladium-based alloys versus some specific metal content is also reported. It is interesting to observe that also the Pd–Cu alloy shows a maximum (at 40 wt% Cu). At this composition this alloy forms an equilibrium phase having body-centered cubic (b.c.c.) structure with a simultaneous increase in diffusivity by two orders of magnitude.

Hydrogen permeance of Pd–Cu alloy membranes over a wide range of temperatures and pressures has recently been studied by Howard *et al.* [120]. Their membranes show some decrease in hydrogen permeance when exposed for long time at 1173 K. In particular, the decline was most pronounced when the 20% Cu.

Another interesting study is the one of Roa and Way [121]. They found that when Pd and Pd–Cu membranes, prepared by electroless deposition on ceramic supports, are exposed to air, at temperatures above 350 °C, the H_2 flux increased substantially immediately after the air exposure and then decreased to a new steady-state value. The authors also found that the air exposure produces some oxidation of the palladium film and produces PdO, suggesting that the common hypothesis stating that air oxidation just cleans the surface of the membrane might not be sufficient to explain all the changes they observed.

Concerning other literature data regarding these alloys can be cited, for example, Nam and Lee [122]. Recently, they were able to fabricate a structurally stable and pinhole-free Pd–Cu alloy composite membranes with a diffusion barrier on mesoporous stainless steel supports by vacuum electrodeposition. Their Pd–Cu alloy composite membrane yielded excellent separation performance for hydrogen: hydrogen permeance of 2.5×10^{-2} cm^3/cm^2 s cmHg and H_2/N_2 selectivity above 70,000 °C at 723 K.

Hoanga *et al.* [123] using a sputtering technique deposited submicron thin Pd–Cu alloy films. Their membrane shows a high hydrogen flux of 1.6 mol/m^2 s and H_2/He selectivity at least 500 at 725 K.

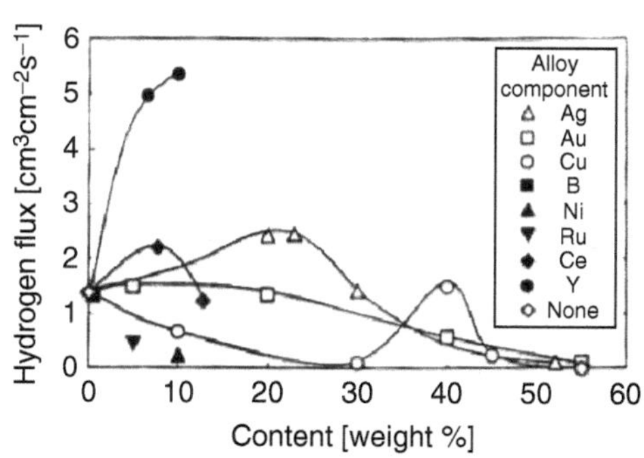

Figure 8.20 Hydrogen flux through Pd alloy membranes at 350 °C and 2.2 MPa [119].

Unfortunately, also the Pd–Cu alloy shows a hydrogen permeation and selectivity reduction in presence in the process streams of aggressive compounds such as H_2S, which practically deactivates the metal membrane [124].

In general, making a Pd-alloy with other metals such as Cu, Ni, Fe, Pt, and Ag able to maintain high hydrogen permeability, but minimizing $\alpha \rightarrow \beta$ PdH phase transition, reduces the chemisorption of poison agents such as CO, sulfur, and H_2O and, minimizing coking in WGS and hydrocarbon steam reforming, is the most commonly used method to improve the stability.

In conclusion and in general, a Pd alloy with other metals such as Cu and Ag maintains very high hydrogen permeability and, minimizing the $\alpha \rightarrow \beta$ PdH phase transition, reduces embrittlement, which is a serious disadvantage characteristic of pure palladium.

4. Palladium Membrane Reactors

4.1. A brief history on the development of membrane reactors

In 1866, Graham (UK) first used palladium to separate H_2 from gaseous mixtures [125]. Pd was also used for the catalytic activation of chlorine—water to hydrochloric acid at room temperature. In 1898, Zelinskii (Moscow University) used the first composite membrane catalyst by combining hydrogen production during hydrochloric acid reaction with zinc and hydrogenation by a palladium sponge covering the zinc granules. In 1915, Snelling (USA) patented hydrogen removal through palladium or platinum tubes from a reactor using a granular catalyst of dehydrogenation. In 1935, Kobosev (Moscow University) investigated the hydrogenation on palladium membranes by hydrogen supplied as a gas to other surfaces. Wagner *et al.* (Germany) in 1939, and Wood *et al.* (USA) in 1966, also studied hydrogenation on palladium membranes by hydrogen supplied as a gas to other surfaces. In 1956, a practical metal membrane was first invented by Hunter, who patented palladium-alloy membranes [126, 127].

Finally, in 1964, the academician Professor Vladimir Gryaznov (Moscow University) first proposed a method for carrying out simultaneously the evolution and the consumption of H_2 in a dense tubular palladium reactor, where palladium is permeable to H_2 only and also serves as a catalyst. The reaction studied was cyclohexane dehydrogenation to benzene [128, 129].

Publications of the dense membrane reactor concept first started appearing in the 1970s.

On the other hand, almost 50 years ago, porous inorganic membranes were industrially developed in order to separate UF_6 using gas phase diffusion processes.

The application of porous ceramic membranes as catalytic reactors started in the 1980s.

In the following years, as summarized in Table 8.6, several works concerning high temperature membrane reactors for different applications were performed, in terms of both experimental studies and mathematical models.

A large number of papers have been published on this multidisciplinary subject at the frontier between catalysis, membrane science, and chemical engineering. A database research shows the still growing interest in catalytic membrane reactors. In particular, the history of the development of supported metal membranes for hydrogen separation is well summarized by Uemiya in a recent paper [20].

Table 8.6 Several high temperature membrane reactors for different applications

Kind of reaction	Membrane	Material
Claus reaction for the desulfurization of gases	Porous	Al_2O_3
Coupling of hydrogenation and dehydrogenation reactions	Dense	Pd
Decomposition of H_2S	Porous	γ-$A1_2O_3$
Decomposition of H_2I	Dense	Pt
Decomposition of RuO_4 to $RuO_2 + O_2$	Dense	Pd/Ag
Decomposition of water	Porous	Fe_2O_3 + molecular sieve
Decomposition of ammonia	Dense	Pd
Decomposition of CO_2	Dense	Y stab. ZrO_2
Dehydrogenation of cyclohexane to benzene	Porous	Al_20_3
	Dense	Pd/Ag
Dehydrogenation of ethylbenzene to styrene	Porous	Al_2O_3/Pd
Dehydrogenation of ethane to ethylene	Porous	Al_2O_3
	Dense	Pd/Ag
Dehydrogenation of propane to propene	Porous	Al_2O_3/silica
Dehydrogenation of propane to propylene	Porous	Al_2O_3
Dehydrogenation of 2- or 3-methyl-1-butene to isoprene	Porous	Al_2O_3
Dehydrogenation of isopropyl alcohol to acetone	Dense	Pd
Dehydrogenation of water gas shift reaction	Dense	Pd, Pd/Ag
	Porous	Pd
Dehydrogenation of *n*-heptane to toluene + benzene	Dense	Pd/Rh
Dehydrogenation of butene to butadiene	Dense	Pd
Dehydrogenation of 1,2-cyclohexanediol	Dense	Pd/Cu
Dehydrogenation of isobutene	Porous	silica
Dehydrogenation of isobutene	Dense	silica
Dehydrogenation of methanol	Porous	γ-$A1_2O_3$
Dehydrogenation of *n*-butane	Porous	γ-$A1_2O_3$
Dehydration of isopropanol to propylene	Porous	$A1_2O_3$
Dry reforming of methane	Dense	Pd-alloy
Hydrogenation of ethylene to ethane	Porous	$A1_2O_3$

(*continued*)

Table 8.6 (*Continued*)

Kind of reaction	Membrane	Material
	Dense	Pd
Hydrogenation of butadiene	Dense	Pd
Hydrogenation of acetylene	Dense	Pd/Ag
Hydrogenation of butenes	Dense	Pd/Sb
Hydrogenation of diene hydrocarbons	Dense	Pd/Ru
Hydrogenation of dehydrolinalool	Porous	stainless steel
Hydrogenation of CO_2 to methanol	Porous	zeolite, mordenite
Hydrogenation of nitrobenzene	Porous	Pt on γ-$A1_2O_3$
Hydrogenation of cyclohexene	–	Al_2O_3-supp. RhS
Hydrogenation of phenol to cyclohexanone	Dense	Pd; $Pd_{93}Ni_7$; $Pd_{93}Ru_7$; $Pd_{77}Ag_{23}$
Hydrogenolysis of ethane to methane	Porous	Al_2O_3
Hydrodealkylation of dimethylnapthalenes	Dense	Pd/Ni
Hydrodesulfurization of triophene	Dense	Al_2O_3-supp. RhS
Fisher—Tropsch Synthesis	Dense	Carbon
Methane conversion into hydrogen and higher hydrocarbons	Porous	Pd-al1oy
Octane reforming	Dense	Pd and Pd-al1oy
Oxidative dehydrogenation of methanol	Porous	Al_2O_3
Oxidative dehydrogenation of propane	Porous	Zeolite
Oxidative dehydrogenation of l-butene to butadiene	Porous	Zr–Y–Ti oxides
Oxidative dehydrogenation of ethanol to acetaldehyde	Porous	Al_2O_3
Oxidative methane coupling	Porous	Zr–Y–Ti oxides/ perovskite-type oxides
	Dense	Ceramic (with catalyst deposited on)
Oxidative dimerization of propylene to benzene and hexadiene	Porous	Ceramic
Oxidation of CO to CO_2	Porous	Al_2O_3
Oxidation of ammonia to nitrogen	Dense	Ag
Oxidation of ethylene to ethylene oxide	Porous	Zr–Y–Ti oxides
Oxidation of propene	Porous	MoO3 on Au/YSZ/Ag
Oxidation of methane to methanol	Porous	ZrO–CaO–Al_2O_3
Oxidation of toluene	Porous	Al_2O_3
Partial oxidation of methane	Porous	Perovskite and Pd-alloy
Steam reforming of ethanol	Dense	Pd and Pd-al1oy
Steam reforming of methane	Dense	Pd-alloy
Steam reforming of methanol	Dense	Pd and Pd-alloy

Figure 8.21 shows the main application opportunities of inorganic membrane reactors. Specifically, (A) and (B) refer to the conversion enhancement of an equilibrium limited reaction by the selective permeation of a reactant product through the membrane and by the coupling of reactions in both sides of the membrane, respectively, while (C) and (D) refer to the selectivity enhancement by selective permeation of an intermediate product and by dosing a reactant through the membrane, respectively.

4.2. Commercial and potential applications of palladium-based membrane reactors

The first commercial application of a metal membrane was hydrogen purification using a 23 wt% palladium–silver membrane by Johnson Matthey in 1964 [130]. The process was used in small-to-medium scale plants, with individual units for the production of up to 56 m^3/h of very high purity hydrogen (impurities <0.1 ppm total) [131].

In the following years, Johnson Matthey also developed a hydrogen generator consisting of a Pd membrane reactor, fed with a methanol/water mixture. The hydrogen production capacity was up to 25 m^3/h. A small scale prototype of this unit was used by the British Antarctic Survey in 1975 [132, 133].

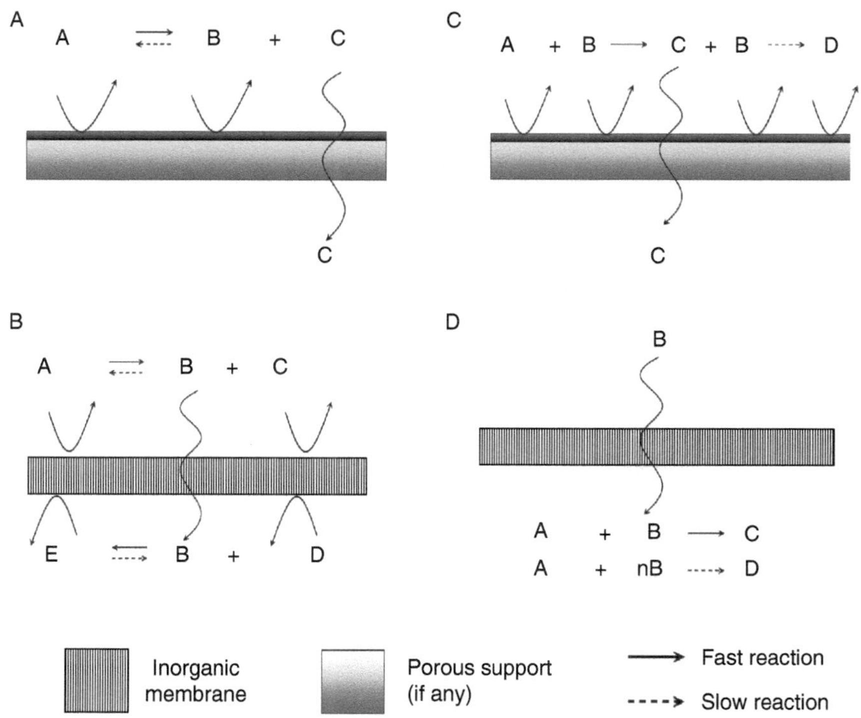

Figure 8.21 Main application opportunities of inorganic membrane reactors.

Most of the work on the development of membrane reactors originated from the Soviet Union, and more recently from Japan, where several small scale systems were commercialized. In all these cases, according to Shu *et al.* [119], it seems that the metal cost was not a limiting step for the application of palladium-based membranes in several small to medium processes.

A short list of the most representative membranes commercialized in the last 10–15 years is shown in Table 8.7.

In the following, different applications of membrane reactors are presented.

4.2.1. The first pilot-scale membrane reformer for hydrogen production

The first pilot-scale membrane reformer for direct ultrapure hydrogen production has been realized by the largest gas company in Tokyo (Tokyo Gas Company Ltd.). This reformer, indicated as *World's first membranes type hydrogen producer* [134, 135] (Fig. 8.22), is a membrane reactor of cylindrical shape with external dimensions: diameter of 0.6 m and length 1 m. It is constituted of a 0.18 m diameter furnace, containing a catalytic bed (height 0.735 m and length 0.038 m) and a certain number of membranes (in the case of tubular membranes they are 24 with length 0.6 m). The reaction of interest is the steam reforming of methane coupled to the water gas shift (WGS) reaction. The feed consists of city gas, having the molar composition: methane 88.5%, ethane 4.6%, propane 5.4%, and butane 1.5%. The hydrogen produced from the reaction is selectively removed by means of composite membranes (tubular or flat: today the flat ones are used; however, both are constituted of a stainless steel support on which a dense palladium layer of 20 μm is deposited with the electroless plating technique).

The purity of hydrogen produced is >99.999% with a flow rate of 40 Nm^3/h. The steam/methane ratio in the feed stream is 3. The reaction temperature is in the range between 500 and 550 °C. In particular, at 550 °C, methane conversion approaches 70%. Among the advantages of this particular system with respect to the traditional ones:

- an increase in methane conversion for the steam reforming of methane (SRM) reaction, thanks to the selective separation of hydrogen by means of palladium membranes;
- pure hydrogen production; lowering of the reaction temperature, from 800 to 500–550 °C.

4.2.2. The Japanese RITE project

The Research Institute of Innovative Technology for the Earth (RITE) project was an interesting research project concerning the CO_2 fixation and utilization based on its catalytic hydrogenation [134]. The Research Institute of Innovative Technology for the Earth (Japan) carried out this project from 1990 to 1999 which consisted of the collection of the CO_2 from the emission gases of power plants and factories and its conversion, by adding H_2 (produced using electrical power generated by natural energy) into substances such as methanol: Fig. 8.23 illustrates the conceptual approach followed for the RITE project. The use of MRs for the methanol production played the most important role for the success of the whole project.

Table 8.7 Some representative membranes commercialized in the last 10–15 years

Manufacturer	Trade Name	Membrane material	Pore size	Geometry
Asahi Glass		Glass	0.1–1.4 μm	Tube
ATECH		SiC/SiC	0.05–1.0 μm	Tube/Monolith
		Al_2O_3		
Carré		ZrO_2/SS		Tube
Ceramem	CERATREX®	Ceramics/Ceramics	0.1 μm	
		Ceramics/Cordierite	0.05–0.5 μm	Honeycomb
CERASIV		Al_2O_3/Al_2O_3	0.1–1.2 μm	Tube/Monolith
		$TiO_2/ZrO_2/Al_2O_3$	5–100 nm	
Du Pont/Carré		$Zr(OH)_4$/SS	0.2–0.5 μm	Tube
Fairey	STRATA-PORE®	Ceramics/Ceramics	1–10 μm	
Fuji Filters		Glass	4–90 nm	Tube
		Glass	0.25–1.2 μm	
Gaston County	UCARSEP®	ZrO_2/C	4 nm	Tube
Hoogovens		Al_2O_3/Al_2O_3	0.1–1 μm	Tube
Inocermic	INOPOR®	α-$Al_2O_3 + TiO_2 + ZrO_2$	3–1000 nm	Monochannel and multichannel tubes with a lenght up to 1200 mm
Koch	SelRO®	SS		Tube
Le Carbone Lorraine	KERASEP®	$TiO_2/Al_2O_3 + TiO_2$	0.1–0.45 μm	Monolith
		$ZrO_2/Al_2O_3 + TiO_2$	15–300 kD	
NKG		$Al_2O_3/Al_2O_3 + SiO_2$	0.5–5 μm	Tube/Monolith
Osmonics	MICROFILTRES®	SS	0.2–1.0 μm	
	HYTREX®	Ag	0.2–5.0 μm	
Schumacher		TiO_2	5 nm	Tube
Steenecker		Al_2O_3	0.4 μm	Tube

(continued)

Table 8.7 *(Continued)*

Manufacturer	Trade Name	Membrane material	Pore size	Geometry
NOK		Al_2O_3/Al_2O_3	0.2–6 μm	Tube
TDK	DYNACERAM®	Zr/Al_2O_3	10 nm	Tube
		Al_2O_3/Al_2O_3	50 nm	
TECH-SEP	CERAFLO®	Al_2O_3/Al_2O_3	0.2–1.0 μm	Monolith
	CARBOSEP®	ZrO_2/C	10–300 kD	Tube
		TiO_2/C	0.14 μm	
TOTO		Al_2O_3/Al_2O_3	0.1–0.2 μm	Tube
		ZrO_2/Al_2O_3	5–30 nm	Monolith
Trideltafiltration		Al_2O_3/Al_2O_3	0.1–7 μm	Tube/Monolith
USF/SCT	MEMBRALOX®	Zr/Al_2O_3	20–100 nm	Monolith
		Al_2O_3/Al_2O_3	5 nm–12 μm	
Whatman	ANOPORE®	Al_2O_3	20 nm–0.2 μm	Disk

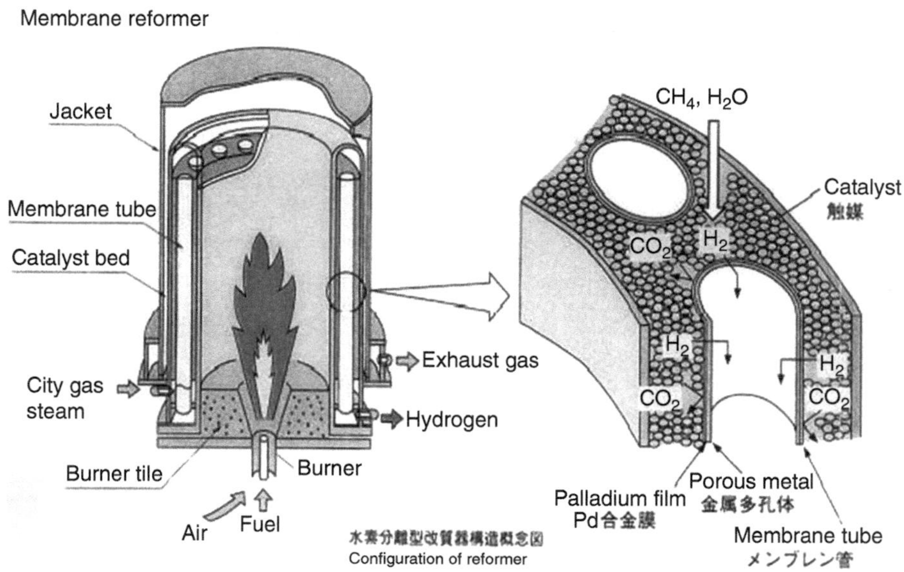

Figure 8.22 World's first membranes type hydrogen producer, by Tokyo Gas company, Ltd.

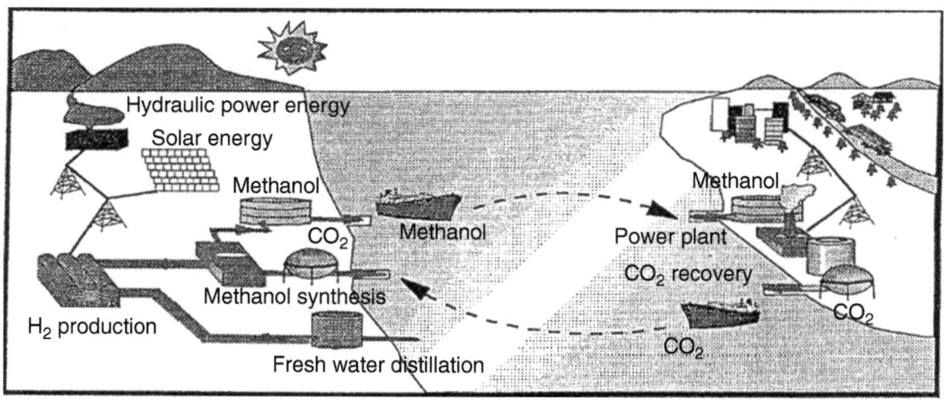

Figure 8.23 The conceptual approach used for the RITE project.

The research included the development of polymeric separation membrane and modules with high CO_2 selectivity.

4.2.3. The water gas shift reaction in the fusion reactors fuel cycle

Another direction for the hydrogen production is WGS reaction. This reaction is present in several industrial processes, such as catalytic steam reforming of hydrocarbons, coal gasification, and production of ammonia. WGS reaction is also particularly interesting from the point of view of the application of membrane reactors in the fuel cycle of the fusion reactors where the hydrogen isotopes are separated and purified from the gaseous mixtures. In the next sections special attention will be paid to the potential applications of membrane reactors to different fusion reactors.

4.2.4. The water gas shift in the NET application

The future fusion reactor fuel will be a mixture of deuterium and tritium: while deuterium can be relatively easily produced using traditional separation technology, tritium must be produced by means of a nuclear reaction between neutrons and lithium atoms within the fusion reactor breeder blanket (made of lithiated material). The tritium produced in the breeder needs a proper extraction process to reach the required purity level.

The Next European Torus (NET) is a process of an experimental fusion reactor developed at the end of 1980s; a scheme of the reactor is shown in Fig. 8.24. In the NET a process based on a chemical reaction between tritium and oxygen has been proposed in order to recover tritium from the blanket in the form of tritiated water. Then, the tritiated water is separated from the gaseous stream and the WGS reaction between this product and CO takes place: tritium and the other hydrogen isotopes are recovered in a dense palladium permeator obtaining CO_2 as by-product. This way was suggested, for example, by Violante *et al.* [136]. Figure 8.25 shows the conceptual scheme suggested by Violante *et al.* to recover tritium from the blanket of NET.

4.2.5. Membrane applications in the tritium extraction system of ITER

ITER is the International Thermonuclear Experimental Reactor to be built in France: it will be the largest world facility for studying the applicability of nuclear fusion by magnetically confined plasmas. In this plant, special modules containing

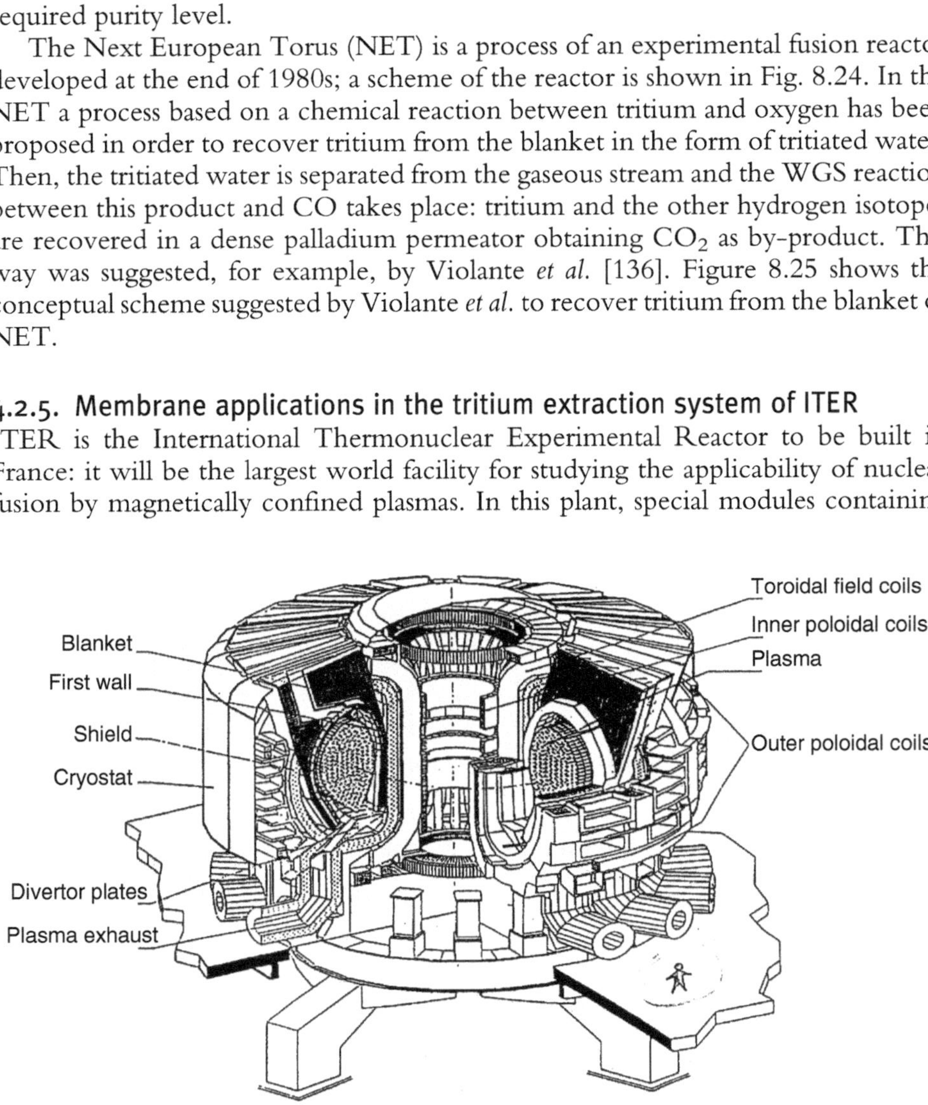

Figure 8.24 A picture of the Next European Torus.

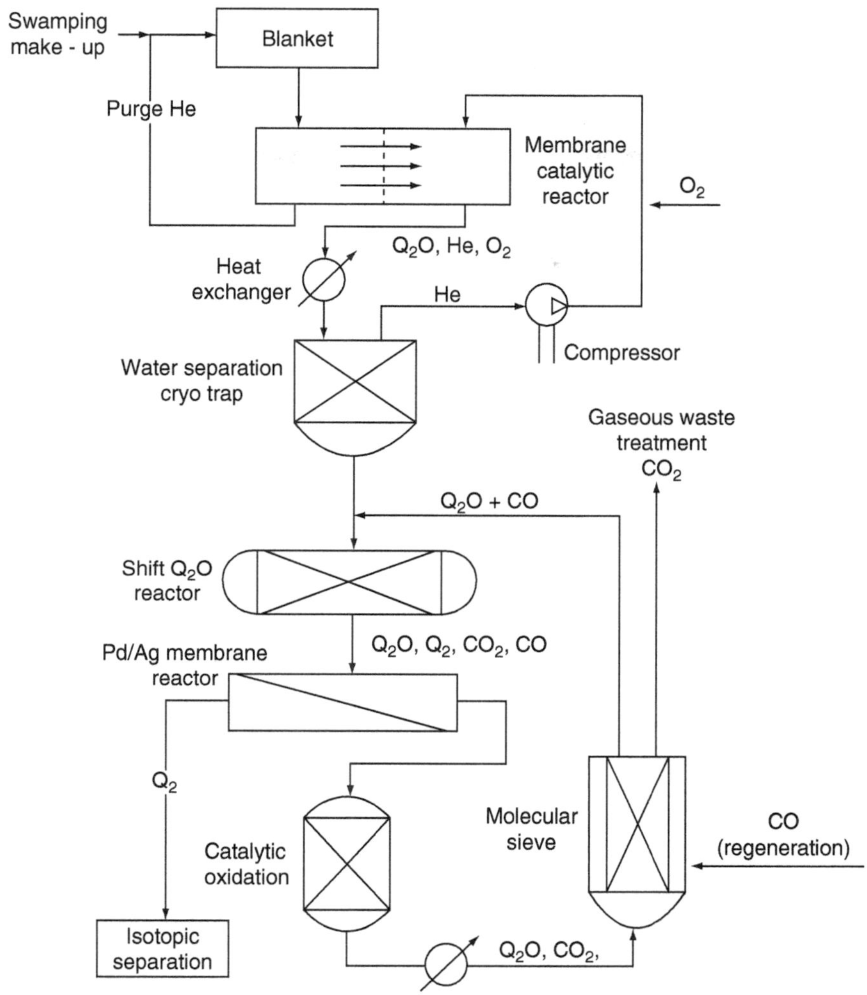

Figure 8.25 The way suggested by Violante *et al.* [136] for carrying out an application of a membrane reactor to an integrated system regarding a fusion cycle.

lithium (Breeding Blankets Test Modules, BTM) will be operated in order to verify their tritium production. Especially, a study of the potential applications of Pd–Ag membrane tubes and membrane reactors has been performed: the conceptual scheme reported in Fig. 8.26 refers to the Tritium Extraction System (TES) where tritium is recovered from the blanket module (TBM) [137] via a He purge stream: it is expected that at least 50% of the tritium generated in the TBM will be collected in the form of tritiated water (HTO). The tritiated water is collected in cold traps (component no. 4 of Fig. 8.26) while molecular sieve beds gather the tritium gas: alternatively, the tritium gas can be separated onto a low temperature adsorption bed.

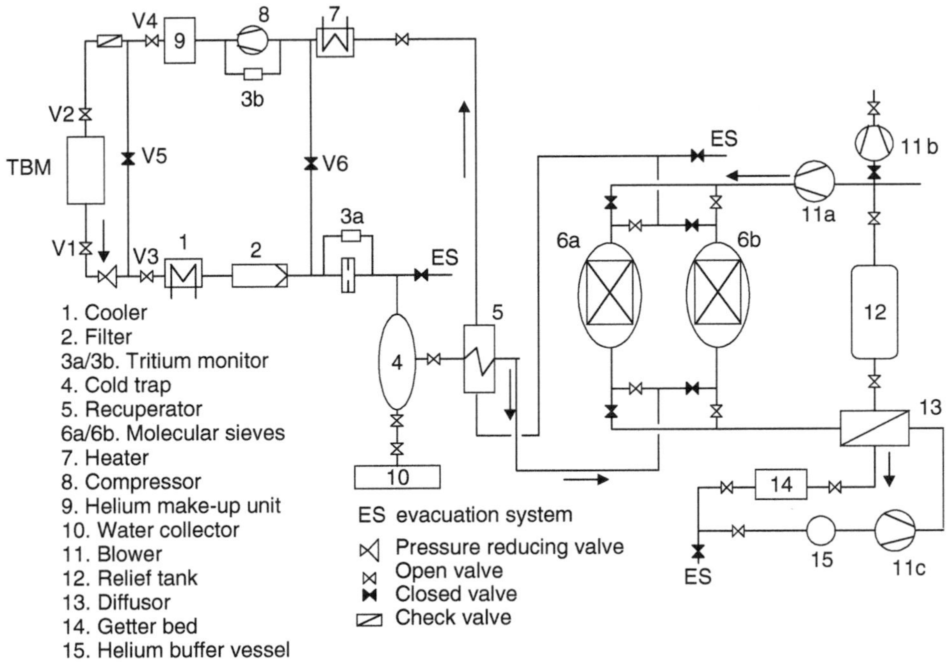

Figure 8.26 Flow sheet of the tritium extraction subsystem.

4.2.6. Tritium recovery from tritiated water by water gas shift reaction

For recovering tritium from the tritiated water collected in the cold traps a membrane reactor operating in a "closed loop" process previously studied and tested has been proposed [138]. The scheme of the process is reported in Fig. 8.27.

This process is based on a batch mode operation by one membrane reactor for both the WGS reaction: $CO_2 + H_2 \Leftrightarrow CO + H_2$ and the reverse one, promoted operating at higher temperatures: $CO + H_2O \Leftrightarrow CO_2 + H_2$.

In the first phase, operating at 325 °C and feed pressure of 100 kPa with a Pd–Ag thin wall membrane tube (thickness 60–70 μm, diameter 10 mm, and length 150 mm) the WGS reaction takes place in the membrane reactor until all the CO is converted into CO_2: the hydrogen isotopes produced by this reaction are selectively separated through the membrane and recovered in the shell side. Then, in a second phase, the temperature is increased (500 °C) and the pure hydrogen is supplied in the shell side of the reactor: in this way, the CO_2 conversion into CO (reverse WGS reaction) is operated.

As a main advantage with respect to the process studied for the NET, this process presents low tritium inventories, moderate operating pressures and temperatures, low dilution of the stream to be processed by isotopic separation. In fact the CO stream feeding the WGS reactor operates in a closed loop avoiding the production of tritiated CO_2 wastes; furthermore, the WGS reactor and the Pd permeator are integrated in one membrane reactor.

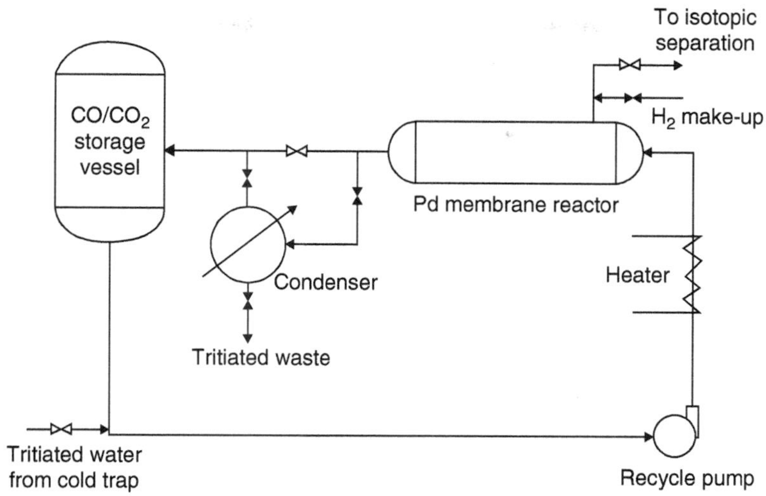

Figure 8.27 Scheme of the closed loop process.

4.2.7. Diffusor of the tritium extraction system

In the tritium extraction system, a Pd–Ag diffusor (the component no. 13 in Fig. 8.26) separates the hydrogen isotopes from the helium carrier gas used to regenerate the zeolite adsorber beds. This component consists of a bundle of 28 Pd–Ag permeator tubes operating at 350 °C and having the following characteristics: thickness 50 μm, diameter 10 mm, and length 300 mm. In the membrane module the Pd–Ag permeator tubes have a finger-like arrangement (avoiding the mechanical stresses due to the elongation as a consequence of hydrogen loading). As an advantage, this design permits reduced size and tritium inventory.

4.2.8. Hydrogen isotope separation from the helium purge gas by permeation (alternative to the process using low temperature adsorber beds)

In this case, the entire helium purge gas stream coming from the cold trap and the recuperative heat exchanger is treated in a Pd–Ag diffusor instead of the zeolite adsorber bed and its diffusor. A set of four permeators in series plus one purged with oxygen is proposed. Each of the four permeators is connected to a vacuum pump (this arrangement reduces the overall vacuum pumping requirements and the permeating surface) while the last permeator, where the hydrogen partial pressure is lower, is purged by a flow of added oxygen (see the scheme reported in Fig. 8.28).

New systems for tritiated inert gas processing in both high hydrogen isotopes partial pressure [117] and low hydrogen isotopes partial pressure [198] were also studied. In these systems, a gas mixture is fed to one side of a Pd–Ag membrane and a vacuum is applied to the other side of the membrane. The difference in hydrogen isotopes partial pressure across the membrane is the driving force, which generates a flow of hydrogen isotopes from one side to the other one. An overall system based on this concept was proposed.

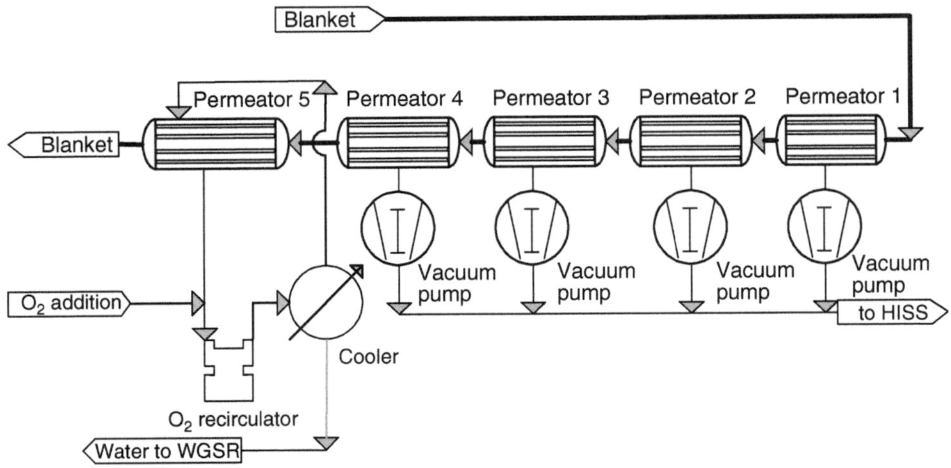

Figure 8.28 Hydrogen recovery by four permeators in series operating by vacuum pumping plus one purged with oxygen.

About one third of the hydrogen isotopes is recovered in oxidized form (tritiated water) in the cooler of the fifth permeator and can be sent to the system described above (suitably upgraded). This solution is competitive with respect to the use of the zeolite adsorber bed in terms of reduced size and lower tritium inventories.

4.2.9. Permcat

As a consequence of their complete hydrogen selectivity and applicability in continuous processes, palladium–silver permeators have also been developed for separating hydrogen isotopes in plasma exhaust clean-up processing. In fact, after the discharge burning the deuterium–tritium mixture, the plasma chamber contains gases (tritiated water, tritiated hydrocarbons, etc.) coming from the interactions between the hydrogen isotopes and the chamber walls.

Since the 1990s at the Tritium Laboratory of Karlsruhe (TLK) Pd–Ag permeators were operated [140]: a membrane reactor, named *PERMCAT* (*permeator* + *catalyst*), was developed at the TLK for detritiating plasma exhaust gases in a continuous process [141, 142]. In this reactor, the tritiated species are sent to a catalyst bed while H_2 is sent in counter-current mode inside a dense Pd–Ag tube Fig. 8.29. On the catalyst, tritium is exchanged with protium by isotopic swamping; as an example, the swamping reactions of tritiated methane and tritiated water (Q is indicating a hydrogen isotope: protium, deuterium, or tritium) are reported:

$$2_{H_2} + CQ_4 \Leftrightarrow CH_4 + 2Q_2$$

$$H_2 + Q_2O \Leftrightarrow H_2O + Q_2$$

The PERMCAT permitted to attain a high decontamination factor in the final step of the Catalytic Purification Experiment (CAPRICE) facility [141]. The finger-type configuration was adopted in order to let the membrane tube free in its

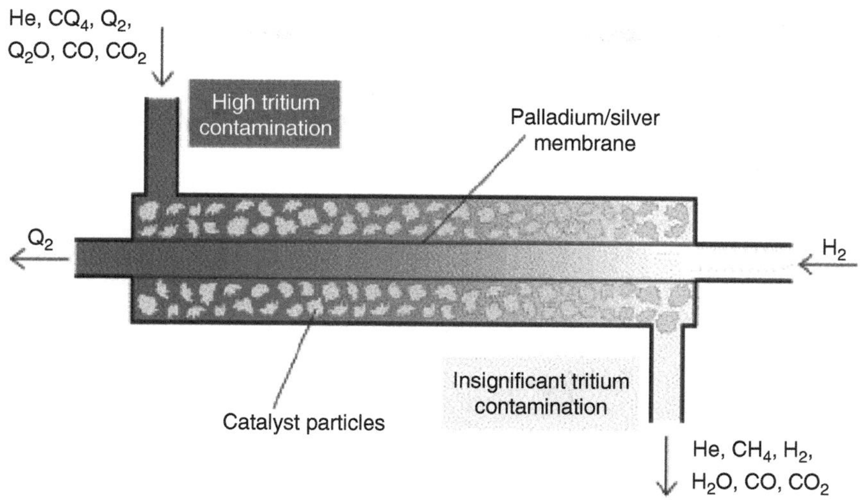

Figure 8.29 Principles of the PERMCAT reactor [142].

elongation, due to hydrogen loading and thermal expansion. The same configuration was applied to the PERMCAT proposed for impurity processing in the Joint Experimental Torus (JET), the largest tokamak today in operation [142]. The PERMCAT was designed in order to operate with a tritiated gas flow rate of at least 1.5 mol/h, containing up to 1% of tritium in different chemical forms, such as water, methane, or molecular hydrogen.

Finally, the PERMCAT was applied in a process developed at the TLK for the final clean-up of the ITER exhaust gases: the results from preliminary tests demonstrated that an overall tritium removal efficiency of about 10^8 can be reached [143].

4.2.10. The water gas shift reaction from the IGCC viewpoint

The WGS reaction in a membrane reactor could also be adopted within an Integrated Gasification Coal Cycle (IGCC). This process is able to convert the chemical energy of coal into electrical energy: coal gasification produces a gas mixture containing H_2, which is separated from CO_2 and burned for producing vapor and finally electrical energy by means of turbines. WGS reaction occurs after coal gasification, for hydrogen enrichment of the gas mixture.

The capture (separation and storage) of the carbon dioxide, emitted by the big fossil fuel combustion plants, is considered as one of the few possible available operations able to stabilize the greenhouse-gas concentrations in the troposphere, at least as long as the energetic system will be fed with new energetic sources. There are many available technologies to capture carbon dioxide in IGCC power plants; membranes offer several advantages: small size, simplicity of operation and maintenance, compatibility, and diversity. Various studies on the techno-economic impact of high temperature membranes to separate CO_2 from H_2, eventually integrated with WGS reactors, indicate that the IGCC conversion efficiency can be increased and investment costs reduced in comparison to physical absorption [144, 145].

As far as metallic membranes are considered, a techno-economic study performed by Kreutz *et al.* [146] regarding the hydrogen production with 60/40 Pd–Cu membranes and microporous membrane versus Selexol and PSA (pressure swing adsorption) processes indicates that cost advantages of membrane separators/reactors are moderate.

A proposal for a process aimed at the application of a WGS membrane reactor in IGCC systems for the CO_2 emission mitigation was named WIHYS (Water Gas Shift With Integrated H_2/CO_2 Separation) process [139]. These authors also presented a process scheme of IGCC with integrated WIHYS process for the CO_2 recovery, a very interesting process from the point of view of greenhouse-gas reduction.

An economic and energetic analysis of IGCC plants with Pd–Ag membranes was recently performed by Amelio *et al.* [147] demonstrating that this plant is more competitive with respect to an IGCC plant with the CO_2 adsorption systems due to the low costs of membrane reactors and to the absence of the CO_2 absorption unit.

4.2.11. Integrated systems involving hydrogen production and recovering for fuel cell applications

An interesting and important research line regarding the integrated systems consisting of MRs producing hydrogen for fuel cell applications. A fuel cell could be viewed as a low temperature membrane reactor where the chemical energy of a fuel is converted into electrical energy. Fuel cells operate at high efficiencies and normally emit very low amounts of hazardous compounds. A fuel cell can run using pure hydrogen or other fuels (methane, methanol, and gasoline); one drawback is due to the fact that each fuel needs suitable conditions to give the maximum efficiency. Today, a polymer electrolyte membrane fuel cell (PEMFC) attracts significant research efforts. It is composed of a number of components, each one made of a composite membrane (a cation-exchange layer) clamped between two current collector plates and two gas diffusion backings. After the reformer, a clean-up system is needed, since it eliminates compounds that might be detrimental to this fuel cell. Finally, the resulting fuel is fed to the fuel cell unit, followed by a catalytic burner for complete combustion of residues.

A palladium membrane unit could be integrated in the reforming step, in order to enhance the fuel conversion and to feed pure hydrogen to the fuel cell. Recently, an integrated system that combines steam reforming of methane and pure hydrogen recovery for fuel cell application using palladium-based membranes was presented by Wieland *et al.* [148]. Pd membrane methane steam reformer promises considerable weight and space savings (particularly important for fuel cell car applications). The cost of a fuel cell vehicle equipped with a multifuel reformer is very high, given the level of technology available today [149]. A rough estimate yields a cost exceeding US$100,000 for a fuel cell vehicle. To reduce the very high costs of both the fuel cell and the reformer system a major research effort within an international cooperative program is necessary.

Ultrapure hydrogen (i.e., <10 ppm CO) is required because of the sensitivity of the anode catalyst in the PEMFC to CO poisoning. A PEMFC can be powered directly by hydrogen stored or by hydrogen that is produced on site from a suitable hydrocarbon feedstock such as gasoline, natural gas, and methanol [150]. The use of pure hydrogen as the energy carrier requires an expensive hydrogen-fueling network, leading to high costs in the fuel delivery system. Moreover, the low volumetric energy density of hydrogen at ambient conditions makes hydrogen storage uneconomical. Therefore, on site hydrogen generation from a hydrocarbon feedstock is preferred.

Figure 8.30 shows a scheme of an integrated system for converting the chemical energy of fuels into electric energy via fuel cell.

A process in which a membrane reactor produces the hydrogen needed to a 500 W polymeric fuel cell has been designed, for example, by Tosti *et al.* [151]. The process foresees the use of a traditional ethanol reformer and a Pd membrane reactors. The membrane reactor consists of a bundle of Pd–Ag thin wall tubes operating in parallel where the WGS reaction takes place at 350 °C and 200 kPa: the hydrogen stream of 4.46×10^{-3} mol/s (about 6 l/min) is recovered in the shell side by means of nitrogen sweep gas at 100 kPa. The assessment of the membrane module carrying out the WGS reaction has been performed on the basis of

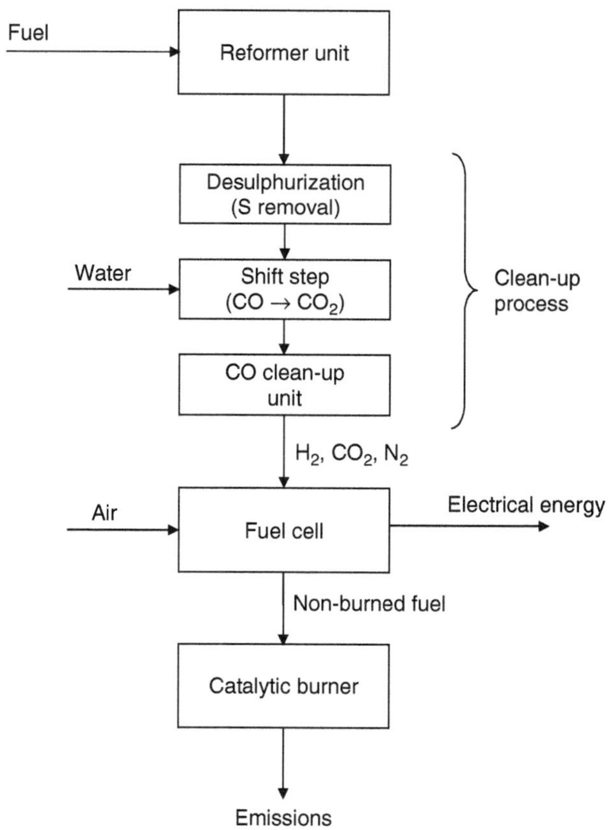

Figure 8.30 Scheme of an integrated system for converting chemical energy of fuels into electric energy via fuel cell.

membrane tubes of diameter 10 mm, wall thicknesses 40 m, and length 180 mm. The membrane reactor needs 19 tubes: their cost based on the Pd quotation of 5 €/g proves to be 250,00 thus much lower than the cost of the polymeric fuel cell.

4.3. The role of the hydrogen gas

Although hydrogen is considered to be the most common element in the known universe, unfortunately it is not present in pure form on our planet. Hydrogen is an odorless, colorless gas, and with molecular weight of 2.016 it is the lightest element. Its density is about 14 times less than air (0.08376 kg/m^3 at standard temperature and pressure). Hydrogen is liquid at temperatures below 20.3 K (at atmospheric pressure) and has the highest energy content per unit mass of all fuels—higher heating value is 141.9 MJ/kg, almost three times higher than gasoline.

It has to be observed that hydrogen gas is not a primary fuel in the same sense as natural gas, oil, and coal. No wells produce hydrogen gas from geologically identified deposits. Rather, hydrogen is an energy carrier like electricity. Hydrogen is a secondary form of energy, produced using other primary energy sources, such as natural gas, coal, or solar technologies. Today, the world demand for hydrogen corresponds to about 50 million metric tons per year of which about 80% is deliberately produced while 20% is a by-product of other chemical processes. In particular, the United States alone each year consumes more than 8 million tons of hydrogen, mainly by the chemical and petroleum industries.

The primary uses of hydrogen today are for producing fertilizers, and for other industrial processes. The need of hydrogen for these purposes is currently growing at the rate of 5–10% per year. Considering an average growth of 7% per year, the use of hydrogen would grow to 100 million metric tons per year in a decade.

In 2003 U.S. President Bush proposed "$ 1.2 billion in research funding so that America can lead the world in developing clean, hydrogen-powered automobiles," while on August 2004, a special section "toward a hydrogen economy" was reported [152].

Nowadays, hydrogen is considered to be the greatest potential energy carrier in the world, whose application area will be expanded to the operation of fuel cells and mobile vehicles far from the traditional petroleum refining and the semiconductor processing. It will be the main future energy carrier [153–156].

However, the production and separation of hydrogen from hydrogen-containing sources such as water, natural gas, and coal is the real precondition for all the realistic applications of hydrogen energy. Even though the hydrogen source of water has greater potential because of its huge abundance, its application is still unrealistic because of its high cost.

In order to produce pure hydrogen or syngas ($CO + H_2$), many chemical reactions have been suggested in the specialized literature (see Table 8.6). Hydrogen is today traditionally produced via multiple reaction steps as a primary product from the steam reforming of hydrocarbons such as methane, naphtha oil, or methanol. On an industrial scale, most of the hydrogen is currently produced by the steam reforming of natural gas. Natural gas production increased from 2.5×10^{12} Nm^3 in 1990 to 3.0×10^{12} Nm^3 in 1999 [2]. The reforming of the abundant natural gas (mainly methane) is today considered one of the most important chemical processes for the production of hydrogen, which has been proved to be more economical than other processes such as coal vaporization, hydrocarbon partial oxidation, and water electrolysis to produce hydrogen.

With steam reforming of methane carried out using traditional reactors high hydrogen yields can be achieved. However, the methane steam reforming reaction is largely endothermic and equilibrium limited, the high temperature (i.e., costly high temperature heat exchange equipments and complex energy integration between a large number of process units, including reformer, high and low temperature shift reactors and a preferential oxidation reactor) must be employed to achieve the satisfactory methane conversion. Moreover, often a pressure swing adsorption unit is used to achieve the desired hydrogen purity. For the production of ultrapure hydrogen for small-scale applications, this route is not preferred because of the large number of process units and the associated uneconomical downscaling.

A high degree of process integration and process intensification can be accomplished by integrating palladium-based membrane reactors in the steam reformer. Using palladium-based membranes, the number of process units can be decreased and the total required reactor volume can be reduced, while higher methane conversion and hydrogen yields beyond thermodynamic equilibrium limitations can be achieved, at lower temperatures and with higher overall energy efficiencies.

4.4. High temperature membrane reactors

In recent years, membrane reactors have been of considerable interest for many gas-phase catalytic reaction processes. Membrane reactors are multifunctional reactors suitable for the combination of molecular separation and chemical conversions in one step. This system shows several interesting applications. In fact, several works concerning high temperature membrane reactors have been performed in terms of both experimental studies and mathematical models describing the main kinetic and fluid-dynamic phenomena occurring in these systems. Excellent reviews have been presented, for example, by Saracco *et al.* [157], Zaman and Chakma [158], Coronas and Santamaria [159], and Julbe *et al.* [160]. More recently, also the works presented by Marcano and Tsotis [161], Paturzo *et al.* [133], Uemiya [20], Dixon [162], Paglieri [103], Specchia *et al.* [199], and Gryaznov *et al.* [200] should be considered. Several high temperature membrane reactors for different applications are summarized in Table 8.7.

Moreover, recently, new patents and experimental plants were introduced for the treatment of large amount of materials. For example, considering palladium-based membrane reactors:

- Amiridis *et al.* [163] patented a process for producing substantially pure hydrogen by contacting a stream of a hydrocarbon gas with a nickel-containing catalyst in a membrane reactor. The electroless plating technique was used to deposit a thin palladium film (1–2-μm thickness) on the surface of a refractory metal such as Nb, Ta, Zr, and V. The invention is applicable for cracking any suitable hydrocarbon, such as natural gas, and hydrocarbons with molecular weights in the gasoline and diesel range.

- Fuder [164] patented a membrane reactor consisting of a hydrogen-permeable diffusion membrane for generating high-purity hydrogen from a hydrocarbon stream. A precious metal alloy is used as the membrane material, such as a palladium–silver alloy that also serve as a catalyst. Additionally, other metals such as rhodium, ruthenium, nickel, cobalt, and iron can also be deposited onto the membrane using standard procedures.

Briefly, membrane reactors offer the possibility of overcoming the thermodynamic limitations of reversible chemical reactions. If at least one of the reaction products can selectively be removed from the reaction system, equilibrium-limited reactions can achieve higher conversions at lower temperatures. This concept, called "equilibrium shifting," is a simple consequence of Le Chatelier–Braun principle: "If an external force leads to the separation as a specified component from a system at equilibrium, this system adjusts so as to minimize the effect of the applied force," and thereby, it produces more of the removed component.

Consider, for example, the following general reversible reaction:

$$\mathbf{A} + \mathbf{B} = \mathbf{C} + \mathbf{D}$$

If only product C is permeable through the membrane (e.g., hydrogen through dense palladium membrane or oxygen through dense silver membrane), the conversion becomes higher as the equilibrium is displaced toward the products side. In the case of palladium-based membrane reactors, the product selectively removed is the hydrogen gas. Higher conversion than a traditional system also means that the membrane reactor can work at the same conversion rate as traditional reactors, but in less severe conditions (temperature and pressure). Of course, it also implies low energy consumption.

It should be well kept in mind that the equilibrium we are speaking about is the *hypothetical* equilibrium of the *closed* system formed by a fixed mass reacting according to the previous reaction. It is not the equilibrium of the actual membrane reactor system, which is an open system, including both the feed and permeate sides. In order to better understand the effect of the equilibrium shifting, some considerations should be made comparing a traditional reactor with a membrane reactor. In a conventional isothermal plug-flow reactor (PFR), two important rates govern its performance: the rate of the reaction and the rate of the reactants fed per catalyst volume. Their ratio gives the so-called Damkohler number:

$$\mathrm{Da} = \frac{(\text{reactor volume}) \times (\text{maximum reaction rate per volume})}{(\text{inlet flow rate})}$$

which also involves reactor tube dimensions.

The membrane reactor brings in at least one additional rate, the permeation rate of the fastest gas. The ratio of these has been labeled differently by various authors.

For example, following Bernstein and Lund [165], we can term it as the Damkohler–Peclet product:

$$\mathrm{Da} \times \mathrm{Pe} = \frac{(\text{maximum reaction rate per volume})}{(\text{maximum permeation rate per volume})} \tag{8.10}$$

Harold *et al.* [166] explained that if the permeation rate is too low (i.e., the product Da × Pe is very high), the membrane has little effect and the reactor behaves like a PFR with the reactant gas feed. Vice versa, if the permeation rate is too high (i.e., the product Da × Pe is very low), the shell and tube sides will equilibrate rapidly. In particular, Bernstein and Lund [165] recommend $0.1 < \mathrm{Da} \times \mathrm{Pe} < 10$ as covering the optimal range.

The activity of the catalyst is also important, as reflected in the Da value. If the reaction rate is slow, equilibrium is not approached and the removal of a product by a membrane will not affect the ultimate yield, as demonstrated by Raich and Foley [167] in the dehydrogenation of cyclohexane to benzene (fast reaction rate). Slower reactions, such as the dehydrogenation of isobutane to isobutene, are not helped by the membrane reactor as they are kinetically limited (low Da).

Following Raich and Foley [167], the selective and continuous removal of hydrogen in palladium membrane reactors is a *necessary* condition for achieving a level of conversion higher than the thermodynamic equilibrium value for traditional system, but it is not *sufficient*. In the case of dehydrogenation reactions, for example, the *necessary* and *sufficient* conditions have to take both thermodynamics and kinetics into consideration. In fact, generally, the rate of the hydrogen permeation is not the only impediment to achieving supra-equilibrium conversions since kinetics of the reaction is also a critical point.

To better understand this important point, let us consider the simple generic dehydrogenation chemical reaction: $A + B = C + H_2$.

Figure 8.31 shows the conversion versus the temperature for this generic (endothermic) reaction.

The *Thermodynamic Equilibrium* curve represents the equilibrium conversion for a traditional reactor. The conversion for a traditional reactor necessarily falls under this curve (region A–B at a fixed temperature). In particular, the conversion depends on the catalyst's activity used for the specific reaction: a very active catalyst bed gives conversion values (curve N. 3) close to the thermodynamic equilibrium curve, while a poor active catalyst gives conversion values (curve N. 1) far from the thermodynamic equilibrium curve.

Vice versa, the *Dynamic Equilibrium* curve reported in Fig. 8.31 is related only to a dense palladium-based membrane reactor and represents the reaction not limited by chemical kinetics and thus the feed gas is assumed to be continuously at equilibrium inside the membrane reactor. As hydrogen permeates through the palladium layer, the feed gas composition changes because of the hydrogen permeation. This means that the reaction rate is not the limiting step: all molecules in contact with the catalyst are considered to react with an infinite velocity.

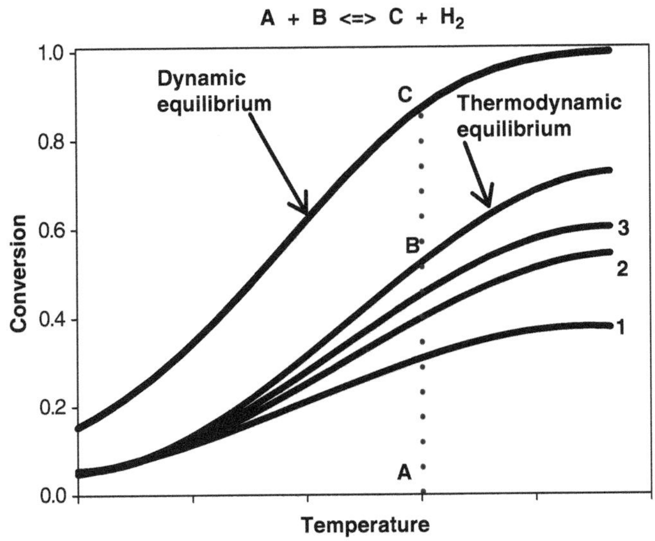

Figure 8.31 Conversion versus temperature for a generic (endothermic) reaction.

The conversion for a membrane reactor necessarily falls under this curve (region A–C at a fixed temperature). Also in this case, the conversion depends on the catalyst's activity used for the specific reaction: a very active catalyst bed gives conversion values possibly higher than the thermodynamic equilibrium curve (and falls in the region B–C at a fixed temperature), while a poor active catalyst bed gives conversion values far from the dynamic equilibrium curve (and can fall in the region A–B at a fixed temperature).

4.5. Some case studies of membrane reactors in the literature

The majority of the reactions studied by membranologists have mainly taken advantage of the ability of palladium membranes to selectively remove product hydrogen, either by diffusion through palladium or palladium alloys or by Knudsen diffusion through microporous glass or ceramic. In the following, some experimental results, presented recently in the literature and concerning hydrogen production using both traditional reactors and membrane reactors, are presented. In particular, focusing our attention on hydrogen and synthesis gas ($CO + H_2$) production, which are mainly obtained in industrial plants by the dehydrogenation of methane or methanol on a nickel catalyst, the following reactions for producing hydrogen, studied by different authors from an experimental point of view, are taken into consideration:

- Steam reforming of methane (Fig. 8.32).
- Dry reforming of methane (Fig. 8.33).
- Partial oxidation of methane (Fig. 8.34).
- Steam reforming of methanol (Fig. 8.35).
- Oxidative steam reforming of methanol (Fig. 8.36).

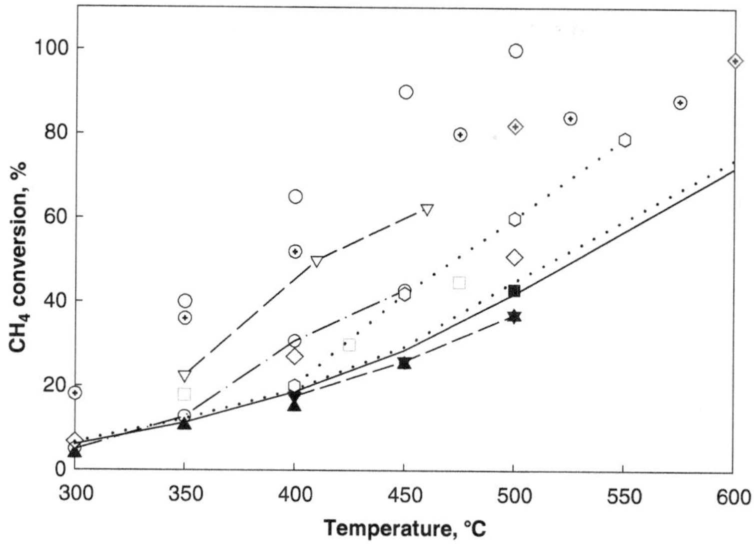

Figure 8.32 Methane steam reforming literature data: conversion versus temperature, $H_2O/CH_4 = 3$, $p_{lumen} = 1.1–1.36$ bar. ○: MR, Kikuchi *et al.* [168]; ⊙: MR, Chai *et al.* [169]; ■: TR, Madia *et al.* [170]; □: MR, Madia *et al.* [170]; ▲: TR, Shu *et al.* [171]; ◇: MR, Shu *et al.* [171]; ◈: MR, Nam *et al.* [172], $H_2O/CH_4 = 4$; . . .: Thermodynamic equilibrium, $p = 1.1$ bar; ——: Thermodynamic equilibrium, $p = 1.36$ bar; —▼—: TR, Gallucci *et al.* [173]; —▽—: MR, Gallucci *et al.* [173] H_2O sweep gas; —⊖ ·: MR, Gallucci *et al.* [173] N_2 sweep gas; · ⬡ ·: MR, Kusakabe *et al.* [174].

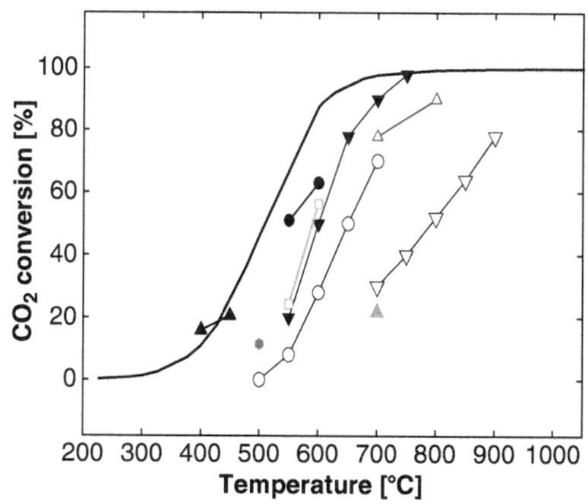

Figure 8.33 Dry reforming of methane literature data in terms of methane conversion versus temperature. —△—: Chang *et al.* [175]; —○—: Wang and Lu [176]; —●—: Erdöhelyi *et al.* [177]; —▲—: Lemonidou *et al.* [178]; —▼—: Ji *et al.* [179]; —▽—: Choudhary *et al.* [180]; —□—: Galuszka *et al.* [181]; ——: Equilibrium conversion ($CH_4/CO_2 = 1/1$; $p = 1$ at; —▲—: MR, Paturzo *et al.* [182]; —●—: MR, Galuszka *et al.*, [181].

In all these cases, the concepts of thermodynamic and dynamic equilibrium are well evidenced. In fact, in all these approaches, it is evident that a membrane reactor has a higher methane or methanol conversion with respect to the traditional one.

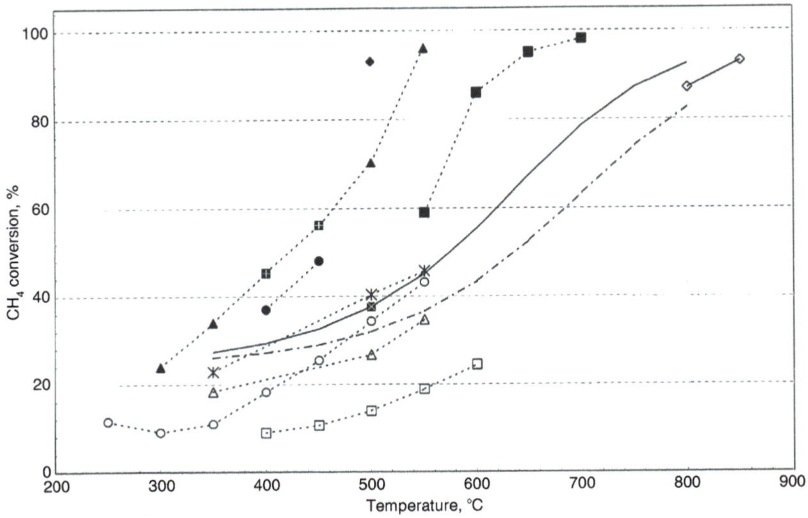

Figure 8.34 Methane conversion versus temperature for membrane reactors and traditional reactors for the partial oxidation of methane (from Basile and Paturzo [183]). ——: Thermod. Equilibrium, $P = 1$ bar; –·–·: Thermod. Equilibrium, $P = 2$ bar; ··■··: Ioannides and Verykos [202], MR; ··△··: Galuszka *et al.*, [181], TR; ··✱··: Galuszka *et al.*, [181], MR; ··⊠··: Kikuchi and Chen [203], TR; --◆--: Kikuchi and Chen [203], MR; ··□··: Nakagawa *et al.*, [204], TR; —◇—: Ruckenstein and Hu [205], TR; ··▲··: Basile *et al.* [110], MR; ··○··: TR, Basile and Paturzo [183]; ··●··: MR, Basile and Paturzo [183]; ··⊞··: PMR, Basile and Paturzo [183].

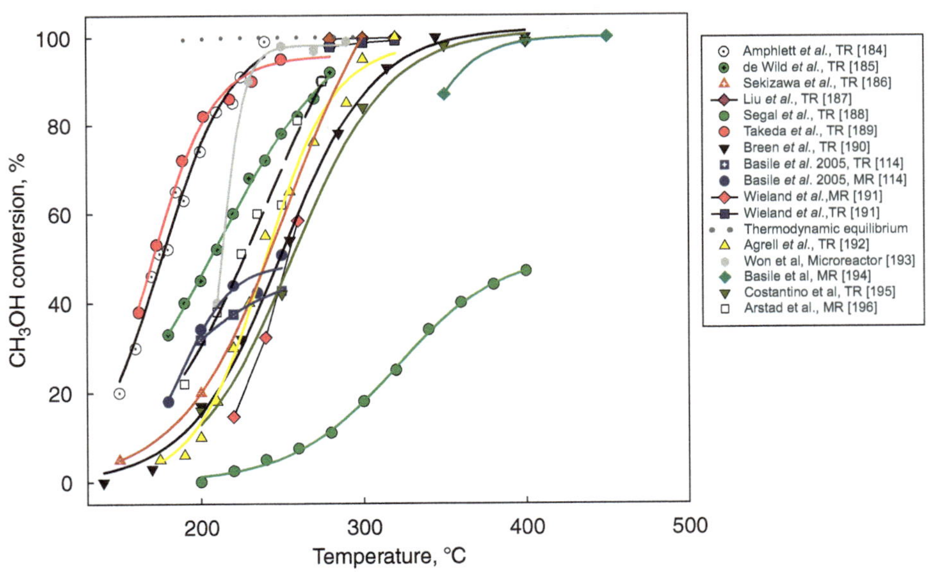

Figure 8.35 Methanol conversion versus temperature for membrane reactors and traditional reactors for the steam reforming of methanol.

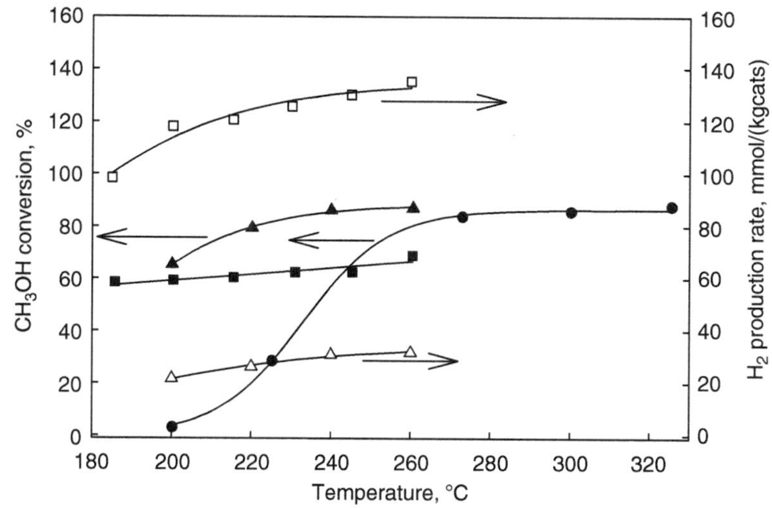

Figure 8.36 Methanol conversion versus temperature for membrane reactors and for traditional reactors for the oxidative steam reforming of methanol (from Basile *et al.*, [63]). •: Murcia-Mascaros *et al.*, [206], TR; ▪: Velu *et al.*, [207], TR; □: Velu *et al.*, [207], TR; △: Basile *et al.*, 2005, MR ($O_2/CH_3OH = 0.17$); : Basile *et al.*, 2005, MR ($O_2/CH_3OH = 0.17$).

In particular, it should be observed that all experimental data carried out in traditional reactors are referred to be below the corresponding thermodynamic equilibrium curve. Vice versa, all experimental data regarding membrane reactors are always above the corresponding thermodynamic equilibrium curve, except for the dry reforming of methane, where carbon deposition does not permit high conversion values to be obtained. This means that palladium membranes enhance the catalytic conversion of both methane and methanol. In other words, in a membrane reformer a greater amount of hydrogen could be produced if working is considered at the same experimental conditions as the traditional one. In principle, in fact, it should be possible to use a membrane reactor for achieving the same methane or methanol conversion as in a traditional reactor, but operating at a lower temperature, thus resulting in energy savings.

The same considerations and figures can be produced for all the reaction systems reported in Table 8.6.

Another very important parameter to be considered when the membrane reactor is used for producing ultrapure hydrogen is the hydrogen recovery factor, which represents the percentage of hydrogen recovered as CO (and CO_2) free stream. It is easy to observe in Fig. 8.37, where the hydrogen recovery by a dense Pd–Ag membrane reactor is reported, that, by using a Pd–Ag dense membrane reactor, up to 40% of the total hydrogen produced in the membrane reactor is recovered as a CO-free stream, when the counter-current mode is considered. This hydrogen rich stream is directly available for its use in a PEM fuel cell system.

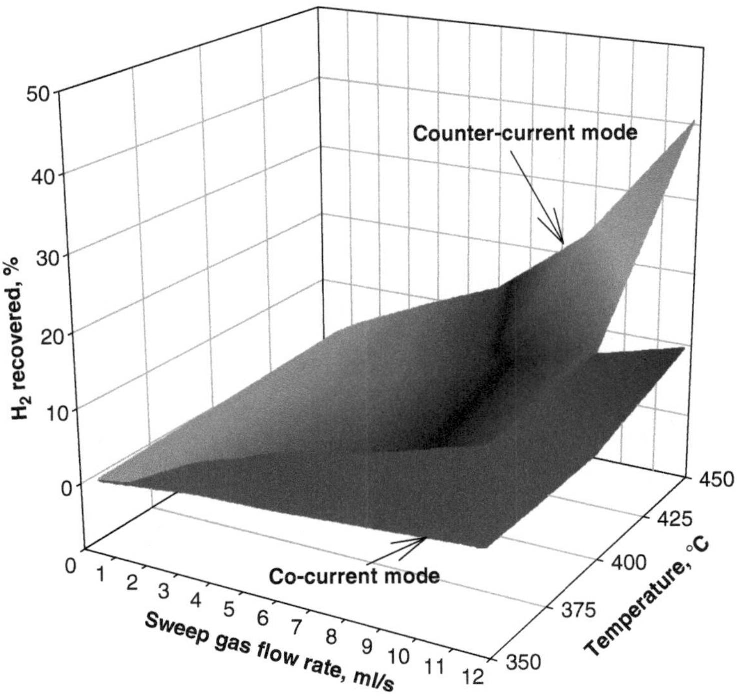

Figure 8.37 Hydrogen recovery versus temperature and versus sweep gas flow rate for membrane reactors in co-current and counter-current mode for the steam reforming of methanol (from Basile *et al.* [194]).

5. Conclusions

In this chapter, an extensive overview concerning Pd-based membranes and membrane reactors is presented. In particular, the preparation of both dense Pd-based membranes and composite ones as well as their characterizations in terms of gas permeation and surface properties are reported in details.

Problems related to the membrane performances in terms of hydrogen selectivity and durability are described: the study of the Pd–H system is reported in order to explain the hydrogen mass transfer mechanisms through palladium and its alloys. The degradation of these materials under the operating conditions used in the membranes and membrane reactors devices is also discussed.

The study of membrane reactors for producing hydrogen is reported from the thermodynamics and kinetics viewpoint: the effect of the membrane in shifting the thermodynamic equilibrium is explained.

Some potential applications, for example the recovery of tritium in the fusion reactors fuel cycle, nearby to the recent literature results related to the hydrogen production in Pd membrane reactors are also shown.

ACKNOWLEDGMENTS

We thank the publishers of the cited figures for giving the permission to reprint the figures here reported.

REFERENCES

[1] McLeary, E. E., Jansen, J. C., and Kapteijn, F. (2006). Zeolite based films, membranes and membrane reactors: progress and prospects. *Microporous Mesoporous Mater.* **90,** 198–220.
[2] Eni (2001). *World and oil gas review* **70**.
[3] Tosti, S., Bettinali, L., Lecci, D., Marini, F., and Violante, V. (2001). Method of Bonding Thin Foils Made of Metal Alloys Selectively Permeable to Hydrogen, Particularly Providing Membrane Devices, and Apparatus for Carrying Out the Same, European Patent EP 1184125.
[4] Tosti, S, and Bettinali, L. (2004). Diffusion bonding of Pd-Ag membranes. *J. Mater. Sci.* **39,** 3041–3046.
[5] Tosti, S., Bettinali, L., Castelli, S., Sarto, F., Scaglione, S., and Violante, V. (2002). Sputtered, electroless, and rolled palladium-ceramic membranes. *J. Membr. Sci* **196,** 241–249.
[6] Tosti, S (2003). Supported and laminated Pd-based metallic membranes. *Int. J. Hydrogen Energy* **28,** 1455–1464.
[7] Bettinali, S., Tosti, L., and Violante, V. (2000). Rolled thin Pd and PdAg membranes for hydrogen separation and production. *Int. J. Hydrogen Energy* **25,** 319–325.
[8] Tosti, S., Bettinali, L., Lecci, D., Rizzello, C., and Basile, A. (2005). Dispositivo a membrana a fascio tubiero per la produzione di idrogeno ultrapuro, Italian Patent n. RM2005A000399.
[9] Basile, A., and Tosti, S. (2005). Dispositivo protatile a membrana intercambiabile per valutazione di processi di permeazione e reazione in fase gassosa, Italian Patent n. RM2005U000107.
[10] Tosti, S., Bettinali, L., Lecci, D., and Violante, V. (2003). Procedimento di saldatura di strutture metalliche di rinforzo a lamine in lega di palladio per la fabbricazione di membrane composite per la separazione di idrogeno e apparato per la sua realizzazione, Italian Patent IT20RM010748.
[11] Shmayda, W. T. (1985). *Tritium Interaction with materials* Report Ontario Hydro Research Division, Sept. 17, 1984–Revision.
[12] Hara, S., Sakaki, K., Itoh, N., Kimura, H.-M., Asami, K., and Inoue, A. (2000). An amorphous alloy membrane without noble metals for gaseous hydrogen separation. *J. Membr. Sci.* **164,** 289–294.
[13] Nishimura, C., Komaki, M., Hwang, S., and Amano, M. (2002). V-Ni alloy membranes for hydrogen purification. *J. Alloys Compd.* **330–332,** 902–906.
[14] Zhang, Y., Ozaki, T., Komaki, M., and Nishimura, C. (2003). Hydrogen permeation of Pd-Ag alloy coated V-Ni composite membrane: effects of overlayer composition. *J. Membr. Sci.* **224,** 81–91.
[15] Ozaki, T, Zhang, Y., Komaki, M., and Nishimura, C. (2003). Hydrogen permeation characteristics of V-Ni-Al alloys. *Int. J. Hydrogen Energy* **28,** 1229–1235.
[16] Hashi, K., Ishikawa, K., Matsuda, T., and Aoki, K. (2004). Hydrogen permeation characteristics of multi-phase Ni-Ti-Nb alloys. *J. Alloys Compd.* **368,** 215–220.
[17] Peachey, N. M., Snow, R. C., and Dye, R. C. (1996). Composite Pd/Ta metal membranes for hydrogen separation. *J. Membr. Sci* **111**(1), 123–133.
[18] Tereschenko, G. F., Ermilova, M. M., Mordovin, V. P., Orekhova, N. V., Gryaznov, V. M., Iulianelli, A. Gallucci, F., and Basile, A. New Ti-Ni dense membranes with low palladium content, Int. J. Hydrogen Energy (2007), **32,** 4016–4022.
[19] Shu, A, Adnot, A., Grandjean, B. P. A., and Kaliaguine, S. (1996). Structurally stable composite Pd-Ag alloy membranes: introduction of a diffusion barrier. *Thin Solid Films* **286,** 72–79.
[20] Uemiya, S (1999). State-of-art-the-art of supported metal membranes for gas separation. *Sep. Purif. Methods* **28**(1), 51–85.
[21] Powell, C. F., Oxley, J. H., and Blocher, J. M. (1966). Vapor Deposition. John Wiley & Sons, New York.

[22] Ilias, S., and Govind, R. (1989). Development of high temperature membranes for membrane reactor: An overview. *AIChE Symp. Ser.* **268**(85), 18–25.
[23] Basile, A, Drioli, E., Santella, F., Violante, V., Capannelli, G., and Vitulli, G. (1996). Study on catalytic membrane reactors for water gas shift reaction. *Gas Sep. Purif.* **10**(1), 53–61.
[24] Li, Z. Y., Maeda, H., Kusakabe, K., Morooka, S., Anzai, H., and Akiyama, S. (1993). Preparation of palladium-silver alloy membranes for hydrogen separation by the spray pyrolysis method. *J. Membr. Sci.* **78,** 247–254.
[25] Konno, M., Shindo, M., Sugawara, S., and Saito, S. (1988). A composite palladium and porous aluminium oxide membrane for hydrogen gas separation. *J. Membr. Sci.* **37,** 193–197.
[26] Gryaznov, V. M., Serebryannikova, O. S., Serov, Y. M., Ermilova, M. M., Karavanov, A. N., Mischenko, A. P., *et al.* (1993). Preparation and catalysis over palladium composite membranes. *Appl. Catal. A.* **96,** 15–23.
[27] Capannelli, G., Bottino, A., Gao, G., Grosso, A., Servida, A., Vitulli, G., *et al.* (1993). Porous Pt/γ-Al_2O_3 catalytic membrane reactors prepared using mesitylene solvated Pt atoms. *Catal. Lett.* **20,** 287–297.
[28] Lowenheim, F. A. (1974). Modern Electroplating. John Wiley & Sons, New York.
[29] Wise, E. M. (1968). Palladium—Recovery, Properties and Uses Academic Press, New York.
[30] Kikuchi, H. (1988). Alloy-Plated Membranes for Hydrogen Separation and Their Manufacture. Japanese Patent N. 88,294,925.
[31] Tong, J., Shirai, R., Kashima, Y., and Matsumura, Y. (2005). Preparation of a pinhole-free Pd-Ag membrane on a porous metal support for pure hydrogen separation. *J. Membr. Sci.* **260,** 84–89.
[32] Uemiya, S., Koseki, M., and Kojrma, T. (1994). Preparation of highly permeable membranes for hydrogen separation using a CVD technique. *In* "Proc. of the 3rd Int. Conf. Inorganic Membranes" (Y. H. Ma, ed.), pp. 545. Worcester (MA).
[33] Bobrov, V. S., Digurov, N. G., and Skudin, V. V. (2005). Propane dehydrogenation using catalytic membrane. *J. Membr. Sci.* **253,** 233–242.
[34] Itoh, N., Akiha, T., and Sato, T. (2005). Preparation of thin palladium composite membrane tube by a CVD technique and its hydrogen permselectivity. *Catal. Today* **104,** 231–237.
[35] Tong, J., Su, C., Kuraoka, K., Suda, H., and Matsumura, Y. (2006). Preparation of thin Pd membrane on CeO_2-modified porous metal by a combined method of electroless plating and chemical vapor deposition. *J. Membr. Sci.* **269,** 101–108.
[36] Han, J., Zeng, Y., Xomeritakis, G., and Lin, Y. L. (1997). Electrochemical vapour deposition synthesis and oxygen permeation properties of dense zirconia-yttria-ceria membranes. *Solid State Ionics* **98,** 63–72.
[37] Han, J., Xomeritakis, G., and Lin, Y. L. (1997). Oxygen permeation through zirconia-yttria membranes prepared by EVD. *Solid State Ionics* **93,** 263–272.
[38] Méndez-Vivar, J., and Mendoza-Serna, R. (2006). SiO_2-TiO_2 membranes by the sol-gel process. *Silicon Chem.* **3,** 59–64.
[39] Schafer, R., Noack, M., Kolsch, P., Thomas, S., Seidel-Morgenstern, A., and Caro, J. (2001). Development of a H_2-selective SiO_2-membrane for the catalytic dehydrogenation of propane. *Sep. Purif. Technol.* **25,** 3–9.
[40] Itoh, S., and Govind, R. (1989). Combined oxidation and dehydrogenation in a palladium membrane. *Ind. Eng. Chem. Res.* **28,** 1554–1557.
[41] Uemiya, S., Kude, Y., and Sugino, K. (1988). A palladium/porous glass composite membrane for hydrogen separation. *Chem. Lett.* **10,** 1687–1690.
[42] Uemiya, S., Matsuda, T., and Kikuchi, E. (1991). Hydrogen permeable palladium-silver alloy membrane supported on porous ceramics. *J. Membr. Sci.* **56,** 315–325.
[43] Altinisik, O., Dogan, M., and Dogu, G. (2005). Preparation and characterization of palladium-plated porous glass for hydrogen enrichment. *Catal. Today* **105,** 641–646.
[44] Lin, W.-H., and Chang, H.-F. (2005). Characterizations of Pd–Ag membrane prepared by sequential electroless deposition. *Surf. Coat. Technol.* **194,** 157–166.
[45] Shu, J., Grandjean, B. P. A., Ghali, E., and Kaliaguinea, S. (1993). Simultaneous depositing of Pd and Ag on porous stainless steel by electroless plating. *J. Membr. Sci.* **77,** 181–195.

[46] Pacheco Tanaka, D. A., Llosa Tanco, M. A., Niwa, S., Wakui, Y., Mizukami, F., Namba, T., *et al.* (2005). Preparation of palladium and silver alloy membrane on a porous α-alumina tube via simultaneous electroless plating. *J. Membr. Sci.* **247,** 21–27.
[47] Volpe, M., Inguanta, R., Piazza, S., and Sunseri, C. (2006). Optimized bath for electroless deposition of palladium on amorphous alumina membranes. *Surf. Coat. Technol.* **200,** 5800–5806.
[48] Malygin, A. A. (1996). Molecular layering technology and some of its applications. *Russ. J. Appl. Chem.* **69**(10), 1419–1426.
[49] Malygin, A. A. (2006). The molecular layering nanotechnology: Basis and application. *J. Ind. Eng. Chem.* **12**(1), 1–11.
[50] Tereshchenko, G. F., Orekhova, N. V., Ermilova, M. M., Malygin, A. A., and Orlova, A. I. (2006). Nanostructured phosphorus–oxide-containing composite membrane catalysts. *Catal. Today* **118,** 85–89.
[51] Aleksovskii, V. B. (1952). Thesis, Technological Institute Leningrad. cited on [49].
[52] Malygin, A. A., Ermilova, M. M., Gryaznov, V. M., Orekhova, N. V., and Malkov, A. A. (2002). The new catalytic membranes with low sized phosphorus oxide structures on a surface. *Desalination* **144,** 433–435.
[53] Ermilova, M. M., Orekhova, N. V., Tereshchenko, G. F., Malygin, A. A., Malkov, A. A., Basile, A., Gallucci, F., and De Luca, G. (2006). Methanol oxidative dehydrogenation on nanostructured composite membranes. *Desalination* **200,** 692–694.
[54] Armor, J. N. (1992). Challenges in membrane catalysis. *Chem. Tech.* **22,** 557–563.
[55] Lewis, F. A. (1967). The Palladium-Hydrogen System. Academic Press, London.
[56] Wang, D., Tong, J., Xu, H., and Matsamura, Y. (2004). Preparation of palladium membrane over porous stainless steel tube modified with zirconium oxide. *Catal. Today* **93–95,** 689–693.
[57] Sato, K., Hanaoka, T., Niwa, S., Stefan, C., Namba, T., and Mizukami, F. (2005). Direct hydroxylation of aromatic compounds by a palladium membrane reactor. *Catal. Today* **104,** 260–266.
[58] van Dyk, L., Miachon, S., Lorenzen, L., Torres, M., Fiaty, K., and Dalmon, J.-A. (2003). Comparison of microporous MFI and dense Pd membrane performances in an extractor-type CMR. *Catal. Today* **82,** 167–177.
[59] Liang, W., and Hughes, R. (2005). The catalytic dehydrogenation of isobutane to isobutene in a palladium/silver composite membrane reactor. *Catal. Today* **104,** 238–243.
[60] Lin, W. H., and Chang, H. F. (2004). A study of ethanol dehydrogenation reaction in a palladium membrane reactor. *Catal. Today* **97,** 181–188.
[61] Lin, Y. M., Liu, S. L., Chuang, C. H., and Chu, Y. T. (2003). Effect of incipient removal of hydrogen through palladium membrane on the conversion of methane steam reforming experimental and modelling. *Catal. Today* **82,** 127–139.
[62] Kleinert, A., Grubert, G., Pan, X., Hamel, C., Morgenstern, A. S., and Caro, J. (2005). Compatibility of hydrogen transfer via Pd-membranes with the rates of heterogeneously catalysed steam reforming. *Catal. Today* **104,** 267–273.
[63] Basile, A., Gallucci, F., and Paturzo, L. (2005). Hydrogen production from methanol by oxidative steam reforming carried out in a membrane reactor. *Catal. Today* **104,** 251–259.
[64] Shimpo, Y., Yamamura, S., Okouchi, H., Nishida, M., Kajita, O., Rimura, H., *et al.* (2004). Hydrogen permeation characteristic of melt-spun $Zr_{60}Al_{15}Co_{2.5}Ni_{7.5}Cu_{15}$ glassy alloy membrane. *J. Alloys Compd.* **372,** 197–200.
[65] Ryi, S. K., Park, J. S., Kim, S. H., Cho, S. H., Park, J. S., and Kim, D. W. (2006). Development of a new porous metal support of metallic dense membrane for hydrogen separation. *J. Membr. Sci.* **279,** 439–445.
[66] Hoang, H. T., Tong, H. D., Gielens, F. C., Jansen, H. V., and Elwenspoek, M. C. (2004). Fabrication and characterization of dual sputtered Pd-Cu alloy films for hydrogen separation membranes. *Mater. Lett.* **58,** 525–528.
[67] Zhang, Y., Gwak, J., Murakoshi, Y., Ikeara, T., Maeda, R., and Nishimura, C. (2006). Hydrogen permeation characteristics of thin Pd membrane prepared by microfabrication technology. *J. Membr. Sci.* **277,** 203–209.

[68] Hou, K., and Hughes, R. (2006). Preparation of thin and stable Pd/Ag composite membranes and simulative analysis of transfer resistance for hydrogen separation. *J. Membr. Sci.* **214,** 43–55.

[69] Tong, J., Suda, H., Haraya, K., and Matsumura, Y. (2006). A novel method for the preparation of thin dense Pd membrane on macroporous stainless steel tube filter. *J. Membr. Sci.* **260,** 10–18.

[70] Wang, D., Flanagan, T. B., and Shanahan, K. L. (2004). Permeation of hydrogen through pre-oxidized Pd membranes in the presence and absence of CO. *J. Alloys Compd.* **372,** 158–164.

[71] Gao, H., Lin, J. Y. S., Li, Y., and Zhang, B. (2005). Electroless plating synthesis, characterization and permeation properties of Pd-Cu membranes supported on ZrO_2 modified porous stainless steel. *J. Membr. Sci.* **265,** 142–152.

[72] Mardilovich, I. P., Engwall, E., and Ma, Y. H. (2002). Dependence of hydrogen flux on the pore size and plating surface topology of asymmetric Pd-porous stainless steel membranes. *Desalination* **144,** 85–89.

[73] Huang, T. C., Wei, M. C., and Chen, H. I. (2003). Preparation of hydrogen-permselective palladium-silver alloy composite membranes by electroless co-deposition. *Sep. Purif. Technol.* **32,** 239–245.

[74] Zhang, J., Liu, D., He, M., Xu, H., and Li, W. (2006). Experiment and simulation studies on concentration polarization in H_2 enrichment by highly permeable and selective Pd membranes. *J. Membr. Sci.* **274,** 83–91.

[75] Yepes, D., Cornaglia, L. M., Irusta, S., and Lombardo, E. A. (2006). Different oxides used as diffusion barriers in composite hydrogen permeable membranes. *J. Membr. Sci.* **274,** 92–101.

[76] Roa, F., Way, J. D., McCormik, R. L., and Paglieri, S. N. (2003). Preparation and characterization of Pd-Cu composite membranes for hydrogen separation. *Chem. Eng. J.* **93,** 11–22.

[77] Wu, L.-Q., Xu, N., and Shi, J (2000). Preparation of a palladium composite membrane by an improved electroless plating technique. *Ind. Eng. Chem. Res.* **39,** 342–348.

[78] Nam, S.-E., Lee, S.-H., and Lee, K-H (1999). Preparation of a palladium alloy composite membrane supported in a porous stainless steel by vacuum electrodeposition. *J. Membr. Sci.* **153,** 163–173.

[79] Yan, S., Maeda, H., Kusakabe, K., and Morooka, S (1994). Thin palladium membrane formed in support pores by metal-organic chemical vapor deposition method and application to hydrogen separation. *Ind. Eng. Chem. Res.* **33,** 616–622.

[80] Dittmeyer, R., Höllein, V., and Daub, K. (2001). Membrane reactors for hydrogenation and dehydrogenation processes based on supported palladium. *J. Mol. Catal. A* **173,** 135–184.

[81] Liang, W., and Hughes, R (2003). The effect of diffusion direction on the permeation rate of hydrogen in palladium composite membranes. *Chem. Eng. J.* **112,** 81–86.

[82] Uemiya, S., Sato, N., Ando, H., Kude, Y., Matsuda, T., and Kikuchi, E. (1991). Separation of hydrogen through palladium thin film supported on a porous glass tube. *J. Membr. Sci.* **56,** 303–313.

[83] Cheng, Y. K., and Yeung, K. L. (1999). Palladium-silver composite membranes by electroless plating technique. *J. Membr. Sci.* **158,** 127–141.

[84] Ma, Y. H., Akis, B. C., Ayturk, M. E., Guazzone, F., Engwall, E. E., and Mardilovich, I. P. (2004). Characterization of intermetallic diffusion barrier and alloy formation for Pd/Cu and Pd/Ag porous stainless steel composite membranes. *Ind. Eng. Chem. Res.* **43,** 2936–2945.

[85] Guazzone, F., Engwall, E. E., and Ma, Y. H. (2006). Effects of surface activity, defects and mass transfer on hydrogen permeance and *n*-value in composite palladium-porous stainless steel membranes. *Catal. Today* **118,** 24–31.

[86] Souleimanova, R. S., Mukasyan, A. S., and Varma, A. (2000). Effects of osmosis on microstructure of Pd-composite membranes synthesized by electroless plating technique. *J. Membr. Sci.* **166,** 249–257.

[87] Li, A., Liang, W., and Hughes, R. (1999). Fabrication of defect-free Pd/α-Al_2O_3 composite membranes for hydrogen separation. *Thin Solid Films* **350,** 106–112.

[88] Li, A., Liang, W., and Hughes, R. (1998). Characterisation and permeation of palladium/stainless steel composite membranes. *J. Membr. Sci.* **149,** 259–268.

[89] Zhang, Y., Komaki, M., and Nishimura, C. (2005). Morphological study of supported thin Pd and Pd-Ag membranes upon hydrogen permeation. *J. Membr. Sci.* **246,** 173–180.

[90] Hunter, J. B. (1963). *In* "Proc. of the Symp. on the Production of Ultra-Pure Hydrogen" Petroleum Division.
[91] Adrover, A., Giona, M., Capobianco, L., and Violante, V. (2004). Effects of self-stress on hydrogen diffusion in Pd membranes in the coexistence of α and β phases. *J. Alloys Compd.* **368,** 287–297.
[92] Grashoff, G. J., Pilkington, C. E., and Corti, C. W. (1983). Purification of hydrogen. *Platinum Met. Rev.* **27,** 157–169.
[93] Tebus, V., Rivkis, L., Arutunova, G., Dmitrievsky, E., Filin, V., Golikov, Y., *et al.* (1999). Investigation of palladium alloy properties degradation during long-time tritium exposure. *J. Nucl. Mater.* **271/272,** 345–348.
[94] Mundschau, M. V., Xie, X., Evenson, C. R., and Sammells, A. F. (2006). Dense inorganic membranes for production of hydrogen from methane and coal with carbon dioxide sequestration. *Catal. Today* **118,** 12–23.
[95] McCool, B., Xomeritakis, G., and Lin, Y. S. (1999). Composition control and hydrogen permeation characteristics of sputter deposited palladium-silver membranes. *J. Membr. Sci.* **161,** 67–76.
[96] Amandusson, H., Ekedahl, L.-G., and Dannetun, H. (2001). Hydrogen permeation through surface modified Pd and PdAg membranes. *J. Membrane Sci.* **193,** 35–47.
[97] Yamakawa, K., Ege, M., Ludescher, B., and Hirscher, M. (2003). Surface adsorbed atoms suppressing hydrogen permeation of Pd membranes. *J. Alloys Compd.* **352,** 57–59.
[98] Tosti, S., Adrover, A., Basile, A., Camilli, V., Chiappetta, G., and Violante, V. (2003). Characterization of thin wall Pd-Ag rolled membranes. *Int. J. Hydrogen Energy* **28,** 105–112.
[99] Krishna, R. (1993). Problems and pitfalls in the use of the fick formulation for intraparticle diffusion. *Chem. Eng. Sci.* **48,** 845–861.
[100] Keil, F. (1993). Diffusion und chemische Reaktion in der Gas/Feststoff-Katalyse. *Chem. Tech.* **45**(6), 437–447.
[101] Schramm, O., and Seidel-Morgenstern, A. (1999). Comparing porous and dense membranes for the application in membrane reactors. *Chem. Eng. Sci.* **54,** 1447–1453.
[102] Hughes, R. (2001). Composite palladium membrane for catalytic membrane reactors. *Membr. Tech.* **131,** 9–13.
[103] Paglieri, S. N., and Way, J. D. (2002). Innovations in palladium membrane research. *Sep. Purif. Methods* **31**(1), 1–169.
[104] Armor, J. (1995). Membrane catalysis: Where is it now, what needs to be done? *Catal. Today* **25,** 199–207.
[105] McCool, B. A., and Lin, Y. S. (2001). Nanostructured thin palladium-silver membranes: Effects of grain size on gas permeation properties. *J. Mater. Sci.* **36,** 3221–3227.
[106] Jayaraman, V., and Lin, Y. S. (1995). Synthesis and hydrogen permeation properties of ultrathin palladium-silver alloy membranes. *J. Membr. Sci.* **104,** 251–262.
[107] Morreale, B. D., Ciocco, M. V., Enick, R. M., Morsi, B. I., Howard, B. H., Cugini, A. V., *et al.* (2003). The permeability of hydrogen in bulk palladium at elevated temperatures and pressures. *J. Membr. Sci.* **212,** 87–97.
[108] Ward, T. L., and Dao, T. (1999). Model of hydrogen permeation behaviour in palladium membranes. *J. Membr. Sci.* **153,** 211–231.
[109] Tosti, S, Bettinali, L., Violante, V., Basile, A. Drioli, E., Chiappetta, M., Criscuoli, A., and Rizzello, C. (1998). "Pd based ultrathin membranes for the tritiated water gas shift reaction in the ITER breeder recovery system", Proc. of the 20th Symposium on Fusion Technology", pp. 1033–1036. Marseille (France), Sept. 7–11.
[110] Basile, A., Paturzo, L., and Laganà, F. (2001). The partial oxidation of methane to syngas in a palladium membrane reactor: Simulation and experimental studies. *Catal. Today* **67,** 65–75.
[111] Koffler, S. A., Hudson, J. B., and Ansell, G. S. (1969). Hydrogen permeation through Alpha-Palladium. *Trans. AIME* **245,** 1735–1740.
[112] Balovnev, Yu. A. (1974). The pattern of hydrogen diffusion in palladium (in Russian). *Russ. J. Phys. Chem.* **48**(3), 719–720.
[113] Itoh, N., and Xu, W.-C. (1993). Selective hydrogenation of phenol to cyclohexanone using palladium-based membranes as catalysts. *Appl. Catal. A* **107,** 83–100.

[114] Basile, A., Gallucci, F., and Paturzo, L. (2005). A dense Pd/Ag membrane reactor for methanol steam reforming: Experimental study. *Catal. Today* **104,** 244–250.

[115] Keuler, J. N., and Lorenzen, L. (2002). Developing a heating procedure to optimise hydrogen permeance through Pd-Ag membranes of thickness less than 2.2 μm. *J. Membr. Sci.* **195,** 203–213.

[116] Ali, J. K., Newson, E. J., and Rippin, D. W. T. (1994). Deactivation and regeneration of Pd-Ag membranes for dehydrogenation reactions. *J. Membr. Sci.* **89,** 171–184.

[117] Wills, R. S., Arzu, P. R., Honnell, K. G., and Brisdell, S. A. (2000). Initial testing of a low pressure permeator for tritium processing. *Fusion Eng. Des.* **49–50,** 963–970.

[118] Uemiya, S., Kajiwara, M., and Kojima, T. (1997). Composite membranes of group VIII metal supported on porous alumina. *AIChE J.* **43,** 2715–2723.

[119] Shu, J., Grandjean, B. P. A., Vanneste, A., and Kaliaguine, S. (1991). Catalytic palladium-based membrane reactors: A review. *Can. J. Chem. Eng.* **89,** 1036–1060.

[120] Howard, B. H., Killmeyer, R. P., Rothenberger, K. S., Cugini, A. V., Morreale, B. D., Enick, R. M., *et al.* (2004). Hydrogen permeance of palladium–copper alloy membranes over a wide range of temperatures and pressures. *J. Membr. Sci.* **241,** 207–218.

[121] Roa, F., and Way, J. D. (2005). The effect of air exposure on palladium–copper composite membranes. *Appl. Surf. Sci.* **240,** 85–104.

[122] Nam, S.-E., and Lee, K.-H. (2001). Hydrogen separation by Pd alloy composite membranes: Introduction of diffusion barrier. *J. Membr. Sci.* **192,** 177–185.

[123] Hoanga, H. T., Tonga, D., Gielensb, F. C., Jansena, H. V., and Elwenspoeka, M. C. (2004). Fabrication and characterization of dual sputtered Pd–Cu alloy films for hydrogen separation membranes. *Mater. Lett.* **58,** 525–528.

[124] Kulprathipanjaa, A., Alptekinb, G. O., Falconera, J. L., and Wayc, J. D. (2005). Pd and Pd–Cu membranes: Inhibition of H_2 permeation by H_2S. *J. Membr. Sci.* **254,** 49–62.

[125] Graham, T, (1866), On the Absorption and Dialytic Separation of Gases by Colloid Septa, Philosophical Transactions of the Royal Society of London, **156,** pp. 399–439.

[126] Hunter, J. B. (1956). Silver-Palladium Films for Separation and Purification of Hydrogen US Patent n. 2,773,561.

[127] Hunter, J. B. (1960). A new hydrogen purification process. *Platinum Met. Rev.* **4,** 130–131.

[128] Gryaznov, V. M., Smirnov, V.S., Ivanova L.K., and Mischenko, A.P. (1970). "Reactions coupling by hydrogen transfer through the catalyst" (in Russian). *Dokl. Akad. Nauk. SSSR* **190,** 144–147.

[129] V. Gryaznov, *Sov. Patent n.* 274 092 (1964).

[130] Booth, J.C.S., Doyle, M.L., Gee, S.M., Miller, J., Scholtz, L.A., and Walker, P.A. "Advanced hydrogen separation via thin supported Pd membranes" Hydrogen Energy Progress XI, vol. 1, pp. 867–878 (1996).

[131] Connor, H. (1962). Palladium alloy diffusion cells—Commercial units for the production of ultrapure hydrogen. *Platinum Met. Rev* **6,** 130–135.

[132] Cole, M. J. (1981). The generator of pure hydrogen for industrial applications. *Platinum Met. Rev.* **25,** 12–13.

[133] Philpott, J. (1985). Hydrogen diffusion technology. Commercial applications of palladium membrane. *Platinum Met. Rev.* **29,** 12–16.

[134] Paturzo, L., Basile, A., and Drioli, E. (2002). High temperature membrane reactors and integrated membrane operations. *Rev. Chem. Eng.* **18**(6), 511–551.

[135] Mori, T. (2005). RD&D for launching the initial market by hydrogen from natural gas. *In* "Proc. Int. Hydrogen Energy Congress and Exhibition IHEC 2005" July 13–15, Istanbul (Turkey).

[136] Violante, V., Basile, A., and Drioli, E. (1993). Membrane separation technologies—Their application to the fusion-reactor fuel-cycle. *Fusion Eng. Des.* **22**(3), 257–263.

[137] European HCPB Test Blanket (June 2001). *Des. Description Doc.* **8.**

[138] Tosti, S., Basile, A., Chiappetta, G., Rizzello, C., and Violante, V. (2003). Pd-Ag membrane reactors for water gas shift reaction. *Chem. Eng. J.* **93,** 23–30.

[139] Alderliesten, P., and Bracht, M, (1997), "An attractive option for CO_2 control in IGCC systems: water gas shift with integrated H_2/CO_2 separation (WIHYS) process. Phase 1: Proof of principle". Final Report, CEC Project Jou2 CT92-0158, pp. 1–53.

[140] Penzhorn, R. D., Berndt, U., Kirste, E., Hellriegel, W., Jung, W., Pejsa, R., *et al.* (1995). Long term permeator experiment PETRA at the tritium laboratory Karlsruhe: Commissioning tests with tritium. *Fusion Techn.* **28,** 723–731.

[141] Glugla, M., Miller, J., Herrmann, P., Iseli, M., and Penzhorn, R.-D. (1996). Tritium tests with a PERCAT reactor for isotopic swamping, Fusion Technology. *In* "Proceedings of the 19th SOFT" September 16–20, pp. 1193–1196. Lisbon, Portugal.

[142] Glugla, M., Perevezentsev, A., Niyongabo, D., Penzhorn, R.-D., Bell, A., and Herrmann, P. (2000). PERMCAT reactor for impurity processing in the JET active gas handling. *Fusion Eng. Des.* **49–50,** 817–823.

[143] Bornschein, B., Glugla, M., Günther, K., Lässer, R., Le, T. L., Simon, K. H., *et al.* (2003). Tritium tests with a technical PERMCAT for final clean-up of ITER exhaust gases. *Fusion Eng. Des.* **69,** 51–56.

[144] Waku, H., Tamura, I., Inoue, M., and Akai, M. (1995). Life cycle analysis of Fossil Power Plant with CO_2 recovery and sequestring system. *Energy Convers. Manage.* **36,** 877–880.

[145] Bracht, M. (1997). Water gas shift membrane reactor for CO_2 control in IGCC systems: Techno-economic feasibility study. *Energy Convers. Manage.* **38,** 159–162.

[146] Kreutz, T. G., Williams, R. H., Socolow, R. H., Chiesa, P., and Lozza, G. (2003). Production of hydrogen and electricity from coal with CO_2 capture. *In* "Sixth International Conference on Greenhouse Gas Control Technologies" (J. Gale and Y. Kaya, eds.), pp. 141–147. Kyoto, Pergamon.

[147] Amelio, M., Morrone, P., Gallucci, F., and Basile, A. (2007). Integrated gasification gas combined cycle plant with membrane reactors: A technological and economical analysis. *Energy Convers. Manage.* **48,** 2680–2693.

[148] Wieland, S., and Melin, T., (2001). Membrane reactors for hydrogen production, CD-rom Proceedings of the 3rd European Congress on Chemical Engineering, Nuremberg (Germany), 26-28 June, 2001, Topic A, Section A.4.

[149] Petterson, L. J., and Westerholm, R. (2001). State of the art of multi-fuel reformers for fuel cell vehicles: Problem identification and research needs. *Int. J. Hydrogen Energy* **26**(3), 243–264.

[150] Gallucci, F., Paturzo, L., and Basile, A. (2004). Hydrogen recovery from methanol steam reforming in a dense membrane reactor: Simulation study. *Ind. Eng. Chem. Res.* **43**(10), 2420.

[151] Tosti, S., Basile, A. Bettinali, L., and Chiaravalloti, F. (2005). Study and characterization of Pd-based permeators used in membrane reactors. *In* "Proc. of the Int. Hydrogen Energy Congress and Exhibition IHEC 2005" July 13-15, Istanbul, Turkey.

[152] Service, R. F. (2004). The hydrogen backlash. *Science* **305,** 958–961.

[153] Ramachandran, R., and Menon, R. K. (1998). An overview of industrial uses of hydrogen. *Int. J. Hydrogen Energy* **23,** 593–598.

[154] Veziroglu, T. N. (1999). Hydrogen energy system as a permanent solution to global energy-environmental problems. *Chem. Ind.* **53,** 383–393.

[155] Steele, B. C. H., and Heinzel, A. (2001). Materials for fuel-cell technologies. *Nature* **414,** 345–352.

[156] Schlapbach, L., and Züttel, A. (2001). Hydrogen-storage materials for mobile applications. *Nature* **414,** 353–358.

[157] Saracco, G., Neomagus, H.W.J.P., Versteeg, G.F., and Van Swaaij, W.P.M. (1999). High temperature membrane reactors: Potential and problems. *Chem. Eng. Sci.* **54,** 1997–2017.

[158] Zaman, J., and Chakma, A. (1994). Inorganic membrane reactors. *J. Membr. Sci.* **92,** 1–28.

[159] Coronas, J., and Santamaria, J. (1999). Catalytic reactors based on porous ceramic membranes. *Catal. Today* **51,** 377–389.

[160] Julbe, A., Farrusseng, D., and Guizard, C. (2001). Porous ceramic membranes for catalytic reactors—Overview and new ideas. *J. Membr. Sci.* **181,** 3–20.

[161] Marcano, J. G. S., and Tsotsis, T. T. (2002). Catalytic Membranes and Membrane Reactors : Wiley-VCH Verlag GmbH Ed.

[162] Dixon, A. G. (1999). Innovations in Catalytic Inorganic Membrane Reactors—Catalysis : Royal Society of Chemistry Ed.

[163] Amiridis, M.D., Zhang, T., and Bernales, C. A. (2001). Use of Membrane Reactor for Hydrogen Production Via Direct Cracking of Hydrocarbons U.S. Patent n. CN1291166.

[164] Fuder, F. (2004). Membrane Reactor and Method for the Production of Highly Pure Hydrogen Gas U.S. Patent n. US20044237406.
[165] Bernstein, L. A., and Lund, C. F. R. (1993). Membrane reactors for catalytic series and series-parallel reactions. *J. Membr. Sci.* **77,** 155–164.
[166] Harold, M. P., Lee, C., Burggraaf, A. J., Keizer, K., Zaspalis, V. T., and de Lange, R. S. A. (1994). Catalysts with Inorganic Membranes. *MRS Bulletin,* Vol. 19, No. 4, April, 34.
[167] Raich, B. A., and Foley, H. C. (1995). Supra-equilibrium conversion in palladium membrane reactors: Kinetic sensitivity and time dependence. *Appl. Catal. A* **129,** 167–188.
[168] Kikuchi, E., Uemiya, S., and Matsuda, T. (1991). Hydrogen production from methane steam reforming assisted by use of membrane reactor. *Nat. Gas Conv.* **61,** 509.
[169] Chai, M., Machida, M., Eguchi, K., and Arai, H. (1993). Promotion of methane steam reforming using tuthenium-dispersed microporous alumina membrane reactor. *Chem. Lett.* **1,** 41–44, Jan 1993.
[170] Madia, G., Barbieri, G., and Drioli, E. (1999). Theoretical and experimental analysis of methane steam reforming in a membrane reactor. *Can. J. Chem. Eng.* **77,** 698.
[171] Shu, J., Grandjean, B. P. A., and Kaliaguine, S. (1994). Methane steam reforming in asymmetric Pd- and Pd-Ag/porous SS membrane reactors. *Appl. Catal. A: Gen.* **119,** 305–325.
[172] Nam, S. W., Yoon, S. P., Ha, H. Y., Hong, S. A., and Maganyuk, A. P. (2000). Methane steam reforming in a Pd-Ru membrane reactor. *Korean J. Chem. Eng.* **17**(3), 288.
[173] Gallucci, F., Paturzo, L., Famà, A., and Basile, A. (2004). Experimental study of the methane steam reforming reaction in a dense Pd/Ag membrane reactor. *Ind. Eng. Chem. Res.* **43,** 928–933.
[174] Kusakabe, K., Fumio, S., Eda, T., Oda, M., and Sotowa, K. I. (2005). Hydrogen production in zirconia membrane reactors for use in PEMfuel cells. *Int. J. Hydrogen Energy* **30,** 989–994.
[175] Chang, J.-S., Park, S. E., and Chon, H. (1996). Catalytic activity and coke resistance in the carbon dioxide reforming of methane to synthesis gas over zeolite-supported Ni catalysts. *Appl. Catal. A* **145,** 111–124.
[176] Wang, S., and Lu, G. Q. (1998). Reforming of methane with carbon dioxide over Ni/Al_2O_3 catalysts: Effect of nickel precursor. *Appl. Catal. A* **169,** 271–280.
[177] Erdöhelyi, A., Cserényi, J., Papp, E., and Solymosi, F. (1992). Catalytic reaction of methane with carbon dioxide over supported palladium. *Appl. Catal. A* **108**(2), 205–219.
[178] Lemonidou, A. A., Goula, M. A., and Vasalos, I. A. (1998). Carbon dioxide reforming of methane over 5 wt.% nickel calcium aluminate catalysts—Effect of preparation method. *Catal. Today* **46,** 175–183.
[179] Ji, L., Tang, S., Zeng, H. C., Lin, J., and Tan, K. L. (2001). CO_2 reforming of methane to synthesis gas over sol-gel-made $Co/\gamma\text{-}Al_2O_3$ catalysts from organometallic precursors. *Appl. Catal. A* **207,** 247–255.
[180] Choudhary, V. R., Uphade, B. S., and Mamman, A. S. (1998). Simultaneous steam and CO_2 reforming of methane to syngas over NiO/MgO/SA-5205 in presence and absence of oxygen. *Appl. Catal. A* **168,** 33–46.
[181] Galuszka, J., Pandey, R. N., and Ahmed, S. (1998). Methane conversion to syngas in a palladium membrane reactor. *Catal. Today* **46,** 83–89.
[182] Paturzo, L., Gallucci, F., Basile, A., Vitulli, G., and Pertici, P. (2003). An Ru-based catalytic membrane reactor for dry reforming of methane—Its catalytic performance compared with tubular packed bed reactors. *Catal. Today* **82,** 57–65.
[183] Basile, A., and Paturzo, L. (2001). An experimental study of multilayered composite palladium membrane reactors for partial oxidation of methane to syngas. *Catal. Today* **67,** 55–64.
[184] Amphlett, J. C., Mann, R. F., and Weir, R. D. (1988). Hydrogen production by the catalytic steam reforming of methanol. Part 3: Kinetics of methanol decomposition using C18HC catalyst. *Can. J. Chem. Eng.* **66,** 950–956.
[185] de Wild, P. J., and Verhaak, M. J. F. M. (2000). Catalytic production of hydrogen from methanol. *Catal. Today* **60**(1–2), 3–10.
[186] Sekizawa, K., Yano, S., Eguchi, K., and Arai, H. (1998). Selective removal of CO in methanol reformed gas over Cu-supported mixed metal oxides. *Appl. Catal. A* **169,** 291–297.

[187] Liu, Y., Hayakawa, T., Suzuki, K., Hamakawa, S., Tsunoda, T., Ishii, T., *et al.* (2002). Highly active copper/ceria catalysts for steam reforming of methanol. *Appl. Catal. A* **223,** 137–145.

[188] Segal, S. R., Anderson, K. B., Carrado, K. A., and Marshall, C. L. (2002). Low temperature steam reforming of methanol over layered double hydroxide-derived catalysts. *Appl. Catal. A* **231,** 215–226.

[189] Takeda, K., Baba, A., Hishinuma, Y., and Chikahisa, T. (2002). Performance of a methanol reforming system for a fuel cell powered vehicle and system evaluation of a PEFC system. *J SAE Rev.* **23,** 183–188.

[190] Breen, J. P., and Ross, J. R. H. (1999). Methanol reforming for fuel-cell applications: Development of zirconia-containing Cu-Zn-Al catalysts. *Catal. Today* **51**(3–4), 521–533.

[191] Wieland, S., Melin, T., and Lamm, A. (2002). Membrane reactors for hydrogen production. *Chem. Eng. Sci.* **57**(9), 1571–1576.

[192] Agrell, J., Birgersson, H., and Boutonnet, M. (2002). Steam reforming of methanol over a Cu/ZnO/Al_2O_3 catalyst: A kinetic analysis and strategies for suppression of CO formation. *J. Power Sources* **106**(1–2), 249–257.

[193] Won, J. Y., Jun, H. K., Jeon, M. K., and Woo, S. I. (2006). Performance of microchannel reactor combined with combustor for methanol steam reforming. *Catal. Today* **111,** 158–163.

[194] Basile, A., Tosti, S., Capannelli, G., Vitulli, G., Iulianelli, A., Gallucci, F., and Drioli, E. (2006). Co-current and counter-current modes for Methanol Steam Reforming Membrane Reactor: Experimental Study. *Catal. Today* **118,** 237–245.

[195] Costantino, U., Marmottini, F., Sisani, M., Montanari, T., Ramis, G., Busca, G., *et al.* (2005). Cu–Zn–Al hydrotalcites as precursors of catalysts for the production of hydrogen from methanol. *Solid State Ionics* **176,** 2917–2922.

[196] Arstad, B., Venvik, H., Klette, H., Walmsley, J., Tucho, W. M., Holmestad, R., *et al.* (2006). Studies of self-supported 1.6 m Pd/23 wt.% Ag membranes during and after hydrogen production in a catalytic membrane reactor. *Catal. Today* **118,** 63–72.

[197] Tosti, S., Basile, A., Bettinali, L., Borgognoni, F., Chiaravalloti, F., and Gallucci, F. (2006). Long-term tests of Pd–Ag thin wall permeator tube. *J. Membr. Sci.* **284,** 393–397.

[198] Brisdell, S. A., Willis, R. S., and Wilhelm, R. C. (June 16–20, 1996). Ulra-High Tritium Decontamination of Simulated Fusion Fuel Exhaust Using A 2-Stage Palladium Membrane Reactor. *In* "Proc. of the ANS 12th Topical Meeting on Technology of Fusion Energy" Reno.

[199] Specchia, S., Fino, D., Saracco, G., and Specchia, V. (2005). Inorganic membrane reactors. *In* "Structured Catalysts and Reactors" (A. Cybulski and J. A. Moulijn, eds.), Taylor&Francis, London.

[200] Gryaznov, V. M., Ermilova, M. M., Orekhova, N. V., and Tereschenko, G. F. (2005). Reactors with metal and metal-containing membranes. *In* "Structured Catalysts and Reactors" (A. Cybulski and J. A. Moulijn, eds.), Taylor&Francis, London(Ch. 17).

[201] Pellin, M. J., Stair, P. C., Xiong, G., Elam, J. W., Birrell, J., Curtiss, L., *et al.* (2005). Mesoporous catalytic membranes: Synthetic control of pore size and wall composition. *Catal. Lett.* **102,** 127–130.

[202] Ioannides, T, and Verykos, X. E. (1996). Application of a dense silica membrane reactor in the reactions of dry reforming and partial oxidation of methane. *Catal. Lett.* **36,** 165–169.

[203] Kikuchi, E., and Chen, Y. (1998). Syngas formation by partial oxidation of methane in palladium membrane reactor, Natural Gas Conversion V, Vol. 119, Elsevier, Amsterdam, 441–446.

[204] Nakagawa, K., Ikenaga, N., Suzuki, T., Kobayashi, T., and Aruta, M. (1998). Partial oxidation of methane to synthesis gas over supported iridium catalysts, *Appl. Catal.* A, 169, 281-290.

[205] Ruckenstein, E., and Hu, Y. H. (1999). Methane partial oxidation over NiO/MgO solid solution catalysts. *Appl. Catal. A.* **183,** 85-92.

[206] Murcia-Mascaros, S., Navarro, R. M., Gomez-Sainero, L., Costantino, U., Nocchetti, M., and Fierro, J. L. G. (2001). Oxidative Methanol reforming reactions on CuZnAl catalysts derived from hydrotalcite-like precursors. *J. Catal.* **198,** 338–347.

[207] Velu, S., Suzuki, K., Kapoor, M. P., Ohashi, F., and Osaki, T. (2000). Selective production of hydrogen for fuel cells via oxidative steam reforming of methanol over CuZnAl(Zr)-oxide catalysts. *Appl. Catal. A.* **213,** 47–63.

CHAPTER NINE

Mathematical Modeling of Pd-Alloy Membrane Reactors

Giuseppe Barbieri,* Francesco Scura,* *and* Adele Brunetti*,†

Contents

List of symbols

A	Surface area, m^2
ACCUMULATION	Accumulation rate in a balance, s^{-1}
C	Concentration, mol/m^3
C_p	Specific heat, J/(mol K)
d	Diameter, m
D	Diffusivity, m^2/s
$\Delta P_{H2}^{Sievert}$	H_2 permeation Sievert's driving force, $Pa^{0.5}$
E	Activation energy, J/mol

* National Research Council - Institute for Membrane Technology (ITM-CNR), The University of Calabria, cubo 17/C, 87030 Rende CS, Italy

† Department of Chemical Engineering and Materials, The University of Calabria, cubo 44/A, 87030 Rende CS, Italy

Membrane Science and Technology, Volume 13

ISSN 0927-5193, DOI: 10.1016/S0927-5193(07)13009-6

E_d	H desorption activation energy, J/molH
F	Molar flow rate, mol/s
h	Enthalpy, J/mol
I	Sweep factor
ID	Inner diameter, m
IN	Rate of entering in a balance, s^{-1}
J	Permeating flux, mol/m^2 s
k	Kinetic constant, see related equation
K_{eq}	Equilibrium constant
K_p	Equilibrium constant in terms of partial pressures
k_r	Radial thermal conductivity, J/(m s K)
k_z	Axial thermal conductivity, J/(m s K)
L	Length, m
m	Reactant feed molar ratio
n	Number of mole
N	Molar flux, mol/(m^2 s)
OD	Outer diameter, m
OUT	Rate of leaving in a balance, s^{-1}
P	Pressure, Pa
Pe_o	Permeability pre-exponential factor, mol/(m s $Pa^{0.5}$)
Permeability	mol/(m s $Pa^{0.5}$)
Permeance	mol/(m^2 s $Pa^{0.5}$)
Permeating flux	mol/(m^2 s)
PRODUCTION	Rate a production in a balance, s^{-1}
Q	Volumetric flow rate, m^3(STP)/s
R	Gas law constant, 82.05 cm^3 atm /(g-mol K)
RI	Recovery index
r	Radial coordinate, m
r_{ij}	*j*th reaction rate for the *i*th species, mol/(m^3 s)
S	Cross section, m^2
T	Temperature, °C or K
t	Time, s
U	Overall heat transfer coefficient, W(m^2 K)
V	Volume, m^3
VI	Volume index,%
$W_{Catalyst}$	Catalyst weight, g
X	Conversion
Y_i	Molar fraction of the species *i*th
$\bar{Y}^j$	Molar fraction vector of the *j*th stream
z	Axial coordinate, m

Greek letters

Φ	Enthalpy flux associated to hydrogen permeation, J/(m^2 s)
Π	Permeance, mol/(m^2 s $Pa^{0.5}$)
Ψ	Heat generated by chemical reactions, J/(m^3 s)
δ	Membrane thickness, m

ε	Porosity
η	Catalyst effectiveness factor, -
$V_{I,\varphi}$	Stoichiometric coefficient with respect to the reference component of ith species in the *j*th reaction
ρ	Density, g/m^3
τ	Space time, s

Superscripts

Actual	Actual conditions referred to
Annulus	Annulus side in a luminal (tubular) MR
Catalyst	Catalyst referred to
Exit	Membrane reactor exit referred to
Feed	Membrane module inlet stream referred to
Input	Membrane reactor input referred to
Lumen	Lumen side in a tubular MR
Membrane	Membrane phase referred to
Output	Membrane reactor output referred to
Permeate	Membrane module permeate stream referred to
Permeating	Membrane module permeating stream referred to
Permeation	Membrane module permeation stream referred to
Reaction	Membrane module stream on the reaction volume referred to
Retentate	Membrane module outlet stream on the reaction referred to
Shell	Membrane module shell side referred to
Sweep	Membrane module inlet stream on the permeate side referred to

Acronyms

B.C.	Boundary condition
CPC	Concentration polarization coefficient
CSTR	Continuous stirred tank reactor
GHSV	Gas hourly space velocity (STP), s^{-1}
I.C.	Initial condition
MR	Membrane reactor
MREC	Membrane reactor equilibrium conversion
MSR	Methane steam reforming
ODE	Ordinary differential equation
PDE	Partial differential equation
STP	Standard temperature (25 °C) and pressure (100 kPa)
TR	Traditional reactor
TREC	Traditional reactor equilibrium conversion
WGS	Water gas shift

Dimensionless numbers

$Pé_{AB}$	Peclet number

1. Introduction

Membrane science and technology has led to significant innovation in both process and products, offering a lot of opportunities for sustainable industrial growth. Membrane science and technology has developed significantly in the last few years related to the possibility of integrating various membrane operations in the same industrial cycle, with overall important benefits to product quality, plant compactness, environmental impact, and energetic aspects [1]. A lot of research carried out in the membrane field is based on the logic of the Process Intensification Strategy, an innovative methodology for process and plant design, development, and implementation. It is a new design philosophy for achieving significant reductions (by factors of 10–100 or more) in plant volume and at the same production capacity [2]. In fact, 20% of plant costs are in the process equipment and the balance in structural steel, piping, etc., as a consequence smaller unit operations mean lower installed plant costs.

Membrane reactors (MRs) have attracted considerable attention in recent years [3–27]. In these systems, reaction and separation are performed in the same vessel. A permselective membrane, dividing the reactor into a reaction volume (where the reaction happens) and a permeate side, allows the selective removal of products from the reaction volume under the effect of a driving force that is a function of the species partial pressures on the reaction and permeate sides. In particular, it can be created by means of an inert sweep gas in the permeate compartment (nitrogen, helium, water, etc.), or with the application of a pressure difference between the retentate and the permeate compartments or by their combination (Fig. 9.1).

The removal of a product from the reaction implies a series of advantages:

- Yield enhancement of equilibrium-limited reactions, owing to the possibility of a continuous removal of one or more reaction products.
- Selectivity enhancement owing to a possible selective permeation.
- Coupling of two or more reactions, for example, dehydrogenation (endothermic) with a hydrogenation (exothermic) on the two membrane sides.
- Modification of reaction mechanisms by the removal of main or secondary products by the means of appropriate configurations of the catalytic membranes.
- Higher residence time due to the product removal.
- Lower-operating conditions (e.g., temperature) having analogous conversion, etc.

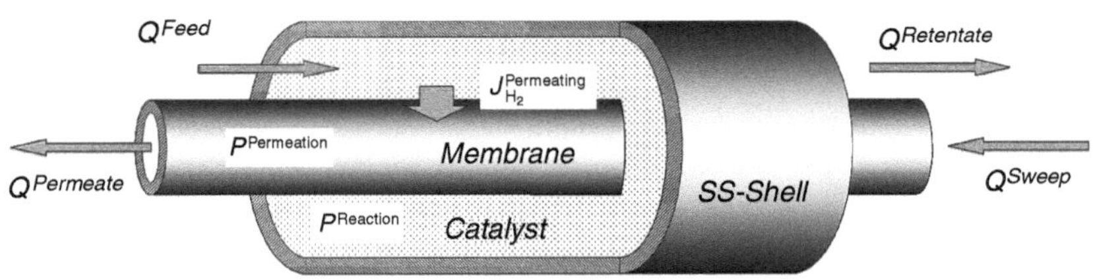

Figure 9.1 MR scheme.

- Controlled addition of a reactant through the membrane in particular for reactions such as partial oxidations where the same oxygen amount can be feed in a distributed way along the whole membrane length.

In the specific case of Pd-alloy MR, other advantages are the following:

- Production of pure H_2 permeate stream (where a pressure difference is used instead of a sweep gas for creating the permeation-driving force).
- Retentate streams concentrate in the other species (e.g., CO_2).
- Positive effect of reaction pressure on the equilibrium conversion also for reactions characterized by an increase in the number of moles.

In particular, the removal of a product from the reaction volume allows the thermodynamic equilibrium limit of a traditional reactor (TR) to be exceeded, obtaining higher conversion in analogous-operating conditions. This allows the MR to achieve, for endothermic reactions, the same conversion as that attained in a TR but at significantly lower temperatures. Instead, a higher temperature favors the kinetics for exothermic reactions.

This also affects the performance of the catalyst in terms of activity and life, also avoiding possible catalyst deactivation with significant advantages in terms of equipment costs. Also the kinetic mechanisms might be modified or controlled by the presence of an appropriate membrane system, which can also act as a reactive interface only with no permselectivity, optimizing phase-transfer catalysis.

Another interesting aspect of MR use is the positive effect that the reaction pressure can have on the process, also for reactions that happen without mole number variation [e.g., water gas shift (WGS)] or with a mole number increase [e.g., methane steam reforming (MSR)]. A classical approach in the chemical process industry is volume reduction by means of a higher pressure also for those reactions characterized by an increase in the mole number.

One of the main application of MRs is hydrogen production by light hydrocarbons (e.g., MSR and WGS); for these applications, the Pd-alloy MRs are the most promising devices; because of Pd-alloy membrane infinite hydrogen selectivity pure H_2 permeate stream can be obtained [28–31].

The term *Process Intensification* is used to describe the strategy of making dramatic reductions in the physical size (Fig. 9.2) of a chemical plant while achieving a given production objective. It is an interesting possible answer to the necessity of redesigning industrial production for sustainable growth, compatible with a high-quality lifestyle, mitigating greenhouse gas emissions, reducing investment and operating costs of chemical plants and thereby increasing profitability.

Pursuing the logic of the Process Intensification Strategy (see Fig. 9.30), MRs provide an interesting answer.

In this context, *membrane engineering* is one of the most interesting possibilities for putting this strategy into practice [1, 2, 32–34]. The synergic effects offered by MRs combining reaction and separation in the same unit, their simplicity and the possibility of advanced levels of automation and control offer an attractive opportunity to redesign industrial processes [35–37].

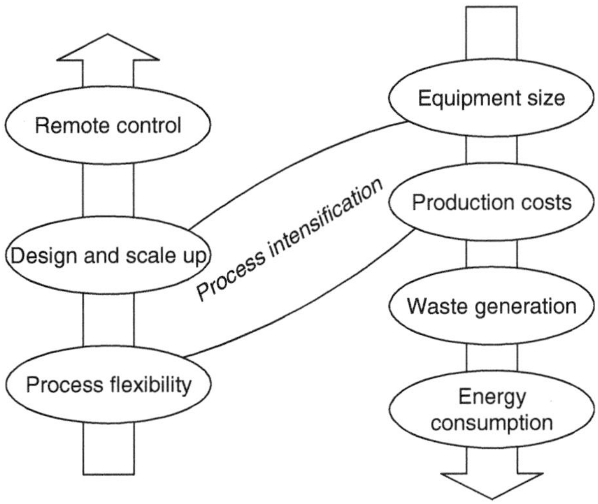

Figure 9.2 The logic of Process Intensification Strategy.

2. Pd and Pd-Alloy Membranes

Pd or better Pd-alloy membranes are the most innovative devices for hydrogen separation [38, 39]. These materials show infinite hydrogen selectivity, and this is the main advantage of their application in MRs. However, the use of pure palladium membranes is hindered by the transition of a metal phase (from the α-phase to the β-phase): at a temperature below 300 °C and a pressure below 200 kPa, which depends on the hydrogen concentration in the metal. Since the lattice constant of the β-phase is 3% higher than that of the α-phase this transition leads to lattice strain and, consequently, after a few cycles to a metal lattice distortion [40] and micro failures inside the metallic bulk are induced. Under the critical temperature (300 °C) in the Pd–H system, the two different phases α and β coexist with two different compositions. The α-phase characterized by a hydrogen/palladium atomic ratio equal to 0.008 (at 20 °C) has a lower presence of hydrogen atoms. The β phase has a hydrogen/palladium atomic ratio equal to 0.607. Both the phases have the same lattice symmetry, but very different specific volumes. Figure 9.3 reports the phase-diagram as function of the hydrogen/palladium atomic ratio.

Pd-alloys are generally used in membrane production in order to reduce embrittlement. The presence of other metals, lowering the coexistence temperature of the two crystalline phases, reduces the internal strength generated and hinders micro failure formation inside the alloy bulk.

Silver is the most-used metal worldwide in Pd-alloy membrane production. Its optimum composition giving greater permeability is 77 and 23 (%wt) for Pd and Ag, respectively.

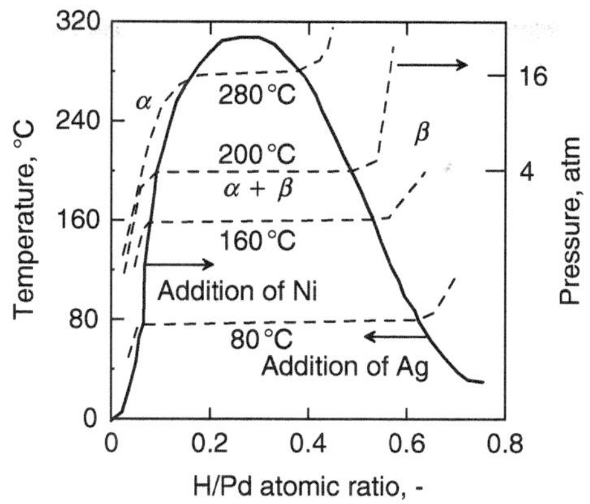

Figure 9.3 Phase diagram Pd-H system [41].

2.1. Elementary step of H_2 permeation and Sievert's law

The H_2 permeation through Pd or Pd-alloy membranes also when supported on porous materials involves several steps in series. They are ranked, in order from the highest H_2 partial pressure side to the lowest one, into:

On the feed side:

1. *Mass transfer in the fluid phase of the film just near the interface.*
2. *Adsorption*: Hydrogen decomposition in atoms on the Pd-alloy surface.

Inside the Pd-alloy layer:

3. *Surface to bulk*: Atomic hydrogen transition from Pd-alloy surface to Pd-alloy bulk.
4. *Diffusion*: Transport in the Pd-alloy bulk.
5. *Bulk to surface*: Atomic hydrogen transition from Pd-alloy bulk to Pd-alloy surface.

On the permeate side:

6. *Desorption*: Hydrogen atomic recombination on the Pd-alloy surface.
7. *Mass transfer in the fluid phase of the multilayered porous support* (for supported membranes).
8. *Mass transfer in the fluid phase of the film just near the interface.*

Each of these steps is characterized by an intrinsic forward and reverse rate. The overall observed permeation rate is governed by the rate-determining steps.

An Arrhenius plot of H_2-permeating flux versus the temperature (Fig. 9.4) is a useful tool for the identification of the rate-determining steps.

Ward and Dao [42] developed a model for H_2 permeation through self-supported Pd membranes considering the elementary steps listed above. They calculated the H_2-permeating flux through Pd membranes in the absence of external mass transfer for three different membrane thicknesses (Fig. 9.4). The flux decreases linearly with reverse of temperature (at a high temperature, left side). Here, the diffusion in the Pd-alloy bulk is the rate-determining step: the permeating flux follows the dashed line calculated for diffusion control condition into Pd bulk.

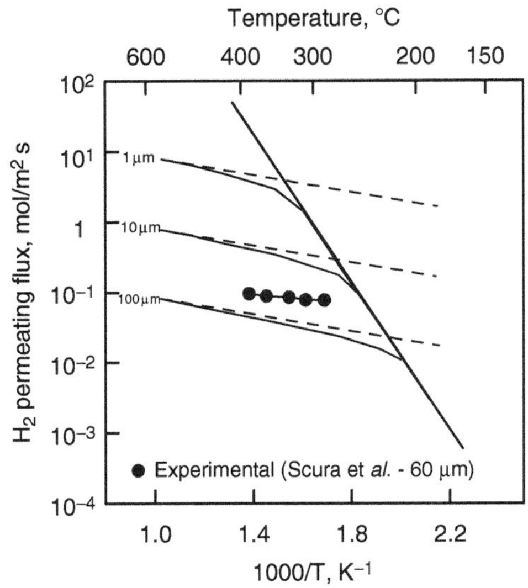

Figure 9.4 Arrhenius plot of H_2-permeating flux as a function of the reverse of temperature through self-supported Pd membrane. Overall permeating flux (solid lines); *diffusion-controlled* H_2-permeating flux (dashed lines); and *desorption-limited* (thick line, for a desorption activation energy of 58 kJ/mol) calculated by Ward and Dao [42] in the absence of external mass transfer resistances, for three different membrane thicknesses (1, 10, and 100 μm), at 100 and 0 kPa of feed and permeation pressures, respectively. Experimental data (symbols) by Scura [43] with a Pd–Ag membrane 60-μm thick.

The slope increases significantly at a lower temperature (right side) where the surface phenomena (adsorption, desorption, etc.) are the rate-determining steps. Here, the permeation does not depend on the membrane thickness. The transition temperature of the rate-determining step for the permeation from diffusion to surface phenomena also depends on the membrane thickness.

The same figure shows a good agreement with some measurements of Scura [43] by using a Pd/Ag membrane of a different thickness (60 μm). These data are in the diffusion-controlled range of the permeation process.

Pd-alloy membranes are generally utilized in a temperature range where the diffusion in the metal bulk is the rate-determining step. Therefore, *Sievert's law* [Eq. (9.1)] is used worldwide for the mathematical description of H_2-permeating flux in these types of membrane. The hydrogen-permeating flux is a linear function of the permeability and driving force and reverse function of the membrane thickness.

$$J_{H_2}^{\text{Permeating}} = \frac{\text{Permeability}_{H_2}}{\text{thickness}} \Delta\sqrt{P_{H_2}} \tag{9.1}$$

$$\Delta\sqrt{P_{H_2}} = \Delta P_{H_2}^{\text{Sievert}} = \sqrt{P_{H_2}^{\text{Reaction side}}} - \sqrt{P_{H_2}^{\text{Permeation side}}} \tag{9.2}$$

The permeation driving force of Sievert's law is the difference of the square root of the hydrogen partial pressure on the two membrane sides [Eq. (9.2)].

The hydrogen permeability is experimentally evaluated measuring the permeating flux as a function of the driving force and the membrane thickness as reported by Eq. (9.3).

$$\begin{aligned}\text{Permeability}_{H_2} &= \frac{\text{Permeating flux}_{H_2}}{\Delta P_{H_2}^{\text{Sievert}}/\text{thickness}} \\ &= \text{Permeability}_{H_2}^0 e^{-E/RT}\end{aligned} \tag{9.3}$$

It is a property of the material (e.g., Pd or Pd–Ag), fabrication procedure, etc., and gives a measure of the permeation fastness through the membrane material. Efforts to increase permeability are devoted to finding new materials, new alloys or alloy compositions [44, 45].

Permeance, another variable widely used to characterize the permeation properties of a membrane, gives information about the permeation of a specific produced membrane with its own specific thickness, separating layer, support, etc. Only the permeation flux and driving force are sufficient for its determination [Eq. (9.4)].

$$\begin{aligned}\text{Permeance}_{H_2} &= \frac{\text{Permeating flux}_{H_2}}{\Delta P_{H_2}^{\text{Sievert}}} \\ &= \text{Permeance}_{H_2}^0 e^{-E/RT}\end{aligned} \tag{9.4}$$

Both permeability and permeance [Eqs. (9.3) and (9.4)] follow an Arrhenius law as a function of temperature; generally they are linear functions of the permeation driving force.

The relationship between the permeability and the permeance is:

$$\text{Permeance}_{H_2} = \frac{\text{Permeability}_{H_2}}{\text{thickness}} \tag{9.5}$$

The permeance will widely be used as a variable for describing the permeation in the mathematical models proposed in this chapter. Permeation modeling requires, in fact, both the permeability and the membrane thickness; both being included in the permeance. Therefore Sievert's law will be hereafter often written in the following form:

$$J_{H_2}^{\text{Permeating}} = \text{Permeance}_{H_2} \Delta\sqrt{P_{H_2}} \tag{9.6}$$

2.2. Literature data of H_2 permeation

Diffusion-limited H_2 permeation rates (permeance and permeability) in Pd-alloy are well documented in the literature. Figures 9.5 and 9.6 give an idea of the typical values and trends of the permeation properties of the Pd-alloy membrane by means of a comparison among the literature data of H_2 permeance or permeability as a function of temperature. In particular, the membranes differ in thickness and in the presence of a support, but in the majority of cases the diffusion in the metallic bulk

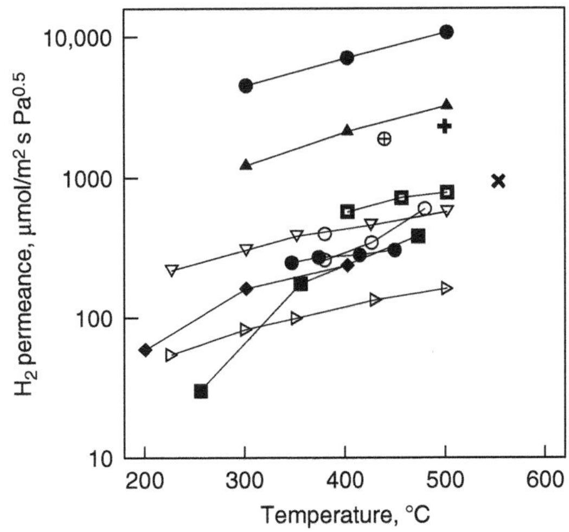

Figure 9.5 Literature data of H_2 permeance as a function of temperature for different thicknesses of Pd-alloy membranes [46–56]. ▲ Shu *et al.* (2 μm) [46], × Collins and Way (20 μm) [47], ◆ Yan *et al.* (5 μm) [48], Uemiya *et al.* (20 μm) [49], • Uemiya *et al.* (3 μm) [50], ∇ Hurlbert and Konecny (27 μm) [51], ▷ Hurlbert and Konecny (106 μm) [51], ■ Xomeritakis and Lin (0.5 μm) [52], ○ Li *et al.* (10 μm) [53], ○ Li *et al.* (10 μm) [54], + Tong *et al.* (1.2 μm) [55], ⊕ Bredesen *et al.* (1.2 μm) [56], • Scura (60 μm) [43].

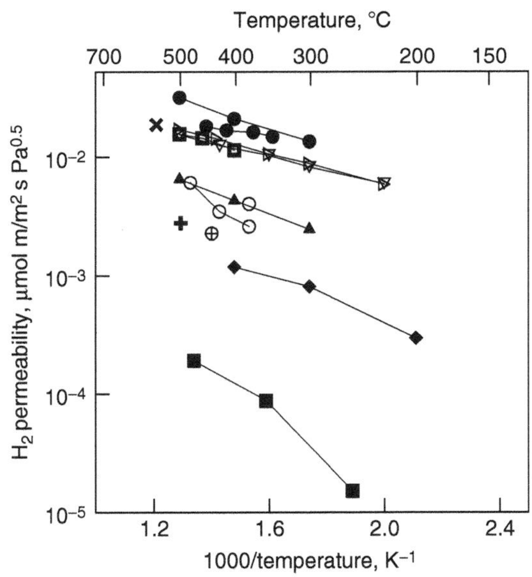

Figure 9.6 Literature data of H_2 permeability as a function of the reverse of temperature for different thicknesses of Pd-alloy membranes. ▲ Shu *et al.* (2 μm) [46], × Collins and Way (20 μm) [47], ◆ Yan *et al.* (5 μm) [48], □ Uemiya *et al.* (20 μm) [49], • Uemiya *et al.* (3 μm) [50], ∇ Hurlbert and Konecny (27 μm) [51], ▷ Hurlbert and Konecny (106 μm) [51], ■ Xomeritakis and Lin (0.5 μm) [52], ○ Li *et al.* (10 μm) [53], ○ Li *et al.* (10 μm) [54], + Tong *et al.* (1.2 μm) [55], ⊕ Bredesen *et al.* (1.2 μm) [56], • Scura (60 μm) [43] .

seems to be the rate-determining step for H_2 permeation. The difference in permeance can be due to different membrane thicknesses (Fig. 9.5), whereas those of permeability (Fig. 9.6) are due to different composition, preparation procedure, etc.

The majority of Pd-alloy membranes are tested in the region of interest of high-temperature reactions where diffusion controls H_2 permeation. Therefore, a lot of mathematical models of Pd-alloy MRs presented in the open literature consider Sievert's law of H_2 permeation.

2.3. Concentration polarization

The permeation driving force is a function of the difference of H_2 partial pressure at the two membrane surfaces. They are equal to those of the bulk phases when the H_2 concentration gradients can be neglected as is shown in Fig. 9.7A. This is the ideal/desired case because it gives the greatest driving force available for the permeation. The actual case is schematized in Fig. 9.7B in which the driving force is reduced by concentration gradients (also generated by the permeation) near the membrane surfaces. The actual driving force, less than the highest one, has to be considered in the mathematical models and simulations of hydrogen permeation.

The concentration polarization more or less significantly affects the performance of all the membrane-based separation processes, producing an accumulation of nonpermeating species in correspondence to the membrane surface and decreasing the separation performance.

Mass transport through the gaseous film produces an actual reduction of the permeating-driving force and a lower-permeating flux with significant reduction of the system performance (less H_2 removed with lower conversion improvement).

Polarization has to be taken into account for a good evaluation of the permeating flux. Thus, the concentration polarization coefficient (CPC) was defined as the ratio of the driving force at the membrane surface to the highest value of the driving force.

$$\mathrm{CPC} = \frac{\Delta\sqrt{P_{\mathrm{H_2}}^{\mathrm{Membrane}}}}{\Delta\sqrt{P_{\mathrm{H_2}}^{\mathrm{bulk}}}} = \frac{\sqrt{P_{\mathrm{H_2}}^{\mathrm{Membrane,\ Feed}}} - \sqrt{P_{\mathrm{H_2}}^{\mathrm{Membrane,\ Permeate}}}}{\sqrt{P_{\mathrm{H_2}}^{\mathrm{bulk,\ Feed}}} - \sqrt{P_{\mathrm{H_2}}^{\mathrm{bulk, Permeate}}}} \tag{9.7}$$

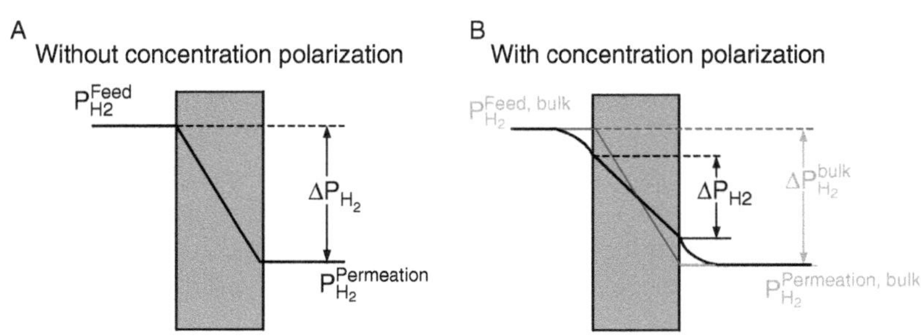

Figure 9.7 Concentration polarization: scheme of the permeation driving force.

The H_2-permeating flux after the introduction of CPC is given by Eq. (9.7).

$$J_{H_2}^{Permeating} = Permeance_{H_2} \quad \Delta\sqrt{P_{H_2}^{Membrane}} \\ = CPC\ Permeance_{H_2} \quad \Delta\sqrt{P_{H_2}^{bulk}} \tag{9.8}$$

The CPC assumes all the values between 0 (complete polarization) and 1 (no polarization). It depends on the membrane properties and fluid dynamics of the system.

Also, the presence of a porous support in Pd-alloy membranes produces H_2 concentration polarization. This was well illustrated by Caravella *et al.* in [57] using the multicomponent approach for the case of a multilayer porous support. The H_2 partial pressure profile from feed to permeate bulk phases was also calculated considering all the elementary permeation steps. The contribution of different steps (porous support, adsorption and desorption, feed and permeate films, and diffusion) on H_2 pressure drop through the membrane can be evaluated for different temperatures, feed composition, and permeation configuration (feed from Pd-membrane or support side, and no support presence). Therefore, the CPC can be estimated from these profiles.

A rough estimation of concentration polarization for the MSR reaction in a Pd-alloy MR was made by Madia *et al.* [58] by means of a comparison between the diffusive flux of H_2 in the radial direction at the equilibrium composition and the maximum permeating flux at extreme driving force conditions. The transmembrane permeating flux, around 5% of the radial diffusive flux, allowed the concentration polarization to be neglected.

Caravella and Barbieri [59] developed a model for H_2 permeation through a supported Pd-alloy membrane with a multicomponent approach. A comparison with the Ward and Dao [42] model shows a significant difference for thin membranes, as the temperature increases because of improved permeation. The polarization phenomenon becomes less and less important at lower temperature and for a higher membrane thickness: the diffusion in the gaseous bulk phases is higher than the permeation through the membrane (Fig. 9.8).

3. Thermodynamic Equilibrium in Pd-Alloy Membrane Reactor

The *nonequilibrium conversion of a TR* depends on several factors, for example the reactor model considered (continuous stirred tank or plug flow or batch), the thermodynamic variables (temperature and pressure), composition and the operating variables (e.g., feed flow rate, catalyst amount and activity, and overall heat exchange coefficient), etc.

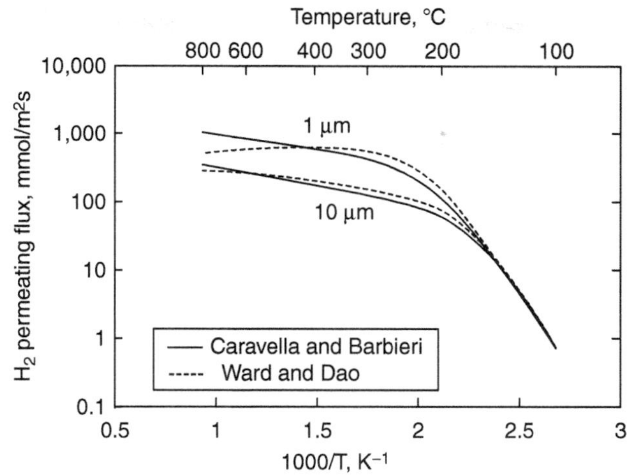

Figure 9.8 Hydrogen-permeating flux calculated considering the concentration polarization phenomena. $P_{H_2}^{Feed} = 101.3kPa, P_{N_2}^{Feed} = 0kPa, P_{H_2}^{Permeate} \approx 0kPa, P_{N_2}^{Permeate} = 101.3kPa.$

$$\text{Nonequilibrium conversion of a TR} = f\left(Kp, T^{Reaction}, P^{Reaction}, F^{Feed}, Y_i^{Feed}, W^{Catalyst}, U^{Overall}\right) \quad (9.9)$$

The upper limit of a chemical reactor is given by the thermodynamics that allows the evaluation of the equilibrium conversion of a TR and of an MR. The calculation of various variables, for example conversion, for a TR is well known starting from reaction equilibrium constants.

$$Kp_j(T) = \prod_i^{N_{species}} P_i^{\nu_{i,j}} \; \forall \text{reaction } j \quad (9.10)$$

The TR equilibrium conversion (TREC) depends only on thermodynamics (temperature, pressure, and equilibrium constant) and on the species initial concentrations.

$$\begin{aligned} \text{TREC} &= \text{TR Equilibrium Conversion} \\ &= f\left(Kp, T^{Reaction}, P^{Reaction}, Y_i^{Feed}\right) \end{aligned} \quad (9.11)$$

It does not depend on the reaction path.

The *nonequilibrium conversion of an MR* depends on temperature and pressure of the permeate side, sweep gas flow rate and composition, membrane properties in addition to the other parameters already cited for a TR.

$$\text{Nonequilibrium conversion of an MR} = f(Kp, T^{\text{Reaction}}, P^{\text{Reaction}}, F^{\text{Feed}}, Y^{\text{Feed}}, W^{\text{Catalyst}}, T^{\text{Permeation}}, P^{\text{Permeation}}, F^{\text{Sweep}}, Y_i^{\text{Sweep}}, U^{\text{Overall}}) \quad (9.12)$$

It depends on the reaction and permeation conditions.

The membrane characteristics and its geometry (permeance, thickness, and surface) influence the permeation rate and hence nonequilibrium conversion of an MR.

The equilibrium of an MR is a relatively new concept [60, 61]. The permeation equilibrium has to be reached in an MR in addition to the reaction equilibrium typical of a TR. This extra constraint can be expressed as no longer permeation.

$$\begin{aligned} J_i^{\text{Permeating}} &= 0 \Leftrightarrow \mathrm{f}(P_i^{\text{Reaction}}, P_i^{\text{Permeation}}) \\ &= 0 \Leftrightarrow P_i^{\text{Reaction}} = P_i^{\text{Permeation}} \;\forall \text{permeable specie } i \end{aligned} \quad (9.13)$$

The condition of no flux is equivalent to 0 permeation driving force, that is when the species partial pressures on both membrane sides are equal. The said equality means that the equilibrium conversion of an MR is also independent of the permeation law.

Therefore, the MR equilibrium conversion **(MREC)** is a function of the thermodynamic variables and initial compositions on both sides of the Pd-alloy membranes.

$$\text{MREC} = \text{MR Equilibrium Conversion} = f(Kp, T^{\text{Reaction}}, P^{\text{Reaction}}, Y_i^{\text{Feed}}, T^{\text{Permeation}}, P^{\text{Permeation}}, F^{\text{Feed}}/F^{\text{Sweep}}, Y_i^{\text{Sweep}}) \quad (9.14)$$

The MREC does not depend on the reaction path as the TREC and permeation rate. It is independent of the membrane permeation properties influencing the time-dependent variables (i.e., the residence time for the plug-flow MR) necessary to reach equilibrium,[1] but the final value reached depends on the extractive capacity of the system. This property is a function of the sweep factor characterizing an MR.

The sweep factor was defined for a closed MR as the ratio of the mole number of the sweep gas to the mole number of the reference component, both evaluated at the starting time:

$$I = \text{Sweep factor} = \left.\frac{n^{\text{Sweep}}}{n^{\text{Feed}}_{\text{Reference component}}}\right|^{\text{Initial time}} \quad (9.15)$$

[1] Theoretically, the time (residence time) required by a batch (plug flow) reactor in order to reach equilibrium conversion is infinite. In calculations it is considered as that time after which any change in the chemical species is lower than a very low set value; practically no change takes place.

In an analogous way the sweep factor can be defined for a flow reactor as the ratio between the flow rate of the sweep gas and the flow rate of a reference component on the feed side as

$$I = \text{Sweep factor} = \frac{F^{\text{Sweep}}}{F^{\text{Feed}}_{\text{Reference component}}} \tag{9.16}$$

The MSR and WGS reactions have been chosen as the reference reactions to calculate the MREC owing to their importance in many industrial processes for hydrogen production.

The MREC of the MSR is an increasing function of temperature [60] as for a TR. It is higher for a greater sweep factor. The increase in the conversion is higher at an intermediate temperature, while at a lower temperature the conversion difference between an MR and a TR decreases owing to low H_2 partial pressure in the reaction chamber (low methane conversion) thus the permeation effect is also very low. At a high temperature (>500 °C), the conversion is higher than 90% (for a sweep factor of 10); a further increase of sweep factor shifts the equilibrium to a total conversion.

Moreover, MSR reaction is characterized by an increase in mole number and thus it is not favored by pressure. Figure 9.9 shows a measure of this reduction; in particular, the TRECs relative to higher pressures are located under the TREC corresponding to atmospheric reaction pressure. The situation is completely different for an MR where an increase of the reaction pressure produces an increase in the driving force for the permeation and then a higher MREC is achieved. The same positive effect is shown by the WGS reaction (Fig. 9.10) characterized by no variation in mole number. However, reaction pressure increases the permeation driving force, and consequently there is a great advantage in the global performance

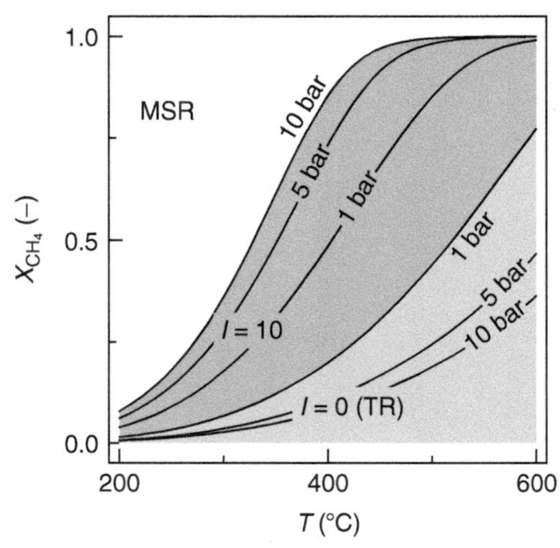

Figure 9.9 MSR reaction. CH_4 equilibrium conversion for both traditional and membrane reactors. *I* is the ratio of the sweep flow rate to the CH_4 feed flow rate. Reprinted from Marigliano *et al.* [60] with permission from Elsevier.

of the reactor. The pressure has a positive effect on the MREC for any reaction type (e.g., MSR). In addition, the permeation reduces the reaction volume for a batch MR or the outlet reaction stream of a plug-flow MR.

In Fig. 9.11 the CH_4 conversion, in the MSR reaction, is reported versus the reaction pressure. The MREC is an increasing function of the pressure at a higher sweep factor (I). When the reaction temperature and I are lower (<500 °C, $I = 1$) the MREC profile presents a minimum at a pressure of a few bars. The negative effect of

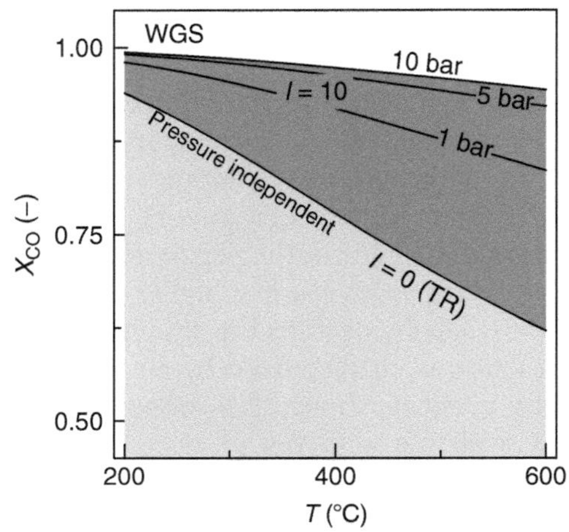

Figure 9.10 WGS reaction. CO equilibrium conversion for both traditional and membrane reactors. I is the ratio of the sweep flow rate to the CO feed flow rate. Reprinted from Marigliano *et al.* [60] with permission from Elsevier.

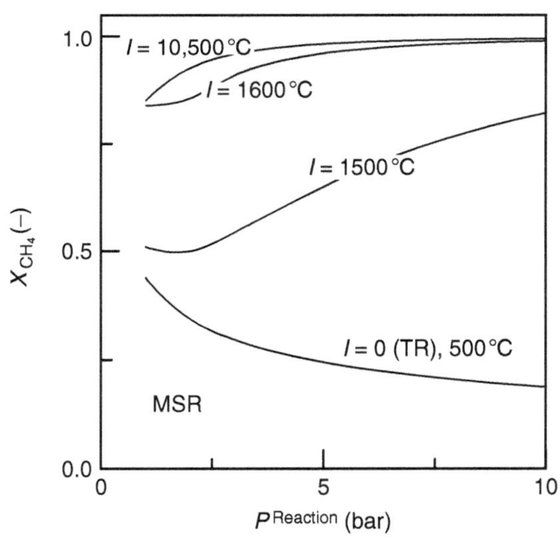

Figure 9.11 CH_4 equilibrium conversion versus reaction pressure for MSR reaction. Reactant (H_2O/CH_4) feed molar ratio = 3. Reprinted from Marigliano *et al.* [60] with permission from Elsevier.

thermodynamics prevails on the permeation before the minimum, after that the permeation shows its efficiency. The TREC decreases over the whole pressure range.

The MREC depends on the reaction pressure owing to two different factors: one negative, owing to thermodynamics, such as in TR, and one positive, owing to an increase of the permeation driving force. At 500 °C and $I = 1$ these two effects are comparable, and they generate a minimum. The two opposite effects on the conversion are equal for an MR at 600 °C with $I = 1$, and the profile presents a flex point (Fig. 9.12).

When the permeation is improved at a higher I, the flex point disappears. The conversion after the minimum increases monotonically. In an MR, H_2 partial pressures on the reaction and permeate sides are equal under the equilibrium condition, by the definition of the equilibrium of an MR. H_2 partial pressure shows the same behavior as the conversion degree, increasing for an MR and decreasing for a TR. In addition, it decreases when I increases, even if the CH_4 conversion and H_2 produced increase. Therefore, a low sweep factor is desired for a high hydrogen concentration in the permeate stream.

The constraint of Eq. (9.13) and the bond with the I variable are shown in Figs. 9.13 and 9.14. In fact, the final conversion reached using different flow rates is the same at the same I value (Fig. 9.13). These profiles were calculated by using a simulation program describing the nonisothermal behavior of a co-current MR. This final value is achieved when the partial pressure of H_2, the product selectively removed from the reaction environment, is the same on both membrane sides as required from Eq. (9.13). The outlet (equilibrium) value of H_2 partial pressure is independent of the membrane thickness, even if the profiles are very different (Fig. 9.14) for the membrane thickness considered (δ and 10δ). They reach the same final value, even if different reactor lengths are required to reach equilibrium. Similar differences in the

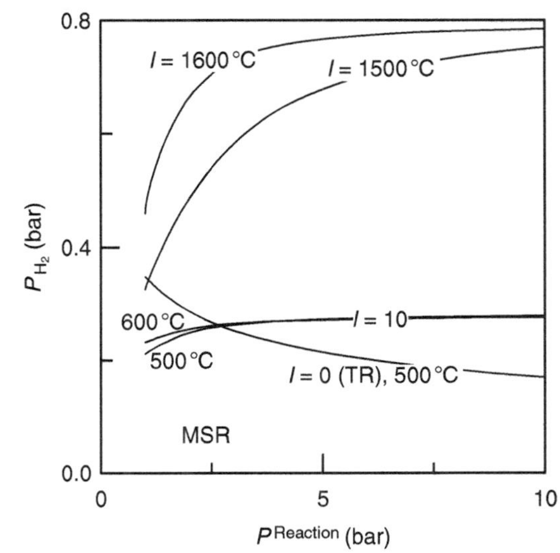

Figure 9.12 H_2 partial pressure under equilibrium condition in the MSR reaction. $T^{Reaction} = 500$ °C, $P^{Reaction} = 100$ kPa, reactant (H_2O/CH_4) feed molar ratio = 3. Reprinted from Marigliano *et al.* [60] with permission from Elsevier.

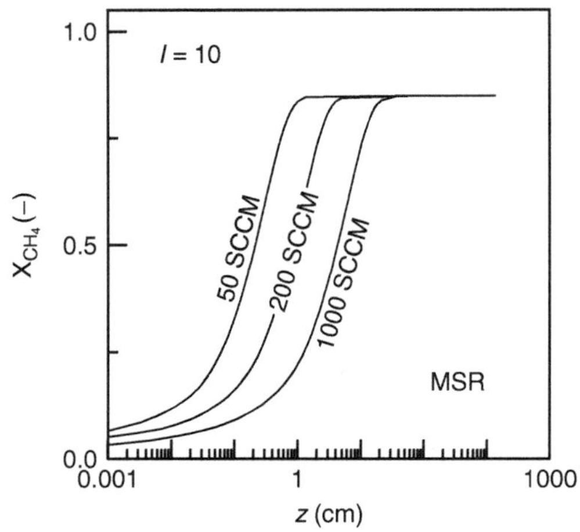

Figure 9.13 CH_4 conversion profile in a plug-flow MR for the MSR reaction at different feed flow rates. $P^{\text{Reaction}} = P^{\text{Permeate side}} = 100$ kPa, $I = 10$, $T = 500\ °C$, reactant (H_2O/CH_4) feed molar ratio = 3. Reprinted from Marigliano *et al.* [60] with permission from Elsevier.

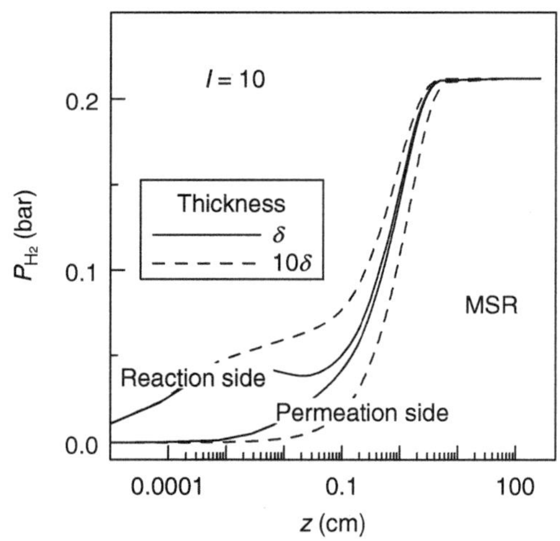

Figure 9.14 H_2 partial pressure in a plug-flow MR for the MSR reaction for two different values of membrane thickness. $\delta = 7.5\ \mu m$, $P^{\text{Reaction}} = P^{\text{Permeate side}} = 100$ kPa, = 10, $T = 500\ °C$, reactant (H_2O/CH_4) feed molar ratio = 3, CH_4 feed flow rate = 200 cm^3(STP)/min. Reprinted from Marigliano *et al.* [60] with permission from Elsevier.

profiles, but not in the final value, are shown when other parameters of the permeation, for example, the membrane area or permeation law parameters are changed. In a similar way, the energy transfer properties of the MR, catalyst activity or amount, etc., can be changed having the same equilibrium as an MR.

In Fig. 9.15, the equilibrium conversion of an MR system is represented by the *I* curves. The equilibrium conversion of an MR is an increasing function of the sweep factor. The MR equilibrium conversion is double that of a TR for $I = 10$, and its value

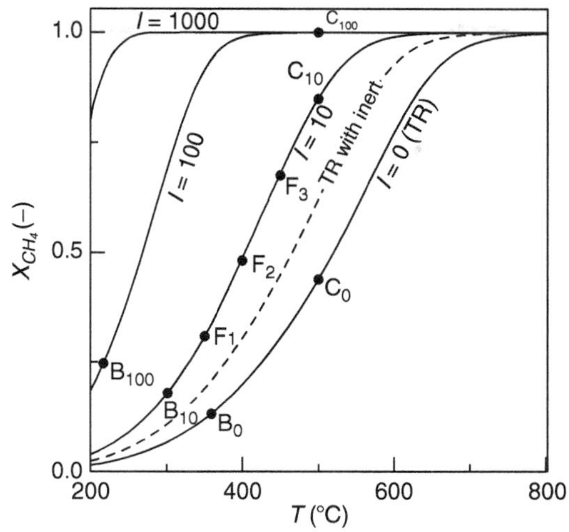

Figure 9.15 Methane conversion versus temperature at several sweep factors (solid lines). Reactant (H_2O/CH_4) feed molar ratio = 3, $P^{Reaction} = P^{ermeation} = 100$ kPa. TR with inert: equilibrium of a TR when an inert stream with a flow rate equal to 10 times that of methane was added to the feed. Points B_I and C_I are the final points of the TR and MRs obtained by simulation in adiabatic, isothermal, and nonisothermal reactors, respectively, with the following conditions: $T^{Reaction} = T^{Permeation} = 500$ °C. F_1, F_2, and F_3 are the MR outlet conversions, calculated at different furnace temperatures (350, 400, and 450 °C), for $I = 10$. Reprinted from Barbieri *et al.* [100] with permission from American Chemical Society.

is equal to 100% for $I = 100$. The addition of an inert gas in a TR has a positive effect on methane conversion in the MSR reaction, but the conversion is lower than that obtained in an MR where the same inert gas is used to sweep. Figure 9.15 reports the conversion of a TR (dashed line) and of an MR (solid line, $I = 10$) when the flow rates of the inert and sweep gases are equal to 10 times the methane feed flow rate.

4. Models of Pd-Alloy Membrane Reactors

Sections 4.1 and 4.2 report mass and energy balances of Pd-alloy MRs. Section 4.3 reports the cross-linking of mathematical models and the various references of scientific journals.

An MR is a chemical device in which reaction and separation occur simultaneously.

Two different sides and four streams are generally identified in an MR (Fig. 9.16). The two sides are:

- *Reaction side* where the reactions involving the hydrogen take place.
- *Permeate side* where, usually, the hydrogen produced is recovered.

This configuration considering H_2 as a reaction product is the most commonly used and will be considered in this chapter. When H_2 is a reactant, such as in hydrogenation reactions, reaction and permeate sides coincide.

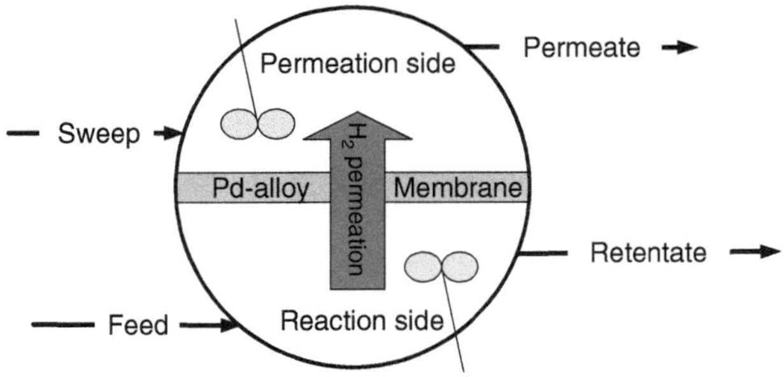

Figure 9.16 Scheme of a Pd-alloy MR.

The four inlet–outlet streams are:

- *Feed* is the inlet stream on the reaction side.
- *Retentate* is the outlet stream from the reaction side.
- *Sweep* is the inlet stream on the permeate side (optional).
- *Permeate* is the outlet stream from the permeate side.

As well as for a TR, a mathematical model of an MR is a set of equations [partial differential equations (PDEs) or ordinary differential equations (ODEs), algebraic equations] able to simulate all the possible evolutions of the variables describing the modeled system in accordance with the initial status and operating conditions: the local variables (temperatures, pressures, and compositions) and the overall ones (e.g., flow rates, conversion, and overall recovery index).

Generally, these equations are the momentum, mass, and energy balances. These balances have also to be written for the permeate side in addition to the reaction volume.

The MRs considered as well as TRs might be distinguished as systems with:

- Distributed parameters, such as *tubular MRs*, in which the state variables for both reaction and permeate sides depend on the axial and/or the radial position.
- Lumped parameters, such as *completely stirred MRs*, in which both reaction and permeate sides are described by global variables.

In a tubular MR species composition, temperatures and pressures change along the reactor length and the radial direction; also the permeation rate through the membrane changes along the reactor length. Therefore, these systems must be described by PDEs.

For batch and continuous stirred MRs global balance equations (ODEs) must be considered focusing on the overall behavior of the two sides of the reactor; in this case, the permeating flux through the membrane is assumed the same for the whole membrane surface.

The presented mathematical models will be developed under the following *hypothesis*, common to most of the models present in the open literature:

- Velocity field for the studied systems will be assumed assigned (only *mass* and *energy balances* will be considered).
- Pd-alloy membranes with ideal behavior: infinite H_2 selectivity.

- H_2 permeation *diffusion-controlled*: H_2-permeating flux expressed by Sievert's law.
- Isobaric conditions on both membrane sides. It is justified by calculations which indicate that the pressure drop is negligible in the catalyst bed of laboratory-scale reactors. Ergun's equation can be used when the pressure drop is not negligible.
- Ideal gas behavior for both membrane sides; the majority of H_2-involving reactions carried out in an MR occur in the high temperature gas phase.
- Pseudo-homogeneous description of heterogeneous catalytic reactions: the void fraction, effectiveness factor, and specific catalytic surface are included into the reaction rate expressions.

4.1. Mass balances

In general, a *mass balance* is an accounting for the net mass flow through a system. The total mass entering (IN) into a system must either leave (OUT) the system or accumulate (ACCUMULATION) in it:

$$\text{ACCUMULATION} = \text{IN} - \text{OUT} \tag{9.17}$$

where ACCUMULATION may be negative or positive.

The presence of chemical reactions, the mass balance of the *i*th component must contain the production term (PRODUCTION):

$$\text{ACCUMULATION}_i = \text{IN}_i - \text{OUT}_i + \text{PRODUCTION}_i \tag{9.18}$$

For the Pd-alloy MRs, the mass balance principle is the same as previously expressed and, for the reaction side, the OUT term also contains the H_2 permeating through the membrane [Eq. (9.19)].

$$\begin{bmatrix} \text{H}_2 \text{ mole} \\ \text{variation} \\ \text{in the time} \\ \text{ACCUMULATION} \end{bmatrix} = \begin{bmatrix} \text{H}_2 \\ \text{entering the system} \\ \text{with input stream} \\ \text{IN} \end{bmatrix} - \left[\begin{pmatrix} \text{H}_2 \\ \text{leaving the system} \\ \text{with output stream} \end{pmatrix} \underset{\text{OUT}}{+} \begin{pmatrix} \text{H}_2 \\ \text{permeated through} \\ \text{themembrane} \end{pmatrix} \right] + \begin{bmatrix} \text{H}_2 \\ \text{produced} \\ \text{by reaction} \\ \text{PRODUCTION} \end{bmatrix} \tag{9.19}$$

The same equation can be employed for the permeate side, where the permeating flux must be considered with the "+" sign in the "IN" term.

$$\begin{bmatrix} H_2\text{mole} \\ \text{variation} \\ \text{in the time} \\ \text{ACCUMULATION} \end{bmatrix} = \left[\begin{pmatrix} H_2 \\ \text{entering the system} \\ \text{with input stream} \end{pmatrix} + \begin{pmatrix} H_2 \\ \text{permeated through} \\ \text{the membrane} \end{pmatrix} \right]_{\text{IN}} - \begin{bmatrix} H_2 \\ \text{leaving the system} \\ \text{with output stream} \\ \text{OUT} \end{bmatrix} \tag{9.20}$$

In the majority of the literature, no reaction occurs in the permeate side; therefore, the PRODUCTION term is nil.

If steady-state condition is considered, the accumulation terms for the reaction and permeate sides are 0, too.

The term related to the permeation through the membrane must be taken into account by means of the permeating flux expression that depends on the membrane type and, therefore, on the permeation mechanism.

As previously reported (par. 2), a Pd-alloy membrane allows only H_2 permeation. This can be expressed by the Eq. (9.21), foreseeing Sievert's permeation for H_2 and the absence of flux for other species:

$$J_i^{\text{Permeating}} = \begin{cases} \text{see Eq.(9.1)for } i = H_2 \\ 0 \text{ for } i \neq H_2 \end{cases} \tag{9.21}$$

where $i = 1 \ldots N_{\text{Species}}$ will be utilized in the mass balance equations.

The convective term in mass balance equations will always be written in terms of molar fluxes (N_i) as reported by Barbieri and Di Maio: the use of concentration or molar fraction requires the velocity field definition (usually the velocity is not a constant for these kinds of system, owing to for example the mole number variation in gas phase reactions, temperature change, and also permeation).

The used relationship between N_i and P_i, T, etc., can be expressed by means of the ideal gas law:

$$N_i(\text{Reactor cross section}) = \frac{P_i}{\text{RT}}(\text{Volumetric flow rate}) \tag{9.22}$$

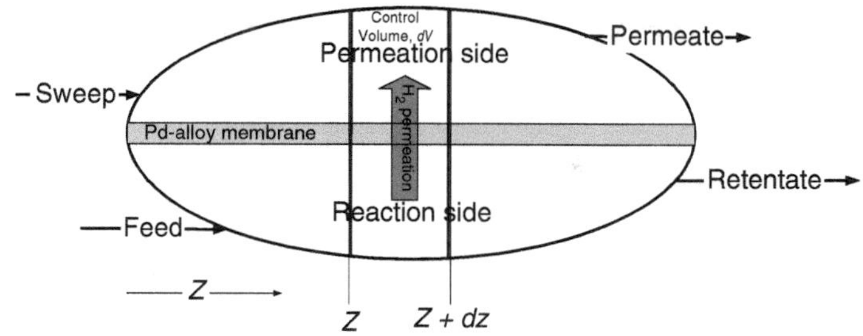

Figure 9.17 Tubular Pd-alloy MR.

4.1.1. Tubular membrane reactor

The majority of Pd-alloy MR simulations reported in the literature refers to tubular systems (Fig. 9.17). This kind of reactor is, in fact, the most used in industrial applications owing to its advantages with respect to the other configuration. In the tubular configurations the state variables describing the system change along the reactor axis and also tubular MRs assure higher conversions with respect to completely stirred MRs.

The mass balances for a tubular MR on both membrane sides must be written for a differential "reference volume," of *dz* length.

Three different tubular MRs were considered:

1. Plug-flow MR (1D-1st-order model).
2. With axial diffusion (1D-2nd-order model).
3. With axial and radial diffusion (2D-2nd-order model).

The mass balance equations and the corresponding initial conditions (I.C.s) and boundary conditions (B.C.s) (all defined at the inlet of the system, "initial value" problem) for the plug-flow MR are reported in Eqs. (9.23) and (9.24) for the reaction and permeate sides, respectively. This equation set can be used indifferently for reaction in the annulus or in the lumen (Table 9.1).

The constitutive equation terms are the following for the reaction side (Table 9.2):

The equations of the permeate side are constituted of the same terms as at the reaction side, without that relative to the chemical reaction. The term of the permeating flux through the membrane has a positive sign because in this case the H_2 enters in the permeation volume [see Eq. (9.20)].

Most of the cases treated in the literature refer to the *steady-state* model and the mass balance of a plug-flow tubular MRs Eqs. (9.25) and (9.26) are the same as (9.23) and (9.24) without the accumulation term [62–96] (Table 9.3).

Table 9.4 reports the model equations for a *counter-current* tubular MR. They are the same on the reaction of the co-current mode. The only difference is the sign of the term related to convective flux on the permeate side [Eq. (9.28)]. The B.C.s for this set of equations are defined at the two ends of the membrane module ("boundary type problem").

Table 9.1 Mass balance of a tubular Pd-alloy MR with cylindrical symmetry

Reaction side:

$$\frac{1}{RT^{\text{Reaction}}}\frac{\partial P_i^{\text{Reaction}}}{\partial t} = -\frac{\partial N_i^{\text{Reaction}}}{\partial z} + \sum_{j=1}^{N_{\text{Reactions}}} v_{i,j} r_j - \frac{A^{\text{Membrane}}}{V^{\text{Reaction}}} J_i^{\text{Permeating}} \tag{9.23}$$

$$I.C.\ P_i^{\text{Reaction}}(z) = P_i^{\text{Reaction,Initial}}$$

$$B.C.\ P_i^{\text{Reaction}}\Big|_{z=0} = P_i^{\text{Feed}}$$

Permeate side:

$$\frac{1}{RT^{\text{Permeation}}}\frac{\partial P_i^{\text{Permeation}}}{\partial t} = -\frac{\partial N_i^{\text{Permeation}}}{\partial z} + \frac{A^{\text{Membrane}}}{V^{\text{Permeation}}} J_i^{\text{Permeating}} \tag{9.24}$$

$$I.C.\ P_i^{\text{Permeation}}(z) = P_i^{\text{Permeation,Initial}}$$

$$B.C.\ P_i^{\text{Permeation}}\Big|_{z=0} = P_i^{\text{Sweep}}$$

Note: Plug-flow MR (1D-1st-order model)—co-current flow configuration—transient state.

Table 9.2 Constitutive terms of mass balance equation

$\frac{1}{RT^{Reaction}}\frac{\partial P_i^{\text{Reaction}}}{\partial t}$	Accumulation term; it is present only in the transient state
$-\frac{\partial N_i^{\text{Reaction}}}{\partial z}$	Convective flux variation of the *i*th species along the reaction side
$+\sum_{j=1}^{N_{\text{Reactions}}} v_{i,j} r_j$	Reaction term involving *i*th species in all the reactions
$-\frac{A^{\text{Membrane}}}{V^{\text{Reaction}}} J_i^{\text{Permeating}}$	Permeation term of the *i*th species through the membrane. See Eq. (9.21) for further details

In general, the axial dispersion can be neglected when Péclet number:

$$Pé_{AB} = \frac{(\text{Volumetric flow rate/Cross section})L}{D_{AB}} \le 1.$$

In the case of nonnegligible axial dispersion a further term related to diffusive transport for the species along the MR axis must be considered in the equations (Table 9.5).

The model mass balances [Eqs. (9.29) and (9.30)] constituted of the same terms as Eqs. (9.27) and (9.28), previously described, are second-order PDEs owing to the presence of axial diffusion. Thus, it requires additional B.C.s (e.g., Danckwerts's condition expressing the absence of the concentration gradient at the reactor exit) on both the reaction and the permeation volumes.

When radial profiles cannot be neglected the compositions, temperatures, and pressures also depend on the radial coordinate and a different differential "reference volumes" must be assumed.

Table 9.3 Mass balance of a tubular Pd-alloy MR with cylindrical symmetry

Reaction side:

$$-\frac{dN_i^{\text{Reaction}}}{dz}+\sum_{j=1}^{N_{\text{Reactions}}}\nu_{i,j}r_j-\frac{A^{\text{Membrane}}}{V^{\text{Reaction}}}J_i^{\text{Permeating}}=0 \quad (9.25)$$

$$\text{B.C. } P_i^{\text{Reaction}}\big|_{z=0}=P_i^{\text{Feed}}$$

Permeate side:

$$-\frac{dN_i^{\text{Permeation}}}{dz}+\frac{A^{\text{Membrane}}}{V^{\text{Permeation}}}J_i^{\text{Permeating}}=0 \quad (9.26)$$

$$B.C.\ P_i^{\text{Permeation}}\big|_{z=0}=P_i^{\text{Sweep}}$$

Note: Plug-flow MR (1D-1st-order model)—co-current flow configuration—*steady state* [69–82, 89–91].

Table 9.4 Mass balance of a tubular Pd-alloy MR with cylindrical symmetry

Reaction side:

$$-\frac{dN_i^{\text{Reaction}}}{dz}+\sum_{j=1}^{N_{\text{Reactions}}}\nu_{i,j}r_j-\frac{A^{\text{Membrane}}}{V^{\text{Reaction}}}J_i^{\text{Permeating}}=0 \quad (9.27)$$

$$B.C.\ P_i^{\text{Reaction}}\big|_{z=0}=P_i^{\text{Feed}}$$

Permeate side:

$$+\frac{dN_i^{\text{Permeation}}}{dz}+\frac{A^{\text{Membrane}}}{V^{\text{Permeation}}}J_i^{\text{Permeating}}=0 \quad (9.28)$$

$$B.C.\ P_i^{\text{Permeation}}\big|_{z=L}=P_i^{\text{Sweep}}$$

Note: Plug-flow MR (1D-1st-order model)—counter-current flow configuration—*steady state* [83–88].

In this case the mathematical model is 2D and a new term related to radial diffusion[2] appears with respect to Eqs. (9.31) and (9.32); in the meantime, the term related to the permeation disappears from the equation. In fact, the membrane permeation is taken into account by means of the B.C. at the membrane surface. This B.C. is the same for both reaction and permeate sides.

Also a second B.C. for the radial coordinate is required (e.g., the absence of radial flux on symmetry axis and on the shell internal surface for the tube and the annulus, respectively).

[2] The *Fick* diffusion is considered as reported by *Vant Sint Annaland et al.* both in the axial and in the radial directions.

Table 9.5 Mass balance of a tubular Pd-alloy MR

Reaction side:

$$\frac{1}{RT^{\text{Reaction}}}\frac{\partial P_i^{\text{Reaction}}}{\partial t} = \frac{1}{RT^{\text{Reaction}}}D_i\frac{\partial^2 P_{\text{i}}^{\text{Reaction}}}{\partial z^2} + -\frac{\partial N_i^{\text{Reaction}}}{\partial z} + \sum_{j=1}^{N_{\text{Reactions}}} \nu_{i,j} r_j - \frac{A^{\text{Membrane}}}{V^{\text{Reaction}}} J_i^{\text{Permeating}} \tag{9.29}$$

$$I.C.\ P_i^{\text{Reaction}}(z) = P_i^{\text{Reaction,Initial}}$$
$$B.C.1\ P_i^{\text{Reaction}}\big|_{z=0} = P_i^{\text{Feed}}$$
$$B.C.2\ \frac{\partial P_i^{\text{Reaction}}}{\partial z}\Big|_{z=L} = 0$$

Permeate side:

$$\frac{1}{RT^{\text{Permeation}}}\frac{\partial P_i^{\text{Permeation}}}{\partial t} = \frac{1}{RT^{\text{Permeation}}}D_i\frac{\partial^2 P_i^{\text{Permeation}}}{\partial z^2} + -\frac{\partial N_i^{\text{Permeation}}}{\partial z} + \frac{A^{\text{Membrane}}}{V^{\text{Permeation}}} J_i^{\text{Permeating}} \tag{9.30}$$

$$I.C.\ P_i^{\text{Permeation}}(z) = P_i^{\text{Permeation,Initial}}$$
$$B.C.1\ P_i^{\text{Permeation}}\big|_{z=0} = P_i^{\text{Sweep}}$$
$$B.C.2\ \frac{\partial P_i^{\text{Permeation}}}{\partial z}\Big|_{z=L} = 0$$

Note: Tubular MR with axial diffusion (1D-2nd-order model)—transient state.

The B.C.s on radial coordinate for the tube or the annulus are (9.33) and (9.34) independent of the MR configuration (reaction in the tube or in the annulus) (Table 9.6).

4.1.2. Batch membrane reactor

For batch or continuous stirred MRs, the balances at both the membrane sides must be written on the integral control volumes; in fact, in this case the composition and temperature are the same at each point of the fluid phases (Fig. 9.18).

Table 9.7 reports the mass balance ODEs for the reaction and permeate sides of a batch MR in the hypothesis of constant volumes for the two membrane sides. This equation set contains the accumulation and the permeation terms and, only for the reaction side, the reaction term. No inlet and outlet streams (terms) are present.

4.1.3. Continuous stirred tank membrane reactor

The mass balance ODEs of continuous stirred MRs (Fig. 9.19) must contain the IN and OUT terms related to the continuous flow of the feed (sweep) and retentate (permeate) streams in addition to the terms presented by the batch MRs.

Table 9.6 Mass balance of a tubular Pd-alloy MR

Reaction side:

$$\frac{\partial P_i^{\text{Reaction}}}{\partial t} = D_{i,z}\frac{\partial^2 P_i^{\text{Reaction}}}{\partial z^2} + D_{i,r}\frac{\partial^2 P_i^{\text{Reaction}}}{\partial r^2} + \\ +RT^{\text{Reaction}}\left(-\frac{\partial N_{i,z}^{\text{Reaction}}}{\partial z} + \sum_{j=1}^{N_{\text{Reactions}}} v_{i,j} r_j\right) \\ I.C.\ P_i^{\text{Reaction}}(z,r)|_{t=0} = P_i^{\text{Reaction, Initial}} \\ B.C.1\ P_i^{\text{Reaction}}|_{z=0} = P_i^{\text{Feed}} \\ B.C.2\ \frac{\partial P_i^{\text{Reaction}}}{\partial z}|_{z=L} = 0 \tag{9.31}$$

Permeate side:

$$\frac{\partial P_i^{\text{Permeation}}}{\partial t} = D_{i,z}\frac{\partial^2 P_i^{\text{Permeation}}}{\partial z^2} + \\ +D_{i,r}\frac{\partial^2 P_i^{\text{Permeation}}}{\partial r^2} - RT^{\text{Permeation}}\frac{\partial N_i^{\text{Permeation}}}{\partial z} \\ I.C.\ P_i^{\text{Permeation}}(z,r)|_{t=0} = P_i^{\text{Permeation, Initial}} \\ B.C.1\ P_i^{\text{Permeation}}|_{z=0} = P_i^{\text{Sweep}} \\ B.C.2\ \left.\frac{\partial P_i^{\text{Permeation}}}{\partial z}\right|_{z=L} = 0 \tag{9.32}$$

(continued)

Table 9.6 *(Continued)*

For the lumen:

$$\begin{cases} N_{H_2,r} = -D_{H_2,} \dfrac{1}{RT^{\text{Lumen}}} \dfrac{\partial P_{H_2}}{\partial r} + \dfrac{1}{RT^{\text{Tube}}} P_{H_2} \left(\dfrac{(-1)^p J_{H_2}^{\text{Permeating}} RT^{\text{Lumen}}}{p^{\text{Lumen}}} \right) \Bigg|_{R=\frac{ID^{\text{Membrane}}}{2}} \quad \text{for } i \neq H_2 \\ \text{B.C.3} \quad \left. \dfrac{\partial P_i}{\partial r} \right|_{R=\frac{ID^{\text{Membrane}}}{2}} = 0 \text{ for } i \neq H_2 \\ \text{B.C.4} \quad \left. \frac{\partial P_i}{\partial r} \right|_{r=0} = 0 \end{cases} \tag{9.33}$$

For the annulus:

$$\begin{cases} N_{H_2,r} = -D_{H_2,} \dfrac{1}{RT^{\text{Annulus}}} \dfrac{\partial P_{H_2}}{\partial r} + \dfrac{1}{RT^{\text{Annulus}}} P_{H_2} \left(\dfrac{(-1)^p J_{H_2}^{\text{Permeating}} RT^{\text{Annulus}}}{p^{\text{Annulus}}} \right) \Bigg|_{R=\frac{OD^{\text{Membrane}}}{2}} \quad \text{for } i \\ \text{B.C.3} \quad \left. \dfrac{\partial P_i}{\partial r} \right|_{r=\frac{OD^{\text{Membrane}}}{2}} = 0 \text{ for } i \neq H_2 \\ \text{B.C.4} \quad \left. \frac{\partial P_i}{\partial r} \right|_{R=\frac{ID^{\text{Membrane}}}{2}} = 0 \end{cases} \tag{9.34}$$

$$\text{Being } p = \begin{cases} 1 \text{ for reaction side} \\ 2 \text{ for permeation side} \end{cases}$$

Note: Tubular MR with axial and radial diffusion (2D-2nd-order model)—transient state [92, 95, 96].

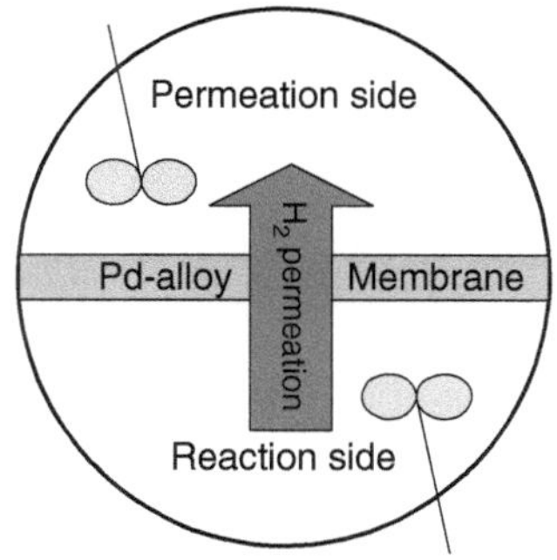

Figure 9.18 Batch Pd-alloy MR.

Table 9.7 Mass balance equations of a Pd-alloy batch (discontinuous) MR

Reaction side

$$\frac{1}{RT^{\text{Reaction}}}\frac{dP_i^{\text{Reaction}}}{dt} = \sum_{j=1}^{N_{\text{Reactions}}} \nu_{i,j} r_j - \frac{A^{\text{Membrane}}}{V^{\text{Reaction}}} J_i^{\text{Permeating}} \tag{9.35}$$

$$I.C.\ P_i^{\text{Reaction}} = P_i^{\text{Reaction, Initial}}$$

Permeate side:

$$\frac{1}{RT^{\text{Permeation}}}\frac{dP_i^{\text{Permeation}}}{dt} = \frac{A^{\text{Membrane}}}{V^{\text{Permeation}}} J_i^{\text{Permeating}} \tag{9.36}$$

$$I.C.\ P_i^{\text{Permeation}} = P_i^{\text{Permeation, Initial}}$$

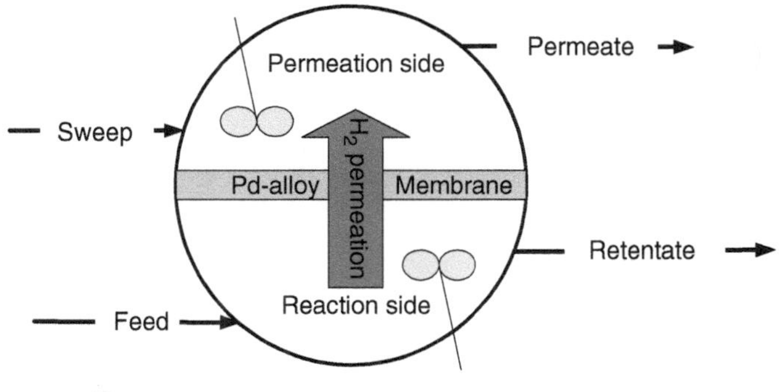

Figure 9.19 Continuous stirred tank Pd-alloy MR.

Table 9.8 reports the mass balance equations for the reaction and permeate sides, in the case of transient state.

The steady-state equations (Table 9.8) can easily be obtained assuming the accumulation term null. Therefore, the ODE set becomes an algebraic equation set that can be solved much more easily (Table 9.9).

Table 9.8 Mass balance equations of a continuous stirred tank Pd-alloy MR *(0D model) — transient state*

Reaction side:

$$\frac{1}{RT^{\text{Retentate}}}\frac{dP_i^{\text{Retentate}}}{dt} = \frac{F_i^{Retentate} - F_i^{Feed}}{V^{\text{Reaction}}} + \sum_{j=1}^{N_{\text{Reactions}}} \nu_{i,j}\, r_j - \frac{A^{\text{Membrane}}}{V^{\text{Reaction}}} J_i^{\text{Permeating}} \quad (9.37)$$

$$I.C.\ P_i^{\text{Retentate}} = P_i^{\text{Reaction, Initial}}$$

Permeate side:

$$\frac{1}{RT^{\text{Permeation}}}\frac{dP_i^{\text{Permeate}}}{dt} = \frac{F_i^{Permeate} - F_i^{Sweep}}{V^{\text{Permeation}}} + \frac{A^{\text{Membrane}}}{V^{\text{Permeation}}} J_i^{\text{Permeating}} \quad (9.38)$$

$$I.C.\ P_i^{\text{Permeate}} = P_i^{\text{Permeation, Initial}}$$

Table 9.9 Mass balance equations of a continuous stirred tank Pd-alloy MR *(0D model)—steady state* [98]

Reaction side:

$$\frac{F_i^{\text{Retentate}} - F_i^{\text{Feed}}}{V^{\text{Reaction}}} + \sum_{j=1}^{N_{\text{Reactions}}} \nu_{i,j}\, r_j - \frac{A^{\text{Membrane}}}{V^{\text{Reaction}}} J_i^{\text{Permeating}} = 0 \quad (9.39)$$

Permeate side:

$$\frac{F_i^{\text{Permeate}} - F_i^{\text{Sweep}}}{V^{\text{Permeation}}} + \frac{A^{\text{Membrane}}}{V^{\text{Permeation}}} J_i^{\text{Permeating}} = 0 \quad (9.40)$$

4.2. Energy balances

The mass balance equations have to be generally solved simultaneously with the *energy balances* in order to describe the behavior of MRs. In fact, the energy transport and the temperature profiles inside an MR have a significant importance on the thermodynamics, the kinetics and the transport properties (through the membrane and inside the reaction and permeation phases). The knowledge of these profiles allows the reaction in the desired conditions to be performed.

The energy balance must contain the terms related to accumulation (ACCUMULATION), input (IN), output (OUT) of energy within the "reference volume" (differential or integral) like the mass balance Eqs. (9.17) and (9.18). These terms are substantially the convective flux of energy, the conductive fluxes through and within the control volumes, respectively, and the term related to heat involved

Table 9.10 Characteristic terms of energy balance in MR

$\Phi = \begin{cases} 0 \text{ on reaction side} \\ J_{H_2}^{\text{Permeating}}\left(h_{H_2}^{T^{\text{Reaction}}} - h_{H_2}^{T^{\text{Permeation}}}\right) \text{on permeation side} \end{cases}$	Temperature variation due to enthalpy flux associated with hydrogen permeation	(9.41)
$\Psi = \begin{cases} \sum_{j=1}^{N^{\text{Raction}}} r_j(-\Delta H_j) \text{on reaction side} \\ 0 = \text{on permeation side} \end{cases}$	Heat produced by chemical reactions	(9.42)

in the chemical reactions. In Pd-alloy MRs, a further term (Φ) related to energy associated with the H_2-permeated flux [Eq. (9.41)] must be considered in the energy balances [69].

All the equations will contain the terms of Table 9.10. In the following sections, the energy balances will be written for the three different Pd-alloys MRs types treated in the previous section.

4.2.1. Tubular membrane reactor

The energy balance equations were written for both the annular and lumenal spaces.

Figure 9.20 schematically reports the different heat transfer resistances in a tubular MR.

The annulus exchanges heat with the furnace and the lumen side, whereas the stream in the lumen exchanges heat only with the annular volume. The heat generated by the reaction, ψ [Eq. (9.42)], will be present in the equation of the annulus or the lumen side, depending on the configuration used. The reaction temperature does not have any variation due to the H_2 permeation: hydrogen leaves the reaction stream at the reaction temperature itself. The system behaves as a splitting point of a stream, where the intensive variables (e.g., temperature) do not have any variations, only the extensive variables (such as flow rate and stream enthalpy) undergo some variations. Instead, hydrogen permeated through the membrane mixes with the permeate stream. The extensive and also intensive variables change on the permeate side. In particular, the temperature of the permeation stream undergoes a variation, depending on the permeate flux of hydrogen. The energy associated with the permeate flux of hydrogen (Φ) is present only in the permeate side equations.

The effect of the energy transport on the performance for an annular and an lumenal MR is different. Energy transport between the furnace and the reaction volume depends on the MR configuration used: for an annular MR the heat transfer resistance is characteristic of the gaseous film near the stainless steel shell; when the reaction takes place in the lumen, heat transfer resistance is located in the membrane

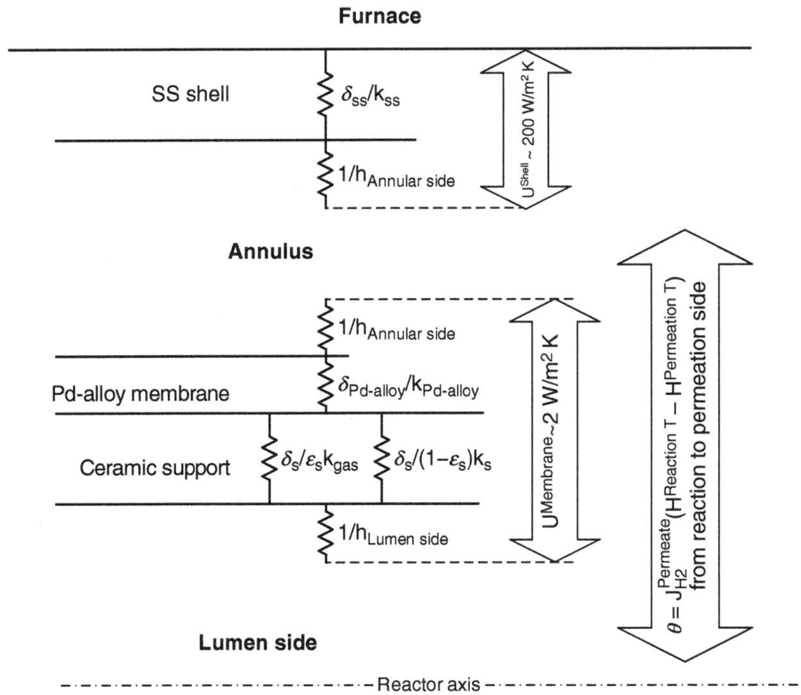

Figure 9.20 Scheme of the overall heat transfer coefficients in an MR and the energy flux associated with the hydrogen permeation through Pd–Ag membrane.

and its support. The overall heat transfer coefficient between the furnace and the annulus and that between the annulus and the lumen are typically of a few hundred and a few units of W/m^2 K, respectively. An annular MR characterized by a greater heat flux exchange with the heat source has a higher performance with respect to a lumenal MR.

The energy balance equations for the three different tubular MRs considered (plug flow, with axial diffusion, and with axial and radial diffusion) will be reported in the following tables.

The *plug-flow* system (T^{Annulus} and T^{Lumen}) temperatures change only along the membrane length. The first-order PDEs (9.43) and (9.44) are the energy balances for the annulus and the lumen, respectively, in the case of a transient state. In Table 9.11 also the I.C.s and B.C.s are reported.

In these equations, the rate of enthalpy increase in the control volume must equal the sum of heat fluxes owing to input–output convection, conduction from the membrane side and the furnace (only for the annulus), reactions (Ψ), and H_2 permeation through the membrane Φ (only for the permeate side). For each equation, I.C.s and B.C.s are required.

The majority of nonisothermal plug-flow MR models present in the literature were developed for steady state. In this case, the energy balance equations are reported in Table 9.12.

Table 9.11 Energy balance equations of a tubular Pd-alloy MR

Annulus:

$$\sum_{i}^{Nspecies} C_i Cp_i \frac{\partial T^{\text{Annulus}}}{\partial t} = -\sum_{i=1}^{Nspecies} N_i Cp_i \frac{\partial T^{\text{Annulus}}}{\partial z} + \\ + \frac{U^{\text{Shell}} A^{\text{Shell}}}{V^{\text{Annulus}}}\left(T^{\text{Furnace}} - T^{\text{Annulus}}\right) - \frac{U^{\text{Membrane}} A^{\text{Membrane}}}{V^{\text{Annulus}}}\left(T^{\text{Annulus}} - T^{\text{Lumen}}\right) + \\ + \Psi + \Phi \frac{A^{\text{Membrane}}}{V^{\text{Annulus}}} \tag{9.43}$$

$$I.C.\ T^{\text{Annulus}}\big|_{t=0} = T^{\text{Annulus, Initial}}$$
$$B.C.\ T^{\text{Annulus}}\big|_{z=0} = T^{\text{Feed}} \text{or}\, T^{\text{Sweep}}$$

Lumen:

$$\sum_{i}^{Nspecies} C_i Cp_i \frac{\partial T^{\text{Lumen}}}{\partial t} = -\sum_{i=1}^{Nspecies} N_i Cp_i \frac{\partial T^{\text{Lumen}}}{\partial z} + \\ + \frac{U^{\text{Membrane}} A^{\text{Membrane}}}{V^{\text{Lumen}}}\left(T^{\text{Annulus}} - T^{\text{Lumen}}\right) + \Psi + \Phi \frac{A^{\text{Membrane}}}{V^{\text{Lumen}}} \tag{9.44}$$

$$I.C.\ T^{\text{Lumen}}\big|_{t=0} = T^{\text{Lumen, Initial}}$$
$$B.C.\ T^{\text{Lumen}}\big|_{z=0} = T^{\text{Feed}} \text{ or } T^{\text{Sweep}}$$

Note: Plug-flow MR (1D-1st-order model)—transient state.

In the case of the counter-current mode, the energy balance equation for the reaction will be the same as the co-current mode, but for the permeate side the sign of convective heat flux must change ("+" instead "−"). Moreover, the B.C. for the permeate side must be defined in the other extremity of the MR ("boundary type problem").

A second-order term referring to heat diffusion along the MR axis must be included in the energy balances. In this case, when the axial dispersion is nonnegligible the equations become (9.47) and (9.48), for the annulus and the lumen, respectively, and require a second B.C. (usually a Danckwerts's one at the exit of MR) (Table 9.13).

The term related to radial heat diffusion inside the reaction and permeation phases must be considered and the second-order energy balance Eqs. (9.49) and (9.50) are necessary to describe the annulus and the lumen profiles, respectively, when the hypothesis of absence of radial temperature profiles is removed. In this case, the "reference volume" will be differential in both axial and radial coordinates and the energy balance equations will not contain the terms related to conductive heat exchange with the membrane and the furnace and the term related to "convective" heat flux owing to H_2 permeation (only for the permeate side) (Table 9.14).

Table 9.12 Energy balance equations of a tubular Pd-alloy MR

Annulus:

$$-\sum_{i=1}^{Nspecies} N_i Cp_i \frac{\partial T^{\text{Annulus}}}{\partial z} + \frac{U^{\text{Shell}} A^{\text{Shell}}}{V^{\text{Annulus}}} \left(T^{\text{Furnace}} - T^{\text{Annulus}}\right) + \\ - \frac{U^{\text{Membrane}} A^{\text{Membrane}}}{V^{\text{Annulus}}} \left(T^{\text{Annulus}} - T^{\text{Lumen}}\right) + \Psi + \Phi \frac{A^{\text{Membrane}}}{V^{\text{Annulus}}} = 0 \quad (9.45)$$

$$I.C.\ T^{\text{Annulus}}\big|_{t=0} = T^{\text{Annulus,Initial}}$$
$$B.C.\ T^{\text{Annulus}}\big|_{z=0} = T^{\text{Feed}} \text{ or } T^{\text{Sweep}}$$

Lumen:

$$-\sum_{i=1}^{N\,species} N_i Cp_i \frac{\partial T^{\text{Lumen}}}{\partial z} + \frac{U^{\text{Membrane}} A^{\text{Membrane}}}{V^{\text{Lumen}}} \left(T^{\text{Annulus}} - T^{\text{Lumen}}\right) + \\ + \Psi + \Phi \frac{A^{\text{Membrane}}}{V^{\text{Lumen}}} = 0 \quad (9.46)$$

$$I.C.\ T^{\text{Lumen}}\big|_{t=0} = T^{\text{Lumen Initial}}$$
$$B.C.\ T\big|_{z=0} = T^{\text{Feed}} \text{ or } T^{\text{Sweep}}$$

Note: Plug-flow MR (1D-1st-order model)—steady state [87, 97].

The 2D energy balance equations require two B.C.s along the radial direction, in addition to those along the reactor length. These further B.C.s are reported in the relationships (9.51) and (9.52) for the annulus and the lumen, respectively. In the second-type B.C.[3] on the membrane surface the term related to the convective heat flux (Φ) is 0 for the reaction side, while the insulation–symmetry condition is imposed for the MR axis ($r = 0$) (Table 9.15).

4.2.2. Batch membrane reactor

The energy balance equations are ODEs in the case of lumped parameter systems (batch or continuous stirred MRs). In this case the control volumes are the whole reaction or permeate ones. Assuming for analogy, the notation (annulus and lumen) already used for the tube-and-shell system also for the 0D MR models, Eqs. (9.53) and (9.54) for the two membrane sides of a batch system can be written (Table 9.16).

4.2.3. Continuous stirred tank membrane reactor

The term related to continuous inlet–outlet flow on the two membrane sides must be added to the previous equations for the continuous stirred MR. The energy balance Eqs. (9.55) and (9.56) are the same as those for the plug-flow MR in which the integral MR control volume substitutes the differential one. The same notation (annulus and lumen) of a tubular MR was used (Table 9.17).

[3] The heat flux is known on the boundaries, instead of the temperature.

Table 9.13 Energy balance equations of a tubular Pd-alloy MR

Annulus:

$$\sum_{i}^{Nspecies} C_i Cp_i \frac{\partial T^{\text{Annulus}}}{\partial t} = -\sum_{i=1}^{Nspecies} N_i Cp_i \frac{\partial T^{\text{Annulus}}}{\partial z} + k_z \frac{\partial^2 T^{\text{Annulus}}}{\partial z^2}$$
$$\frac{U^{\text{Shell}} A^{\text{Shell}}}{V^{\text{Annulus}}}\left(T^{\text{Furnace}} - T^{\text{Annulus}}\right) - \frac{U^{\text{Membrane}} A^{\text{Membrane}}}{V^{\text{Annulus}}}\left(T^{\text{Annulus}} - T^{\text{Lumen}}\right)$$
$$+\Psi + \Phi \frac{A^{\text{Membrane}}}{V^{\text{Annulus}}} \tag{9.47}$$
$$I.C.\ T^{\text{Annulus}}\big|_{t=0} = T^{\text{Annulus, Initial}}$$
$$B.C.1\ T^{\text{Annulus}}\big|_{z=0} = T^{\text{Feed}} \text{ or } T^{\text{Sweep}}$$
$$B.C.2\ \frac{\partial T^{\text{Annulus}}}{\partial z}\Big|_{z=L} = 0$$

Lumen:

$$\sum_{i}^{Nspecies} C_i Cp_i \frac{\partial T^{\text{Lumen}}}{\partial t} = -\sum_{i=1}^{Nspecies} N_i Cp_i \frac{\partial T^{\text{Lumen}}}{\partial z} + k_z \frac{\partial^2 T^{\text{Lumen}}}{\partial z^2} +$$
$$+\frac{U^{\text{Membrane}} A^{\text{Membrane}}}{V^{\text{Lumen}}}\left(T^{\text{Annulus}} - T^{\text{Lumen}}\right) + \Psi + \Phi \frac{A^{\text{Membrane}}}{V^{\text{Lumen}}} \tag{9.48}$$
$$I.C.\ T^{\text{Lumen}}\big|_{t=0} = T^{\text{Lumen, Initial}}$$
$$B.C.1\ T^{\text{Lumen}}\big|_{z=0} = T^{\text{Feed}} \text{ or } T^{\text{Sweep}}$$
$$B.C.2\ \frac{\partial T^{\text{Lumen}}}{\partial z}\Big|_{z=L} = 0$$

Note: MR with axial diffusion (1D-2nd-order model)—transient state.

4.3. Literature models

The main characteristics of some mathematical models present in the literature and developed for simulating, describing, or predicting different H_2-involving reactions carried out in Pd-alloy MRs are summarized in Table 9.18.

The majority of the papers on Pd-alloy MR modeling refer to tubular plug-flow MRs for steady-state operations, but models considering completely stirred MRs or radial diffusion effects in the tubular system were also present. For this reason the models were grouped into three main categories on the basis of model dimensions:

- *0D models:* completely stirred MRs
- *1D models:* tubular MRs with or without axial dispersion
- *2D models:* tubular MRs with radial (and axial) dispersion

Table 9.18 reports the year of publication of the proposed model (for those authors that published more than one paper, the year of the first publication is

Table 9.14 Energy balance equations of a tubular Pd-alloy MR

Annulus:

$$\sum_{i}^{N\text{species}} C_i Cp_i \frac{\partial T^{\text{Annulus}}}{\partial t} = -\sum_{i=1}^{N\text{species}} N_i Cp_i \frac{\partial T^{\text{Annulus}}}{\partial z} + k_z \frac{\partial^2 T^{\text{Annulus}}}{\partial z^2} + k_r \frac{\partial^2 T^{\text{Annulus}}}{\partial r^2} + \Psi$$

$$I.C.\ T^{\text{Lumen}}\big|_{t=0} = T^{\text{Lumen, Initial}}$$
$$B.C.1\ T^{\text{Lumen}}\big|_{z=0} = T^{\text{Feed}} \text{ or } T^{\text{Sweep}} \quad (9.49)$$
$$B.C.2\ \frac{\partial T^{\text{Lumen}}}{\partial z}\big|_{z=L} = 0$$

Lumen:

$$\sum_{i}^{N\text{species}} C_i Cp_i \frac{\partial T^{\text{Lumen}}}{\partial t} = -\sum_{i=1}^{N\text{species}} N_i Cp_i \frac{\partial T^{\text{Lumen}}}{\partial z} + k_z \frac{\partial^2 T^{\text{Lumen}}}{\partial z^2} + k_r \frac{\partial^2 T^{\text{Lumen}}}{\partial r^2} + \Psi$$

$$I.C.\ T^{\text{Lumen}}\big|_{t=0} = T^{\text{Lumen, Initial}}$$
$$B.C.1\ T^{\text{Lumen}}\big|_{z=0} = T^{\text{Feed}} \text{ or } T^{\text{Sweep}} \quad (9.50)$$
$$B.C.2\ \frac{\partial T^{\text{Lumen}}}{\partial z}\big|_{z=L} = 0$$

Note: MR with axial and radial diffusion (2D-2nd-order model)—transient state [92].

indicated), the balance equations considered (especially mass or energy, but also momentum or pressure drop equations), and the state considered (transient or stationary). All the different terms (convective and conductive heat flux, heat generated by reactions, etc.) are specified for the energy balances.

4.3.1. 0D models

Raich and Foley proposed an isothermal and isobaric CSTR-CSTR model of a Pd-alloy MR for the transient analysis of dehydrogenation reactions. Sievert's law for H_2 permeation was utilized. Specifically, this model considers two additional parameters taking into account the catalyst and membrane deactivation (the activity functions variation were experimentally determined).

4.3.2. 1D models

Barbieri and Di Maio proposed a nonisothermal steady-state model of a plug-flow tubular MR for the MSR reaction in co-current and counter-current mode (see Tables 9.3, 9.4, and 9.11). In particular, Barbieri *et al.* studied the behavior of a Pd-alloy MR changing the operating conditions and comparing the annular and the lumenal configuration. In fact, the two configurations showed different MR

Table 9.15 B.C.s for radial coordinate

For the annulus:

$$\begin{cases} B.C.3 & q\Big|_{r=\frac{ID^{\text{Shell}}}{2}} = -U^{\text{Shell}}\left(T^{\text{Annulus}} - T^{\text{Furnace}}\right) \\ B.C.4 & q\Big|_{r=\frac{OD^{\text{Membrane}}}{2}} = -U^{\text{Membrane}}\left(T^{\text{Lumen}} - T^{\text{Annulus}}\right) + \Phi \end{cases} \quad (9.51)$$

For the lumen:

$$\begin{cases} B.C.3 & q\Big|_{r=\frac{ID^{\text{Membrane}}}{2}} = -U^{\text{Membrane}}\left(T^{\text{Lumen}} - T^{\text{Annulus}}\right) + \Phi \\ B.C.4 & \dfrac{\partial T^{\text{Lumen}}}{\partial r}\Big|_{r=0} = 0 \end{cases} \quad (9.52)$$

Table 9.16 Energy balance equations of a Pd-alloy batch (discontinuous) MR

Annulus

$$\sum_{i=1}^{N\text{species}} C_i Cp_i \frac{dT^{\text{Annulus}}}{dt} = \frac{U^{\text{Shell}} A^{\text{Shell}}}{V^{\text{Annulus}}}\left(T^{\text{Furnace}} - T^{\text{Annulus}}\right) + \\ - \frac{U^{\text{Membrane}} A^{\text{Membrane}}}{V^{\text{Annulus}}}\left(T^{\text{Annulus}} - T^{\text{Lumen}}\right) + \Psi + \Phi \frac{A^{\text{Membrane}}}{V^{\text{Annulus}}} \quad (9.53)$$

$$I.C.\, T^{\text{Annulus}}\Big|_{t=o} = T^{\text{AnnulusInitial}}$$

Lumen

$$\sum_{i=1}^{N\text{species}} C_i Cp_i \frac{dT^{\text{Lumen}}}{dt} = \frac{U^{\text{Membrane}} A^{\text{Membrane}}}{V^{\text{Lumen}}}\left(T^{\text{Annulus}} - T^{\text{Lumen}}\right) + \Psi \\ + \Phi \frac{A^{\text{Membrane}}}{V^{\text{Lumen}}} \quad (9.54)$$

$$I.C.\; T^{\text{Lumen}}\Big|_{t=o} = T^{\text{Lumen, Initial}}$$

performance because of thermal effects. The energy transport was analyzed for annular and lumenal configurations taking into account also the energy transfer associated with H_2 permeation through the membrane.

Table 9.17 Energy balance equations of a Pd-alloy continuous stirred tank MR (0D model)

Shell (as annulus for tubular MR)

$$\sum_{i}^{N\mathrm{species}} C_i Cp_i \frac{dT^{\mathrm{Annulus,\ Exit}}}{dt} = \frac{\left(\sum_{i=1}^{N\mathrm{species}} F_i Cp_i T^{\mathrm{Annulus}} \Big|^{\mathrm{Exit}} - \sum_{i=1}^{N\mathrm{species}} F_i Cp_i T^{\mathrm{Annulus}} \Big|^{\mathrm{Input}} \right)}{V^{\mathrm{Annulus}}} + \frac{U^{\mathrm{Shell}} A^{\mathrm{Shell}}}{V^{\mathrm{Annulus}}} \left(T^{\mathrm{Furnace}} - T^{\mathrm{Annulus,\ Exit}} \right) - \frac{U^{\mathrm{Membrane}} A^{\mathrm{Membrane}}}{V^{\mathrm{Annulus}}} \left(T^{\mathrm{Annulus,\ Exit}} - T^{\mathrm{Lumen,\ Exit}} \right) + + \Psi + \Phi \frac{A^{\mathrm{Membrane}}}{V^{\mathrm{Annulus}}} \tag{9.55}$$

$$I.C.\ T^{\mathrm{Annulus}} \big|_{t=0} = T^{\mathrm{Annulus,Initial}}$$

Core (as lumen for tubular MR)

$$\sum_{i}^{N\mathrm{species}} C_i Cp_i \frac{dT^{\mathrm{Lumen,Exit}}}{dt} = \frac{\left(\sum_{i=1}^{N\ \mathrm{species}} F_i Cp_i T^{\mathrm{Lumen}} \Big|^{\mathrm{Exit}} - \sum_{i=1}^{N\ \mathrm{species}} F_i Cp_i T^{\mathrm{Lumen}} \Big|^{\mathrm{Input}} \right)}{V^{\mathrm{Lumen}}} + \frac{U^{\mathrm{Membrane}} A^{\mathrm{Membrane}}}{V^{\mathrm{Lumen}}} \left(T^{\mathrm{Annulus,\ Exit}} - T^{\mathrm{Lumen,\ Exit}} \right) + \Psi + \Phi \frac{A^{\mathrm{Membrane}}}{V^{\mathrm{Lumen}}} \tag{9.56}$$

$$I.C.\ T^{\mathrm{Tube}} \big|_{t=0} = T^{\mathrm{Lumen,\ Initial}}$$

Hughes *et al.* developed a steady-state isothermal 1D model for the comparison of dense and microporous MRs for MSR reaction. The co-current model neglects pressure drops, axial, and radial diffusion phenomena. The mass balance ODEs written in terms of molar flux for the species in both membrane sides considers the Sievert's law for the permeation.

Basile *et al.* proposed 1D steady-state model for the partial oxidation of methane in dense Pd MR (energy balance was also proposed), WGS in both Pd and Pd–Ag MR considering also axial diffusion, MSR in dense Pd MR [84].

Using a 1D isothermal model Harold *et al.* simulated a packed-bed annular MR surrounding a supported Pd membrane tube in steady state. The model was used focusing on reaction and flow effects for the methanol decomposition (direct and oxidative), methanol steam reforming, and methanol partial oxidation. The effect of the Pd membrane surface coverage by CO on the permeating flux term is also considered, and Sievert's law is assumed to describe the permeating flux through the Pd membrane. They also proposed a comparison between TR- and MR-based fuel processors for the autothermal hydrocarbon reforming in terms of hydrogen recovery and overall efficiency.

Table 9.18 Overview on the literature mathematical models of Pd-alloy MRs

			Balance									
							Energy					
									Heat exchange			
References	Year of first publication	Reference number	Mass	Momentum	Continuity	Ergun's equation	Convective flux	Reaction	With furnace	Through the membrane	Due to H_2 permeation	State
						0D						
Raich and Foley	1995	[98]	•									Steady
This work	2007		•				•	•	•	•	•	Steady
This work	2007		•				•	•	•	•	•	Transient
						1D						
Barbieri *et al.*	1996	[58, 69, 70, 83, 99–103]	•				•	•	•	•	•	Steady
Basile *et al.*	1996	[71, 72, 84, 97, 7–104, 105]	•				•	•	•			Steady
Harold *et al.*	2003	[85, 86]	•				•	•		•		Steady
Dittmeyer *et al.*	1997	[73]	•			•	•	•				Steady
Johannessen and Jordal	2005	[87]	•			•	•	•	•		•	Steady
Itoh *et al.*	1997	[74–78]	•									Steady
Elnashaie *et al.*	2003	[79, 88, 106, 107, 108, 88, 109]	•			•	•	•	•			Steady
Dixon *et al.*	2001	[80, 110, 111]	•									Steady

(continued)

Table 9.18 (*Continued*)

References	Year of first publication	Reference number	Balance									State
							Energy					
									Heat exchange			
			Mass	Momentum	Continuity	Ergun's equation	Convective flux	Reaction	With furnace	Through the membrane	Due to H_2 permeation	
Fernandes and Soares Jr.	2006	[81, 82, 89]	•				•	•	•			Steady
Alfadhel and Kothare	2005	[90, 91]	•									Steady
This work	2007		•				•	•	•	•	•	Transient
						2D						
Hughes *et al.*	1995	[112–115]	•									steady
Basile *et al.*	2003	[72]	•									Steady
Clarizia *et al.*	2006	[116, 117]	•					•	•	•		Steady
Van Sint Annaland *et al.*	2006	[92, 93, 96, 118]	•	•	•	•	•	•		•		Transient
This work	2007		•				•	•	•	•	•	Transient

Hermann *et al.* proposed a 1D nonisothermal mathematical model taking also into account the pressure drops along the catalytic bed (Ergun's equation) for the catalytic dehydrogenation of ethylbenzene to styrene in a composite MR. In the low H_2 partial pressure side the proposed model considers the oxidation of the permeated hydrogen by air, which promotes a higher dehydrogenation.[4] Multicomponent Stefan–Maxwell diffusion in the stagnant gas film on both membrane faces, combined Knudsen diffusion and viscous flow in the macroporous support, Knudsen diffusion in the microporous layer, and Sievert's law for H_2 transport through the Pd–Ag film describes the transmembrane flux in radial direction.

Johannessen and Jordal proposed a nonisothermal 1D steady-state model for a catalytic MR for MSR reaction; the pressure drop in the catalytic bed by means of Ergun's equation was also taken into account.

Itoh *et al.* developed a mathematical model for isothermal, isobaric, steady-state, co-current Pd-alloy packed bed MR simulation. They coupled the endothermic cyclohexane dehydrogenation with the exothermic H_2 oxidation. The dimensionless mass balance equation takes the mixing diffusion of hydrogen in the radial direction into account. The decline in hydrogen permeation owing to the concentration polarization and the hindrance by carbon monoxide were considered in the simulation of methanol decomposition.

Elnashaie *et al.* proposed a simultaneous steam and oxidative reforming of heptane in a circulating fast fluidized-bed membrane reformer equipped with dense Pd and perovskite membranes. The authors developed a 1D, steady state, nonisothermal, isobaric, plug-flow (for both reaction and permeate sides) reactor model, for co-current configuration. The mass balance equations take into account the two different permeating terms related to H_2 through Pd membrane (Sievert's law) and O_2 through the perovskite membrane. A model of the simultaneous dehydrogenation of ethylbenzene and hydrogenation of benzene was also proposed. Mass, energy, and pressure drop equations and Sievert's H_2-permeating flux were considered in the steady-state analysis.

A general (only one species permeating) steady-state 1D model was elaborated by Dixon *et al.* to compare TR and MR. In this model the membrane type is taken into account by the permeating flux expression. Both dimensional and dimensionless equations were presented for mass and energy balance, assuming no axial dispersion and isobaric conditions.

Fernandes and Soares jr. developed a 1D steady-state model for MSR simulation in a Pd tube-and-shell MR (with reaction taking place in the annulus). Plug flow on both reaction and permeate volumes and no boundary layer on the membrane surfaces were assumed. H_2 permeation through the membrane was modeled by Sievert's law. In the energy balance only the term related to heat exchange with furnace is reported, besides those related to convective flux and heat generated by reaction.

Alfadhel and Kothare studied the modeling of multicomponent concentration profiles in a membrane-based microreactor. Using the continuum approach, they derived a 1D mathematical model for a generic steady-state, isothermal Pd-alloy

[4] In this case the permeate side is a volume for a coupled reaction.

micro-MR, incorporating chemical reaction on a catalytic boundary of rectangular microchannel and permeation through a selective micromembrane. The model is valid for high-speed flows for which the diffusive fluxes are ignored (first-order model). The ODE describing the molar fraction profiles into microchannels are written in dimensionless form. The model considers an excess of sweep in operating conditions and no equations are provided for the permeate side.

4.3.3. 2D models

Hughes *et al.* proposed a 2D, isothermal, and isobaric model for co- and counter-current mode in ceramic supported Pd/Ag MR for the ethylbenzene dehydrogenation. The model neglects axial dispersion, but considers the radial dispersion in both reaction side and ceramic support.[5] Sievert's H_2 permeation was considered and, owing to the excess of sweep gas, no permeate side mass balances were considered. Also the dimensionless equations are proposed.

Basile *et al.* also proposed a 2D, isothermal, and steady-state model for WGS in both co- and counter-current MRs consisting in the mass balances for the reaction and permeate sides, written in terms of species concentrations; Sievert's H_2 permeation is taken into account in B.C.s defined on the membrane surface.

Van Sint Annaland *et al.* developed a mathematical model for the transient analysis of a packed bed MR for the autothermal reforming of methane and pure H_2 production. The 2D pseudo-homogeneous packed bed MR model takes into account the continuity and momentum equations in addition to the component mass and energy balances. The model is based on a standard dispersion model that describes the gas phase mass and energy transport as convective flow with superimposed radial and axial dispersion. A power law equation is used (H_2 partial pressure exponent equal to 0.72) to describe the overall permeation rate of hydrogen through the membrane. Also mass and energy intraparticle balances for the reaction were solved in the model.

Clarizia *et al.* proposed a 2D mathematical model for a co-current, nonisothermal, and steady-state Pd MR for WGS reaction in which radial diffusion phenomena are considered. Sievert's H_2 permeation through the Pd film was considered.

5. Simulation

The simulations of MRs behavior are useful to predict and analyze the effect of some parameters on the performance of a Pd-alloy MR. In particular, the important variables are:

- Temperatures on both membrane sides.
- Reaction and permeation pressures.
- Reactor length.
- Feed molar ratio.

[5] Nevertheless, since in the same authors showed that radial species concentration gradients were flat and could be neglected, only the simulation results of a simplified 1D model were proposed.

- Sweep factor.
- Operating configuration (co- or counter-current mode).
- Space velocity, space time, or time factor.

The main variables expressing the MR performance are:

- Conversion.
- H_2 recovery.
- Reaction selectivity (in the case of competitive reactions).

In this section, some results of Pd-alloy MRs of models presented above will be reported.

Table 9.19 reports a summary of the simulation results present in the open literature. It distinguishes the results by the model dimension (0D, 1D, or 2D) and balances (mass, energy) utilized.

5.1. Isothermal model results of Pd-alloy MRs

In this section basic simulation results of isothermal MRs obtained solving the mass balance equations is presented. The isothermal operating condition is ideal and fundamental since the reactor behavior can be analyzed considering some main variables such as reaction and permeation pressures, flow rates, molar ratio, overcoming for a while the nonlinearity introduced by the variation of the temperature.

Table 9.19 Simulation results of Pd-alloy MRs reported in considered literature

Reference (principal investigator)	Mass results	Also energy balance solution
0D		
Raich and Foley	•	
1D		
Barbieri *et al.*	•	•
Basile *et al.*	•	
Harold *et al.*	•	•
Dittmeyer *et al.*	•	•
Alfadhel and Kothare	•	
Johannessen and Jordal	•	•
Itoh *et al.*	•	
Elnashaie *et al.*	•	•
Dixon *et al.*	•	
Fernandes and Soares jr.	•	•
Hughes *et al.*	•	
Clarizia *et al.*	•	
2D		
Van Sint Annaland *et al.*	•	•

5.1.1. Tubular membrane reactor

The species profiles along the reactor axis give interesting indications in order to understand the benefits of Pd-alloy MRs with respect to TRs. In fact, the H_2-permeating flux depends on the permeation driving force and, thus, on H_2 partial pressures.

The H_2 partial pressure profiles in a tubular MR obtained for a steady-state 1D model (for the MSR reaction) shows that in co-current mode [Eqs. (9.25) and (9.26)] the H_2-driving force is always positive (H_2 partial pressure is higher in reaction than the permeate side), becoming nil for high-reactor lengths when the permeation equilibrium[6] is reached (Fig. 9.21A). A higher H_2 partial pressure difference, higher methane conversion, and H_2 recovery are available in the MR, and equilibrium is reached in lower lengths (Fig. 9.22A) for a higher temperature.

The H_2 partial pressure difference becomes negative in the first part of the module, giving a reverse H_2 permeation and shifting the MSR reaction to the left side and not to the right, in the case of the counter-current mode (Fig. 9.21B and Fig. 9.22B). The reactor length involved also depends on the operating conditions.

In general, the most important aspect to be considered in the comparison between co- and counter-current mode is related to the driving force behavior along the MR. For the co-current mode the hydrogen driving force along the reactor axis tends to zero, that is, the hydrogen partial pressure in the permeate side tends to reach the hydrogen partial pressure in the reaction side. Therefore, they can have the same value at the end of the MR. The inlet abscissa of the reaction (i.e., $z = 0$) corresponds to the outlet one for the counter-current flow configuration. In this case, the hydrogen partial pressure in the permeate side exceeds the hydrogen partial pressure in the reaction side, thereby giving reverse hydrogen permeation.

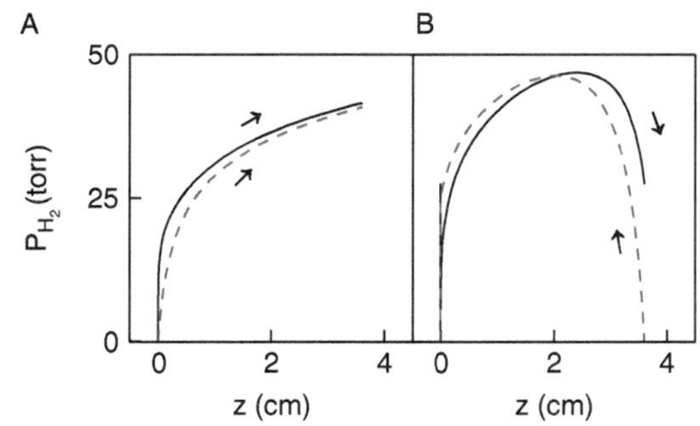

Figure 9.21 H_2 partial pressures against reactor length, on the reaction (solid line) and permeation (dashed line) sides, both (A) parallel flow and (B) counter-current flow. Temperature = 300 °C, reaction pressure = 136 kPa, overall feed flow rate = 162 cm^3(STP)/min, permeate side pressure = 101 kPa, and sweep flow rate = 40 cm^3(STP)/min. Reprinted from Barbieri and Di Maio [83] with permission from American Chemical Society.

[6] An infinite reactor length is required to reach permeation equilibrium. However, from a practical point of view, equilibrium can be considered reached when the difference between the actual and the theoretical values is lower than a fixed limit.

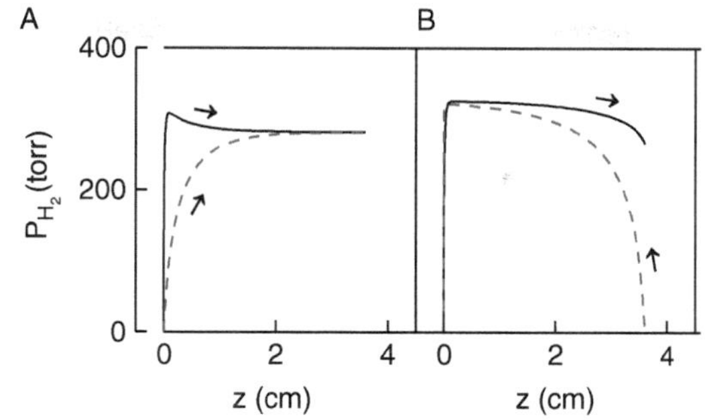

Figure 9.22 H_2 partial pressures against reactor length, for the reaction (solid line) and permeation (dashed line) sides, both (A) co-current flow and (B) counter-current flow. Temperature = 500 °C, reaction pressure = 136 kPa, overall feed flow rate = 162 cm^3(STP)/min; permeate side pressure = 101 kPa, and sweep flow rate = 40 cm^3(STP)/min. Reprinted from Barbieri and Di Maio [83] with permission from American Chemical Society.

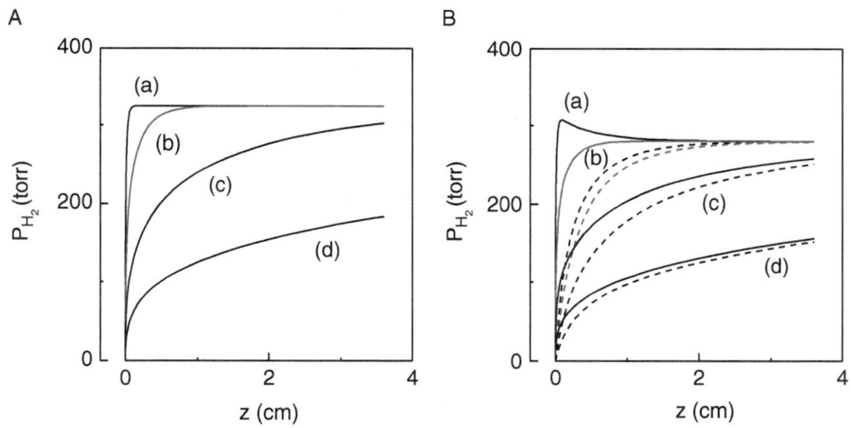

Figure 9.23 Calculated axial profiles of H_2 partial pressures of (A) TR and (B) MR, for different normalized space velocities: (a) 1, (b) 10, (c) 100, and (d) 1000. Temperature = 500 °C, reaction pressure = 136 kPa, feed flow rate = 162 cm^3(STP)/min, permeate side pressure = 101 kPa, and sweep flow rate = 40 cm^3(STP)/min. Reprinted from Barbieri and Di Maio [83] with permission from American Chemical Society.

The modeling explains the effect of space velocities on a co-current MR behavior and its difference with respect to a TR (Fig. 9.23). In particular, the lower H_2 partial pressure in the MR (right side) with respect to the TR (left side) is due to the H_2 permeation that lowers the H_2 partial pressure in the reaction promoting a further conversion. In both TR and MR, the equilibrium conversion (and also the permeating one) is reached before the reactor outlet, for low space velocities [(a) and (b)].

For higher space velocities [(c) and (d)], in both reactors models, the increase of H_2 partial pressure along the reactor axis denotes that a longer membrane is required in order to obtain reaction and permeation *equilibria*.

The major advantage of the MRs is the enhancement in conversion of the equilibrium limited reaction as a result of selective H_2 removal. As is shown in

Fig. 9.24B, the MR length required for reaching equilibrium is higher than that of TR (Fig. 9.24A) owing to H_2 removal from the reaction volume that shifts the reaction towards the products, with a higher methane conversion. The figure also shows an increase in CO and CO_2 partial pressures in MR with respect to TR.

In the counter-current mode, the presence of a "reverse permeation region" at the inlet of MR mainly depends on the permeation rate; the species profiles in an MR for the WGS reaction are illustrated, focusing on the effect of membrane permeability. Figures 9.25 and 9.26 show the variation of the hydrogen molar fraction in the permeate side versus reactor length analyzed in co-current and counter-current modes, respectively, for different H_2 permeance values. With

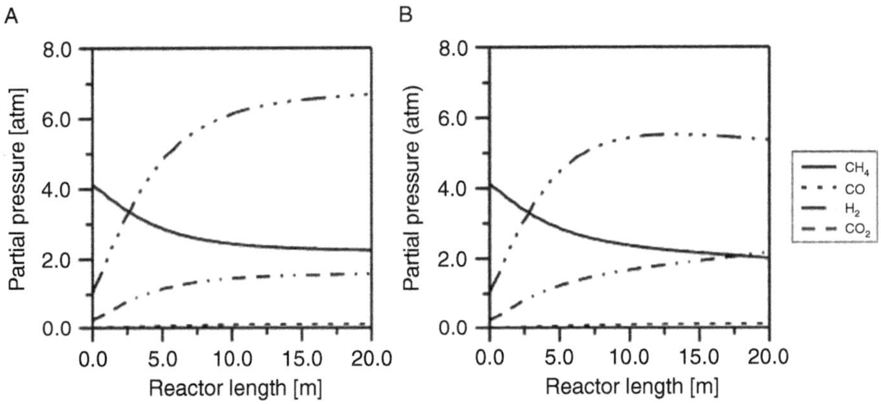

Figure 9.24 Partial pressures along the reactor axis of (A) TR and (B) MR for the MSR reaction. Reprinted from Fernandes and Soares [81] with permission from Elsevier.

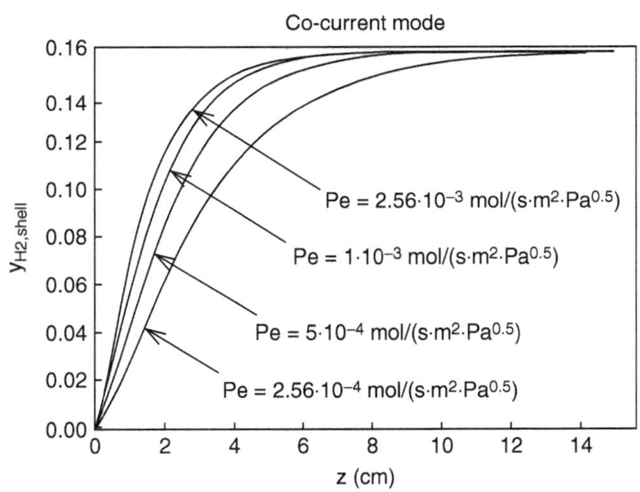

Figure 9.25 Hydrogen molar fraction in the permeate side versus the axial coordinate at different hydrogen permeability through the membrane; co-current mode, $T = 600$ K. Reprinted from Basile *et al.* [72] with permission from Elsevier.

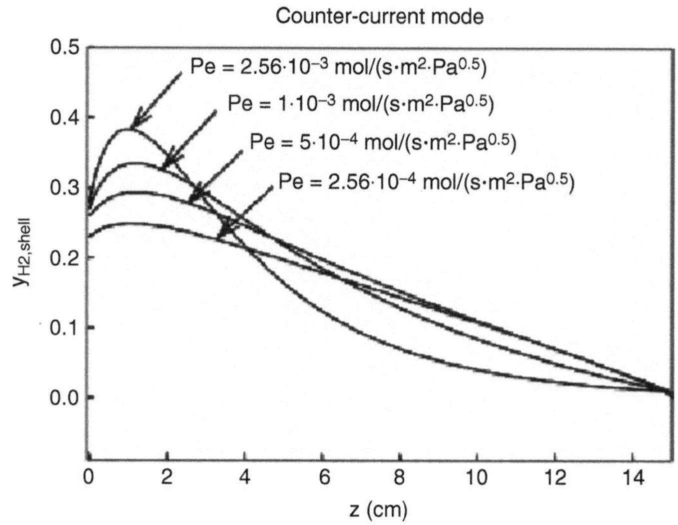

Figure 9.26 Hydrogen molar fraction in the permeate side versus the axial coordinate at different hydrogen permeances through the membrane; counter-current mode, $T = 600$ K. Reprinted from Basile *et al.* [72] with permission from Elsevier.

regard to the co-current mode (Fig. 9.25), at a fixed axial coordinate, increasing the hydrogen permeance (or by decreasing the membrane thickness), the hydrogen molar fraction in the permeate side increases owing to the higher hydrogen flux through the membrane.

The same trend is observed for the counter-current mode (Fig. 9.26) up to a maximum, after that H_2 partial pressure drops down more or less rapidly depending on permeability.

The temperature effect on the performance of counter- and co-current reactor in terms of, for example, conversion increase is also an interesting aspect of MR understanding. The calculated methane conversion for MSR reaction as a function of temperature is illustrated in Fig. 9.27 for co-current (continuous line) and counter-current mode (dashed line). Furthermore, a validation of the mathematical model is made comparing numerical results for the co-current mode with experimental data in a stainless steel supported Pd-alloy MR ([119]). There is a good agreement between the co-current mode (solid line) and experimental data at a low temperature. In the same figure the positive effect of sweep gas flow rate increase on methane conversion for both operating modes is shown.

A comparison between modeling results of MR and TR with experimental data as a function of temperature is reported in Fig. 9.28. The methane conversion of MSR reaction was evaluated at different temperatures integrating the isothermal mass balances equations. A higher increase with respect to the TR is achieved at higher temperatures.

The design equations typically used for catalytic reactors were applied to the WGS reaction, including a subtractive term (recovery index) related to the H_2 permeation through the membrane (Fig. 9.29).

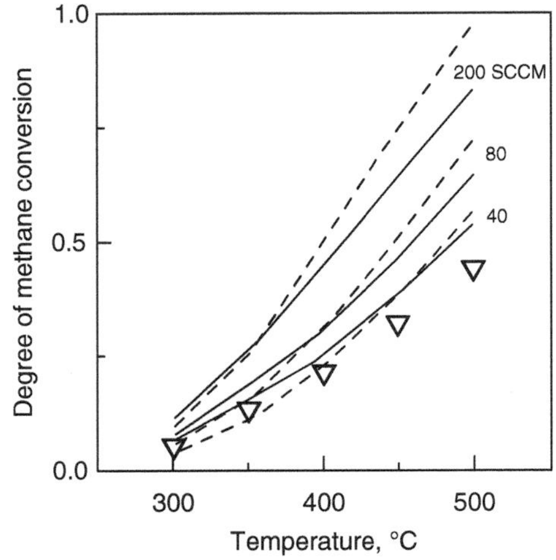

Figure 9.27 Methane conversion of co- and counter-current MRs against temperature. Comparison of the experimental data (symbols) and the model results (lines) for 40 cm^3(STP)/min sweep flow rate. The solid line represents the results for the parallel-flow configuration, whereas the dashed line is the simulation results of the counter-current. Calculated conversions for 80 and 200 cm^3(STP)/min sweep flow rate are also reported. Reaction side: pressure = 136 kPa, H_2O/CH_4 feed molar ratio = 3, feed flow rate = 162 cm^3(STP)/min. Permeate side: pressure = 101 kPa. Reprinted from Barbieri and Di Maio [83] with permission from American Chemical Society.

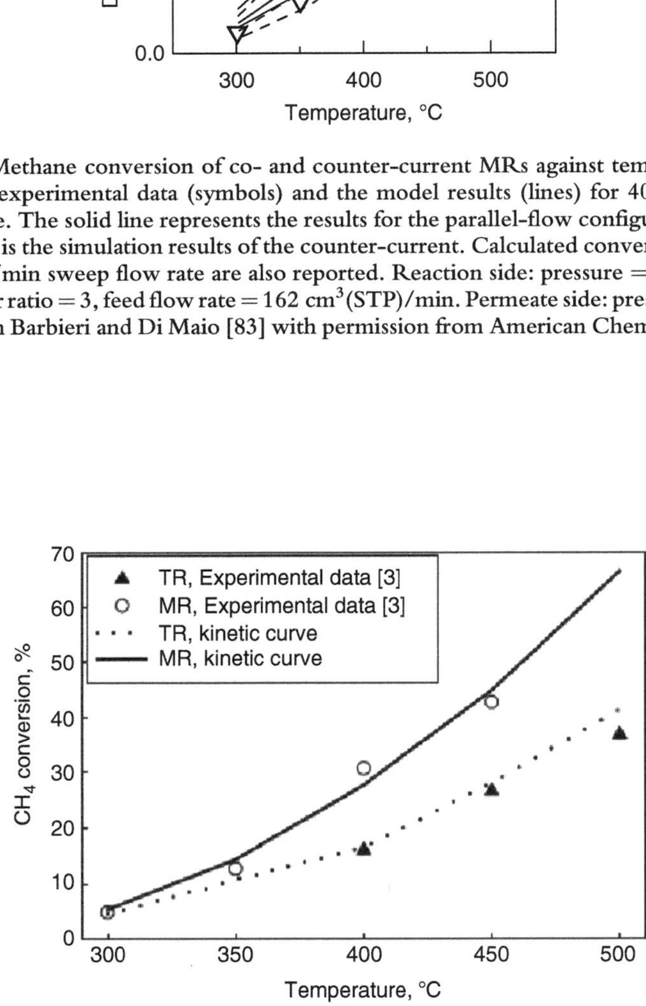

Figure 9.28 Methane conversion versus temperature for TR and MR: comparison between experimental data and simulation results; sweep gas flow rate = 36.2 μmol/s; (MR), P^{Tube} = 122 kPa, $P^{Annulus}$ = 110 kPa, H_2O/CH_4 feed molar ratio = 3. Reprinted from Gallucci *et al.* [84] with permission from Elsevier.

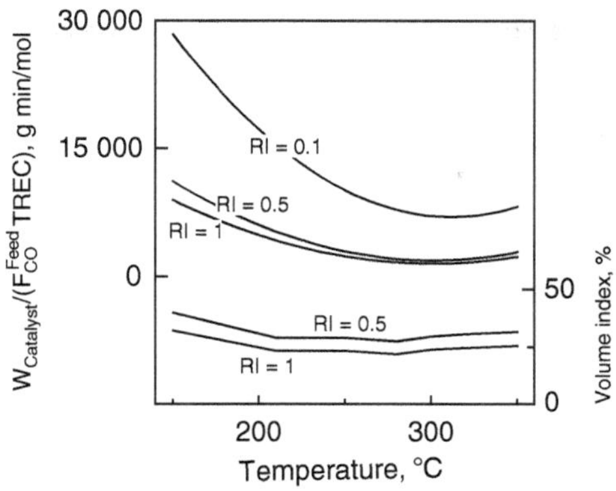

Figure 9.29 Calculated $W_{\text{Catalyst}}/(F_{\text{CO}}^{\text{Feed}}\ \text{TREC})$ and space time ratio between MR and TR as a function of temperature at different *RI* values. Reaction pressure = 100 kPa.

The recovery index (RI) is estimated from the reactor outlet streams, representing the overall hydrogen fraction permeated through the membrane.

$$\text{Recovery index} = \text{RI}_{\text{H}_2} = \frac{(F\gamma)_{H_2}^{\text{Permeate}}}{(F\gamma)_{\text{H}_2}^{\text{Permeate}} + (F\gamma)_{\text{H}_2}^{\text{Retentate}}} \tag{9.57}$$

Furthermore, volume index was introduced as a useful parameter to summarize the advantage in the MR use with respect to TR:

$$\text{Volume index} = \text{VI} = \frac{\tau_{\text{Catalyst}}^{\text{MR}}}{\tau_{\text{Catalyst}}^{\text{TR}}} \tag{9.58}$$

Figure 9.29 reports the factor $W_{\text{Catalyst}}/(F_{\text{CO}}^{\text{Feed}}\ \text{TREC})$ necessary to obtain the 99% of MR equilibrium conversion (MREC) and the volume index, as a function of temperature for different values of H_2 recovery index (percentage of the produced hydrogen recovered in permeate). As temperature increases the catalyst necessary decreases rapidly because the kinetics is faster. The volume index shows the advantage in MR use (in this case) for the WGS; in fact, the low value indicates the low reaction volume (catalyst amount) required by an MR against that necessary in a TR for the same conversion. For instance, the MR volume to reach 95% of equilibrium conversion is about 1/3 of the TR volume required at the same operating conditions (1000 kPa, 280 °C) when H_2 recovery index is 50%. This also affects the required catalyst amount and implies significant advantages in terms of reduced equipment costs. Same advantages were shown by a silica MR requiring about 1/2 of the TR volume at the same operating conditions. Figure 9.30 reports the reduction volume achieved using a Pd-alloy or a silica MR.

The main advantage of an MR with respect to a TR is the positive effect of the pressure on the conversion also for reactions with mole number increase, such as

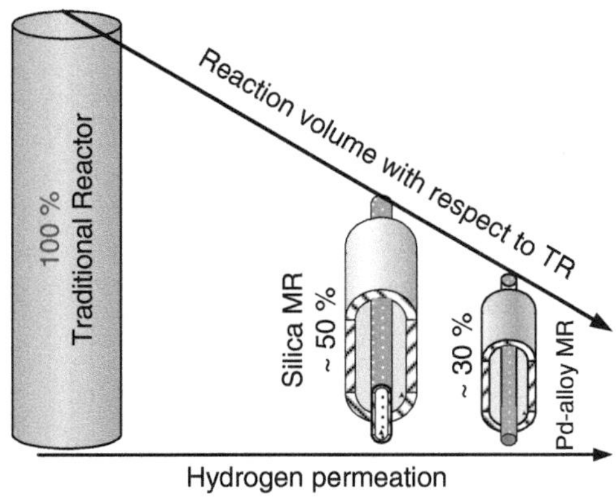

Figure 9.30 MR reduction volume due to the H_2 permeation for Pd-alloy and silica membranes.

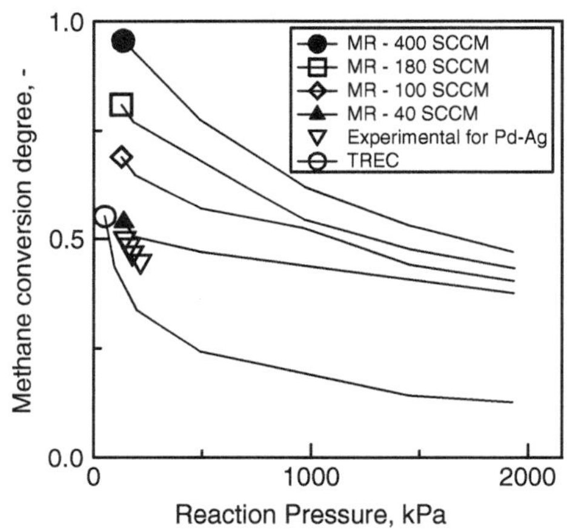

Figure 9.31 Methane conversion dependence on reaction pressure at different sweep flow rates. Temperature 500 °C, feed flow rate, 162 cm^3(STP)/min, and permeate side pressure 101 kPa. Experimental data of Shu *et al.* [119] for Pd–Ag MR are also reported. Reprinted from Barbieri and Di Maio [83] with permission from American Chemical Society.

MSR or methanol steam reforming reactions. Therefore, the *reaction pressure* has a greater importance in MRs than in a TR. The methane conversion in MSR reaction as a function of pressure on the reactor side calculated at different sweep flow rates is reported in Fig. 9.31. The conversion decreases increasing the reaction pressure, but more slowly than the TREC does, reported in the same figure. Therefore, a global positive effect in terms of methane conversion owing to pressure increase is obtained; nevertheless, the MSR reaction produces an increase of the mole number.

Figure 9.32 shows methane conversion versus the reaction pressure for MRs of different reactor lengths. For a high reactor length, a minimum for the methane conversion appears because of a trade-off between two competitive effects. The first

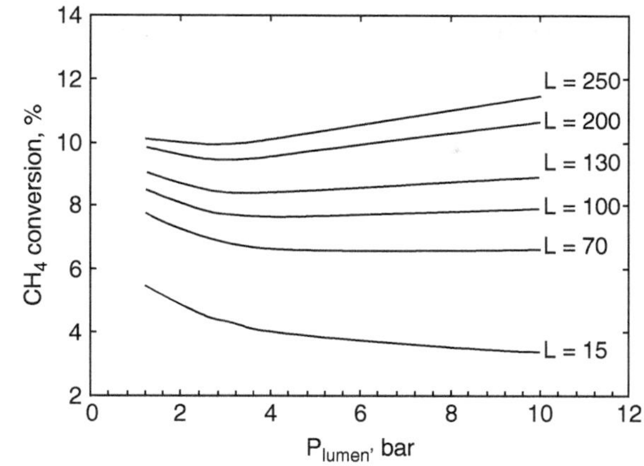

Figure 9.32 Methane conversion versus reaction (lumen) pressure for MR, at different values of the reactor length (L, expressed in centimeter), membrane thickness = 50 μm, permeation pressure = 110 kPa, T = 300 °C, H_2O/CH_4 feed molar ratio = 3. Reprinted from Gallucci *et al.* [84] with permission from Elsevier.

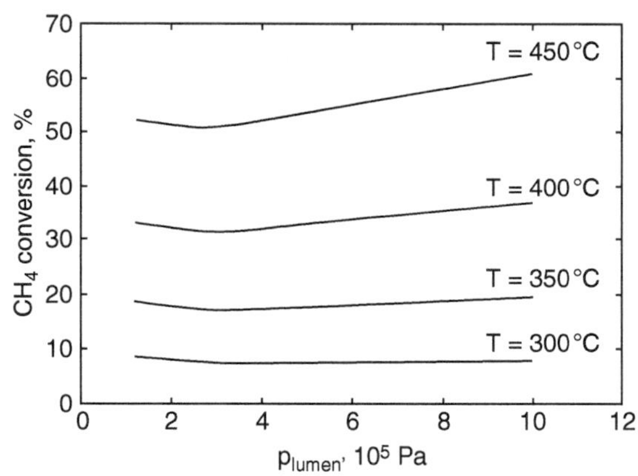

Figure 9.33 Methane conversion versus lumen pressure for MR, at different temperatures, membrane thickness = 50 μm, $P^{Annulus}$ = 110 kPa, H_2O/CH_4 feed molar ratio = 3, membrane length = 1 m. Reprinted from Gallucci *et al.* [84] with permission from Elsevier.

one is the reduction of the methane conversion owing to the increasing pressure. The second is a higher methane conversion promoted by the higher H_2 permeation driving force.

Figure 9.33 shows methane conversion versus the reaction pressure for an MR at different temperature values, at a fixed reactor length of 1 m. Also in this case, two coexisting and competitive effects produce a minimum in CH_4 conversion; in fact, higher values of both temperature and reaction pressure improve H_2 removal.

The effect of sweep factor (*I*, molar ratio between sweep to feed flow rates, [see Eq. (9.15)]) on MSR reaction for a Pd MR performance was also analyzed in Fig. 9.34, where steam was fed at different flow rates such as sweep gas (in addition,

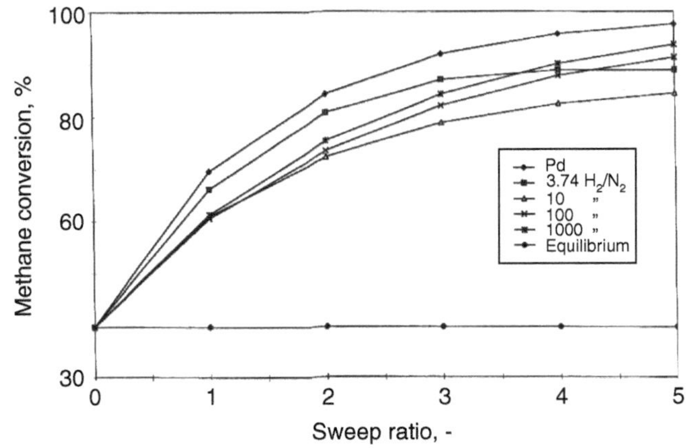

Figure 9.34 Pd/Ag and microporous membrane performance as a function of the sweep/feed flow. Steam as sweep. Hydrogen separation factor as parameter for microporous membranes. Reprinted from Oklany *et al.* [114] with permission from Elsevier.

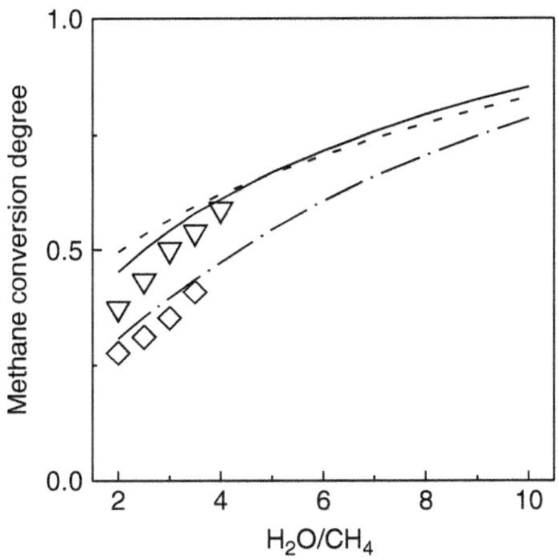

Figure 9.35 Methane conversion against H_2O/CH_4 feed molar ratio. *Squares:* TR experimental data. *Triangles*: MR experimental data (Pd–Ag, $\delta = 10\ \mu m$) for parallel flow. (–.–) TR simulation results. *Continuous line:* simulation results of co-current MR. *Dashed line:* simulation results of counter-current MR. $T = 500\ °C$, reaction pressure = 136 kPa, feed flow rate = 162 cm^3(STP)/min, permeate side pressure = 101 kPa, and *Sweep flow rate* = 40 cm^3(STP)/min. Reprinted Barbieri and Di Maio [83] with permission from American Chemical Society.

in the figure the performance of different microporous MRs were reported). The methane conversion increases as well as the capacity of the sweep to remove H_2 from the reaction side.

Another parameter influencing the conversion is the reactant molar feed ratio (*m*).

The methane conversion increases with the H_2O/CH_4 molar feed ratio (*m*) as is shown in Fig. 9.35. The model results of co- and counter-current MRs are in good

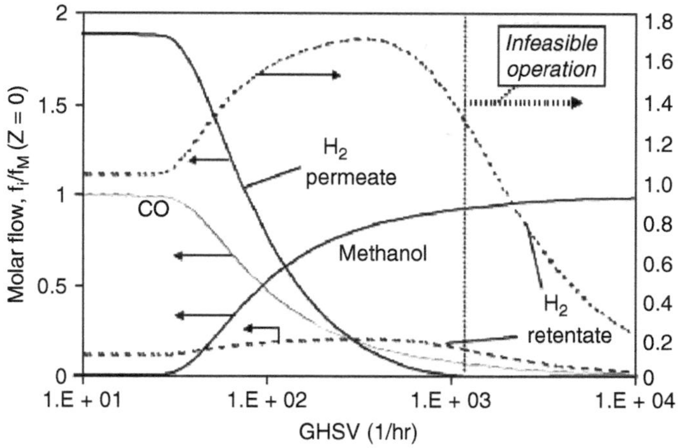

Figure 9.36 Simulation of methanol decomposition in the packed bed MR. Species molar flow rates and retentate hydrogen pressure are shown as a function of the space velocity. Complete conversion of methanol is achieved. Reprinted from Harold *et al.* [85] with permission from Elsevier.

agreement with experimental data ([119]). In the same figure, the calculated methane conversion of the TR is also reported.

The effect of space velocity on Pd-alloy MR is illustrated for direct methanol decomposition (Fig. 9.36). The species profiles over a wide range of space velocities show an increase of methanol effluent molar flow and a conversion decrease with Gas Hourly Space Velocity (GHSV), while in the meantime H_2 and CO flow rates decrease. At a sufficiently low space velocity (approaching 30 h^{-1}), methanol conversion is nearly complete and the permeate hydrogen flow rate approaches its theoretical limit, that is, twice the methanol feed flow rate.

A comparison of simulated results for a Pd/Ag MR and experimental data of a TR is proposed (Fig. 9.37) for the catalytic dehydrogenation of isobutane to isobutene. The reactor performance was evaluated as a function of the space time, that is, the ratio of the reaction volume to the reference component molar flow rate (V/F_{A0}). The advantage of employing an MR is quite clear, since the yield of isobutene in MR is compared (for different membrane permeability) with TR experimental results.

5.1.2. 2D models

Some authors, (e.g., Itoh *et al.* [120]), have demonstrated the existence of H_2 radial profiles in MRs owing to the radial mixing diffusion of H_2, which can unfavorably affect the MRs performances; in particular, the developed "radial diffusion model" provides better prediction than the "ideal plug-flow mode," mostly when a high sweep factor promotes a high H_2 recovery index (high-H_2-permeating flux). However, the PDEs describing the 2D system, even if more complete in terms of capacity in systems description, often require more complex algorithms and considerable calculus efforts. This is the main reason for the difference between the 2D and the 1D model numbers available in the open literature.

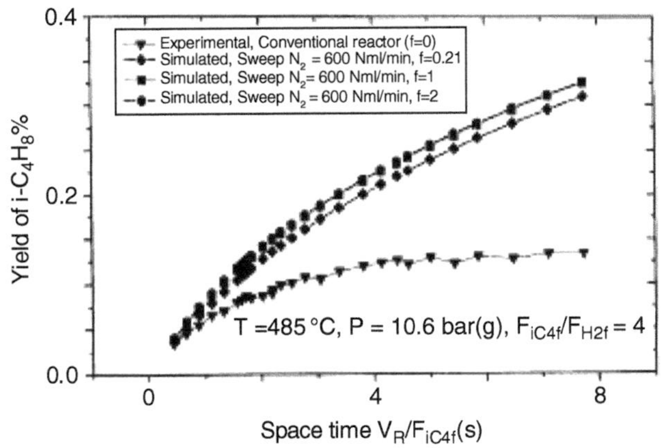

Figure 9.37 Yield of isobutene as a function of space-time. Calculated results for a Pd/Ag MR and experimental data of a TR. Reprinted from Liang and Hughes [115] with permission from Elsevier.

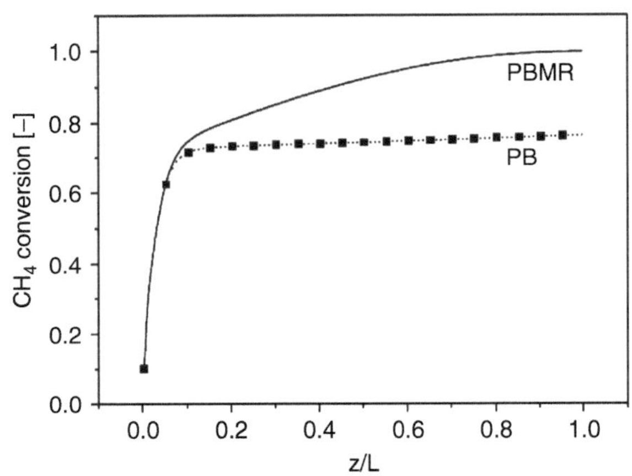

Figure 9.38 Calculated methane conversion in the packed bed TR (PB) and packed bed MR (PBMR) in isothermal operations (base case). Reprinted from Tiemersma *et al.* [92] with permission from Elsevier.

The simulation of auto-thermal reforming of methane in a packed bed MR and a packed bed TR using a 2D is proposed in Fig. 9.38. MR isothermal simulations show an increasing methane conversion along the reactor length and it is higher than that of the packed bed TR at each abscissa.

When a relative low H_2-permeating flux occurs, the methane conversion is restricted by H_2 removal rate through the membrane and radial concentration gradients are very small (Fig. 9.39).

In an MR the permeation of the species through the membrane, as well as the mole number variation owing to gas phase reactions and temperature changes, produces a gradient of velocity. The momentum balance PDE solution produces the velocity profiles in the MR, as is displayed in Fig. 9.40A and B for the radial and

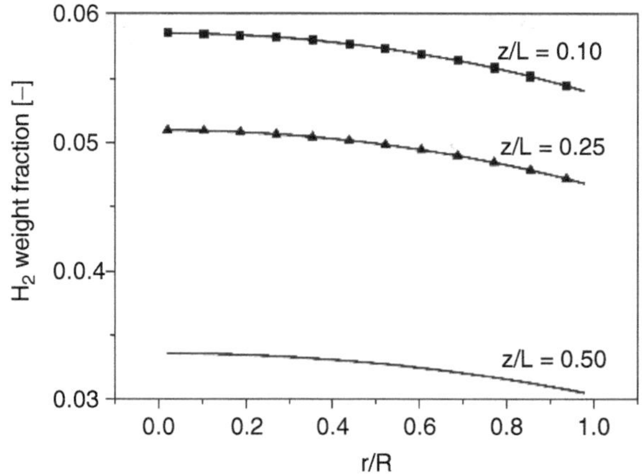

Figure 9.39 Calculated H_2 radial profile for the isothermal operation mode (base case). Reprinted from Tiemersma *et al.* [92] with permission from Elsevier.

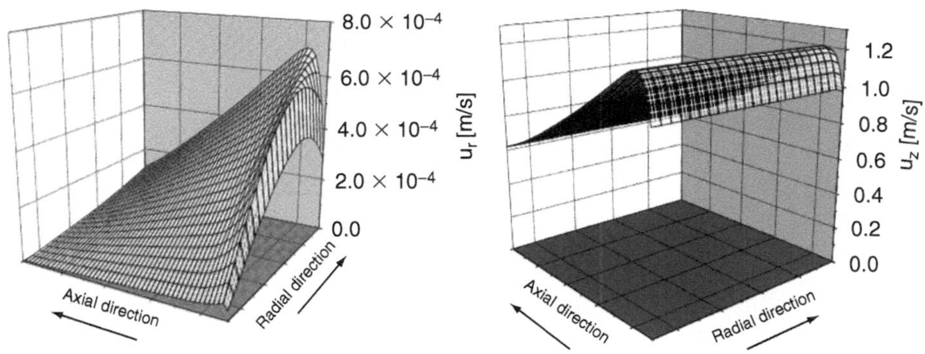

Figure 9.40 (A) Radial and (B) axial velocity profiles for the isothermal operation mode (base case). Reprinted from Tiemersma *et al.* [92] with permission from Elsevier.

axial directions, respectively. The radial velocity profile shows a nearly linear increase toward the membrane due to the selective removal of hydrogen. At the reactor inlet the axial velocity increases owing to the high reaction rate, which results in a net production of moles; the downstream axial velocity decreased owing to hydrogen extraction.

5.1.3. Continuous stirred tank membrane reactor

In order to understand the specific effect of some parameters on MRs performance, it can be useful to consider a lumped parameter system (batch or continuous).

The simulation of cyclohexane and isobutane dehydrogenation in a Pd MR (Fig. 9.41) modeled by means of 0D mass balance equations shows (for both reactions) that a conversion increase takes place by increasing the sweep factor, because this allows a larger removal of H_2 from the reaction volume. Moreover, for cyclohexane dehydrogenation the total amount of H_2 produced is higher than that

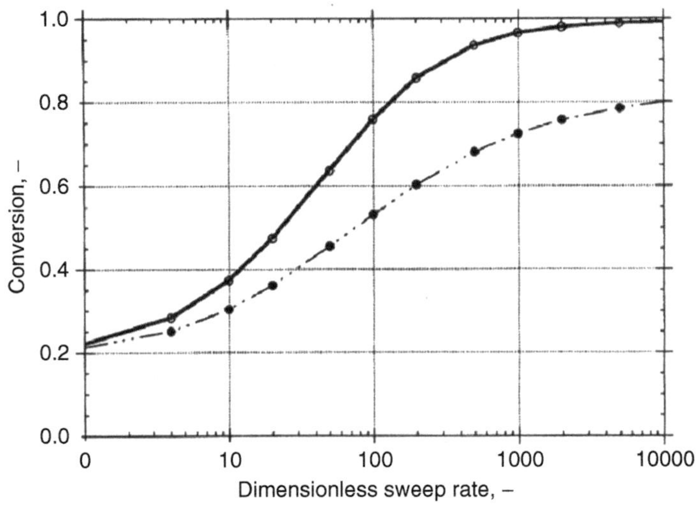

Figure 9.41 Comparison of cyclohexane (empty circles) and isobutane (filled circles) dehydrogenation with a 20% equilibrium conversion basis. Reprinted from Raich and Foley [98] with permission from Elsevier.

produced in the other reaction owing to the stoichiometric coefficients. A reaction producing more H_2 moles per mole of reactant is more sensitive to the effect of H_2 removal through the Pd-membrane and more readily achieves the MR equilibrium conversion.

5.2. Nonisothermal models of Pd-alloy MRs

In this section, the energy balance is introduced in the differential equation set and thus resolved simultaneously. The interaction between mass and energy balances and the related nonlinear behavior will be contained in the simulation results. These results are more interesting than the basic isothermal one, since they are closer to laboratory experiments and industrial applications. Therefore, temperature profile analysis will be proposed in addition to that of other variables (e.g., species partial pressure and conversion) already presented in the basic isothermal section.

5.2.1. Tubular membrane reactor

The effect of the energy transport in a Pd-alloy MR for an annular and a lumenal configuration was proposed by means of integrating a 1D mathematical model including the energy balance Eqs. (9.43) and (9.44).

Figure 9.42 reports calculated temperature profiles for an annular, a lumenal, and an adiabatic MR and a TR for the MSR reaction. The reaction temperature decreases (in the first part) along the reaction length because of the endothermicity of the reaction but, after a minimum, rises until the furnace temperature is reached when the methane conversion reaches the equilibrium value. This minimum is higher for the annular MR than for the lumenal one (the supply of heat is easier for

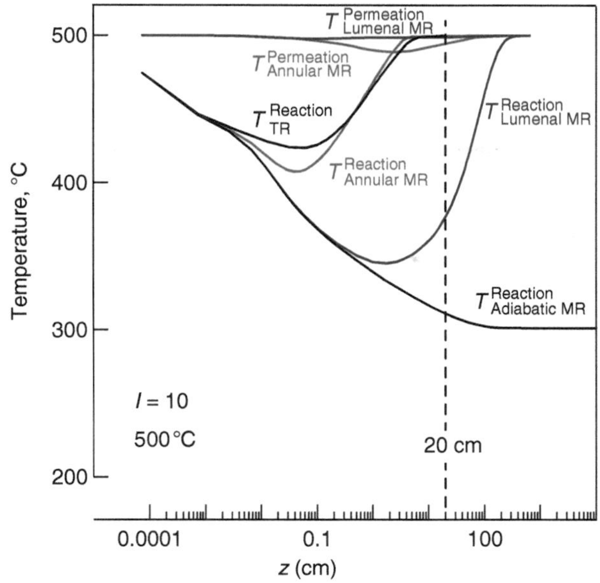

Figure 9.42 Temperature profiles versus reactor length for TR and annular, lumenal, and adiabatic MRs. I.C.s: reaction pressure = permeate side pressure = 100 kPa, $I = 10$, $m = 3$ and CH_4 feed flow rate = 200 cm^3(STP)/min. Reprinted from Marigliano *et al.* [69] with permission from Elsevier.

the annular configuration). The temperature profile for the TR has a minimum a few degree higher than that of an annular MR. Also heat fluxes are higher for the annular MR owing to a lower resistance and as a consequence a short reactor is necessary to reach equilibrium (with lower amount of catalyst). The reaction temperature profile for the adiabatic MR decreases continuously as the methane conversion increases "cooling the reaction stream." The permeation temperature profile for the lumenal MR has a minimum higher than the annular MR because the heat exchange between the furnace and the permeation stream in the annulus is high.

Figure 9.43 reports methane conversion along the reactor axis that is related to the different temperature profiles of Fig. 9.42. In particular, the annular and the lumenal MRs allow the conversion to be the highest (a shorter reactor length is enough for the annular configuration owing to the highest temperature profile), while the adiabatic MR conversion is the lowest one (the stream is "cooled" owing to the endothermic reaction). The difference between methane conversions of MR and TR depends on the equilibrium conversion value: highest in the MR.

Figure 9.44 reports H_2 partial pressure profiles on the reaction and the permeate side related to the previous figures, for both annular and lumenal configurations and adiabatic operation.

Hydrogen partial pressure on the reaction side increases with the reactor length until a maximum corresponding also to the maximum hydrogen permeation flux of an annular MR. Then, it decreases reaching a minimum at the same reactor length of the minimum reaction temperature, after that it increases, because the temperature

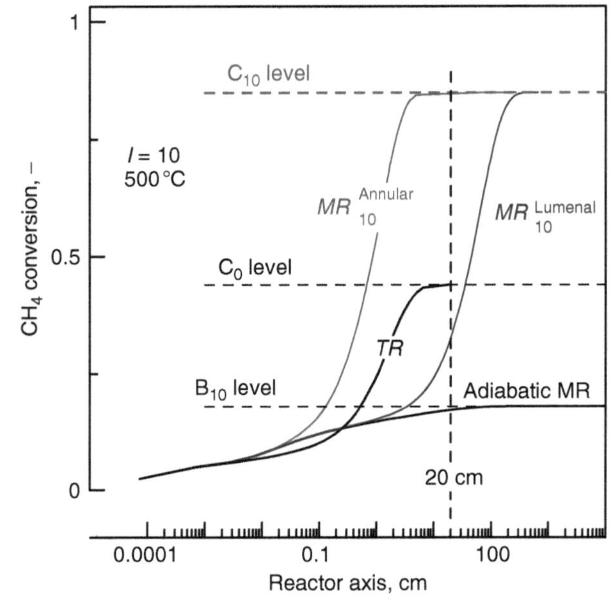

Figure 9.43 Conversion degree versus reactor length for TR and annular, lumenal (tubular), and adiabatic MRs. Initial conditions: reaction pressure = permeate side pressure = 100 kPa, $I = 10$, $m = 3$, and CH_4 feed flow rate = 200 cm^3(STP)/min. Reprinted from Marigliano *et al.* [69] with permission from Elsevier.

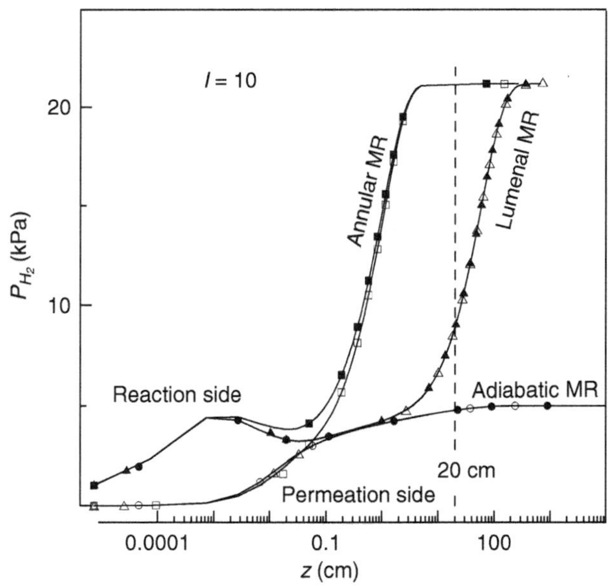

Figure 9.44 Hydrogen partial pressures on the reaction for (filled squares) annular, (filled triangles) lumenal, (filled circles) adiabatic and on the permeate side for (empty squares) annular, (empty triangles) lumenal, (empty circles), and adiabatic MRs. Operating conditions: reaction pressure = permeate side pressure = 100 kPa, $I = 10$, $m = 3$, and CH_4 feed flow rate = 200 cm^3(STP)/min. Reprinted from Marigliano *et al.* [69] with permission from Elsevier.

increases achieving its own equilibrium value. The permeation has a higher weight than the reaction on the hydrogen concentration, also owing to the reduced production rate as a consequence of the negative effect of the temperature fall on the thermodynamics and kinetics. On the permeate side hydrogen partial pressure increases monotonically; it achieves the same value of hydrogen partial pressure on the reaction at a short reactor length (<20 cm). Their difference is equal to zero before the thermal equilibrium is reached as it is possible to see from Fig. 9.42, where the permeation temperature is not yet equal to the furnace temperature. Hydrogen partial pressures of lumenal and adiabatic MRs show about the same maximum value reached by the annular MR on the reaction side. Also on the permeate side they have the same behavior until 0.5 cm of the reactor length where the difference in the partial pressures is very little. The system variation happens in conditions of permeating equilibrium: the rate-determining step is the heat transfer. Hydrogen partial pressure profiles of lumenal and adiabatic MRs follow different paths in the last reactor part. The outlet hydrogen partial pressure of a lumenal MR is the same as the annular MR (about 20 kPa), whereas an adiabatic MR reaches a lower value.

Temperature profiles for lumenal and annular MRs with higher sweep factors have similar types of behavior (Fig. 9.45, ten times higher), but, owing to a higher conversion promoted by a higher H_2 removal capacity of the system, the energy requirements are higher and therefore the minimum temperatures are lower. Reaction temperature reaches that of the furnace at a lower reactor length than is the case

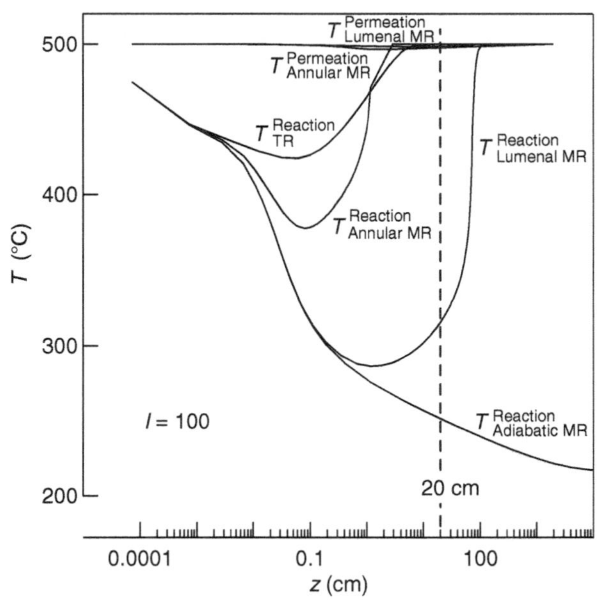

Figure 9.45 Temperature profiles versus reactor length for annular, lumenal, and adiabatic MRs. I.C.s: reaction pressure = permeate side pressure = 100 kPa, $I = 10$, $m = 3$, and CH_4 feed flow rate = 200 cm^3(STP)/min. Reprinted from Marigliano *et al.* [69] with permission from Elsevier.

of the lower sweep factor due to the reduced heat capacity of the reaction stream, which is a consequence of the higher hydrogen permeate. Permeation temperature for the annular reactor is quite constant because of the higher heat capacity of the sweep stream.

Simulations at different furnace temperatures show that in all cases the annular MR has greater efficiency than the lumenal MR (Fig. 9.46). In fact, an annular MR reaches equilibrium conversion before 20 cm of the reactor length for all furnace temperatures, whereas the lumenal MR reaches the same values at a higher reactor length. The energy required by the reaction increases with the temperature owing to an increase in the equilibrium conversion. Also the difference in the behavior of the two proposed configurations increases with the temperature owing to the difference in the overall heat transfer coefficient between the furnace and the reaction side. At the same reactor length, for example 20 cm, the difference in the conversion between the annular and the lumenal MRs operating at 600 °C is about 0.5 much higher than the difference (0.3) of the same variable for an MR operating at 400 °C.

Moreover, in Fig. 9.47 profiles of the energy required by the reaction are reported. In the first part the profile of the annular MR is higher than the corresponding profile of the lumenal MR, but it reaches a null value at a shorter reactor length. In fact, the heat required by the reaction is an increasing function of the sweep factor (a higher conversion can be achieved increasing the H_2 removal capacity). Therefore, the profile of heat required by the reaction, for an annular MR

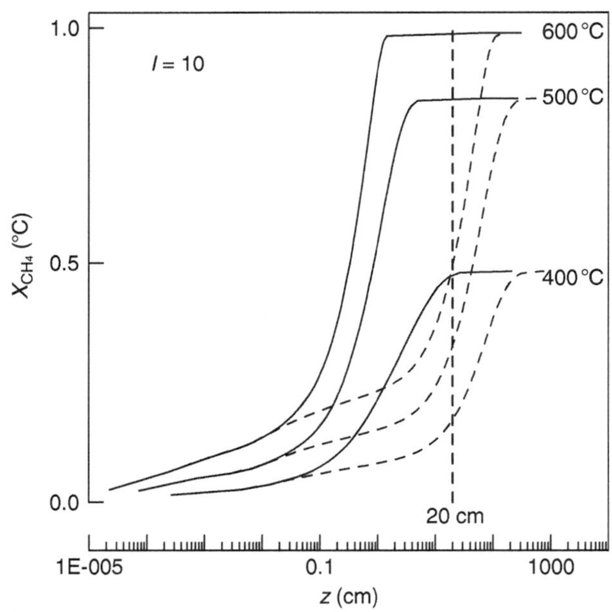

Figure 9.46 Methane conversion for annular (solid lines) and lumenal (dashed lines) MRs at several furnace temperatures. Operating conditions: reaction pressure = permeate side pressure = 100 kPa, $I = 10$, $m = 3$, and CH_4 feed flow rate = 200 cm^3(STP)/min. Reprinted from Marigliano *et al.* [69] with permission from Elsevier.

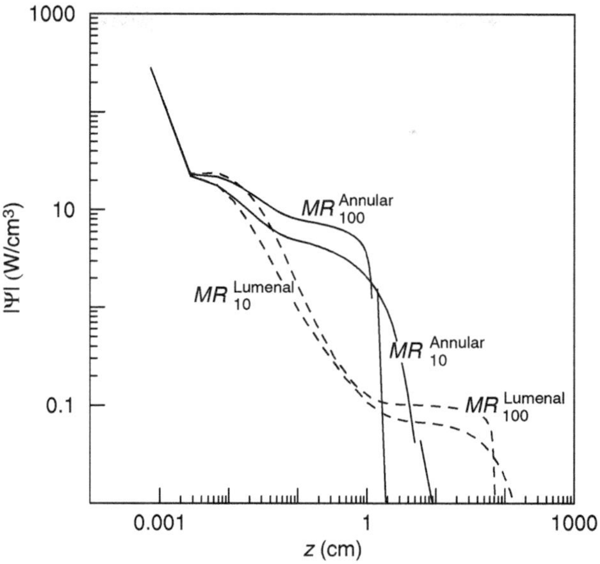

Figure 9.47 Heat required by reaction versus reactor length for annular and lumenal MR at two different sweep factors ($I = 10$ and 100). Feed conditions: reaction pressure = permeate side pressure = 100 kPa, $m = 3$, and CH_4 feed flow rate = 200 cm^3(STP)/min. Reprinted from Marigliano *et al.* [69] with permission from Elsevier.

with $I = 100$, is higher compared with the same profile for $I = 10$. However it fell to zero at shorter reactor length.

The lumenal reactors present the same behavior even if the heat required at a specific reactor length is lower compared with the annular MRs. However, for a lumenal MR greater reactor length is necessary to reach the equilibrium conditions.

The relative importance of the different terms along the reactor length, which contribute to the energy balance of the system, is reported in Figs. 9.48 and 9.49 for an annular MR and a lumenal MR. The term related to hydrogen permeation is much lower than the other terms. The heat exchange between the annulus and the furnace increases, and it exceeds the value assumed by the reaction requirements; its relative importance decreases following the absolute profile of the heat required by reaction (dashed line). Simultaneously, the relative importance of the heat exchanged between the annulus and the furnace and between the annulus and the lumen increases up to the same value.[7]

A closed region delimited by the two ideal modes (*isothermal* and *adiabatic*) and the equilibrium conversion curve is identified in the conversion–temperature plane using the simulation results of simultaneous mass and energy balance integration. This region includes all reaction paths with the same I.C. (e.g., temperature, feed

7 All the differences in equilibrium condition must be equal to zero, but in the numerical simulations $|T^{Furnace} - T^{Reaction}| < \varepsilon$ and $|T^{Furnace} - T^{Permeation}| < \varepsilon$ etc. with, e.g., $\varepsilon = 0.00001$, were considered the limits to achieve the equilibrium and to stop the simulations. For this reason all the terms of the relative importance are numerically different from zero, at the reported reactor exit, even if their absolute values are practically zero.

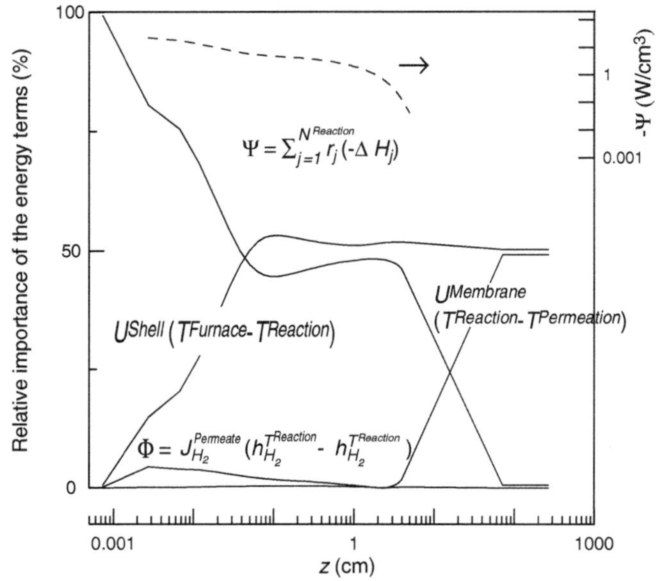

Figure 9.48 Relative importance of the energy terms for an annular MR. Operating conditions: reaction pressure = permeate side pressure = 100 kPa, $I = 100$, $m = 3$, and CH_4 feed flow rate = 200 cm^3(STP)/min. Reprinted from Barbieri *et al.* [100] with permission from American Chemical Society.

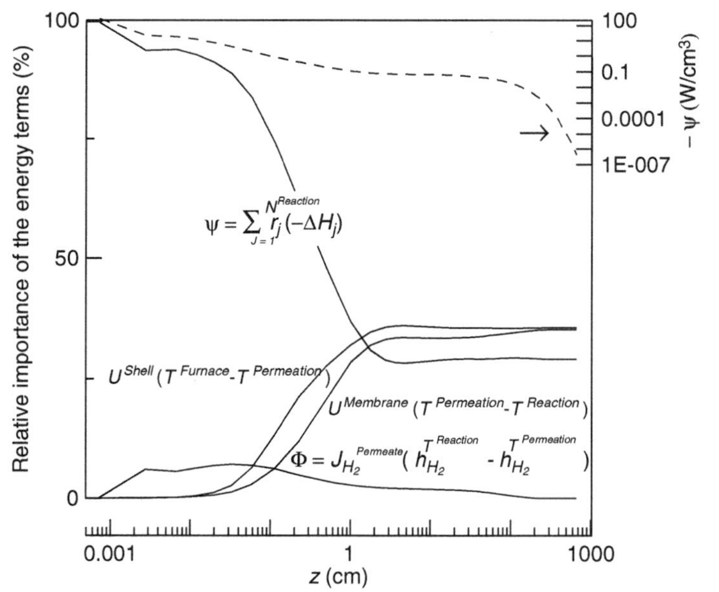

Figure 9.49 Relative importance of the energy terms for lumenal (tubular) MR. Operating conditions: reaction pressure = permeate side pressure = 100 kPa, $I = 100$, $m = 3$, and CH_4 feed flow rate = 200 cm^3(STP)/min. Reprinted from Barbieri *et al.* [100] with permission from American Chemical Society.

molar ratio, and sweep factor). Different reaction paths are followed when flow rates, membrane thickness, and overall heat transfer coefficient are changed, but in any case these paths belong to this closed region: the *existence region*, the overall operative limits in which any MR can/has to operate when the I.C.s are fixed.

For a nonisothermal counter-current MR the H_2 partial pressure and temperature profiles along the membrane axis for both membrane sides are reported in Figs. 9.50 and 9.51, respectively, in the case of MSR reaction. H_2 partial pressure profiles are substantially similar to those of the isothermal case (see Fig. 9.21B).

The temperature profiles in a hybrid TR- counter-current MR for the autothermal reforming of hydrocarbons is reported in Fig. 9.52. The temperature peak in

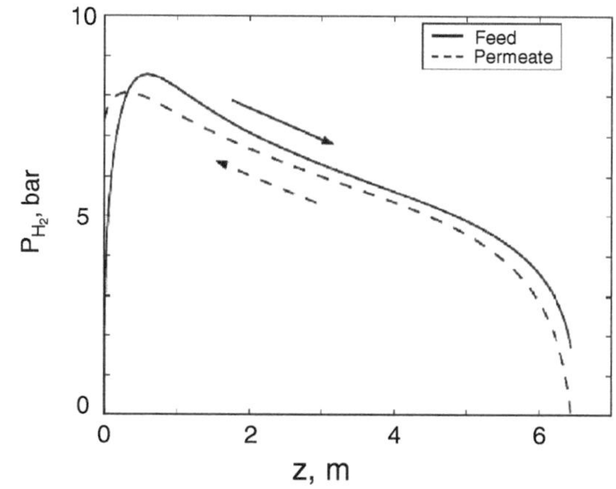

Figure 9.50 The partial pressure of H_2 on the feed and the permeate side. The flow directions are indicated. Reprinted from Johannessen and Jordal [87] with permission from Elsevier.

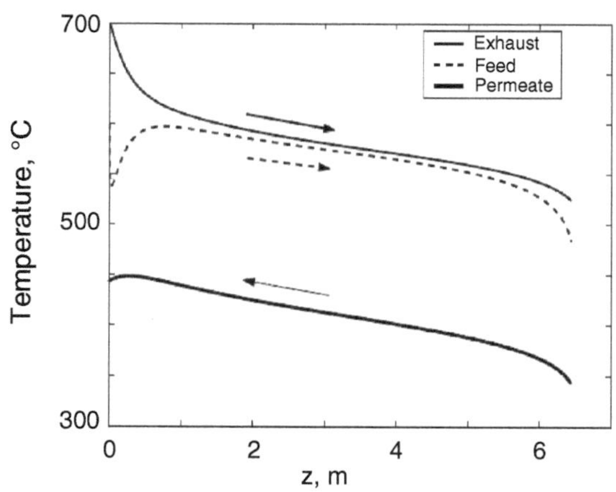

Figure 9.51 Temperature profiles. The flow directions are indicated. Reprinted from Johannessen and Jordal [87] with permission from Elsevier.

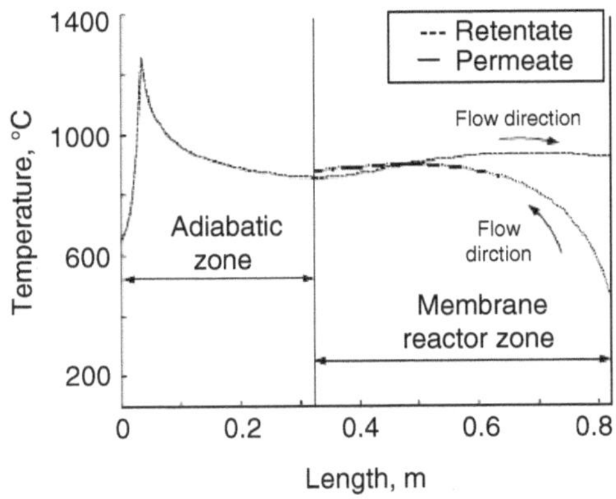

Figure 9.52 Temperature profiles in adiabatic TR zone and in MR zone (retentate and permeate) for autothermal reforming methane. Reprinted from Lattner and Harold [86] with permission from Elsevier.

the first part of the reactor depends on the fast oxidative reaction, while the adiabatic reaction produces a decreasing temperature profile until the MR zone where the counter-current flow and the permeation produce the temperature profile intersection (see Figs. 9.50 and 9.21).

In Fig. 9.53, the reaction paths for the isothermal, adiabatic, and nonisothermal luminal (tubular) TR and MR for MSR are reported. The isothermal path is a vertical line that starts from initial point A and reaches point C_I, corresponding to the maximum conversion that can be obtained with an MR at the specified conditions. The curve AB_I is calculated for the adiabatic conditions. Also this reaction path ends on the corresponding equilibrium curve because point B_I represents an equilibrium condition at the adiabatic reactor temperature. The last path is included between the previous two equilibrium points B_I and C_I. In particular, the existence regions for different sweep factors, indicated with D_I, for TR and MR (*a*, *b*, *c* and *d*) are shown in the conversion–temperature diagram, increasing *I* from 0 (TR) to, for instance, 1, 10, 100 the existence region widens including the existence regions at a lower *I*:

$$\mathbf{D}_0 \subset \mathbf{D}_1 \subset \mathbf{D}_{10} \subset \mathbf{D}_{100}$$

Figure 9.54 shows the behavior of a TR ($TR^{2.4}$ and TR^{227}) with two different overall heat transfer coefficients; the behavior of an annular MR (MR^{227}) and lumenal MRs ($MR^{2.4}$, $MR^{0.1}$, and $MR^{0.01}$) with different values of the overall heat transfer coefficient is also reported. Other $U_{\text{Membranes}}$ values [0.1 and 0.01 W/(m^2 K)] are hypothetical and were used to approach adiabatic conditions. The reacting stream of an annular MR receives heat quickly from the furnace and slowly a small contribution from the sweep stream. In a lumenal MR, only the sweep stream offers heat to the reactants. The heat flux exchanged by the lumenal MR is less than that exchanged for MR^{227} because of the lower values of the alumina thermal conductivity, and so $T^{\text{Permeate}} = T^{\text{Furnace}}$.

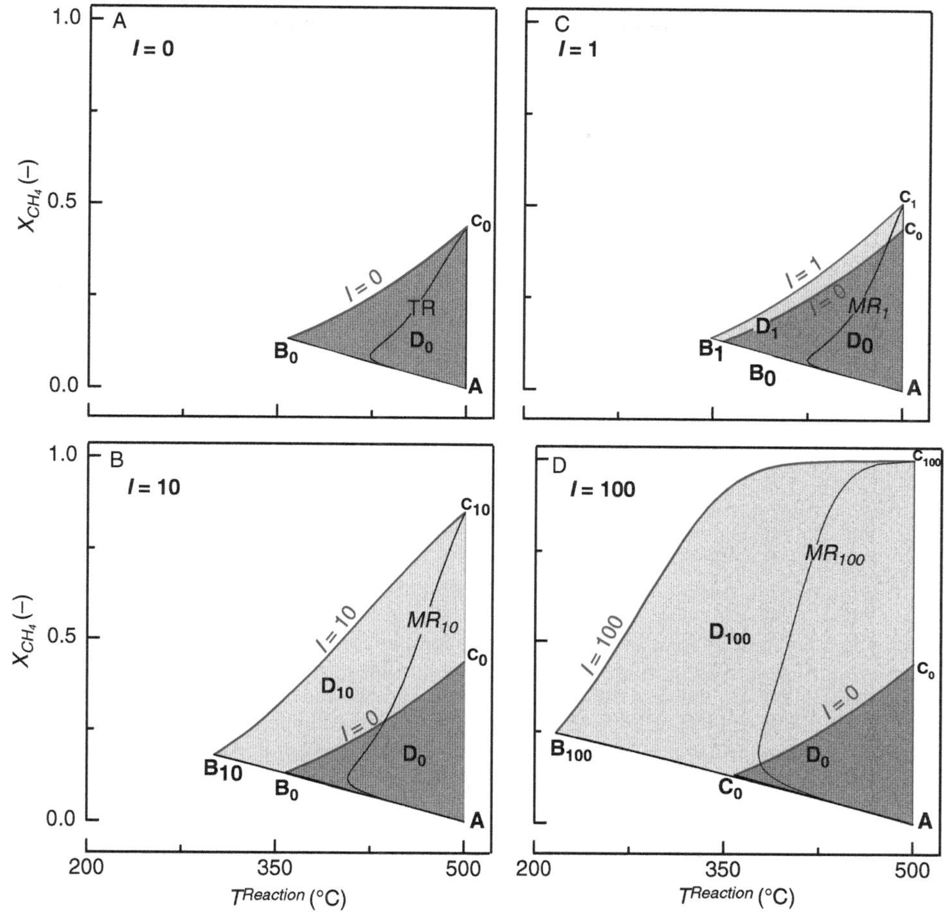

Figure 9.53 Existence regions for an MR and a TR. $P^{\text{Reaction}} = P^{\text{Permeation}} = 100$ kPa, $T^{\text{Feed}} = T^{\text{Sweep}} = 500\ °\text{C}$, $Q^{\text{Feed}}_{\text{CH}_4} = 200\text{cm}^3(\text{STP})/\text{min}$, $m = 3$. (A) Traditional reactor with $I = 0$. AC_0 and AB_0 are the isothermal and adiabatic behavior, respectively; B_0C_0 is the equilibrium locus for $I = 0$. $D_0 \leftrightarrow AB_0C_0$ is the existence region of a TR. *TR* is the path of a nonisothermal TR [$U = 227$ W/(m^2 K)]. (B), (C), and (D) MRs with $I = 1$, 10, and 100, respectively. AC_I and AB_I are the isothermal and adiabatic behavior, respectively; B_IC_I is the equilibrium locus for $I = 1$, 10, and 100. $D_I \leftrightarrow AB_ICI$ is the existence region of the MR. *MR* is the path of a nonisothermal annular MR [$U^{\text{Shell}} = 227$ W/(m^2 K)]. Reprinted from Barbieri *et al.* [100] with permission from American Chemical Society.

The rate-limiting step between the reaction and the heat exchange can be determined by following a generic point on the reaction path. The driving force promoting the reaction is a function of the difference between the conversion reached by the reactor and the corresponding point, at the same temperature, on the equilibrium curve. The difference between the reaction and the furnace temperatures is the driving force for heat exchange (it is evaluated on the horizontal line). The conversion difference of a TR is less than that of an MR. In the first part, the energy consumed by the reaction is greater than that supplied by thermal

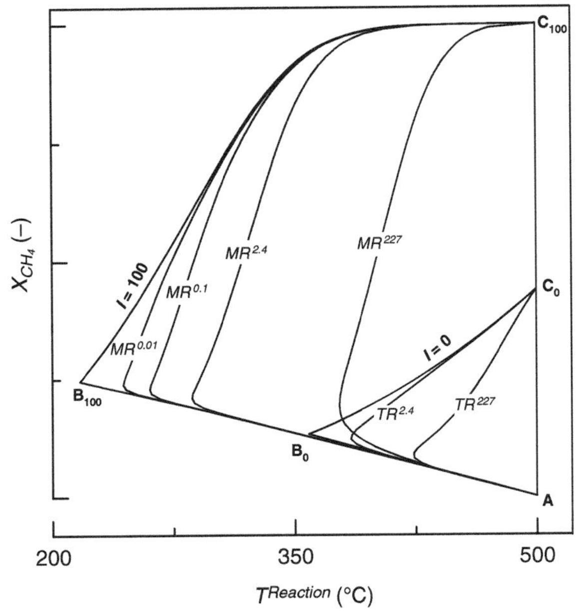

Figure 9.54 Methane conversion versus reaction temperature for the traditional and membrane reactors. TR^{227} and $TR^{2.4}$ are the TR conversion profiles for $U = 227$ and 2.4 W/(m^2 K), respectively. MR^{227} is the annular MR conversion profile for $U^{\text{Shell}} = 227$ W/(m^2 K) and $U^{\text{Membrane}} = 2.4$ W/(m^2 K). $MR^{2.4}$, $MR^{0.1}$, and $MR^{0.01}$ are the lumenal (tubular) MR conversion profiles for $U^{\text{Shell}} = 227$ W/(m^2 K) and $U^{\text{Membrane}} = 2.4$, 0.1, and 0.01 W/(m^2 K), respectively. $I = 0$ and $I = 100$ are the equilibrium curves without sweep gas and with a sweep factor equal 100, respectively. $m = 3$, $T^{\text{Feed}} = T^{\text{Sweep}} = 500$ °C, $P^{\text{Reaction}} = P^{\text{Permeation}} =$ 100 kPa, $Q^{\text{Feed}}_{\text{CH}_4} =$, 200cm^3(STP)/min, $Q^{\text{Sweep}} = 100\,Q^{\text{Feed}}_{\text{CH}_4}$. Reprinted from Barbieri *et al.* [100] with permission from American Chemical Society.

exchange. From the adiabatic trajectory, the reaction paths of the different reactors branch out along different paths. The reaction path reaches a minimum temperature when the amounts of energy required by the reaction and the heat exchanged are equal. Above the minimum temperature, the energy supplied is greater than that required by the reaction; the system tends to reach thermal equilibrium. For lower U values, the minimum temperature decreases, and after the minimum is reached, the reaction path is very close to the equilibrium line and more distant from thermal equilibrium. For the TR, the minimum temperature in reactor $TR^{2.4}$ was about 385 °C, 40 °C lower than the value of 425 °C for TR^{227}, and the path for the lowest overall heat transfer coefficient value is close to the equilibrium line. Both reactors reach the same final equilibrium achieved by the isothermal reactor. The MR path follows, after the minimum, the equilibrium line for lower overall heat transfer coefficient values. Under these conditions, the system is near the reacting and the permeating equilibrium but far from thermal equilibrium. For lower overall heat transfer coefficient values, the generic reaction path departs from isothermal behavior, the optimal path, and approaches the equilibrium curve. This behavior goes to a lower temperature and, as a consequence, implies both a reduction in

$(\Delta X)_T$ and a significant decrease in the reaction rate than that possible under isothermal (optimal) conditions for both MR and TR at the respective *equilibria*.

Membrane thickness variations have different effects on the reaction paths of the annular and the lumenal MRs. Figure 9.55 reports the results for both configurations when the membrane thickness was varied from 0.1 to 100 $\delta(\delta=7.5\ \mu m)$. The thickness variation has a small effect on the reaction path of the lumenal MR. The ten-fold reduction of the membrane thickness did not result in any variation in the reaction path. When the thickness of the palladium film was 10δ, the reaction path is near the previous reaction path. This is due to the low hydrogen permeation flux that is a consequence of the closeness of the reaction path to equilibrium (curve $I = 10$). In this case, the rate-determining step is the heat transfer between the furnace and the reaction volume. Thickness reduction of the palladium film shows a lower minimum temperature as a consequence of the better permeation in the annular MR. Otherwise, the reactor follows the same path as the MR with a membrane thickness equal to δ. The situation is very different when the membrane thickness is increased to 10δ. A temperature minimum higher than that of the TR is present: the conversion is analogous, but the heat capacity of the reaction stream is

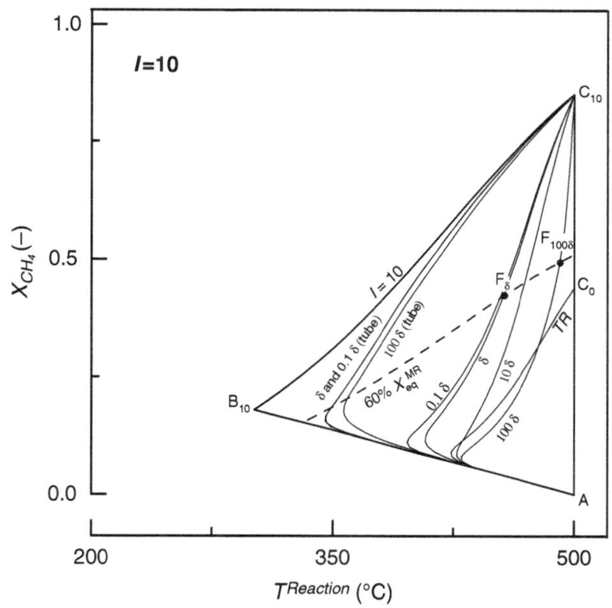

Figure 9.55 Effect of membrane thickness on the MR path. The curves indicated with 0.1δ, δ, 10δ, and 100δ are the reaction paths of the annular MR with different values of the membrane thickness. The curves indicated with δ and 0.1δ (lumen (tube)) and 100δ (lumen (tube)) are the reaction paths of the lumenal (tubular) MR with different values of the membrane thickness. *TR* is the TR path. The dashed curve represents the conditions of 60% of the equilibrium of an MR system. $\delta = 7.5\ \mu m$; $T^{Feed} = T^{Sweep} = 500\ °C$, $P^{Reaction} = P^{Permeation} = 100$ kPa, $m = 3$, $Q_{CH_4}^{Feed} = 200\ cm^3(STP)/min$; $Q^{Sweep} = 10Q_{CH_4}^{Feed}$, $U^{Shell} = 227\ W/(m^2\ K)$, and $U^{Membrane} = 2.4\ W/(m^2\ K)$. Reprinted from Barbieri *et al.* [100] with permission from American Chemical Society.

reduced because of hydrogen permeation. Changes in the reaction path are greater when the membrane thickness is 100 δ. The effects of membrane thickness variation are more evident for the annular MR, where the hydrogen permeation flux has greater changes. Heat exchange and reaction-permeation have the same importance for the system with a membrane thickness equal to δ. When the membrane is thicker, for example, 100 times, the importance of heat exchange is reduced, so the reaction and the permeation increase their own effects on the system performance. The temperature difference between the furnace and the reaction stream for a membrane thickness equal to 100δ (distance of the point $F_{100\ \delta}$ from the line AC_{10}, Fig. 9.55) is much lower than the corresponding value for a membrane thickness equal to δ (distance of the point F_{δ} from the line AC10). Both points, F_{δ} and $F_{100\ \delta}$, are at the same distance from the equilibrium of the membrane system. The reaction path does not tend toward isothermal (optimal) behavior but toward that of the TR when the membrane thickness is increased. In fact, the system performances are not better: it is necessary to have a greater membrane area and catalyst bed to achieve the final point C_{10}.

The effect of reaction pressure on MRs performance is proposed by means of a simulation of catalytic dehydrogenation of ethylbenzene to styrene in a Pd MR. Figure 9.56 shows the calculated conversion of ethylbenzene as a function of the reaction pressure and the sweep factor. Increasing both parameters the ethylbenzene conversion increases, but at higher reaction pressure the increase with I is higher.

Figure 9.57 reports styrene selectivity as a function of the reaction pressure and the sweep factor. Increasing the ethylbenzene sweep gas flow rate the selectivity increases, while it decreases increasing the reaction pressure.

It has to be emphasized that in Figs. 9.56 and 9.57 with increasing reaction pressure the pressure difference between retentate and permeate, and hence the driving force of hydrogen removal, also increases. However, this is not the main reason for the observed increase of conversion/yield. Figure 9.58 shows the calculated conversion of ethylbenzene as a function of both the reaction and permeate sides pressure, in the case of evacuation of permeate gas. Increasing the reaction pressure and decreasing the permeate pressure (in other words, increasing the transmembrane pressure difference) the ethylbenzene conversion increases.

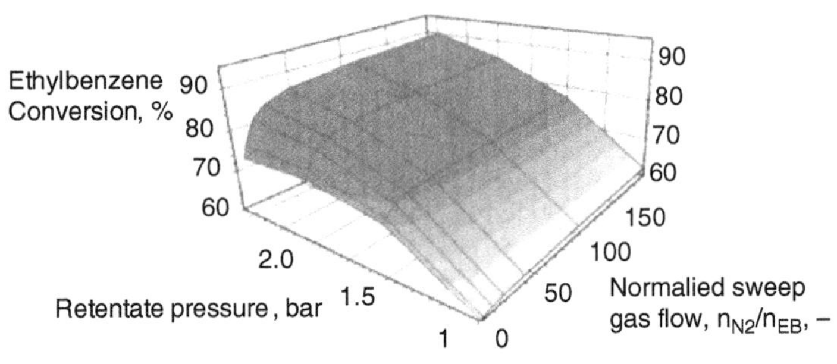

Figure 9.56 Ethylbenzene conversion versus sweep gas flow and reaction pressure. Reprinted from Hermann et al. [73] with permission from Elsevier.

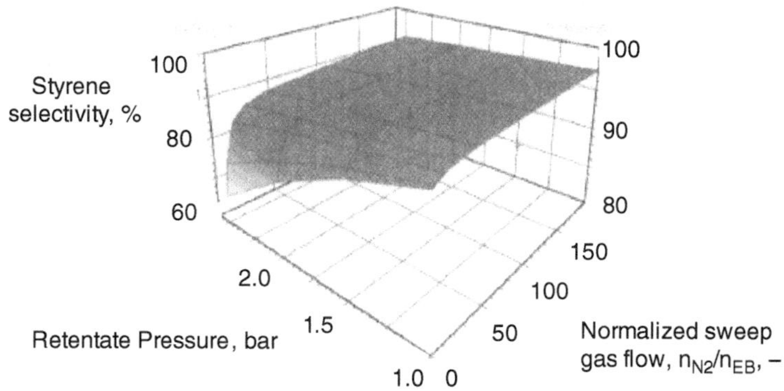

Figure 9.57 Styrene selectivity versus sweep gas flow and reaction pressure. Reprinted from Hermann *et al.* [73] with permission from Elsevier.

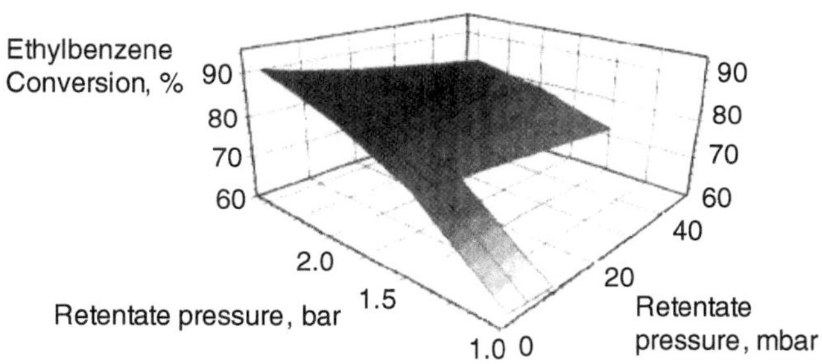

Figure 9.58 Ethylbenzene conversion versus permeate and reaction pressure in the case of evacuation of the permeate gas. Reprinted from Hermann *et al.* [73] with permission from Elsevier.

Figure 9.59 reports styrene selectivity as a function of both the reaction and the permeate side pressure, in the case of the evacuation of permeate gas. The selectivity increases reducing the permeate side, whereas no significant selectivity changes are observed changing the reaction pressure.

In Fig. 9.60, ethylbenzene conversion is plotted versus the reaction pressure and compared with TR, in the case of excess of sweep gas (partial pressure of H_2 in the permeate gas is zero). The plot also reports styrene selectivity for both MR and TR, showing the positive effect of the selective removal of H_2 in the presence of competitive reactions.

5.2.2. 2D models

Regarding nonisothermal 2D mathematical models, the behavior of an adiabatic/wall-cooled MR was also proposed by means of some experimental results. Simulations in the adiabatic mode of operation showed large temperature gradients along the reactor length. Because of the relatively faster methane combustion reaction rates compared with the reforming reaction rate, a temperature overshoot of almost 200–300 K over a length <1 cm results. By cooling/heating the reactor wall with

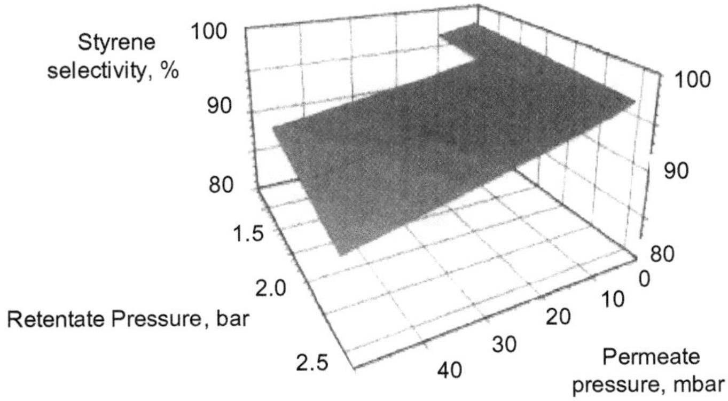

Figure 9.59 Styrene selectivity versus permeate and reaction pressure in the case of evacuation of the permeate gas. Reprinted from Hermann *et al.* [73] with permission from Elsevier.

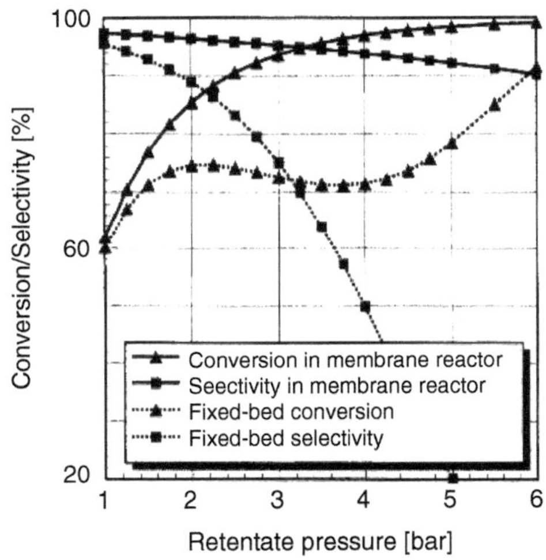

Figure 9.60 Conversion and selectivity reaction as a function of reaction pressure. Reprinted from Hermann *et al.* [73] with permission from Elsevier.

a high sweep gas rate, the membrane wall could be maintained at a constant temperature. The effect of a constant membrane wall temperature on the overall reactor performance was investigated. If the membrane wall temperature is maintained at a constant value of 873 K large radial temperature gradients occur, as is shown in Fig. 9.61, particularly at the inlet of the reactor, because of the faster methane combustion reaction rate compared with MSR. These large temperature gradients are undesirable from a catalyst stability point of view. The reactor performance in terms of methane combustion, H_2 yield, and CO selectivity in the exhaust are comparable to the isothermal operation mode, because the hydrogen permeation was comparable in these cases. Nevertheless, the simulation results show that the

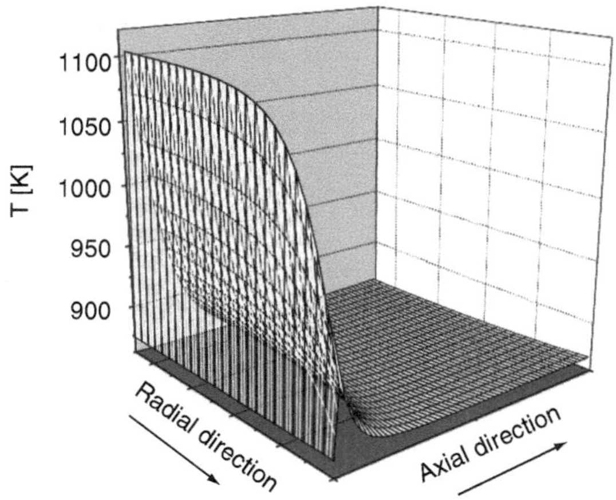

Figure 9.61 2D temperature distribution in the packed bed MR for cooled membrane wall at 873 K. Reprinted from Tiemersma *et al.* [92] with permission from Elsevier.

large temperature gradients cannot be avoided, not even with an infinitely high sweep gas rate (infinite heat capacity → isothermal model).

ACKNOWLEDGMENTS

The "Ministero degli Affari Esteri, Direzione Generale per la Promozione e la Cooperazione Culturale" and "Ministero dell' Istruzione, Università e Ricerca" Progetto "FIRB-CAMERE RBNE03JCR5 – Nuove membrane catalitiche e reattori catalitici a membrana per reazioni selettive come sistemi avanzati per uno sviluppo sostenibile" are gratefully acknowledged for the financial support.

REFERENCES

[1] Drioli, E., and Romano, M. (2001). Progress and new perspectives on integrated membrane operations for sustainable industrial growth. *Ind. Eng. Chem. Res.* **40,** 1277–1300.
[2] Stankiewicz, A., and Moulijn, J. A. (2002). Process intensification. *Ind. Eng. Chem. Res.* **41,** 1920–1924.
[3] Kikuchi, E., Uemiya, S., Sato, N., Inoue, H., Ando, H., and Matsuda, T. (1989). Membrane reactor using microporous glass-supported thin film of palladium. Application to the water–gas shift reaction. *Chem. Lett.* **3,** 489–492.
[4] Uemiya, S., Sato, N., Inoue, H., Ando, H., and Kikuchi, E. (1991). The water–gas shift reaction assisted by palladium membrane reactor. *Ind. Eng. Chem. Res.* **30,** 585–589.
[5] Basile, A., Violante, V., Santella, F., and Drioli, E. (1995). Membrane integrate system in the fusion reactor cycle. *Catal. Today* **25,** 321–326.
[6] Violante, V., Basile, A., and Drioli, E. (1995). Composite catalytic membrane reactor analysis for the water gas shift reaction in the tritium fusion fuel cycle. *Fusion Eng. Des.* **30,** 217–223.
[7] Criscuoli, A., Basile, A., and Drioli, E. (2000). An analysis of the performance of membrane reactors for the water–gas shift reaction using gas feed mixtures. *Catal. Today* **56,** 53–64.
[8] Barbieri, G., and Bernardo, P. (2004). Experimental evaluation of hydrogen production by membrane reaction. CarbonDioxide Capture for Storage in Deep Geologic Formations—Results from the CO_2 Capture Project. Vol. 1, pp. 385–408. Chapter 22. Elsevier, Oxford, UK.

[9] Onstot, W. J., Minet, R. G., and Tsotsis, T. T. (2001). Design aspects of membrane reactors for dry reforming of methane for the production of hydrogen. *Ind. Eng. Chem. Res.* **40,** 242–251.
[10] Sjardin, M., Damen, K. J., and Faaij, A. P. C. (2006). Techno-economic prospects of small-scale membrane reactors in a future hydrogen-fuelled transportation sector. *Energy* **31**(14), 2187–2219.
[11] Bracht, M., Alderliesten, P. T., Kloster, R., Pruschek, R., Haupt, G., Xue, E., *et al.* (1997). Water gas shift membrane reactor for CO_2 control in IGCC systems: Techno-economic feasibility study. *Energy Convers. Manage.* **38,** S159–S164.
[12] Yeung, K. L., Zhang, X., Lau, W. N., and Martin-Aranda, R. (2005). Experiments and modeling of membrane microreactors. *Catal. Today* **110**(1–2), 26–37.
[13] Dudukovic, M. P. (1999). Trends in catalytic reaction engineering. *Catal. Today* **48**(1–4), 5–15.
[14] Sun, J., Desjardins, J., Buglass, J., and Liu, K. (2005). Noble metal water gas shift catalysis: Kinetics study and reactor design. *Int. J. Hydrogen Energy* **30**(11), 1259–1264.
[15] Majumder, D., and Broadbelt, L. J. (2006). A multiscale scheme for modeling catalytic flow reactors. *AIChE J.* **52**(12), 4214–4228.
[16] Collins, J. P., Way, J. D., and Kraisuwansarn, N. (1993). A mathematical model of a catalytic membrane reactor for the decomposition of NH_3. *J. Membr. Sci.* **77**(2–3), 265–282.
[17] Chmielewski, D., Ziaka, Z., and Manousiouthakis, V. (1999). Conversion targets for plug flow membrane reactors. *Chem. Eng. Sci.* **54,** 2979–2984.
[18] Belgued, M., Pareja, P., Amariglio, A., and Amariglio, H. (1991). Conversion of methane into higher hydrocarbons on platinum. *Nature* **352,** 789–790.
[19] Koerst, T., Deelen, M. J. G., and Van Santen, R. A. (1992). Hydrocarbon formation from methane by a low-temperature two-step reaction sequence. *J. Catal.* **138,** 101–114.
[20] Gryaznov, V. (1999). Membrane catalysis. *Catal. Today* **51,** 391–395.
[21] Armor, J. N. (1998). Applications of catalytic inorganic membrane reactors to refinery products. *J. Membr. Sci.* **147,** 217–233.
[22] Coronas, J., and Santamaria, J. (1999). Catalytic reactors based on porous ceramic membranes. *Catal. Today* **51,** 377–389.
[23] Sirkar, K. K., Shanbhag, P. V., and Kovvali, A. S. (1999). Membrane in a reactor: A functional perspective. *Ind. Eng. Chem. Res.* **38,** 3715–3727.
[24] Drioli, E., Basile, A., and Criscuoli, A. (2000). High temperature membrane reactors for clean productions. *Clean Prod. Processes* **2,** 179–186.
[25] Julbe, A., Farrusseng, D., and Guizard, C. (2001). Porous ceramic membranes for catalytic reactors—Overview and new ideas. *J. Membr. Sci.* **181,** 3–20.
[26] Tosti, S., Bettinali, L., and Violante, V. (2000). Rolled thin Pd and Pd–Ag membranes for hydrogen separation and production. *Int. J. Hydrogen Energy* **25,** 319–325.
[27] Belgued, M., Amariglio, A., Pareja, P., and Amariglio, H. (1996). Oxygen-free conversion of methane to higher alkanes through an isothermal two-step reaction on ruthenium. *J. Catal.* **161,** 282–291.
[28] She, Y., Dardas, Z., Gummalla, M., Vanderspurt, T., and Emerson, S. (2005). Integrated water gas shift (WGS) Pd membrane reactors for compact hydrogen production systems from reforming of fossil fuels. *ACS Div. Fuel Chem. Prepr.* **50**(2), 561–569.
[29] Xu, Z. Q., Chen, Q. L., and Lu, G. Z. (1999). The use of membrane reactor in the reaction of ethylbenzene dehydrogenation to styrene. *Petrochemical Technol.* **28**(6), 362–371.
[30] Xu, Z., Chen, Q., and Lu, G. (1999). Use of membrane reactor in reaction of ethylbenzene dehydrogenation to styrene. *Shiyou Huagong/Petrochemical Technology* **28**(6), 358–362.
[31] Assaf, E. M., Jesus, C. D. F., and Assaf, J. M. (1998). Mathematical modelling of methane steam reforming in a membrane reactor: An isothermic model. *Braz. J. Chem. Eng.* **15**(2), 160–166.
[32] Lin, S. T., Chen, Y. H., Yu, C. C., Liu, Y. C., and Lee, C. H. (2006). Dynamic modeling and control structure design of an experimental fuel processor. *Int. J. Hydrogen Energy* **31**(3), 413–426.
[33] Lin, Y. M., Liu, S. L., Chuang, C. H., and Chu, Y. T. (2003). Effect of incipient removal of hydrogen through palladium membrane on the conversion of methane steam reforming: Experimental and modelling. *Catal. Today* **82**(1–4), 127–139.

[34] Capobianco, L., Del Prete, Z., Schiavetti, P., and Violante, V. (2006). Theoretical analysis of a pure hydrogen production separation plant for fuel cells dynamical applications. *Int. J. Hydrogen Energy* **31,** 1079–1090.
[35] Abashar, M. E. E., Alhumaizi, K. I., and Adris, A. M. (2003). Investigation of methane-steam reforming in fluidized bed membrane reactors. *Chem. Eng. Res. Des.* **81**(2), 251–258.
[36] Tsotsis, T. T., Champagnie, A. M., Vasileiadis, S. P., Ziaka, Z. D., and Minet, R. G. (1992). Packed bed catalytic membrane reactors. *Chem. Eng. Sci.* **47**(9–11), 2903–2908.
[37] Adris, A. M., Lim, C. J., and Grace, J. R. (1997). The fluidized-bed membrane reactor for steam methane reforming: Model verification and parametric study. *Chem. Eng. Sci.* **52**(10), 1609–1622.
[38] Saracco, G., Neomagus, H. W. J. P., Versteeg, G. F., and Van Swaaij, W. P. M. (1999). High-temperature membrane reactors: Potential and problems. *Chem. Eng. Sci.* **54,** 1997–2017.
[39] Paturzo, L., Basile, A., and Drioli, E. (2002). High temperature membrane reactors and integrated membrane operations. *Rev. Chem. Eng.* **18**(6), 511–551.
[40] Shu, J., Grandjean, B. P. A., Van Neste, A., and Kaliaguine, S. (1991). Catalytic palladium-based membrane reactors: A review. *Can. J. Chem. Eng.* **69,** 1036–1060.
[41] Lewis, F. A. (1967). Palladium Hydrogen System London, New York: Academic Press, London, New York.
[42] Ward, T. L., and Dao, T. (1999). Model of hydrogen permeation behavior in palladium membranes. *J. Membr. Sci.* **153,** 211–231.
[43] Scura, F. (2007). Private communication, material of PhD thesis still in progress.
[44] Arstad, B., Venvik, H., Klette, H., Walmsley, J. C., Tucho, W. M., Holmestad, R., *et al.* (2006). Studies of self-supported 1.6 μm Pd/23 wt.% Ag membranes during and after hydrogen production in a catalytic membrane reactor. *Catal. Today* **118**(1–2), 63–72.
[45] Klette, H., and Bredesen, R. (2005). Sputtering of very thin palladium-alloy hydrogen separation membranes. *Membr. Technol.* **2005**(5), 7–9.
[46] Shu, J., Grandjean, B. P. A., Kaliaguine, S., Ciavarella, P., Giroir-Fendler, A., and Dalmon, J. A. (1997). Gas permeation and isobutene dehydrogenation over very thin Pd/ceramic membranes. *Can. J. Chem. Eng.* **75,** 712.
[47] Collins, J. P., and Way, J. D. (1993). Preparation and characterization of a composite palladium-ceramic membrane. *Ind. Eng. Chem. Res.* **32,** 3006–3013.
[48] Yan, S., Maeda, H., Kusakabe, K., and Morooka, S. (1994). Thin palladium membrane formed in support pores by metallorganic chemical vapor deposition method and application to hydrogen separation. *Ind. Eng. Chem. Res.* **33,** 616–622.
[49] Uemiya, S., Sato, N., Ando, H., Kude, Y., Atsuda, T., and Kikuchi, E. (1991). Separation of hydrogen through palladium thin film supported on a porous glass tube. *J. Membr. Sci.* **56,** 303–313.
[50] Uemiya, S., Sato, N., Ando, H., Matsuda, T., and Kikuchi, E. (1991). Steam reforming of methane in a hydrogen-permeable membrane reactor. *Appl. Catal.* **67,** 223–230.
[51] Hurlbert, R. C., and Konecny, J. O. (1961). Diffusion of hydrogen through palladium. *J. Chem. Phys.* **34,** 655–658.
[52] Xomeritakis, G., and Lin, Y. S. (1996). Fabrication of a thin palladium membrane supported in a porous ceramic substrate by chemical vapor deposition. *J. Membr. Sci.* **120,** 261–272.
[53] Li, A., Liang, W., and Hughes, R. (1998). Characterisation and permeation of palladium/ stainless steel composite membranes. *J. Membr. Sci.* **149,** 259–268.
[54] Li, A., Liang, W., and Hughes, R. (2000). The effect of carbon monoxide and steam on the hydrogen permeability of a Pd/stainless steel membrane. *J. Membr. Sci.* **165,** 135–141.
[55] Tong, J., Suda, H., Haraya, K., and Matsumura, Y. (2005). A novel method for the preparation of thin dense Pd membrane on macroporous stainless steel tube filter. *J. Membr. Sci.* **260,** 10–18.
[56] Bredesen, R., Jordal, K., and Bolland, O. (2004). High-temperature membranes in power generation with CO_2 capture. *Chem. Eng. Proc.* **43,** 1129–1158.
[57] Caravella, A., Barbieri, G., and Drioli, E. (2006). H_2 permeation through thin Pd-alloy membranes deposited on a multilayer porous support. *In* "9th International Conference on Inorganic Membranes, ICIM9, " June 25–29, Lillehammer, Norway.

[58] Madia, G., Barbieri, G., and Drioli, E. (1999). Theoretical and experimental analysis of methane steam reforming in a membrane reactor. *Can. J. Chem. Eng.* **77,** 698–706.
[59] Caravella, A., and Barbieri, G. (2006). Modelling and simulation of hydrogen permeation through supported Pd-Alloy membranes with a multicomponent approach. *Priv. Commun* pp.112–159.
[60] Marigliano, G., Barbieri, G., and Drioli, E. (2003). Equilibrium conversion for a Pd-alloy membrane reactor. Dependence on the temperature and pressure. *Chem. Eng. Proc.* **42**(3), 231–236.
[61] Barbieri, G., Scura, F., and Drioli, E. (2006). Equilibrium of a Pd-alloy membrane reactor. *Desalination* **200**(1–3), 679–680.
[62] Kumar, S., Shankar, S., Shah, P. R., and Kumar, S. (2006). A comprehensive model for catalytic membrane reactor. *Int. J. Chem. Reactor Eng.* **4,** 1–29.
[63] Tayakout, M., Bernauer, B., Toure, Y., and Sanchez, J. (1995). Modelling and simulation of a catalytic membrane reactor. *Simul. Pract. Th.* **2**(4–5), 205–219.
[64] Koukou, M. K., Papayannakos, N., Markatos, N. C., Bracht, M., and Alderliesten, P. T. (1998). Simulation tools for the design of industrial-scale membrane reactors. *Chem. Eng. Res. Des.* **76** (A8), 911–920.
[65] Markatos, N. C., Vogiatzis, E., Koukou, M. K., and Papayannakos, N. (2005). Membrane reactor modelling: A comparative study to evaluate the role of combined mass and heat dispersion in large-scale adiabatic membrane modules. *Chem. Eng. Res. Des.* **83**(10 A), 1171–1178.
[66] Ettouney, H. M., Masiar, B., Bouhamra, W. S., and Hughes, R. (1996). High temperature CO shift conversion (HTSC) using catalytic membrane reactors. *Chem. Eng. Res. Des.* **74**(6), 649–657.
[67] Birdsell, S. A., and Willms, R. S. (1995). Modeling and data analysis of a palladium membrane reactor for tritiated impurities cleanup. *Fusion Technol.* **28,** 530–537.
[68] Chan, P. P. Y., Vanidjee, K., Adesina, A. A., and Rogers, P. L. (2000). Modeling and simulation of non-isothermal catalytic packed bed membrane reactor for H_2S decomposition. *Catal. Today* **63**(2–4), 379–385.
[69] Marigliano, G., Barbieri, G., and Drioli, E. (2001). Effect of energy transport on a palladium-based membrane reactor for methane steam reforming process. *Catal. Today* **67,** 85–99.
[70] Barbieri, G., Brunetti, A., Granato, T., Bernardo, P., and Drioli, E. (2005). Engineering evaluations of a catalytic membrane reactor for the water gas shift reaction. *Ind. Eng. Chem. Res.* **44,** 7676–7683.
[71] Basile, A., Chiappetta, G., Tosti, S., and Violante, V. (2001). Experimental and simulation of both Pd and Pd/Ag for a water gas shift membrane reactor. *Sep. Purif. Technol.* **25,** 549–571.
[72] Basile, A., Paturzo, L., and Gallucci, F. (2003). Co-current and counter-current modes for water gas shift membrane reactor. *Catal. Today* **82,** 275–281.
[73] Hermann, C. H., Quicker, P., and Dittmeyer, R. (1997). Mathematical simulation of catalytic dehydrogenation of ethylbenzene to styrene in a composite palladium membrane reactor. *J. Membr. Sci.* **136,** 161–172.
[74] Itoh, N., and Govind, R. (1989). Combined oxidation and dehydrogenation in a palladium membrane reactor. *Ind. Eng. Chem. Res.* **28,** 1554–1557.
[75] Itoh, N., and Wu, T. H. (1997). An adiabatic type of palladium membrane reactor for coupling endothermic and exothermic reactions. *J. Membr. Sci.* **124,** 213–222.
[76] Itoh, N., and Sathe, A. M. (1997). Hydrogen transport from gas to liquid phase through a palladium membrane. *J. Membr. Sci.* **137,** 251–259.
[77] Hara, S., Sakaki, K., and Itoh, N. (1999). Decline in hydrogen permeation due to concentration polarization and CO hindrance in a palladium membrane reactor. *Ind. Eng. Chem. Res.* **38,** 4913–4918.
[78] Itoh, N. (1992). Development of a one-side uniform model for palladium membrane reactors. *J. Chem. Eng. Jap.* **25**(3), 336–338.
[79] Chen, Z., Yan, Y., and Elnashaie, S. S. E. H. (2003). Modeling and optimization of a novel membrane reformer for higher hydrocarbons. *AIChE J.* **49**(5), 1250–1265.

[80] Tsai, C. Y., Ma, Y. H., Moser, W. R., and Dixon, A. G. (1995). Modeling and simulation of a nonisothermal catalytic membrane reactor. *Chem. Eng. Comm.* **134,** 107–132.

[81] Fernandes, F. A. N., and Soares,, A. B., Jr. (2006). Methane steam reforming modeling in a palladium membrane reactor. Short Communication. *Fuel* **85,** 569–573.

[82] Fernandes, F. A. N., and Soares,, A. B. (2006). Modeling of methane steam reforming in a palladium membrane reactor. *Lat. Am. Appl. Res.* **36**(3), 155–161.

[83] Barbieri, G., and Di Maio, F. P. (1997). Simulation of the methane steam reforming process in a catalytic pd-membrane reactor. *Ind. Eng. Chem. Res.* **36,** 2121–2127.

[84] Gallucci, F., Paturzo, L., and Basile, A. (2004). A simulation study of the steam reforming of methane in a dense tubular membrane reactor. *Int. J. Hydrogen Energy* **29,** 611–617.

[85] Harold, M. P., Nair, B., and Kolios, G. (2003). Hydrogen generation in a Pd membrane fuel processor: Assessment of methanol-based reaction systems. *Chem. Eng. Sci.* **58,** 2551–2571.

[86] Lattner, J. R., and Harold, M. P. (2004). Comparison of conventional and membrane reactor fuel processors for hydrocarbon-based PEM fuel cell systems. *Int. J. Hydrogen Energy* **29,** 393–417.

[87] Johannessen, E., and Jordal, K. (2005). Study of a H_2 separating membrane reactor for methane steam reforming at conditions relevant for power processes with CO_2 capture. *Energy Convers. Manage.* **46,** 1059–1071.

[88] Elnashaie, S. S. E. H., Moustafa, T., Alsoudani, T., and Elshishini, S. S. (2000). Modeling and basic characteristics of novel integrated dehydrogenation-hydrogenation membrane catalytic reactors. *Comp. Chem. Eng.* **24**(2–7), 1293–1300.

[89] Fernandes, F. A. N., Sousa, J. F., Souza, C. P., and Rodrigues, S. (2005). Modeling of partial oxidation of methane in a membrane reactor. *In* "7th World Congress of Chemical Engineering, GLASGOW2005, incorporating the 5th European Congress of Chemical Engineering" p. 258.

[90] Alfadhel, K., and Kothare, M. V. (2005). Modeling of multicomponent concentration profiles in membrane microreactors. *Ind. Eng. Chem. Res.* **44,** 9794–9804.

[91] Alfadhel, K., and Kothare, M. V. (2005). Microfluidic modelling and simulation of flow in membrane microreactors. *Chem. Eng. Sci.* **60,** 2911–2926.

[92] Tiemersma, T. P., Patil, C. S., van Sint Annaland, M., and Kuipers, J. A. M. (2006). Modelling of packed bed membrane reactors for autothermal production of ultrapure hydrogen. *Chem. Eng. Sci.* **61,** 1602–1616.

[93] Wieland, S. A., Melin, T., and Lamm, A. (2002). Membrane reactors for hydrogen production. *Chem. Eng. Sci.* **57**(9), 1571–1576.

[94] Rahimpour, M. R., and Ghader, S. (2003). Theoretical investigation of a Pd-membrane reactor for methanol synthesis. *Chem. Eng. Technol.* **26**(8), 902–907.

[95] Smit, J., van Sint Annaland, M., and Kuipers, J. A. M. (2005). Feasibility study of a reverse flow catalytic membrane reactor with porous membranes for the production of syngas. *Chem. Eng. Sci.* **60,** 6971–6982.

[96] Deshmukh, S. A. R. K., Heinrich, S., Mörl, L., van Sint Annaland, M., and Kuipers, J. A. M. (2007). Membrane assisted fluidized bed reactors: Potentials and hurdles. *Chem. Eng. Sci.* **62**(1–2), 416–436.

[97] Basile, A., Paturzo, L., and Laganà, F. (2001). The partial oxidation of methane to syngas in a palladium membrane reactor: Simulation and experimental studies. *Catal. Today* **67,** 65–75.

[98] Raich, B. A., and Foley, H. C. (1995). Supra-equilibrium conversion in palladium membrane reactors: Kinetic sensitivity and time dependence. *Applied Catalysis A.: General* **129,** 167–188.

[99] Barbieri, G., Violante, V., Di Maio, F. P., Criscuoli, A., and Drioli, E. (1997). Methane steam reforming analysis in a Palladium-based catalytic membrane reactor. *Ind. Eng. Chem. Res.* **36,** 3369–3374.

[100] Barbieri, G., Marigliano, G., Perri, G., and Drioli, E. (2001). Conversion-temperature diagram for a Palladium membrane reactor. Analysis of an endothermic reaction: Methane steam reforming. *Ind. Eng. Chem. Res.* **40,** 2017–2026.

[101] Hara, S., Haraya, K., Barbieri, G., and Drioli, E. (2007). Reaction rate profiles in long palladium membrane reactors for methane steam reforming. Desalination, accepted.

[102] Brunetti, A., Barbieri, G., and Drioli, E. (2006). Engineering approach for the WGS reaction in a catalytic membrane reactor. *In* "4th Chemical Engineering Conference for Collaborative Research in Mediterranean Countries," January 9–11, 2006. *Israel.*

[103] Brunetti, A., Caravella, A., Barbieri, G., and Drioli, E. (2006). Simulation study of water gas shift reaction in a catalytic membrane reactor. *In* "First Mediterranean Congress Chemical Engineering for Environment," October 4–6, San Servolo, Venice.

[104] Basile, A., Paturzo, L., and Vazzana, A. (2003). Membrane reactor for the production of hydrogen and higher hydrocarbons from methane over Ru/Al_2O_3 catalyst. *Chem. Eng. J.* **93,** 31–39.

[105] Gallucci, F., Comite, A., Capannelli, G., and Basile, A. (2006). Steam reforming of methane in a membrane reactor: An industrial case study. *Ind. Eng. Chem. Res.* **45**(9), 2994–3000.

[106] Elnashaie, S., and Abdelhady, F. (2004). Plant efficiency: To hydrogen efficiency and beyond. *Chem. Eng.* **755,** 33–34.

[107] Elnashaie, S. S. E. H., Abdallah, B. K., Elshishini, S. S., Alkhowaiter, S., Noureldeen, M. B., and Alsoudani, T. (2001). On the link between intrinsic catalytic reactions kinetics and the development of catalytic processes. Catalytic dehydrogenation of ethylbenzene to styrene. *Catal. Today* **64**(3–4), 151–162.

[108] Moustafa, T. M., and Elnashaie, S. S. E. H. (2000). Simultaneous production of styrene and cyclohexane in an integrated membrane reactor. *J. Membr. Sci.* **178**(1–2), 171–184.

[109] Prasad, P., and Elnashaie, S. S. E. H. (2003). Coupled steam and oxidative reforming for hydrogen production in a novel membrane circulating fluidized-bed reformer. *Ind. Eng. Chem. Res.* **42**(20), 4715–4722.

[110] Becker, Y. L., Dixon, A. G., Moser, W. R., and Ma, Y. H. (1993). Modelling of ethylbenzene dehydrogenation in a catalytic membrane reactor. *J. Membr. Sci.* **77**(2–3), 233–244.

[111] Dixon, A. G. (2001). Analysis of intermediate product yield in distributed-feed nonisothermal tubular membrane reactors. *Catal. Today* **67,** 189–203.

[112] Gobina, E., Hou, K., and Hughes, R. (1995). Mathematical analysis of ethylbenzene dehydrogenation: Comparison of microporous and dense membrane systems. *J. Membr. Sci.* **105,** 163–176.

[113] Gobina, E., Hou, K., and Hughes, R. (1995). Ethane dehydrogenation in a catalytic membrane reactor coupled with a reactive sweep gas. *Chem. Eng. Sci.* **50,** 2311–2319.

[114] Oklany, J. S., Hou, R., and Hughes, R. (1998). A simulative comparison of dense and microporous membrane reactors for the steam reforming of methane. *Appl. Catal. A: General* **170,** 13–22.

[115] Liang, W., and Hughes, R. (2005). The catalytic dehydrogenation of isobutane to isobutene in a palladium/silver composite membrane reactor. *Catal. Today* **104,** 238–243.

[116] Chiappetta, G., Clarizia, G., and Drioli, E. (2006). Analysis of safety aspects in a membrane reactor. *Desalination* **193,** 267–279.

[117] Chiappetta, G., Clarizia, G., and Drioli, E. (2006). Design of an integrated membrane system for a high level hydrogen purification. *Chem. Eng. J.* **124,** 29–40.

[118] Deshmukh, S. A. R. K., Laverman, J. A., Van Sint Annaland, M., and Kuipers, J. A. M. (2005). Development of a membrane-assisted fluidized bed reactor. 2. Experimental demonstration and modeling for the partial oxidation of methanol. *Ind. Eng. Chem. Res.* **44**(16), 5966–5976.

[119] Shu, J., Grandjean, B. P. A., and Kaliaguine, S. (1994). Methane steam reforming in asymmetric Pd- and Pd-Ag/porous SS membrane reactor. *Appl. Catal. A* **119,** 305–325.

[120] Itoh, N., Xu, W. C., and Haraya, K. (1994). Radial mixing diffusion of hydrogen in a packed-bed type of palladium membrane reactor. *Ind. Eng. Chem. Res.* **33,** 197–202.

CHAPTER TEN

Oxygen and Hydrogen Separation Membranes Based on Dense Ceramic Conductors

Marie-Laure Fontaine,* Truls Norby,† Yngve Larring,* Tor Grande,‡ *and* Rune Bredesen*

Contents

* SINTEF Materials and Chemistry, NO-0314 Oslo, Norway
† Department of Chemistry, University of Oslo, Centre for Materials Science and Nanotechnology, Gaustadalleen 21, NO-0349 Oslo, Norway
‡ Department of Material Science and Engineering, Norwegian University of Science and Technology (NTNU) Trondheim, Norway.

Membrane Science and Technology, Volume 13
ISSN 0927-5193, DOI: 10.1016/S0927-5193(07)13010-2

1. Introduction

Oxygen and hydrogen are two of the world's most important chemicals and as such have great commercial interest. Hence, their methods of production and traditional use in industry, for example in petrochemistry, are well established. In recent years, however, attention about sustainable development has grown and consequently the production technology for these major chemicals is desirable to improve. Furthermore, a shift in focus from traditional use to application in sustainable power generation has added a new dimension to the interest for these chemicals. On all continents we see that research and development programs devoted to clean energy now aim at bringing to the market new technology for sustainable oxygen and hydrogen production. Inorganic membranes that selectively separate oxygen and hydrogen from gas streams could be the basis for such energy efficient clean technology. Inorganic hydrogen-selective membranes are currently developed along several lines, that is, as microporous, metallic, and dense ceramic type. This diversity opens up for applications over a broad window of operation. Differently from hydrogen-selective membranes, research on inorganic oxygen-selective membranes are only focused on the dense ceramic type.

In this presentation, we aim at giving the reader an introduction to the field of dense ceramic membranes for oxygen and hydrogen separation. Our starting point is the presentation of this subject to students participating at the European Membrane Society Summer School held in Jaca, Spain, 2005. First we introduce the theoretical basis for understanding the transport properties in dense oxides. This treatment starts from basic defect chemistry and the equations for flux of oxygen ions and protons are derived. For clarification, we devote sufficient space to the somewhat less commonly treated case of protons in oxides. We then introduce the most important families of oxides and the specific compositions that demonstrate high ionic diffusion rates. We divide between *electrolytes*, which have high ionic diffusion rate only, and the *mixed conductors*, exhibiting high diffusion rates of both ions and electrons. The important issue of materials stability under operation is then discussed taking into consideration the typical membrane working condition characterized by high temperature, the presence of significant chemical, mechanical, and thermal gradients, as well as aggressive chemical components. In the final part of the presentation, several examples of membrane applications are described. These are taken from the main areas of gas separation/production, membrane reactors, solid oxide fuel cell (SOFC), and water vapor electrolysis. The possibility of mass and heat integration in novel process designs and the potentials of dense ceramic membranes in power generation with CO_2 capture are demonstrated. Finally, we survey the latest developments in SOFC and prospects toward high-temperature water electrolysis. The chapter ends with some concluding remarks and suggestions about future research in the field.

2. Theory

This section will deal with the defect chemistry and transport theory of ion-conducting ceramic materials. Emphasis will be on mixed ionic-electronic conduction in oxides for gas separation. Proton and electron mixed conductors—a relatively new field—will be given particular attention.

2.1. Defects

Dense ceramic gas separation membranes can in principle be envisaged based on molecular or atomic transport of oxygen (as O_2 or O), hydrogen (as H_2 or H), and water (H_2O). These would all require defects, namely the molecules or atoms occupying lattice or interstitial sites. However, here we will mostly restrict ourselves to the more commonly encountered and important charged species, and only in oxides. Oxygen permeation then requires so-called ambipolar transport of oxide ions and electrons (in opposite directions). The oxide ions will move by vacancy or interstitial mechanisms, or both. Hydrogen transport requires transport of hydrogen ions and electrons (in the same direction). The hydrogen ions are normally protons existing as hydroxide ions but moving interstitially by jumping from oxide ion to oxide ion. The electron transport can take place by conduction band electrons and valence band holes. It is possible to foresee mixed transport of oxide ions and protons, which—when they move in the same direction—constitutes transport of water. The mobile species we need to consider are thus O^{2-}, H^+, and e^- and their corresponding defects. In addition we have dopants (acceptors or donors) that we dissolve in order to affect the concentrations of the charged mobile defects. From this, the electroneutrality condition comprising the defects of importance reads

$$2\left[v_O^{\bullet\bullet}\right] + \left[OH_O^{\bullet}\right]] + p + [D^{\bullet}] = [A'] + n + 2[O_i''] \qquad (10.1)$$

Here, the parentheses denote volume concentration or molar fraction (but not site fraction), and the defect symbols follow Kröger-Vink notation in which the main symbol denotes a chemical species or a vacancy (v) or a donor (D) or an acceptor (A), the subscripts denote lattice site identified by its host or interstitial site (i), and the superscript denotes positive (•) or negative (/) effective charge.

Many factors may complicate the defect structure compared to the simple defects shown above. These comprise electronic defects being also valence defects, that is, electrons or holes localized at ambivalent ions (normally cations). They further comprise disorder of natively incomplete sublattices. Finally, we mention defect–defect interactions and structural distortions that tend to accompany large defect concentrations. Here we will treat only simple cases, with idealized defects and equilibria.

In general, electronic defects are much more mobile than ionic point defects, and thus, we need high concentrations of the mobile point defect, but we can do with minority concentrations of electronic defects. From the electroneutrality we may extract a number of simplified limiting cases comprising only two defects, and we will analyze some of these in the following paragraphs.

Under reducing conditions oxides may become dominated by oxygen vacancies and electrons:

$$\mathrm{O^{x}o} = \mathrm{v_O^{\bullet\bullet}} + 2\mathrm{e}' + \frac{1}{2}\mathrm{O_2(g)}\ K_2 = K_2^0 \exp\frac{-\Delta H_2^0}{RT} = \frac{[\mathrm{v_O^{\bullet\bullet}}]n^2 p_{\mathrm{O_2}}^{1/2}}{[\mathrm{O_O^x}]} \tag{10.2}$$

$$2[\mathrm{v_O^{\bullet\bullet}}] = n = [\mathrm{O_O^x}]2^{1/3}K_2^{1/3}p\mathrm{o}_2^{-1/6} \tag{10.3}$$

The enthalpy change of the reaction is positive (endothermic), so that the defect concentrations increase with increasing temperature. The oxygen vacancies and electrons provide ambipolar diffusion. Oxides with this defect structure have reducible cations and comprise fluorites like ceria, $CeO_{2-\delta}$, and perovskites like $LaFeO_{3-\delta}$. Chemical expansion due to the varying concentration of oxygen vacancies prevents the use of many such materials in membranes.

Under oxidizing conditions, some oxides dissolve interstitial oxide ions, compensated by electron holes:

$$\frac{1}{2}\mathrm{O_2(g)} = \mathrm{O''_i} + 2\mathrm{h^\bullet}\ K_3 = K_3^0 \exp\frac{-\Delta H_3^0}{RT} = \frac{[\mathrm{O''_i}]p^2}{p_{\mathrm{O_2}}^{1/2}} \tag{10.4}$$

$$2[\mathrm{O''_i}] = p = 2^{1/3}K_3^{1/3}p_{\mathrm{O_2}}^{1/6} \tag{10.5}$$

Examples comprise Ruddlesden-Popper-series perovskite-related oxides $A_{n+1}B_nO_{3n+1}$ such as $La_2NiO_{4+\delta}$ in which extra oxygen may be incorporated in the relatively open AO layers in-between the perovskite-like ABO_3 layers. The defect formation is exothermic in this case, and the defect concentrations increase with decreasing temperature as long as kinetics of oxygen uptake permits. Ambipolar diffusion occurs by the oxygen interstitials and electron holes. Such materials appear to exhibit less chemical expansion than vacancy-dominated systems.

One may also consider systems dominated by oxygen vacancies and interstitials (anion-Frenkel disorder). These would be permeable by the dominant ionic point defects and the minority electronic defects. Examples of this defect structure in candidate materials for membrane applications are not known to us. A similar situation occurs when a natively oxygen-deficient system disorders—the disordered sublattice—may be seen as full of both vacancies and interstitials, while other nomenclatures and viewpoints have also been put forth [1]. Examples comprise fluorite-related δ-Bi_2O_3 and perovskite-related $Ba_2In_2O_5$ and $Sr_2Fe_2O_5$.

The oxygen-deficient case above may be acceptor-doped, which enhances the oxygen vacancy concentration and decreases the electron concentration. We shall simply refer to the acceptors as A, and, in the ionized state, A′. The electroneutrality becomes

$$2[v_O^{\bullet\bullet}] = [A'] = \text{constant} \tag{10.6}$$

By inserting this into Eq. (10.2) we obtain a $p_{O_2}^{-1/4}$ dependency for the minor electron concentration and via that a $p_{O_2}^{1/4}$ dependency for the minor electron hole concentration:

$$n = \frac{K_i}{p} = \frac{2^{1/2}[O_O^x]^{1/2}K_2^{1/2}}{[A']^{1/2}p_{O_2}^{1/4}} \tag{10.7}$$

where $K_i = n \cdot p$ is the equilibrium constant of the intrinsic ionization of electrons over the band gap. Acceptor-doped oxides constitute the largest class of ionic and mixed conducting oxides. Well-known examples comprise fluorites such as ZrO_2 doped with Y or Sc and CeO_2 doped with Sm or Gd, and perovskites such as $LaGaO_3$, $LaFeO_3$, or $LaCoO_3$ in which La is substituted by Ca or Sr. The close to constant concentration of oxygen vacancies reduces the chemical expansion and allows these materials to be used in membranes over large spans in oxygen activity. The ambipolar diffusion couple is oxygen vacancies compensated by transport of electrons or holes or both, depending on p_{O_2} and temperature. In order to have a reasonable concentration of minor electronic defects the materials need to have small band gaps.

In principle, open-structured oxides may be donor-doped to obtain a high concentration of oxygen interstitials, but this is not a well-proven route. Instead, attempts of this most often result in large concentrations of cation vacancies, with unwanted cation transport as result.

In atmospheres containing water or hydrogen, protons will be taken up and dissolve mainly as effectively positive substitutional hydroxide defects, $OH_O^\bullet$. The protons move mostly by jumping between the stationary oxide ions hosts (Grotthuss mechanism), and it is thus often useful and convenient to describe the protons simply as interstitial defects, $H_i^\bullet$, although their positions are not ordinary interstitial sites, but located in the electron cloud of its host oxide ion.

Defect-structure-wise there are two main routes to a mixed proton–electron conducting oxide. The simplest would be to dissolve protons compensated by electrons originating directly from hydrogen gas:

$$H_2(g) + 2O_O^x = 2OH_O^\bullet + 2e'; \; K_4 = \frac{[OH_O^\bullet]^2 n^2}{p_{H_2}[O_O^x]^2} \tag{10.8}$$

This reaction was reported for ZnO already in the 1950s [2], and it was found that protons and electrons were the dominating defects in hydrogen atmospheres, so that the electroneutrality condition and concentrations are expressed as

$$[\mathrm{OH}_\mathrm{O}^\bullet] = n = K_4^{1/4} p_{\mathrm{H}_2}^{1/4} [\mathrm{O}_\mathrm{O}^x]^{1/2} \tag{10.9}$$

It may be shown that one may only hope to achieve this situation when the material already is dominated by electrons (and some positive point defects, such as oxygen vacancies) under hydrogen-free reducing conditions. By increasing p_{H_2}, the native positive defects are replaced by protons. Protons compensated by electrons are, however, hardly known in systems other than ZnO.

Instead one tends to use acceptor-doped oxides in which the concentrations of protons and all other positive defects are enhanced by the doping. The acceptors are most often substitutionally dissolved lower-valent cations, while, in principle, interstitial anions or substitutional higher-valent anions (e.g., N^{3-} substituting O^{2-}) can also be used. Assuming that the acceptors are compensated by oxygen vacancies in the absence of hydrogen-containing gases, the reaction with water vapor is given by

$$\begin{gathered} \mathrm{H_2O(g)} + \mathrm{v}_\mathrm{O}^{\bullet\bullet} + O_o^x = 2\mathrm{OH}_\mathrm{O}^\bullet; \\ K_5 = \exp\frac{\Delta S_5^0}{R}\exp\frac{-\Delta H_5^0}{RT} = \frac{[\mathrm{OH}_\mathrm{o}^{\bullet\bullet}]^2}{[\mathrm{v}_\mathrm{O}^{\bullet\bullet}][\mathrm{O}_\mathrm{O}^x]p_{\mathrm{H_2O}}} \end{gathered} \tag{10.10}$$

The equilibrium constant, K_5, for the reaction, and the limiting electroneutrality condition

$$2[\mathrm{v}_\mathrm{O}^{\bullet\bullet}] + [\mathrm{OH}_\mathrm{O}^\bullet] = [\mathrm{A}'] = \text{constant} \tag{10.11}$$

yield (assuming $[\mathrm{v}_\mathrm{O}^{\bullet\bullet}] + [\mathrm{OH}_\mathrm{O}^\bullet] << [\mathrm{O}_\mathrm{O}^x] \approx [\mathrm{O}]$) a simplified expression for $[\mathrm{OH}_\mathrm{O}^\bullet]$:

$$[\mathrm{OH}_\mathrm{o}^\bullet] = \frac{[\mathrm{O}]K_5 p_{\mathrm{H_2O}}\left(-1 + \sqrt{1 + (8[\mathrm{A}']/[\mathrm{O}]K_5 p_{\mathrm{H_2O}})}\right)}{4} \tag{10.12}$$

where [O] is the concentration of oxygen sites. From this, and knowing the equilibrium constant K_5 (or better its ΔS^0 and ΔH^0), the dependences on temperature, $p_{\mathrm{H_2O}}$ and acceptor concentration [A′] can be calculated. Further, concentrations of other defects can be calculated from the electroneutrality condition or from equilibria with protons. Rearranging Eq. (10.11), we have

$$[\mathrm{v}_\mathrm{O}^{\bullet\bullet}] = \frac{[\mathrm{A}'] - [\mathrm{OH}_\mathrm{O}^{\bullet\bullet}]}{2} \tag{10.13}$$

in which $[OH_O^{\bullet\bullet}]$ can be inserted from Eq. (10.12). The concentration of defect electrons is in turn obtained, for example, from their equilibrium (10.7) with hydrogen and protons, to become

$$n = \frac{K_4^{1/2} p_{H_2}^{1/2} [O]}{[OH_O^{\bullet\bullet}]} = \frac{K_4^{1/2} p_{H_2O}^{1/2} [O]}{K_{H_2O}^{1/2} p_{O_2}^{1/4} [OH_O^{\bullet\bullet}]} \tag{10.14}$$

where, again, $[OH_O^{\bullet\bullet}]$ is inserted from Eq. (10.12), and K_{H_2O} is the equilibrium constant for formation of water vapor. Finally, the concentration of electron holes, p, is found from the intrinsic electronic ionization equilibrium as $p = K_i/n$.

In some acceptor-doped materials, for example, many acceptor-doped $LaMO_3$ ($M =$ transition metal) perovskites with small band gaps, electron holes may become dominating defects at high oxygen activities:

$$p = [A'] = \text{constant} \tag{10.15}$$

These may be mixed conductors through minor concentrations of oxygen vacancies or protons. For instance, via Eq. (10.7) and $K_i = n \cdot p$, the minor concentration of protons becomes

$$[OH_O^{\bullet}] = \frac{K_4^{1/2} [A'][O] p_{H_2}^{1/2}}{K_i} = \frac{K_4^{1/2} [A'][O] p_{H_2O}^{1/2}}{K_i K_{H_2O}^{1/2} p_{O_2}^{1/4}} \tag{10.16}$$

2.2. Diffusivity, mobility, and conductivity: The Nernst–Einstein relation

We have introduced concentrations and relations between concentrations of species of interest in the above sections. The area-specific flux density of a species i resulting from a driving force F is proportional to its concentration and to its mechanical mobility, $j_i = c_i \cdot B_i \cdot F_i$, as we come back to in the next section. For species with an activated diffusion, the self-diffusion coefficient D, mechanical mobility B, charge mobility u, and conductivity σ are linked through the Nernst–Einstein relation:

$$\sigma_i = z_i F c_i u_i = (z_i F)^2 c_i B_i = \frac{(z_i F)^2 c_i D_i}{RT} \tag{10.17}$$

where $D = D_0 \exp - Q_m/RT$ defines the temperature dependence and activation energy Q_m. Note that diffusivity and mobility terms do not contain concentration, while conductivity does. We have here used the Faraday constant $F = 96{,}485$ C/mol, expressing the charge per mole of charges, and the defect concentration c_i must thus be brought into the form of a molar volume concentration. Thus, from molar fractions, $[i]$, as typically derived in the defect chemistry, one needs to multiply by the molar volume density, d_M (mol/cm^3 or mol/m^3): $c_i = [i]d_M$.

For species that migrate by other mechanisms than diffusion (notably itinerant electrons and holes), the diffusion coefficient is undefined, and instead one uses mobilities with temperature dependencies typical of band transport with phonon-, impurity-, or large polaron-limited transport [3].

2.3. General expressions for transport in dense mixed conducting membranes

A force, expressed as the negative (down-hill) gradient in a potential P_i acting on the species i, gives rise to a net flux density j_i (in moles per unit area per second) of that species which is proportional to its mechanical mobility B_i. By letting the potential acting to be the electrochemical potential, we obtain in the one-dimensional case

$$j_i = B_i c_i \frac{-dP_i}{dx} = B_i c_i \frac{-d\eta_i}{dx} = -B_i c_i \left[\frac{d\mu_i}{dx} + z_i F \frac{d\phi}{dx}\right] \tag{10.18}$$

For charged species, we commonly substitute conductivity from Eq. (10.17) and obtain

$$j_i = \frac{-\sigma_i}{(z_i F)^2}\left[\frac{d\mu_i}{dx} + z_i F \frac{d\phi}{dx}\right] \tag{10.19}$$

The flux density for the species i gives rise to a partial current density, i_i:

$$i_i = z_i F j_i = \frac{-\sigma_i}{z_i F}\left[\frac{d\mu_i}{dx} + z_i F \frac{d\phi}{dx}\right] \tag{10.20}$$

The net current density in the sample is obtained by summing the partial current densities over all the species, k, and for a membrane (that does not have electrodes or an external circuit) it is necessarily zero (open circuit):

$$i_{\text{tot}} = \sum_k z_k F j_k = -\sum_k \frac{\sigma_k}{z_k F}\left[\frac{d\mu_k}{dx} + z_k F \frac{d\phi}{dx}\right] = 0 \tag{10.21}$$

By using the definitions of total conductivity, $\sigma_{\text{tot}} = \sum_k \sigma_k$, and transport number, $t_k = \sigma_k/\sigma_{\text{tot}} = \sigma_k/\sum_k \sigma_k$, we obtain the following important expression for the electrical potential gradient through the sample in terms of the transport number and chemical potential gradient of all charge carriers:

$$\frac{d\phi}{dx} = -\sum_k \frac{t_k}{z_k F}\frac{d\mu_k}{dx} \tag{10.22}$$

2.4. From charged to neutral species: The electrochemical equilibrium

The chemical potentials of charged species are not well defined, and we need to represent them instead by chemical potentials of neutral species. For this purpose, we may assume equilibria between neutral and charged species and electrons in the electrochemical redox reaction:

$$E = E^z + ze^- \tag{10.23}$$

where E is a neutral chemical entity and z may be positive or negative. The equilibrium condition for this is expressed in terms of the chemical potentials of products and reactants:

$$d\mu_{E^z} + zd\mu_{e^-} - d\mu_E = 0 \text{ or } d\mu_{E^z} = d\mu_E - zd\mu_{e^-} \tag{10.24}$$

We may now insert this for all ionic species, n, in the expression for the electrical potential gradient. The entry (among k) for electrons is left unsubstituted.

When required, we may substitute the chemical potential of the neutral species with activities and partial pressures. For isothermal conditions and for ideal gases we obtain

$$d\mu_E = RTd\ln a_E = RTd\ln p_E \tag{10.25}$$

2.5. The voltage over a sample

By inserting for the chemical potentials of ions and by using $\sum_k t_k = 1$, we obtain from Eq. (10.22)

$$\frac{d\phi}{dx} = -\sum_n \frac{t_n}{z_n F}\frac{d\mu_n}{dx} + \frac{1}{F}\frac{d\mu_{e^-}}{dx} \tag{10.26}$$

where n represents all the ionic species and for which the chemical potentials μ_n now refer to the neutral form. We now integrate the electrical potential gradient over the thickness of the sample, from side I to side II, in order to obtain the voltage over the sample:

$$U_{\mathrm{II-I}} = \int_{\mathrm{I}}^{\mathrm{II}} d\phi = -\sum_n \int_{\mathrm{I}}^{\mathrm{II}} \frac{t_n}{z_n F} d\mu_n + \int_{\mathrm{I}}^{\mathrm{II}} \frac{1}{F} d\mu_{e^-} \tag{10.27}$$

$$U_{\mathrm{II-I}} = \phi_{\mathrm{II}} - \phi_{\mathrm{I}} = -\sum_n \int_{\mathrm{I}}^{\mathrm{II}} \frac{t_n}{z_n F} d\mu_n + \frac{1}{F}(\mu_{e^-}^{\mathrm{II}} - \mu_{e^-}^{\mathrm{I}}) \tag{10.28}$$

We further assume that the voltage is measured using the same inert metal on both sides, for example, platinum. This eliminates the difference between chemical potentials of electrons on the two sides, and the voltage measured between the two sides becomes

$$U_{\text{II}-\text{I}} = -\sum_n \int_{\text{I}}^{\text{II}} \frac{t_n}{z_n F} d\mu_n \tag{10.29}$$

This is the basis for establishing transport numbers based on open circuit voltage measurements of cells exposed to a well-defined gradient in chemical activities. If one manages to establish a gradient in the chemical potential of only one species at a time, the transport number of the ion of that species can be determined. If the gradient is small, then the transport numbers may be taken to be constant and equal to an average value so as to simplify the integration.

2.6. Flux of a particular species

One of the general expressions for the electrical potential gradient can now be inserted into the expression for the flux or current density of a particular species. Inserting Eq. (10.22) into Eq. (10.19) and rearranging yields

$$j_i = \frac{-\sigma_i}{(z_i F)^2} \left[\frac{d\mu_i}{dx} - z_i \sum_k \frac{t_k}{z_k} \frac{d\mu_k}{dx} \right] \tag{10.30}$$

This is a general expression, which we can use to calculate flux densities of one charged species in the company of many other species. However, it reflects the flux density and gradients and properties at a particular point in the membrane. The gradients will adjust according to the varying materials' properties so as to maintain a constant flux density everywhere—*steady state*. In order to implement this, we integrate the flux density expression over the thickness of the membrane and require that the flux density remains constant:

$$\int_{\text{I}}^{\text{II}} j_i dx = j_i L = -\int_{\text{I}}^{\text{II}} \frac{\sigma_i}{(z_i F)2} \left[d\mu_i - z_i \sum_k \frac{t_k}{z_k} d\mu_k \right] \tag{10.31}$$

or

$$j_i = \frac{-1}{(z_i F)^2 L} \int_{\text{I}}^{\text{II}} \sigma_i \left[d\mu_i - z_i \sum_k \frac{t_k}{z_k} d\mu_k \right] \tag{10.32}$$

where L is the thickness of the membrane and k now represents all charge carriers. Again, we may substitute the chemical potentials of all ions with that of the neutral species, so that in the following expression, n represents all the ions considered in the total transport system, μ_n and $\mu_{n=i}$ are the chemical potentials of the corresponding neutral species:

$$j_i = \frac{-1}{(z_i F)^2 L} \int_{I}^{II} \sigma_i \left[d\mu_{n=i} - z_i \sum_n \frac{t_n}{z_n} d\mu_n \right] \tag{10.33}$$

2.7. Fluxes in a mixed proton, oxide ion, and electron conductor

We can now write the flux of protons (species i) through an oxide with mixed conductivity of protons, oxide ions, and electrons (species k). In this case, the neutral species (n) are hydrogen and oxygen, and we obtain

$$j_{H^+} = \frac{-1}{F^2 L} \int_{I}^{II} \sigma_{H^+} \left[(t_{O^{2-}} + t_{e^-}) d\mu_H + \frac{1}{2} t_{O^{2-}} d\mu_O \right] \tag{10.34}$$

We substitute to obtain an expression containing partial pressures of molecular neutral species:

$$j_{H^+} = \frac{-RT}{4F^2 L} \int_{I}^{II} \sigma_{H^+} \left[2(t_{O^{2-}} + t_{e^-}) d\ln p_{H_2} + t_{O^{2-}} d\ln p_{O_2} \right] \tag{10.35}$$

From this equation, we see that there are three sources of flux of protons, all proportional to the proton conductivity. A gradient in hydrogen activity drives the protons, and we get a flux if there is a compensating current from fluxes in oxide ions or electrons. Similarly, a gradient in oxygen activity drives oxide ions if they have sufficient transport number and gives a compensating current of protons. In both cases in which oxide ions are involved, the hydrogen comes through as water, whereas with electrons the hydrogen comes through as elemental hydrogen.

It is often practical to insert the equilibrium between hydrogen, oxygen, and water in order to express the flux density of protons in terms of the water vapor activity gradient:

$$j_{H^+} = \frac{-RT}{2F^2 L} \int_{I}^{II} \sigma_{H^+} \left[t_{e^-} d\ln p_{H_2} + t_{O^{2-}} d\ln p_{H_2O} \right] \tag{10.36}$$

This more clearly shows how the proton flux comes about as fluxes of hydrogen and water.

We may also similarly derive the flux density of oxide ions:

$$j_{O^{2-}} = \frac{-RT}{8F^2L}\int_{I}^{II} \sigma_{O^{2-}}\left[(t_{e^-} + t_{H^+})d\ln p_{O_2} + 2t_{H^+}d\ln p_{H_2}\right] \qquad (10.37)$$

It has contributions from ambipolar transport of oxide ions and electrons in an oxygen activity gradient, which corresponds to flux of oxygen through a mixed conducting membrane without proton transport, and from ambipolar transport of oxygen and protons in an oxygen or hydrogen gradient, that is, transport of water. Rewritten in terms of water vapor pressure, it becomes

$$j_{O^{2-}} = \frac{RT}{8F^2L}\int_{I}^{II} \sigma_{O^{2-}}\left[t_{e^-}d\ln p_{O_2} + 2t_{H^+}d\ln p_{H_2O}\right] \qquad (10.38)$$

2.8. Fluxes in a mixed oxide ion and electron conductor

If the transport number for protons is negligibly small, Eqs. (10.37) and (10.38) both end up as the general expression for oxide ion flux density through a mixed oxide ion electron conducting membrane of thickness L:

$$j_{O^{2-}} = \frac{RT}{8F^2L}\int_{I}^{II} \sigma_{O^{2-}}t_{e^-}d\ln p_{O_2} \qquad (10.39)$$

The solution of the integral depends on the oxygen pressure dependencies of the ionic and electronic conductivities.

If we assume that the material contains oxygen vacancies compensated by electrons, we have earlier shown that the defect concentrations are given by $n = 2[v_O^{\bullet\bullet}] = (4K'_{v_O^{\bullet\bullet}})^{1/3}p_{O_2}^{-1/6}$. Since the mobility of electrons is usually much higher than that of ionic defects, and with similar concentrations of electrons and oxygen vacancies, we may assume that the material is mainly an n-type electronic conductor. Therefore, $t_{e^-} \approx 1$. Furthermore, $\sigma_{O^{2-}} = \sigma_{v_O^{\bullet\bullet}} = 2e[v_O^{\bullet\bullet}]u_{v_O^{\bullet\bullet}} = \sigma_0 p_{O_2}^{-1/6}$, and when we insert these in the flux expression we obtain

$$j_{O^{2-}} = -\frac{RT\sigma_0}{8F^2L}\int_{I}^{II} p_{O_2}^{-1/6}d\ln p_{O_2} \qquad (10.40)$$

We substitute $d\ln p_{O_2} = 1/p_{O_2} dp_{O_2}$ and obtain

$$j_{O^{2-}} = -\frac{RT\sigma_0}{8F^2L} \int_{I}^{II} p_{O_2}^{-7/6} dp_{O_2} \tag{10.41}$$

which integrates into

$$j_{O^{2-}} = \frac{6RT\sigma_0}{8e^2L} \{p_{O_2}^{II-1/6} - p_{O_2}^{I-1/6}\} \tag{10.42}$$

One may note that we here have utilized the Wagner approach for electrochemical transport of two defect species and combined it with the solution of the defect chemical equilibria to obtain an accurate description of the flux of oxygen ions. Numerical values for the flux of course require that one knows σ_0, which contains many parameters relating to formation and mobility of the defects. In the right-hand parenthesis, the negative exponent $-1/6$ has the consequence that the smaller of the two partial pressures of oxygen has the biggest influence. The larger partial pressure makes a comparatively smaller difference and may be neglected to a first approximation for large gradients. For instance, consider a gas separation membrane exposed to air on the one side and a very reducing gas (e.g., methane) on the other. The flux would then be essentially proportional to $p_{O_2}^{red-1/6}$, while it would largely be independent of $p_{O_2}^{ox}$. It would thus not help much to compress the air to 100 atm, while removal of oxidized products like H_2O on the reducing side would lower $p_{O_2}^{red}$ and have a large effect.

Similarly, equations can be derived for materials with other defect structures in which the concentration of oxygen vacancies and thus oxygen ion conductivity is independent of p_{O_2} (typical of acceptor-doped materials) or in which the oxygen ion conductivity is proportional to $p_{O_2}^{1/6}$ (typical of oxygen excess materials with oxygen interstitials). The way fluxes vary with high and low p_{O_2} differs from the situation discussed above in these cases.

2.9. Fluxes in a mixed proton and electron conductor

By assuming that the transport number for oxide ions is zero [cf. Eq. (10.35)], the only flux of hydrogen remaining is that of ambipolar transport of protons and electrons:

$$j_{H^+} = \frac{-RT}{2F^2L} \int_{I}^{II} \sigma_{H^+} t_{e^-} d\ln p_{H_2} \tag{10.43}$$

This is a good point to look a little closer at the ambipolar transport term that we have mentioned earlier in passing. The materials property of interest here is the protonic–electronic ambipolar conductivity term, which can be written in many ways: $\sigma_{H^+} t_{e^-} = \sigma_{total} t_{H^+} t_{e^-} = \sigma_{e^-} t_{H^+} = \sigma_{H^+} \sigma_{e^-} / \sigma_{total}$. It may be noted that if—and only if—the material conducts no other species than protons and electrons can it also be expressed in the commonly used form $\sigma_{H^+} \sigma_{e^-} / \sigma_{H^+} + \sigma_{e^-}$.

The hydrogen partial pressure gradient is the driving force of the flux in Eq. (10.43). In order to integrate the equation, we need to know how the proton conductivity and electron transport number vary with the hydrogen activity. If we further assume that the material exhibits dominating electronic conduction, then the electronic transport number is unity, $t_{e^-} \approx 1$. In the following, we describe three limiting cases of this, which allow us to integrate the above equation.

If protons are minority defects, their concentration and the proton conductivity are proportional to $p_{H_2}^{1/2}$, for example Eq. (10.16), and we obtain

$$j_{H^+} = \frac{-RT\sigma_{H^+,0}}{2F^2L} \int_{I}^{II} p_{H_2}^{1/2} d\ln p_{H_2} = \frac{-RT\sigma_{H^+,0}}{2F^2L} \int_{I}^{II} p_{H_2}^{-1/2} dp_{H_2}$$
$$= \frac{-RT\sigma_{H^+,0}}{F^2L} [p_{H_2}^{II-1/2} - p_{H_2}^{I-1/2}] \tag{10.44}$$

where $\sigma_{H^+,0}$ is the proton conductivity at standard hydrogen pressure [10^5 Pa $\approx$ 1 atm $H_2(g)$].

If protons are majority defects and compensated by electrons, then their concentration and the proton conductivity are proportional to $p_{H_2}^{1/4}$, Eq. (10.9), and we obtain

$$j_{H^+} = \frac{-RT\sigma_{H^+,0}}{2F^2L} \int_{I}^{II} p_{H_2}^{1/4} d\ln p_{H_2} = \frac{-RT\sigma_{H^+,0}}{2F^2L} \int_{I}^{II} p_{H_2}^{-3/4} dp_{H_2}$$
$$= \frac{-2RT\sigma_{H^+,0}}{F^2L} [p_{H_2}^{II-1/4} - p_{H_2}^{I-1/4}] \tag{10.45}$$

If protons are majority defects and compensated by acceptor dopants, then their concentration and the proton conductivity are independent of p_{H_2} [limiting case of Eqs. (10.11) and (10.12)]:

$$j_{H^+} = \frac{-RT\sigma_{H^+}}{2F^2L} \int_{I}^{II} d\ln p_{H_2} = \frac{-RT\sigma_{H^+}}{2F^2L} \int_{I}^{II} \frac{1}{p_{H_2}} dp_{H_2}$$
$$= \frac{-RT\sigma_{H^+}}{2F^2L} [\ln p_{H_2}^{II} - \ln p_{H_2}^{I}] \tag{10.46}$$

In the first two cases (protons in minority or compensated by electrons), the hydrogen partial pressure on the high-pressure side has a large effect on the flux, while how small the pressure is on the other side has less influence. In the last of the three cases, however, the differences in the logarithms (representing the logarithm of the ratio of the two hydrogen pressures) make the small and the large pressures become equally important.

In the above cases, we assumed that the transport number of electronic defects was unity. Especially in the case for which protons dominate and compensate acceptors, the electronic defects may be sufficiently minor, in terms of concentration, that they do not dominate the conductivity. If we consider as an alternative that the transport number of protons is unity, then we may write

$$\begin{aligned} j_{H^+} &= \frac{-RT}{2F^2L}\int_{I}^{II} \sigma_{H^+} t_{e^-} d\ln p_{H_2} = \frac{-RT}{2F^2L}\int_{I}^{II} \sigma_{e^-} t_{H^+} d\ln p_{H_2} \\ &= \frac{-RT}{2F^2L}\int_{I}^{II} \sigma_{e^-} d\ln p_{H_2} \end{aligned} \tag{10.47}$$

If protons compensate acceptors, then electronic conductivity is proportional to $p_{H_2}^{1/2}$ and $p_{H_2}^{-1/2}$, respectively, in the case of electrons and holes if p_{O_2} is assumed constant. This gives

$$\begin{aligned} j_{H^+} &= \frac{-RT\sigma_{n,0}}{2F^2L}\int_{I}^{II} p_{H_2}^{1/2} d\ln p_{H_2} = \frac{-RT\sigma_{n,0}}{2F^2L}\int_{I}^{II} p_{H_2}^{-1/2} dp_{H_2} \\ &= \frac{-RT\sigma_{n,0}}{F^2L}\left[p_{H_2}^{II1/2} - p_{H_2}^{I1/2}\right] \end{aligned} \tag{10.48}$$

for limiting n-type conductivity, and

$$\begin{aligned} j_{H^+} &= \frac{-RT\sigma_{p,0}}{2F^2L}\int_{I}^{II} p_{H_2}^{-1/2} d\ln p_{H_2} = \frac{-RT\sigma_{p,0}}{2F^2L}\int_{I}^{II} p_{H_2}^{-3/2} dp_{H_2} \\ &= \frac{RT\sigma_{p,0}}{F^2L}\left[p_{H_2}^{II-1/2} - p_{H_2}^{I-1/2}\right] \end{aligned} \tag{10.49}$$

for limiting p-type conductivity, where σ_n and σ_p are electron and hole conductivities, and $\sigma_{n,0}$ and $\sigma_{p,0}$ are conductivities at standard hydrogen pressure [10^5 Pa $\approx$ 1 atm H_2(g)].

There are, of course, other limiting situations that may be considered, for example, if protons are compensated by negative point defects such as metal vacancies or oxygen interstitials. These are, however, mainly dominant under oxidizing conditions, not commonly considered for hydrogen-separation processes.

Before we move on to more special cases, we take this opportunity to remind the reader of the many different pressure dependencies we have got for fluxes of mixed conducting membranes. This is why permeabilities in dense mixed conducting membranes most often cannot be given as simply and in the same units as permeability in other types of membranes. We will return to the consequences of this when we quote example literature values later on.

2.10. Fluxes in a mixed proton and oxide ion conductor

If the material is a mixed proton and oxide ion conductor and has a negligible transport number for electrons, we have

$$j_{H^+} = \frac{-RT}{2F^2L} \int_{I}^{II} \sigma_{H^+} t_{O^{2-}} d\ln p_{H_2O} \qquad (10.50)$$

that is, the water vapor pressure gradient is the driving force. Typically, the oxide is acceptor-doped, and protons and oxygen vacancies compete to be the compensating defect. If oxygen vacancies are dominant, then their concentration is independent of p_{H_2O}, while the proton concentration increases with $p_{H_2O}^{1/2}$. If on the other hand, the protons are dominant and therefore constant, the concentration of oxygen vacancies decreases (proportional to $p_{H_2O}^{-1}$). Depending on whether the material has a transport number of any of the two that can be set equal to unity, these dependencies can be inserted into the remaining limiting conductivity, and the equation easily integrated.

2.11. Fluxes in a mixed proton, oxide ion, and electron conductor revisited

In ambipolar diffusion of protons and oxide ions, the two types of ions move in the same direction for charge compensation, and the net result is permeation of water. If we also have electronic transport, then the transport of hydrogen and oxide ions may be decoupled. They may move in the same direction or in opposite directions and at different ratios.

Consider, for instance, a mixed conducting membrane exposed to wet hydrogen on one side and wet argon on the other. The intention might be to measure the hydrogen permeability by measuring the amount of hydrogen in the argon side permeate gas. If the water vapor pressures are equal, there will be no ambipolar transport of water in any direction. Hydrogen will permeate from the high hydrogen partial pressure side to the low. The flux will be proportional to the ambipolar proton–electron conductivity, and hydrogen will appear in the permeate. But at the same time, oxide ions will move in the opposite direction, from the higher oxygen activity in the wet argon to the lower oxygen activity in the wet hydrogen, and the oxygen is taken from the water vapor in the argon. This flux is limited by the ambipolar oxide ion–electron conductivity. As a result of this flux, hydrogen is formed in the argon, and it may be difficult experimentally to distinguish whether the hydrogen arises from proton transport or oxide ion transport. Only analysis of the water vapor balance on one or both sides—usually difficult enough in itself—can distinguish one from the other or give a correct balance between proton and oxide ion conductivity. The two opposite fluxes of protons and oxide ions both give charge transport in the same direction, and they are compensated by a flux of electrons being the sum of the two.

In the case discussed here, where three or more species (e.g., A, B, C) provide charge transport, it may as mentioned earlier be important to keep the pairwise ambipolar conductivities in the correct form and to be aware of the inequality: $\sigma_A t_B = \sigma_{total} t_A t_B = \sigma_A \sigma_B / \sigma_{total} \neq \sigma_A \sigma_B / \sigma_A + \sigma_B$.

2.12. Permeation of neutral hydrogen species

Diffusion of neutral species cannot be excluded, especially in the case of hydrogen permeation, and we include this short treatment of it for completeness and for its growing interest and importance. Flux densities of hydrogen atoms and hydrogen molecules can be expressed in terms of diffusivities and chemical potential gradients:

$$j_H = \frac{-D_H c_H}{RT}\frac{d\mu_H}{dx} = \frac{-D_H c_H}{2}\frac{d\ln p_{H_2}}{dx} \tag{10.51}$$

$$j_{H_2} = \frac{-D_{H_2} c_{H_2}}{RT}\frac{d\mu_{H_2}}{dx} = -D_{H_2} c_{H_2}\frac{d\ln p_{H_2}}{dx} \tag{10.52}$$

Integration over the membrane thickness yields

$$j_H = \frac{-1}{2L}\int_{I}^{II} D_H c_H d\ln p_H \tag{10.53}$$

$$j_{H_2} = \frac{-1}{L}\int_{I}^{II} D_{H_2} c_{H_2} d\ln p_{H_2} \tag{10.54}$$

Assuming that the diffusivities are independent of concentration and that the concentrations are given by the equilibria with hydrogen partial pressure treated earlier, we obtain

$$\begin{aligned} j_H &= \frac{-D_H K_{2H}^{1/2}}{2L}\int_{I}^{II} p_{H_2}^{1/2} d\ln p_{H_2} = \frac{-D_H K_{2H}^{1/2}}{2L}\int_{I}^{II} p_{H_2}^{-1/2} dp_{H_2} \\ &= \frac{-D_H K_{2H}^{1/2}}{L}[p_{H_2}^{II1/2} - p_{H_2}^{I1/2}] \end{aligned} \tag{10.55}$$

$$\begin{aligned} j_{H_2} &= \frac{-D_{H_2} K_{H_2}}{L}\int_{I}^{II} p_{H_2} d\ln p_{H_2} = \frac{-D_{H_2} K_{H_2}}{L}\int_{I}^{II} dp_{H_2} \\ &= \frac{-D_{H_2} K_{H_2}}{L}[p_{H_2}^{II} - p_{H_2}^{I}] \end{aligned} \tag{10.56}$$

in which K_{2H} and K_{H_2} are the equilibrium constants of the dissolution of two atoms or one molecule, respectively, from one hydrogen molecule. It may be noted that K_{2H} is equal to the square of the Sieverts' law constant, $K_{2H} = K_S^2$.

We note that the molecular flux is proportional to the difference in hydrogen pressure (as in gas phase transport through a porous membrane), whereas atomic transport has a square root dependency, and as such would behave similar to ambipolar transport of protons and electrons limited by the conductivity of a minority concentration of protons.

2.13. Surface kinetics issues in mixed conductors

Apart from the material itself, the gas pressure gradient, and the temperature, the membrane thickness is one of the major parameters involved in the enhancement of gas permeation rates. As the membrane becomes sufficiently thin, however, surface kinetics will become important and diffusion of ions through the membrane will no longer be rate determining for the overall gas flux.

A critical thickness L_c is defined as the thickness where the loss in the driving force is split equally between the bulk diffusion and the two surface reactions occurring at the permeate and feed sides of the membrane. Decreasing a membrane's thickness below, L_c gives essentially no increase in the flux. L_c decreases with the roughness or porosity of the surface (actual area over the superimposed area). An enhancement of up to 100 times can be expected theoretically. In practice, however, the enhancements for a variety of membranes modified with porous layers are only up to approximately three times. For the best oxide ion and electron conducting materials, the enhancement increases with decreasing temperature, while the opposite is reported for materials with worse properties. The results also suggest that the larger enhancement is obtained when the modification is done on the permeate side. The increase in surface exchange rate may also be enhanced by depositing a catalytically active substance (e.g., Pt, Ag, or Pr oxide) onto the surface of the membrane.

Generally, one may divide the overall surface reaction into adsorption dissociation of the reacting gas molecule and incorporation of the dissociated species. Again these may consist of more elementary reaction sequences (adsorption, dissociation, surface diffusion, charge transfer, several incorporation steps). Hence, the overall series of reactions describing the process from a gas molecule to an ionic defect in the oxide may be complex. However, one step is assumingly much slower than the others and is named the rate-determining step (RDS).

For a mixed oxide ion electron conductor operating between air and hydrogen, the overall reactions at the two interfaces may be written as

$$O_2(g) + 4e^- = 2O^{2-} \tag{10.57}$$

$$H_2(g) + O^{2-} = H_2O(g) + 2e^- \tag{10.58}$$

The electrons can be supplied to or from an electrode or a mixed conductor surface. Thus, basically, the reactions are the same in the electrolyte/electrode as in the mixed conductor, although the physical presence of the electrode can promote (by catalysis) or obstruct the reaction. Some possible reaction steps in the case of the reduction of oxygen gas are given by the following reaction sequence:

$$O_2(g) = O_{2,ads} \quad (10.59)$$

$$O_{2,ads} + e^- = O^-_{2,ads} \quad (10.60)$$

$$O^-_{2,ads} + e^- = 2O^-_{ads} \quad (10.61)$$

$$O^-_{ads} + e^- = O^{2-}_{ads} \quad (10.62)$$

$$O^{2-}_{ads} = O^{2-}_{O} \quad (10.63)$$

The last of these steps may involve bulk defects (e.g., a vacancy that the adsorbed ion may occupy) but here we will not attempt an analysis of surface or electrode kinetics in terms of defect chemistry. Determination of the nature of the rate-limiting step of a reaction is a difficult task. Studies of surface exchange kinetics using ^{18}O or by relaxation methods can give information about the temperature and pressure dependence of the RDS. However, it should be pointed out that information from such methods do not give direct information about the exact mechanism and which intermediate species that are involved. Each step will have its own specific rate and resistance to the mass transport across the membrane. In lack of detailed knowledge of the rate-limiting step(s) we normally have to relate to the overall reaction.

At equilibrium the overall reaction takes place back and forth at equal rates r_0 as a result of thermal energy. Similar to the random diffusion coefficient connected with these thermal disorders in bulk ($D_r = 1/6\ r_0\ s^2$), we introduce an exchange coefficient for interfaces, $k_i = r_0 s_i$, which says something about the thermal fluctuations of the reaction across the interface of thickness s_i. We can alternatively express this as an exchange flux density

$$j_0 = k_i c = r_0 s_i c \quad (10.64)$$

If we now apply a small force over the interface, we get a perturbation and a net flux, given by

$$j = -ck_i \frac{dP}{kT} = -ck_i \frac{d\mu_i}{kT} \quad (10.65)$$

where c is the volume concentration of species in the interface and dP is the potential (chemical, electrical, or electrochemical) step over the interface.

In order to approach an interpretation of transport measurements, we take as starting point that the flux(es) through the interface must be the same as that in the bulk of the membrane next to the interface in all cases (and in the entire membrane in the steady-state cases). This is done in reality and in the mathematical solutions by adjusting the potential at the border between the interface and the bulk so as to equalize the two fluxes. However, this soon becomes difficult, as it is not trivial to choose the proper set of potentials to apply for different interface processes.

We resort first to chemical diffusion and assume that the interface process is one which can be described as species flowing from a high to a low concentration. Then

$$-D\left(\frac{dc}{dx}\right)_{bulk} = -k\Delta c_{\text{interface}} \tag{10.66}$$

and D as well as k must refer to chemical coefficients. This requirement can be included in the solution of Fick's second law for transients, and D and k can in principle be found independently by fitting transient data to this solution.

In order to analyze the effect of interface kinetics on fluxes through a membrane under steady-state conditions, we may take an approach similar to that for fuel cells and electrolyzers. Principally, no electrodes are involved, and voltages are therefore not directly appropriate as in the other cases. Instead we may sum-up chemical potential changes of oxygen (or hydrogen) through the membrane and its interfaces. We start out with the membrane interior, and from earlier we have for a mixed oxygen ion and electron conductor of thickness L:

$$j_{\text{O}^{2-}} = -\frac{1}{L}\int_{\text{I}}^{\text{II}} \frac{\sigma_{\text{O}^{2-}} t_{e^-}}{8e^2} d\mu_{\text{O}_2} \tag{10.67}$$

If we introduce ambipolar diffusivities and conductivities and assume these constant, the equation may be integrated to obtain

$$j_{\text{O}^{2-}} = \frac{D_{\text{amb}} c_{\text{O}^{2-}}}{2kT}\frac{\Delta\mu_{\text{O}_2}}{L} = \frac{\sigma_{\text{amb}}}{8e^2}\frac{\Delta\mu_{\text{O}_2}}{L} \tag{10.68}$$

and, by rearrangement, the chemical potential drop over the bulk of the membrane is

$$\Delta\mu_{\text{O}_2} = \frac{-2 j_{\text{O}^{2-}} kTL}{D_{\text{amb}} c_{\text{O}^{2-}}} = \frac{-j_{\text{O}^{2-}} 8e^2 L}{\sigma_{\text{amb}}} \tag{10.69}$$

This drop over the membrane bulk is given by the external full gradient minus the losses at the interfaces. In the present terms, the flux through an interface is, from Eq. (10.68), given by

$$j_{\text{O}^{2-}} = \frac{-c_{\text{O}^{2-}} k_i \Delta\mu_{\text{O}_2}^{\text{interface}}}{2kT} \tag{10.70}$$

such that, by rearrangement

$$\Delta\mu_{O_2}^{\text{interface}} = \frac{-2jO^{2-}kT}{k_i cO^{2-}} \tag{10.71}$$

Now, we sum-up the total gradient in chemical potential of oxygen over the bulk of the membrane and the two interfaces

$$\Delta\mu_{O_2}^{\text{tot}} = \Delta\mu_{O_2}^{\text{bulk}} + 2\Delta\mu_{O_2}^{\text{interface}} \tag{10.72}$$

and after insertion of the above expressions for bulk and interface, we can solve for the flux of oxygen ions through a membrane with two equally contributing interfaces under the conditions given:

$$j_{O^{2-}} = \frac{-\Delta\mu_{O_2}^{\text{tot}} c_{O^{2-}}/2kT}{(L/D_{\text{amb}}) + (2/k_i)} \tag{10.73}$$

In lack of full understanding of the relations involved in interface kinetics of membranes, we may refer to a critical thickness L_c, the thickness of the bulk membrane where the interface and bulk impose equally large restrictions on the flux. In any membrane application, going much below this thickness is not of much use. We see from the above that this thickness is $L_c = D_{amb}/k_i$ for the thickness of the part of a membrane relating to one interface, or $L_c = 2D_{amb}/k_i$ for thickness of a membrane between two equally limiting surfaces.

Experimentally the value of L_{crit} has been found to be of the order of 100 μm for many polished samples of different fluorite and perovskite materials, indicating that there is a close relationship between, for example, vacancy concentration in bulk and the surface exchange. Roughing the surface should increase relatively the surface flux, and this has been shown to work.

In a review of hydrogen permeable dense ceramic membranes, Norby and Haugsrud [4] state that there is yet no example in literature where surface kinetics has been proven to limit the hydrogen permeation across a mixed proton–electron conducting oxide, and seemingly membrane thicknesses below L_C have thus not been reached. Thus, rates of hydrogen dissociation on mixed proton–electron conductors have up to now hardly been studied. H_2/D_2 dissociation on metallic surfaces has, on the other hand, been extensively investigated as a measure for hydrogen activation. In the simplest mechanism, hydrogen is dissociatively adsorbed on the metallic surface forming two independent hydrogen atoms. Recombination and desorption accordingly result in exchange of partner atoms. It has also been suggested that exchange may occur via triatomic intermediates. These two mechanisms are usually referred to as the Bonhoeffer–Farkas and the Rideal–Eley mechanisms [5, 6]. The former has been observed to be operative usually at higher temperatures whereas the latter, involving intermediates, is operative only at low temperatures.

In the case of oxides, the catalytic activity toward H_2 dissociation has been investigated for several binary oxides including Cr_2O_3, Co_3O_4, NiO, ZnO, Al_2O_3, and some of the rare earth sesquioxides [7–10]. Most of these studies have been performed below room temperature and are, as such, not directly relevant for the high-temperature mixed proton–electron conductors. One may conclude, however, that for higher temperatures, hydrogen dissociation occurs through dissociative adsorption and desorption following Bonhoeffer–Farkas-type mechanisms. On this basis, hydrogen dissociation can be treated according to the Langmuir equation. The simplest route from hydrogen gas to protons dissolved in the oxide may be written as

$$H_2(g) + 2v_{ads} = 2H_{ads} \tag{10.74}$$

$$H_{ads} + O_O^x = OH_O^{\bullet} + e' + v_{ads} \tag{10.75}$$

As an alternative to forming electrons, hydrogen may be oxidized by annihilation of electron holes. These reactions may be divided again into different possible intermediate steps, for example, including charged adsorbed hydrogen species (e.g., H_2^+).

Surface defects are likely to act as catalytically active centers at which hydrogen dissociation takes place. The electronic states of these defects may be located within the band gap of the semiconducting oxide and are likely to be involved in the electron transfer between hydrogen and the membrane. In the reaction in [Eq. (10.74)], we have not given any indication as to which sites are active for adsorption. However, many of these materials are acceptor-doped, and one may expect the dopants or their compensating defects to be involved. One might speculate, for instance, whether oxygen vacancies act as electron sinks as an intermediate step, here shown with two-electron steps:

$$H_2(g) + v_O^{\bullet\bullet} + 2O_O^x = 2OH_O^{\bullet} + v_O^x \tag{10.76}$$

$$v_O^x = v_O^{\bullet\bullet} + 2e' \tag{10.77}$$

The reaction stages and species involved may be expected to be similar on the feed and permeate sides of a hydrogen membrane. If the flux is driven by reaction with oxygen on the permeate side, this side may naturally be quite different, both in terms of ionic species and electronic (holes vs electrons). The formation or presence of water vapor may also play a significant role in hydrogen exchange kinetics.

3. Materials Properties

Key properties determining the use of a material for electrolyte or selective gas separation membrane applications are the partial ionic and electronic conductivities, and suitable thermomechanical and chemical properties with respect to other cell's or membrane device's components. In the electrolyte-type materials, only oxygen anions or protons have significant conductivity. In this case, external electrodes are required on both sides of the electrolyte to lead the electronic current through an outer closed electrical circuit. Selective gas separation membranes exhibit both sufficient ionic and electronic conductivities. The electrical circuit is closed by electronic charge transport inside the membrane and, as a consequence, the selective membrane requires neither electrodes nor external electric connectors. The materials may be mixed ion and electron conductors or, if the electronic conductivity is not sufficiently high, designed as a composite by addition of an electronically conducting second phase. Further development of stable efficient materials for high-temperature applications still represents an important issue and will be discussed in a later paragraph. This paragraph addresses recent development in the field of single-phase oxygen ion or proton conductors and mixed conductors, through doping strategies and discovery of new materials, with special emphasis on transport properties.

3.1. Oxygen membranes

3.1.1. Electrolyte materials

3.1.1.1. Fluorite related oxides MO_2 fluorite-type oxides often exhibit high concentration of oxygen vacancies, either intrinsically or formed by dissolving lower-valent metal oxides in the lattice. They have therefore been widely investigated as base materials for oxygen ion-conducting electrolytes.

ZrO_2-based oxides are the most common electrolytes for SOFC. The typical dopant being yttrium, usually in the range of 8–10 mol % Y_2O_3, gives a large electrolytic domain (defined as the oxygen partial pressure region where $t_{ion} >$ 0.5), and an oxygen ion conductivity of ~0.1 S/cm at 1000 °C [11, 12].

CeO_2-based membranes show high conductivity $>10^{-2}$ S/cm at 500 °C and good chemical stability in connection with commonly used cathode materials [13–17]. These oxides have a smaller electrolytic domain than Yttrium Stabilized Zirconia (YSZ) due to the appearance of electronic conductivity at low oxygen partial pressure [18]. Doping with 15–25% of lower-valent cations broadens the electrolytic domain and a maximum oxygen ion conductivity is found with Gd (commonly referred to as CGO) and Sm (CSO) which also have the lowest ionic size mismatch [19–21].

δ-Bi_2O_3 can also be described as a fluorite phase, where each fourth oxygen site is vacant [22]. This oxide has the highest oxygen ion conductivity of all known fluorite oxides, 2.3 S/cm at 800 °C. The narrow stability range between ~729 and 824 °C can be enlarged by doping with rare earth dopants (Y, Dy, Er) in combination with higher-valent cations, such as W or Nb. This stabilizes the conducting phase to lower temperature and oxygen partial pressure, but stability under reducing conditions still appears questionable. A maximum oxygen ion conductivity of 10^{-1} S/cm at 550 °C has been reported for $Bi_{0.8}Er_{0.2}O_{1.5}$ [12], while electrochemical pumping using $(Bi_2O_3)_{0.75}(Er_2O_3)_{0.25}$ has shown oxygen fluxes up to 180 ml/m^2 s, the latter result being limited by slow surface kinetics [22]. Application of these oxides is, however, complicated due to high corrosion activity, low mechanical strength and volatilization of bismuth oxide at moderate temperatures.

The pyrochlores ($A_2B_2O_7$) relate to the fluorite structure by having one vacant oxygen site per two formula units. The most conductive materials of this family are based on $Gd_2Ti_2O_7$, with a maximum oxygen ion conductivity of ~5.10^{-2} S/cm at 1000 °C obtained by doping with 8 mol % Ca [23]. Other doping may result in appearance of electronic conduction.

3.1.1.2. Perovskites Perovskites (ABO_3) and related phases exhibit a variety of electrical properties, with some showing high oxygen ion conductivity. For $La_{1-x}Sr_xGa_{1-y}Mg_yO_{3-\delta}$ (LSGM) solid solution series with $x = 0.10$–0.20 and $y = 0.15$–0.20, an almost pure oxygen ion conductivity of about 10^{-1} S/cm at 800 °C has been demonstrated [24]. Other advantages of LSGM also include a large electrolytic domain and a moderate thermal expansion coefficient in air ($11 \times 10^{-6}\,K^{-1}$). The high gallium oxide price and possible volatilization are potential drawbacks of these oxides [25].

3.1.1.3. BIMEVOX The Aurivillius $(Bi_2O_2)(A_{n-1}B_nO_x)$ intergrowth phases, particularly those based on γ-$Bi_4V_2O_{11}$ with partial substitution of vanadium with transition metal cations, such as Cu, Ni, or Co, have high oxygen ion conductivity in the temperatures range 300–700 °C [26]. These oxides, termed BIMEVOX (ME being Cu, Ni, etc.), show an oxygen ion conductivity of above 10^{-1} S/cm at 600 °C. A sluggish detrimental phase ordering appears with time at temperature around 500 °C for materials doped with Me^{II}, such as Co, Cu, Ni. However, it has been demonstrated that doping with Me^{V}, such as Ta, Sb, and Nb, limits this evolution without significantly affecting the conductivity [26]. Electrochemical pumping has given flux of about 3–4 ml/cm^2 min, mainly restricted by slow surface exchange kinetics. The use of these materials in electrochemical devices is challenging due to their reactivity, poor stability in reducing atmosphere, and low mechanical strength [12, 27].

3.1.1.4. LAMOX The LAMOX family of oxide conductors [28], which is based on the parent compound $La_2Mo_2O_9$, has an oxygen ion conductivity of ~0.03 S/cm at 720 °C. A vacancy order–disorder phase transition, occurring at ~580 °C from a monoclinic α form to a cubic β form results in the high anionic conductivity [28]. This phase transition leads to an abrupt volume expansion, which may be suppressed by a minor doping of $La_2Mo_2O_9$. The maximum conductivity is reported for the compositions $La_{1.7}Bi_{0.3}Mo_2O_{9-\delta}$, $La_2Mo_{1.7}W_{0.3}O_{9-\delta}$, and $La_2Mo_{1.95}V_{0.05}O_{9-\delta}$ [12]. The stability in reducing atmosphere is an issue for these materials as $La_2Mo_2O_9$ may loose about 1.14 of its 9 oxygen per unit formula. Substitution of lanthanum by other rare earths (Nd, Gd) increases the oxygen loss, thus, further investigation about appropriate doping should be pursued to improve stability and performance.

3.1.1.5. Apatite In the apatite $A_{10-x}(SiO_4)_6O_{2\pm\delta}$ lattice, A-site cations are located in the cavities created by MO_4 tetrahedra with four distinct oxygen positions, while additional oxygen sites form channels through the structure. The ion transport in $Ln_{10-x}Si_6O_{26+y}$ (Ln = La, Pr, Nd, Sm, Gd, Dy) oxides increases with increasing radius of Ln^{3+}, with a maximum conductivity achieved for the

La-containing phase. Despite that high oxygen ion conductivity has been demonstrated for these materials, reported literature data can vary from 8.4×10^{-5} to 4.3×10^{-3} S/cm at 773 K for undoped ceramics. This is mainly ascribed to the difficulty of sintering these materials and to the anisotropy of the anionic transport in the lattice. Increasing oxygen ion conductivity is reported for oxides with more than 26 oxygen ions per unit formula, suggesting an interstitial conduction mechanism [11, 12], and A cation deficiency, as observed for $La_{9.33+x/3}Si_{6-x}Al_xO_{26}$ oxides. Doping these oxides with Fe up to 30% of the Si sites results in a better sinterability and an increasing ionic conductivity. However, as for many electrolyte materials, the stability in reducing atmospheres is an issue to be addressed.

3.1.1.6. Grain boundary and microstructural effects In materials with high bulk ionic conductivity the microstructure may play an important role. Grain boundaries may have a higher resistivity than the grain bulk, which often can be attributed to the combination of an *extrinsic* and *intrinsic blocking effect*. The *extrinsic blocking effect* relates to the segregation of pores, secondary phases, and non- (or poorly) conducting impurity phases in the grain boundary. This leads to a reduction of the ionic transport pathway and constriction of current, as the transport only takes place through direct grain-to-grain contacts. This effect is often encountered when sintering aids are used during manufacturing of bulk ceramics. Hence, abundant impurities, such as SiO_2, have been shown to have detrimental effects on grain boundary conduction in several electrolyte materials such as ceria-, zirconia-, and apatite-based materials due to formation of glassy phase during sintering. The *intrinsic blocking effect* can even be encountered in high purity crystals and is a direct consequence of grain boundary space-charge effects that may affect the conductivity due to electrical polarization and segregation of defects and dopants.

3.1.2. Mixed conductors

3.1.2.1. Fluorites By codoping ZrO_2-based electrolytes with ambivalent cations, such as Ce, Ti, and Tb, mixed ionic and electronic conduction can be obtained. Due to the high-temperature stability, these materials represent an interesting option for membranes operating above 1000 °C. The surface exchange rate coefficient is, however, seemingly low [29], which may limit the flux. The Tb–YSZ membrane has been reported with a high permeation rate of 6.5×10^{-6} mol O_2/m^2 s at 1100 °C. Mixed conduction in CeO_2-based oxides can be achieved by codoping with cations such as Pr, Co, and Mn. Membranes exhibiting a permeability of $\sim 3 \times 10^{-9}$ mol/cm^2 s under small differential oxygen gradients (air/He gradient) at 850 °C have been reported. This flux is higher than those reported for composite membranes of CGO containing good electronic conductors, such as $La_{0.8}Sr_{0.2}Fe_{0.8}Co_{0.2}O_{3-\delta}$, $La_{0.7}Sr_{0.3}MnO_{3-\delta}$, and $Gd_{0.7}Ca_{0.3}CoO_{3-\delta}$ [30–34]. The lower performance of these composite membranes is apparently attributed to blocking effects of the grain boundaries arising from interdiffusion of cations between the two phases.

3.1.2.2. Perovskite-related materials Particularly well studied among the perovskite-type materials are compositions where the A-site constituents are alkali earth metals (Ca, Sr, Ba) and/or rare earth La, and the B site contains transition metals, such as Fe, Co. Typical compositions like $La_{0.2}Sr_{0.8}Fe_{0.2}Co_{0.8}O_{3-\delta}$ (LSCF), $SrCo_{0.8}Fe_{0.2}O_{3-\delta}$ (SCF), and $Ba_{0.5}Sr_{0.5}Co_{0.8}Fe_{0.2}O_{3-\delta}$ (BSCF) are investigated for applications as oxygen transport membranes and cathode materials for SOFC. Interests in these materials stem from their high oxygen permeability in the temperature range 700–1200 °C, which increases in the order LSCF < SCF ~ BSCF [35–37]. Recently, Vente *et al.* [35] showed for flat self-supported membranes of 200-μm thickness (5 × 5 cm^2 surface area), an oxygen flux in the range 10–13 ml/cm^2 min at 1275 K in 1 bar O_2 and He on the feed and permeate side, respectively. However, as also pointed out by these authors, problems related to long-term stability and temperature cycling still exist. Sirman *et al.* [38] and more recently, Nagai *et al.* [39] have investigated the effects of doping similar oxides on transport properties and some mechanical and chemical properties. As a general rule, large cations and a high degree of lower valency doping on the A site are beneficial for the oxygen diffusion, while for doping on the B site, the diffusivity and the oxygen surface exchange rate change according to the sequence Co>Fe>Ga>Mn>Cr>Ti. On this basis, $SrCo_{0.9}Nb_{0.1}O_{3-\delta}$ membrane of 1-mm thickness was shown to exhibit an oxygen flux of 4.23 ml/cm^2 min at 1173 K, in (0.21 atm/0.01 atm) oxygen partial pressure gradient. Further investigation must be addressed to confirm the long-term stability of this promising material [40].

The $SrFeCo_{0.5}O_x$ [41–45] consisting mainly of $Sr_4Fe_{6-x}Co_xO_{13\pm\delta}$ exhibits appreciable oxygen permeability. A 1-mm-thick membrane demonstrated ~0.2 ml O_2/cm^2 min at 900 °C under an air/helium gradient [46–49].

Doping strategies to increase electronic conductivity in LSGM have shown promise when introducing >10% transition metal cations, such as Fe, Co, or Ni, into the Ga sublattice. The compositions $La_{0.90}Sr_{0.10}Ga_{0.65}Mg_{0.15}Ni_{0.20}O_{3-\delta}$ and $La_{0.40}Sr_{0.60}Ga_{0.40}Fe_{0.60}O_{3-\delta}$ exhibit the highest fluxes with moderate thermal expansion coefficient (~12 × 10^{-6} K^{-1}) and sufficient phase stability. A 1-mm-thick membrane shows a permeation rate of the order 2 × 10^{-7} mol/s cm^2 at 900 °C under (21 kPa/2.1 kPa) oxygen partial pressure gradient [50–52].

Despite that $CaTi_{1-x}Fe_xO_{3-\delta}$ exhibits moderate oxygen ion and electronic conductivities, the stability, a nearly composition- and temperature-independent moderate thermal expansion (~12 × 10^{-6} K^{-1}), and low cost, makes it interesting as membrane material [53–59]. Smith et al. [60] found that for membranes containing 10% Fe, bulk diffusion controls oxygen fluxes in the thickness range 1.5 mm–30 mm. At 900 °C an oxygen flux of ~0.8 ml/min cm^2 was demonstrated under an oxygen/argon gradient, with low activation energy of 50 kJ/mol. $BaCe_xFe_{1-x}O_{3-\delta}$ (BCF) oxides exhibit high oxygen permeability with reported good structural stability under both oxidizing and reducing conditions [61, 62]. At 900 °C, the flux of a 1.4-mm-thick membrane under an air/helium gradient is 0.5 ml/min cm^2. Further increase in flux may to some extent be achieved by reduction in thickness. For the conditions investigated, a transition from bulk diffusion to surface exchange limitation occurs at fairly thick membranes of 0.7–0.8 mm. Further flux increase would require an increase in the active surface area and/or use of catalysts. It should be mentioned that BCF has high-temperature-dependent thermal expansion coefficients in air of about 26×10^{-6} K^{-1} in the range 600–1200 °C, which may prove difficult to integrate with other materials.

Rare earth nickelates $Ln_2NiO_{4+\delta}$ have the perovskite-related Ruddlesden-Popper structure and dominant p-type electronic conductivity with oxygen ion conductivity 10^2–10^4 times lower [63, 64]. The oxygen ion transport is suggested to occur by a complex mechanism, combining vacancy diffusion in the perovskite planes and interstitial anion migration in the rock salt layers, the diffusion in the *ab* plane being much faster than in the *c*-direction [65]. The oxygen diffusion coefficients $D^{\star}$ in $Ln_2NiO_{4+\delta}$-based oxides are among the highest reported for perovskite-type oxides (e.g., around 6×10^{-8} cm^2/s for $Pr_2NiO_{4+\delta}$ at 700 °C) and the oxygen ion conductivity reaches $\sim 2 \times 10^{-2}$ S/cm at 700 °C, with an activation energy of 0.9 eV [36, 65–67]. Boehm et al. [67] recently demonstrated that the insertion of A-site vacancies in the lattice results in enhanced oxygen diffusion coefficient and total conductivity, in particular at intermediate temperature.

Li et al. [68, 69] evaluated the oxygen permeation rate at intermediate temperatures of thin supported $La_2NiO_{4+\delta}$ and $La_2Ni_{0.8}Fe_{0.2}O_{4+\delta}$ with membranes 40 nm thickness. The oxygen flux of the undoped membrane was 0.3 ml/cm^2 min at 750 °C under an air/argon gradient [68] and was significantly increased by doping with iron to about 1 ml/cm^2 min at 650 °C [69]. High-temperature permeation rates of $La_2NiO_{4+\delta}$ membranes have been found by Smith *et al.* [70]. The membranes exhibit stable oxygen flux of 4 ml/cm^2 min at 900 °C under an oxygen/argon gradient. High flux and moderate thermal expansion coefficient of $\sim 14 \times 10^{-6}$ K^{-1} render these membranes potentially viable in oxygen production. Under reducing conditions, however, these membranes are not stable. The low stability of $La_2NiO_{4+\delta}$ under reducing conditions was demonstrated by Zhu *et al.* [71] in partial oxidation of methane. At 900° C, methane conversion reached 89%, CO selectivity 96%, H_2/CO ratio 1.5, with corresponding equivalent oxygen flux of 6.8 ml/min for a tubular membrane with an inner surface area of 4.58 cm^2. However, as expected from the phase stability limits ($pO_2 \sim 10^{-9}$ Pa at 900 °C) [72], it was observed that the membrane surface exposed to reducing atmosphere decomposed into La_2O_3 and Ni.

3.2. Hydrogen membranes

3.2.1. Ceramic proton conductors

Protons in oxides have been particularly studied in the II-IV-type perovskite oxides like (Ca,Sr,Ba)(Ce,Zr,Ti)O_3 [73–119]. Protons are also found in I-V-type perovskites, for example, KTaO3 [120, 121], II_2-(III/V)-type, for example, Sr_2ScNbO_6 [122–124], II3-(II/V2)-type, for example, $Ba_3(Sr_{1.18}Nb_{1.82})O_{9-x}$ [125–129], (III-III)-type, for example, $LaErO_3$ [130–132], $LaYO_3$ [133–136], $LaScO_3$ and mixtures of $LaScO_3$ and $SrTiO_3$, for example, $La_{1-y}Sr_ySc_{1-x}Ti_xO_3$ [137–143]. The investigations indicate proton conductivities in the range 10^{-2}–10^{-5} S/cm. The mobility of protons depends on the softness of the lattice and is favored by large basic atoms, for example, Ba on the A site in the perovskite. The proton conducting oxides have an exothermic reaction with steam; the enthalpy for this reaction is often within −50 to −250 kJ/mol. The trend indicates that increasing negative enthalpy of formation gives increasing negative enthalpy for incorporation of protons. These enthalpies mean that the oxides will be dominated by protons at low temperature and start to loose protons around 500–900 °C in most cases. Since oxygen vacancies are getting more energetically preferable for many of perovskites above these temperatures, oxygen permeation may become dominating. For Y_2O_3 and rare earth sesquioxides [144–161] systematical studies show that the enthalpy for incorporation of water gets more negative along the Ln series.

In the field of high-temperature proton conductors, a few new materials have been reported the last couple of years. One of the new classes of proton conductors are the acceptor-doped rare earth *ortho*-niobates $LnNbO_4$ and *ortho*-tantalates $LnTaO_4$ described by Haugsrud and Norby [162–164]. These materials have a moderate proton conductivity approaching 10^{-3} S/cm at 900 °C and are presumably stable in CO_2 and steam containing atmospheres. Examples of other advances in this field mainly rely on detailed understanding of protons motion in perovskites structure by IR spectroscopy [165, 166], and on investigations of structural variations in doped systems [167], positions of deuterons in $BaZr_{0.5}In_{0.5}O_3$ [168] and $BaSn_{0.5}In_{0.5}O_{2.75}$ [169–170], thermoelectric power [171], thermodynamic H/D isotope effects [172], etc.

3.2.2. Hydrogen permeation in ceramic proton conductors

The number of publications on mixed proton–electron conducting membranes is fairly modest. Cheng et al. [173] have measured fluxes through $SrCe_{0.95}Tm_{0.05}O_3$ membranes with thickness in the range 150–800 μm deposited on a porous support of similar composition. The 150-μm-thick membrane demonstrated the highest flux of 0.13 ml/cm^2 min at 950 °C using 10% H_2 as feed (upstream) and 20% O_2 as sweep (downstream) gases. Moreover, Hamakawa *et al.* [174] showed that the hydrogen flux is proportional to inverse thickness down to 2 μm for a supported $SrCe_{0.95}Yb_{0.05}O_3$ membrane, indicating bulk diffusion control in the investigated thickness range.

Balachandran *et al.* [175–177] investigated mixed proton–electron conducting materials based on $BaCe_{1-x}Y_xO_3$ and $SrCeO_3$ with substitution by Yb, Sm, or Eu on the B site. Hydrogen permeabilities of $SrCe_{0.95}M_{0.05}O_3$ membranes are compared in Fig. 10.1 [175–178]. Doping results in increasing flux, which nevertheless, remains lower than the flux obtained in a short-circuited (SC) cell, indicating that electronic conductivity is rate limiting. These results also highlight the importance of humidifying the permeate gas to keep up the proton concentration at this side.

Larring et al. [178] recently reported hydrogen permeabilities for a number of oxide materials: La_6WO_{12}, $SrCe_{0.95}Yb_{0.05}O_3$, $Er_{1.99}Ca_{0.01}Ti_2O_7$, $Nd_{1.95}Ca_{0.05}$-Ti_2O_7, $Nd_{1.9}Ca_{0.1}O_3$, and $La_{1.99}Ca_{0.01}NbO_4$ (Fig. 10.2). Among these, La_6WO_{12}-based materials are reported with high ambipolar proton–electron conductivity and hydrogen permeability [179–181], similar to perovskite, such as $SrCe_{0.95}Yb_{0.05}O_3$. Again, a low electronic conduction seems to be the main drawback.

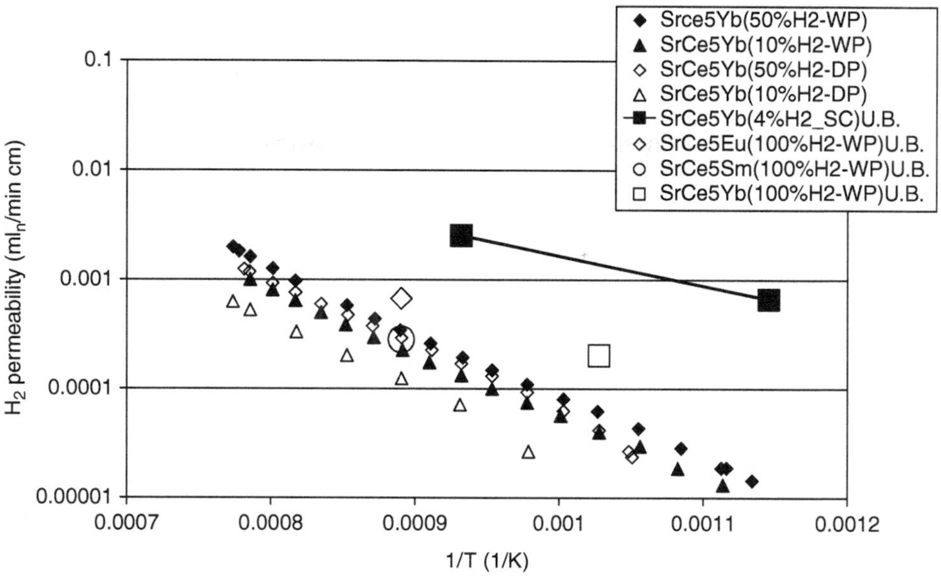

Figure 10.1 Permeability of hydrogen through $SrCe_{0.95}M_{0.05}O_3$ using humidified hydrogen on the feed side and dry argon on permeate (DP) or humidified argon on permeate (WP).

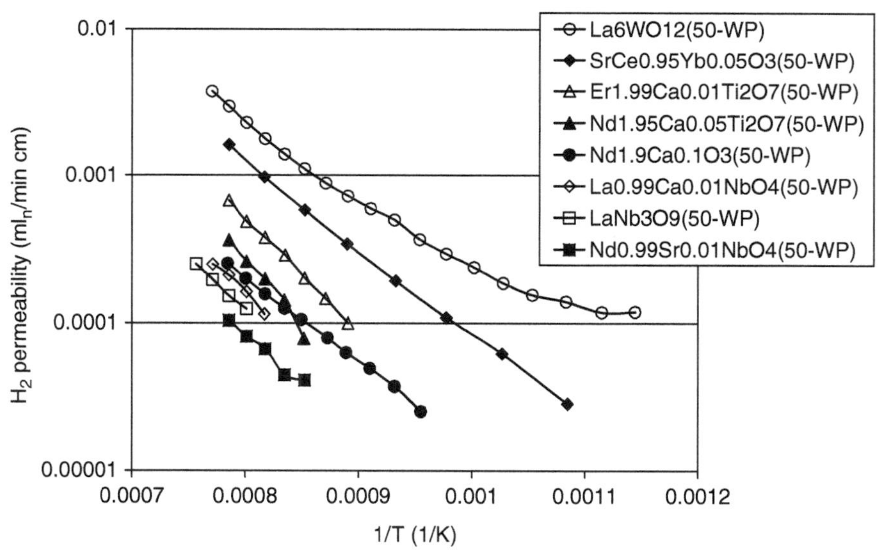

Figure 10.2 Hydrogen permeability of some selected materials using 50% humidified H_2 on feed side and humidified argon on permeate side.

In general, protons are found in perovskite materials where A site contains alkaline earths or rare earth elements, and the B site contains rather basic elements, such as rare earth elements, and Zr, Ce, Nb, Ta. It is therefore believed that addition of transition metals with possibility of exhibiting valence transition may increase the electronic conductivity. However, attempts to measure hydrogen flux in perovskite materials having transition metals like Co, Mn, and Fe on the B site, have so far, been unsuccessful.

Alternative routes to increase the electronic conductivity are explored through the development of composite ceramic metal (cermet) membranes [182–184]. However, proton transport in Ni-based cermets is reported to be restricted by dilution of the ceramic phase in the composite. Despite that high hydrogen flux was demonstrated for Pd-based cermets, it is believed that this mainly results from hydrogen diffusion in the Pd phase [185].

In permeability measurement, a sweep gas such as Ar or an inert gas containing a low amount of hydrogen is often used on the permeate side. This may give flux restriction on the permeate side, since the n-type conductivity here is low (Fig. 10.3). Several proton conducting perovskites have a certain n-type contribution in reducing atmosphere, however, this diminishes fast at higher oxygen pressure limiting the flux considerably. Going to expected process conditions will decrease this restriction, since the n-type conductivity increases as a function of the partial pressure of hydrogen.

4. Stability

The purpose of this paragraph is to discuss potential causes that can lead to membrane failures during operation. Despite their importance, sealing, connections, and housing materials will not be included in this overview on stability issues.

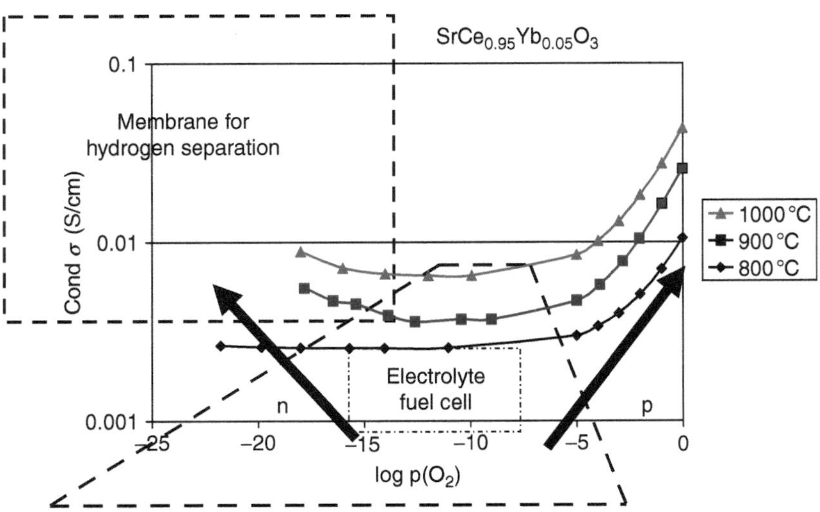

Figure 10.3 Process window for ceramic membranes and fuel cells as a function of oxygen partial pressure for $SrCe_{0.95}Yb_{0.05}O_3$.

4.1. Thermodynamic stability

Well-known perovskites contain transition metal oxides in unusually high oxidation states. For example, Co may take the valence III or IV rather than the more common valence II (though Co_3O_4 is also stable up to about 900 °C in air). The stabilization of the high oxidation state of Co in perovskite materials can be understood from the Gibbs energy, for example, for the simplified reaction

$$La_2O_3 + CoO + \frac{1}{2}O_2(g) \leftrightarrow LaCoO_3 \tag{10.78}$$

The Gibbs energy of this reaction, $\Delta_{f,\mathrm{ox}}G^{\mathrm{o}}_{\mathrm{m}}$, corresponding to the Gibbs energy of formation from the component oxides may be used to estimate the stability of the membrane material. Higher stability is anticipated by a more negative $\Delta_{f,\mathrm{ox}}G^{\mathrm{o}}_{\mathrm{m}}$ [186]. The main contribution to $\Delta_{f,\mathrm{ox}}G^{\mathrm{o}}_{\mathrm{m}}$ is the oxidation of Co(II) to Co(III). The enthalpy of oxidation of the first row transition metals from II to III is becoming less exothermic from Ti to Cu [187]. This means that materials containing Cu, Ni, and Co are not stable under conditions corresponding to partial oxidation of natural gas, while materials containing Ti, Cr, and Mn are expected to be stable. Materials based on Fe fall in-between these two categories. Finally, proton conducting oxides based on $BaZrO_3$, $SrCeO_3$, or similar materials are also stable under reducing conditions due to the stability of Zr(IV) and Ce(IV).

Many good ionic and mixed conducting oxides contain Ba or Sr as main components. These are vulnerable to decomposition by reacting with H_2O, CO_2, H_2S, or SO_2/SO_3 in the process gases, forming $Ba(OH)_2$, $BaCO_3$, BaS, or $BaSO_4$, or the Sr equivalents. Even low levels of such gas components may react and modify the flux properties. Compounds stable at operating temperatures may sometimes be unstable during cooling to room temperature and react during cycling or storage. In this context, water is as harmful as CO_2. While Sr may seem to be less vulnerable to reaction with acidic gases, this may be counteracted by a smaller lattice energy of the Sr-based oxide so that Sr- and Ba-based oxides often come out with comparable stabilities (or rather reactivities). La-based compounds are more stable toward acidic gases than Sr- and Ba-based compounds. One should still not underestimate the stability of lanthanum sulphate, $La_2(SO_4)_3$, and oxycarbonate, $La_2O_2CO_3$, and reaction with water vapor at low temperature. Taking $LaCoO_3$ as an example, the reactions with CO_2 and H_2O can be described as

$$LaCoO_3 + \frac{1}{2}CO_2(g) \leftrightarrow \frac{1}{2}La_2O_2CO_3 + \frac{1}{3}Co_3O_4 + \frac{1}{12}O_2(g) \tag{10.79}$$

and

$$LaCoO_3 + \frac{3}{2}H_2O(g) \leftrightarrow La(OH)_3 + \frac{1}{3}Co_3O_4 + \frac{1}{12}O_2(g) \tag{10.80}$$

The enthalpies of these two reactions at ambient temperature are −31.1 and −119.7 kJ/mol, respectively, based on data from Refs. [188–190]. When we know that the transport properties of membrane materials are tailored by alkali earth substitution of La site, membrane materials may become thermodynamically unstable due to formation of carbonates or hydroxides. In a similar way, decomposition of proton conducting materials based on $BaZrO_3$ and $SrCeO_3$ due to reaction with CO_2 (g) is well documented in the literature [191].

4.2. Kinetic stability

When a membrane is exposed to various gradients (chemical, mechanical, electrical, thermal) fluxes of mobile components that are generated may lead to different materials degradation phenomena:

- the multicomponent material, which was originally homogeneous, becomes chemically inhomogeneous (so-called kinetic demixing)
- formation of new phases might take place, that is, the initially single-phase material might decompose into new phases (so-called kinetic decomposition)
- the original morphology of the material might become unstable and a new morphology might be established (so-called morphological instability)

All three phenomena have a purely kinetic origin, that is, if the applied thermodynamic potential gradient is removed, the direct fluxes will disappear, but instability may remain and the original material might not be reestablished through diffusion. As the properties of the membrane are strongly dependent on their composition and their microstructure, these phenomena might therefore result in a loss of performance during operation and a reduced longevity.

Kinetic demixing or decomposition is anticipated after long operation times if the mobility of cations in a material are substantially different. The gradient in the oxygen chemical potential, μ_O, across an oxide $A_{1-x}A'_xB_{1-y}B'_yO_{3-\delta}$ induces a corresponding, but inverse gradient in the cation chemical potentials, as described by Gibbs-Duhem equation [187]. The accumulation of the fastest diffusion cation at the side with the highest partial pressure of oxygen as observed by Lein et al. [192] is therefore expected (Fig. 10.4). The cation diffusion in perovskite materials is many orders of magnitude lower than oxygen diffusion, and the cations do not contribute to the electrical conductivity. Cation diffusion is therefore only interesting with regard to the long-term stability and changes in performance of the membrane due to changes caused by difference in cation mobility. For $LaBO_3$-based materials (B = Fe, Co, Mn) it is reported that B-cation diffusion is dominating [193–195], which may explain the accumulation of Fe and Co at the high oxygen pressure side after 1 month operation of $La_{0.5}Sr_{0.5}Fe_{1-x}Co_xO_{3-\delta}$ ($x = 0$, 0.5, 1) membranes [190]. However, it was observed that A cations (Sr, La) were accumulated at the high-pressure side by changing the secondary phases in the materials [190]. A possible explanation of the change in the relative mobility of A and B cations in this case calls for fundamental studies of cation diffusion, which takes into account both bulk and grain boundary diffusion.

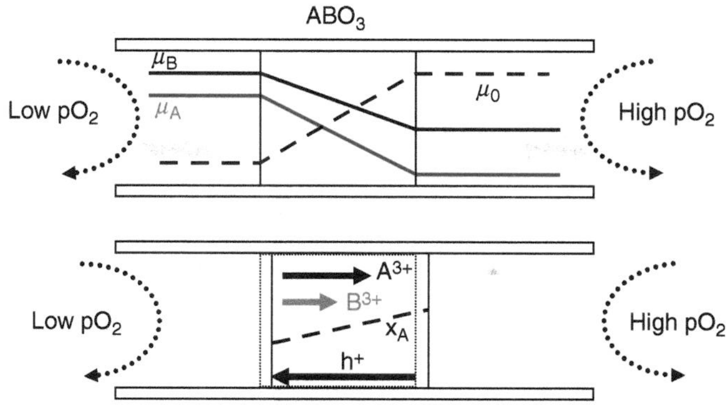

Figure 10.4 Schematic illustration of chemical potential gradients in an oxygen permeable membrane and transport of A and B cations.

4.3. Mechanical stability

The stresses to which ceramic components are subjected in operation arise from various causes as manufacturing (residual stresses, presence of flaws, and their distribution in the functional layers and to the interfaces . . .), thermal and chemical expansion, gradients in temperature, oxygen (hydrogen) activity, and mechanical loading. The magnitude and distribution of stress depend on the material properties, operating conditions, and both geometry and architecture of the membrane devices. Thus, to make a detailed analysis of the mechanical behavior of membrane devices is not straightforward. In the following paragraphs, some basic features related to mechanical stability will be discussed.

Thermal cycling during fabrication or operation of membranes will induce stresses. The thermal shock resistance can be described by

$$\sigma_{\text{th}} = \frac{E\alpha\Delta T}{1 - v} \tag{10.81}$$

where E is the Young's modulus, α is the coefficient of linear thermal expansion (TEC), v is the Poisson's ratio, and $\Delta\ T$ represents a thermal gradient. Thermal stress due to $\Delta\ T = 100$ is ~220 MPa for a typical perovskite material (using the following values: $E = 130$ GPa, $v = 0.3$, $\alpha = 1.2 \times 10^{-5}\ \text{K}^{-1}$), which is of the order of the highest strength reported, see below.

Chemical expansion of membrane materials is a well-known phenomenon [196, 197]. In a first approximation, the stress induced by a chemical potential gradient may be expressed as follows:

$$\sigma_{\text{th}} = \frac{E\frac{1}{3}\beta_c\Delta\delta}{1 - v} \tag{10.82}$$

where $\frac{1}{3}\beta_c$ is the linear chemical expansion [197] and Δ_δ represents a change in oxygen nonstoichiometry (δ) due to a gradient in partial pressure of oxygen. Typical values for $\frac{1}{3}\beta_c$ of perovskite materials are in the range 0.06–0.15. A chemically induced stress of the order of 250 MPa for $\frac{1}{3}\beta_c = 0.1$ corresponds to $\Delta\delta = 0.0135$. Membrane materials with large chemical expansion will therefore suffer due to chemically induced stresses in an oxygen activity gradient. It is therefore important to search for materials with low chemical expansion in the pO_2 range the membrane is designed to operate in.

Reports about mechanical properties of membrane materials to withstand induced stress are limited. During the last decade some studies on the bending strength, E-modulus, and fracture toughness of perovskite membrane materials have occurred. For the room temperature, fracture strength of $LaMO_3$ (M = Fe, Co, Mn, Cr) based materials measured by 4-point bending, reported values are typically in the range 100–200 MPa [198–204]. The corresponding fracture toughness of the materials are in the range 1.2–2.5 $MPa.m^{1/2}$. There are conflicting reports on the Young's modulus of these materials, but data from the most reliable techniques give values in the range 120–190 GPa. The relatively low mechanical strength reported for many of these materials can be attributed to material flaws in the range ~10–100 μm introduced during fabrication and tensional stresses resulting from chemical contraction during cooling.

Non-cubic perovskites may possess ferroelastic behavior [196, 197], and reorientation of the ferroelastic domains under mechanical stress may lead to toughening of materials. The toughening effect has been shown to decrease with increasing temperature for rhombohedral materials [196], but for orthorhombic perovskites it may also increase with temperature [197]. The fracture mode has also been shown to change from transgranular at ambient temperature to intergranular at elevated temperatures [198]. It may be noted that the release of frozen-in tensional stresses may gradually disappear with increasing temperature, and this has given an apparent increase in strength and fracture toughness of materials with large chemical expansion. Moreover, at elevated temperature creep during bending may give systemic errors in the measurements resulting in an overestimation of the strength. High creep resistance is preferred to maintain the desired dimensions of a membrane, but creep may also offer a possible mechanism for stress relaxation [205]. Only a few studies of the creep behavior of mixed ionic conductors $La_{1-x}Sr_xFe_{1-y}A_yO_{3-\delta}$ (A = Co, Cr) have been reported [199, 206–208]. High creep rates at the operation temperature of oxygen permeable membranes have been reported, particularly for materials with high Sr-content [186, 187]. Sirman [38] rates B elements for improving creep resistance according to Ti > Ga > Cr > Mn > Fe > Co in typical perovskites used as oxygen membranes. Diffusion-controlled creep has been suggested and the creep resistance decreases with both decreasing grain size and partial pressure of oxygen. The activation energy for creep reported varies in the range 260–566 kJ/mol, showing that there is no single simple mechanism controlling the mass transport.

5. Applications

Industry and academia have during the past 40 years devoted much effort to the development of SOFC, membrane reactors, and membrane gas separation technology based on high-temperature ceramic mixed conductors and solid electrolytes. This chapter will summarize some main research and developments in these areas.

5.1. Oxygen production

Large-scale oxygen production is currently done by cryogenic air separation which is highly energy consuming. Alternative technology based on oxygen conducting membranes may potentially replace cryogenic methods, since the oxygen purity and energy efficiency can be improved considerably. Air compression to typically 10–20 bars to obtain sufficient oxygen flux requires energy; however, the main component, N_2, is kept at the high-pressure retentate side after separation enabling recovery of compression work with high energy efficiency. Air Products and partners report on membrane module technology for production of 0.5 ton O_2/day [38]. The module is designed as a stack of flat membranes where pure oxygen is collected in a center tube. This design gives a low sealing area and good stress distribution which are important for operation at the high feed pressure.

Large-scale production of oxygen can benefit from process integration as, for example, in power generation and synthesis gas (syngas) production. Dyer *et al.* [209] describe the integration of mixed conducting oxygen separation membranes in a gas turbine cycle: a part of the oxygen in the compressed air is taken out through the membrane before combustion, enabling coproduction of power and oxygen. For the emerging integrated gasification combined cycle (IGCC) power plant, oxygen may be produced with further integration of membrane technology, see Fig. 10.5 [210]. Ceramic membrane technology in oxygen blown gasification of coal has shown promise in terms of cost reduction and energy efficiency improvement in the range of 30–40% [211, 212].

In the advanced zero emission power plant (AZEP) concept, the integrated membrane produces oxygen that reacts with natural gas [213, 214]. The heat from this reaction is transferred to the oxygen-depleted pressurized air, which is expanded in a gas turbine to produce electricity, see Fig. 10.6.

The AZEP module design is based on extruded square channel monoliths to produce highly compacted modules, see Fig. 10.7. Modules produced are $\sim$70 mm^2, with 24 × 24 channels of $\sim$2.2 mm width, separated by porous walls of $\sim$0.5 mm, coated with a $\sim$50-μm-dense membrane. This gives an effective membrane area of 540 m^2/m^3 monolith volume. The length of the monolith is about 170 mm. All module parts, except for the sealant, are made from the same material. Results from module testing under close to realistic process conditions give an extrapolated oxygen production rate of around 37 mol O_2/m^3 s, or 15 MW/m^3 power density. Process calculations have shown that for a 50-MW combined cycle gas turbine, the AZEP concept would result in less than 5% points loss of efficiency

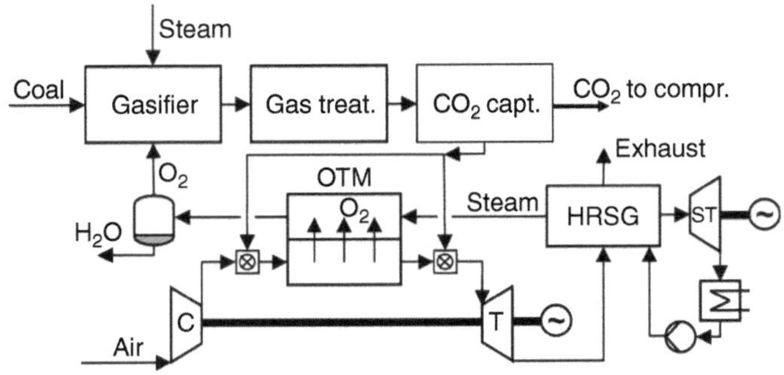

Figure 10.5 Oxygen transport membrane (OTM) integrated in IGCC process. Steam is used as purge at the permeate side. After [211].

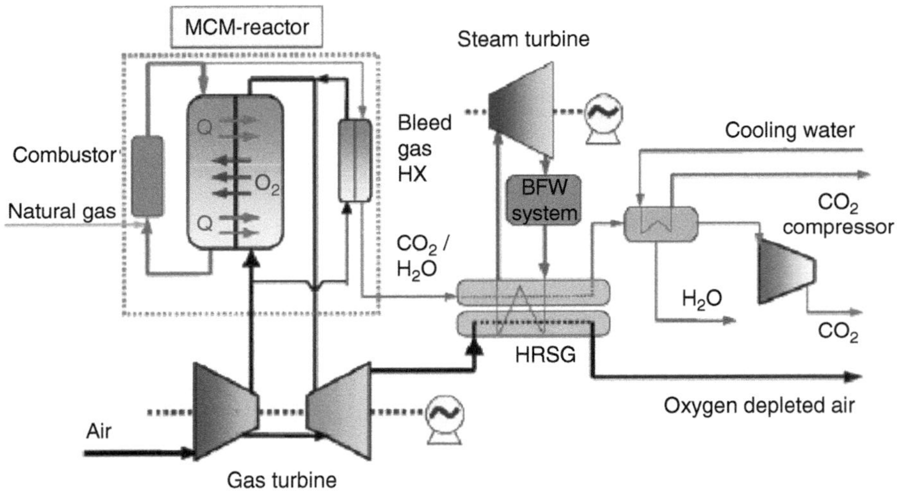

Figure 10.6 The oxygen membrane reactor combustor and heat exchanger integrated with the gas turbine in the AZEP concept. After [214, 215]. This figure was published in *Int. J. Greenhouse Gas Control* [215], Copyright Elsevier (2007).

including CO_2 capture, compared to a conventional gas turbine combined cycle (GTCC) process without CO_2 capture [216].

In processes integrating oxygen production, air is normally used as the feedstock, however, any (hot) oxygen containing gas chemically compatible with the membrane system could in principle be used. Hot exhaust gas from combustion of fossil fuels containing excess oxygen may therefore be the feedstock in integrated processes. A comment on small-scale oxygen production should also be made. For such applications, electrochemical pumping employing electrolyte membranes could be an alternative to mixed conductors. This technology enables direct production of pressurized oxygen from air. Such on-site oxygen production can be envisaged in metal cutting, medical treatment, or in electronics fabrication.

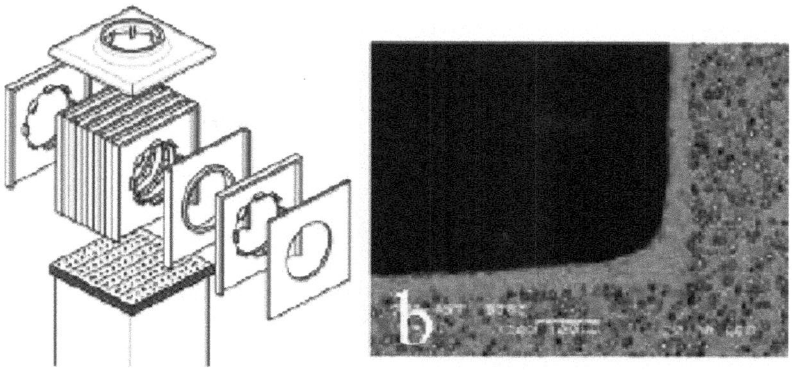

Figure 10.7 Expanded view of a module showing a 24 × 24 channel monolith, flow converter and the different parts of the manifold. Porous monolith with 50-μm dense coating fully sintered [216]. This figure was published in *Int. J. Greenhouse Gas Control* [215], Copyright Elsevier (2007).

5.2. Membrane reactors

Mixed conducting oxygen separation membranes are particularly advantageous in processes where one side is maintained at highly reducing conditions by a chemical reaction, as in partial oxidation of natural gas to syngas. In autothermal reforming, oxygen is commonly used for production of syngas being an intermediate for Fischer-Tropsch products, methanol and hydrogen. Compared to conventional syngas production the integrated oxygen membrane technology appears competitive within reasonable flux requirements (>10 Nml/cm^2 min) and membrane module production costs ~US$1500–3000, depending on process [217]. The challenges involved with respect to materials and module design, and upscaling of the technology are huge as long-term stability towards high total pressure differences and large oxygen partial pressure gradient is required [38, 41]. Steady syngas production (rate 60 Nml/cm^2 min) at 900 °C for 1 year at atmospheric pressure has been reported to yield no material instability [218]. Air Products and partners are also developing planar ceramic membranes for partial oxidation to produce synthesis gas [219]. The membrane modules are manufactured using all-ceramic joining method allowing the module to be made of the same ceramic composition throughout (Fig. 10.8).

The Hydrogen Membrane Reformer (HMR) concept combines three membrane processes at high temperature [220], see Fig. 10.9.

In the first reactor, a ceramic mixed conducting hydrogen separation membrane takes out H_2 from the methane steam reforming reaction, which subsequently reacts with air on the permeate side. This generates heat for the steam reforming and a permeate gas of $N_2 + H_2O$ with an adjusted amount of air flow. In the two next reactors, the syngas from the first reactor is converted to more H_2 and CO_2 by the water gas shift (WGS) reaction. The produced mixture of $H_2 + N_2 + H_2O$ may be fed to a gas turbine to produce electricity. Compared to a conventional GTCC power plant the HMR-integrated power plant has only 5% point lower efficiency including CO_2 capture and compression to 150 bars.

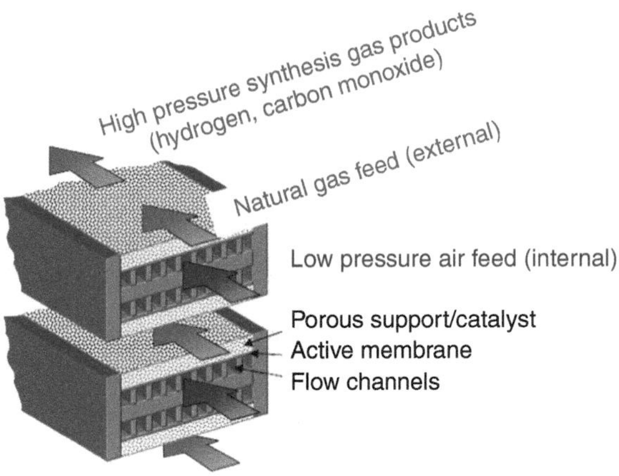

Figure 10.8 Reactor design of the ITM Syngas concept developed by Air Products and partners [219].

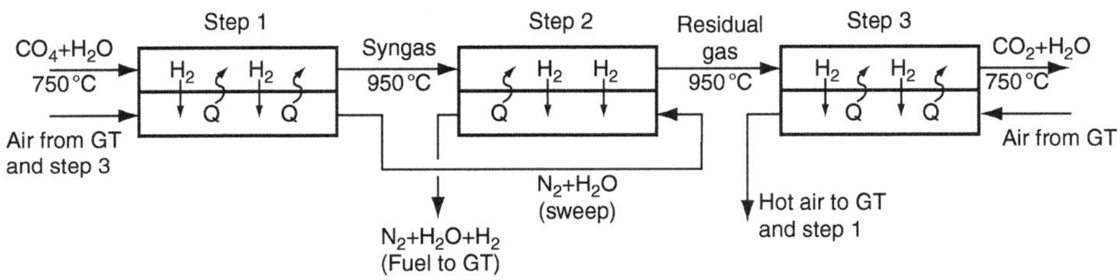

Figure 10.9 Three-stage membrane reactor system in the HMR concept. After [221].

Successful improvement by employing ceramic oxygen membranes in oxidation reaction processes depends on appropriate catalyst formulation and suitable reactor design. In oxidative dehydrogenation of light hydrocarbons, the ceramic membrane is used to feed in a controlled manner activated oxygen species to the catalyst, leading to a different catalytic behavior than under conventional oxygen-hydrocarbon cofed operations. This concept has therefore been largely investigated for the production of olefins from alkanes in oxidative dehydrogenation and methane coupling reactions [221].

Ethylene is a key intermediate product in industrial chemistry mainly produced by steam cracking of ethane, naphtha, or heavier feed stocks [222]. An alternative is oxidative dehydrogenation of ethane (ODHE) in ceramic oxygen membrane reactors. This process involving an exothermic reaction controls the contact between reactants and limits coke formation. To increase selectivity, a catalyst layer is commonly applied at the surface of the dense membrane. Reported performance on ethylene selectivity and yield of reaction for different catalytic membrane reactor (CMR) configuration is gathered in Table 10.1. It should be mentioned that different processing operations have been applied with respect to membrane design, gas mixtures and flow, and pressure gradient across the membrane wall.

Table 10.1 Currently reported performance for the ODHE using CMR with dense MIEC membrane

Membrane formulation and design	Selectivity C_2H_6 (%)	Yield of reaction (%)	Temperature (°C)	References
Disk $Ba_{0.5}Sr_{0.5}Co_{0.8}Fe_{0.2}O_{3-\delta}$ + V/MgO	82	75	767	2005 [223]
$Ba_{0.5}Sr_{0.5}Co_{0.8}Fe_{0.2}O_{3-\delta}$ disk	85	65	800	2002 [223]
Tubular fluorite-type $Bi_{1.5}Y_{0.3}SmO_3$	75	18	825	2002 [223]
Tubular fluorite-type $Bi_{1.5}Y_{0.3}SmO_3$	76	54	875	2002 [224]
$BaCo_xFe_yZr_zO_{3-\delta}$ hollow fiber ($x + y + z = 1$)	64	40	800	2006 [224]

Hence, these data are meant to illustrate increasing achievement while using the CMR concept compared to conventional steam cracking processes. Further improvement in CMR design is expected to increase performance.

Application of proton conductors for the direct conversion of methane to ethane or ethylene has also been investigated using either mixed proton–electron conducting membranes or hydrogen pumps [225]. The reaction proceeds from methane dimerization via dehydrogenation rather than partial oxidation, as encountered with oxygen ion-conducting membranes and produces advantageously hydrogen:

$$2CH_4 \leftrightarrow C_2H_6 + 2H^+ + 2e^- \text{or } C_2H_4 + 4H^+ + 4e^-$$
$$2H^+ + 2e^- \leftrightarrow H_2$$

Despite several of the studied materials (including SCYb) are very efficient methane coupling catalysts, low C_2 yields were reported so far (less than 2%), mainly due to a competitive effect of methane pyrolysis reaction ($CH_4 \rightarrow C + 2H_2$). This process, which allows chemical conversion of fuel to high value-added C_2 products with simultaneous separation and production of pure hydrogen, is, however, considered as a promising application.

5.3. Solid oxide fuel cells

Fuel cells are electrochemical devices which convert directly the chemical energy of a fuel to electricity and surplus heat. SOFCs integrating a solid oxygen ion-conducting electrolyte present great potential with respect to their expected energy efficiency rating above 80% through concurrent generation of heat and electricity, as well as their compatibility with a wide range of fuels, expected long life, and low environmental impact. Hence, SOFCs are drawing considerable attention for their adaptability to a range of applications, from large-scale stationary power systems to small power sources for residential use and transport.

In SOFC operation, oxygen reduction occurs at the cathode side, while at the anode side, oxygen from the electrolyte reacts with the fuel (e.g., CO, H_2) and forms CO_2 and/or H_2O. State-of-the-art SOFCs are made of porous $La_{1-x}Sr_xMnO_{3-\delta}$ (LSM) cathode, dense YSZ electrolyte, and porous Ni-YSZ cermet anode. It has been widely investigated and exhaustive understanding of electrodes and electrolyte behavior with respect to manufacturing and processing of the cell has been gained. Nevertheless, still fairly low power density is achieved with this cell, mainly due to the high cathode overpotential. As LSM is almost a pure electron conducting material with poor ion conductivity, electrochemical reactions are mainly restricted to the triple-phase boundaries (Fig. 10.10). Composite electrodes consisting of an electronic and an ionic conducting phase or a single-phase mixed conducting oxide offer the advantages to enlarge the active area in the electrode volume, and to transport oxide ions within the electrode material. Therefore, advanced cathode architectures are currently being developed to reduce resistance at the electrode/electrolyte interface. Most of them are designed as multilayered architectures made of a functional composite layer at the electrolyte/electrode interface (e.g., LSM/YSZ), and a current collector layer at the oxidant/electrode interface (e.g., LSM) [226]. Other configurations use CGO at the interface between YSZ and LSCF to prevent the formation of poorly conducting $SrZrO_3$ perovskite [227]. Metal-organic deposition or sol-gel made nearly dense interfacial layers of mixed conducting oxide are also developed to improve electrolyte/electrode adhesion and increase electrochemical reaction kinetics [227, 228].

Development of suitable electrolyte materials depends on the cell geometries (most mature ones are tubular and planar) and concepts (electrolyte- or electrode-supported cell). Tubular and planar electrolyte-supported cells often use YSZ

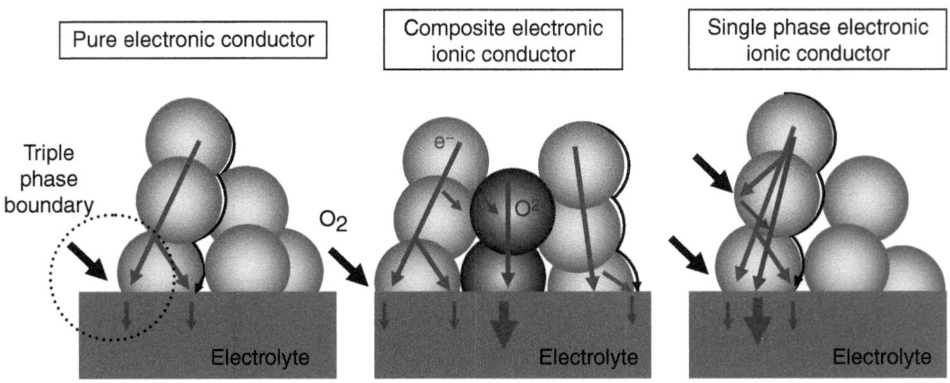

Figure 10.10 Schematic representation of the oxygen reduction for different cathodes. After [227].

electrolyte with a thickness of about 100–200 μm, and therefore require high working temperature (850–1000 °C) due to the fairly high ohmic resistance of the thick electrolyte. Planar electrode-supported cells can feature thinner electrolytes, and thereby being processed at lower temperature, typically 600–800 °C. As the production of thin electrolytes is a manufacturing challenge, another possibility is to use alternative electrolyte materials with improved performance at these intermediate temperatures. For instance, $Ce_{1-x}M_xO_2$ fluorite-type oxides (M = Gd or Sc) and LSGM perovskite-type oxides demonstrate similar or even better conductivity than YSZ at ~200 °C lower temperature [230]. These materials are however, known to suffer from partial reduction of ceria-oxides in reducing atmosphere, and diffusion of Ni into LSGM resulting in catastrophic drop of cell performance. Solutions may be found through the addition of a thin YSZ protective layer on the ceria-electrolyte and new anodes containing MgO in LSGM-based electrolyte cells [229]. Recently reported performance of these SOFCs is gathered in Table 10.2.

High-temperature proton conductors are also promising alternative electrolyte materials for SOFC. In comparison to their oxygen ion-conducting counterpart, protonic fuel cells (here termed PCFC) would offer higher flexibility to various fuels and higher intrinsic thermodynamic efficiency as water vapor is produced at the oxygen electrode and the fuel is not diluted [230–232]. Most extensively studied proton conducting electrolytes are based on perovskite-type materials, such as acceptor-doped $BaCeO_3$ and $BaZrO_3$ [233]. Combining high proton conductivity with sufficient thermodynamic stability is, however, a key issue for these materials due to the formation of alkali earth hydroxides and carbonates, as well as high grain boundary resistance at low temperatures. To date, best performance has been reported for a 0.7-μm-thick $BaCe_{0.8}Y_{0.2}O_{2.90}$ electrolyte integrated in a Pd supported cell showing a power density of 1.4 W/cm^2 at 600 °C, while typical performance hardly exceeds 100 mW/cm^2 [231, 234–237]. As lower ohmic loss is obtained with the development of thinner electrolytes, this low performance is mainly caused by cathode overpotential. One may expect that current development of advanced cermet or ceramic-ceramic composite electrode designs will eventually lead to breakthroughs in this innovative field [16, 238]. Ammonia fueled and reformed fuel gas PCFCs are currently investigated as promising new fuel cells concepts [239].

In SOFC stacks the unit cell components are connected in electrical series through the interconnect material to build up the voltage. Stacks are connected in various ways to increase the current and power output of modules to suitable values. Recently reported performance of stacks developed by several companies is reported in Table 10.3. SOFCs using gasified coal, natural gas, or biomass as hydrogen source are foreseen as promising technology for power generation in the future, as they can operate at high electrical efficiency >60% (see Table 10.4) [244]. SOFCs are currently being investigating in hybrid systems, such as integration of SOFC-gas turbines [245–247] and thermal integration with gasification systems [248, 249]. The inherent advantage of SOFC with respect to N_2-CO_2 gas separation during operation increases the competitiveness compared to other power generation technologies. It is, however, necessary to develop exhaust gas solutions to convert residual fuel to power by oxidation. Both ceramic oxygen and hydrogen membranes may be integrated with the hot exhaust gas in such tail solutions to give electrical efficiency of about 65% [245, 246].

Table 10.2 Recently reported characteristics of single cells in 2003–2007 [227, 228, 230, 240, 241, 234, 242]

Institution/Company	Design of the cell	Cell power density (W/cm^2) at 0.7 V, cathode material, and operating temperature*	Degradation (%/1000 h), time of operation	Maximum active cell size (cm^2)
SWPC	Tubular	0.20, LSM, 1000 °C	0.1, 40,000 h	850
Rolls Royce	Anode substrate	0.25, x, 900 °C	0.1, 700 h	550
Tokyo gas	Anode substrate	0.45, LSCF, 750 °C	0.3, 10,000 h	100
TOTO	Anode substrate	0.08, LSM, 900 °C	0.3, 4000 h	360
FZJ (Forschungszentrum Jülich)	Anode substrate	0.54, YSZ/LSM + LSM, 800 °C	1, 8150 h (2 cell stack)	400
FZJ	Anode substrate	0.60, LSM, 750 °C	0.3, 6000 h	575
Acumentrics	Tubular	0.10, *x*, 900 °C	0.5, 6000 h	175
Risøe	Anode substrate	0.80, LSCF, 750 °C	0.6, 4000 h	670
Cummings/SOFCs	Electrolyte substrate	*X*	1.0, 1000 h	85
Global	Anode substrate	1.90, *x*, 750 °C	1.75, 20000 h	360
GE (Honeywell Allied Signal)	Anode substrate	0.95, LSCF, 750 °C	*x*	90
HTCeramix	Anode substrate	0.55, LSCF, 750 °C	*x*	*X*
Sulzer Hexis	Electrolyte substrate	0.20, LSM, 900 °C	*x*	100
Sulzer Hexis	Electrolyte substrate	0.21, (3YSZ)CGO/LSM, 950 °C	*x*	*X*
FZJ	Anode substrate	1.2, CGO/LSCF, 800 °C	0.5, 3000 h	25
Univ. Connecticut	Tubular	0.48, (LSGM, 220 μm)-LSC, 800 °C (SDC/Ni anode)	*x*	*X*
Univ. Connecticut	Tubular	0.22, LSM, 850 °C	*x*	*X*
Ceramatec Inc.	Tubular	0.16–(LSGM–300 μm), LSC, 800 °C (Ni-MgO-CeO_2 anode)	*x*	*X*

* When not indicated in bracket, electrolyte is YSZ (8 mol % Y_2O_3). 3YSZ, YSZ with 3 mol % Y_2O_3; LSGM, $La_{0.8}Sr_{0.2}Ga_{0.8}Mg_{0.2}O_3$; CGO, $Ce_{1-x}Gd_xO_2$; LSM, $La_{1-x}Sr_xMnO_3$; LSCF, $La_{1-x}Sr_xCo_{0.2}Fe_{0.8}O_3$; LSC, $La_{1-x}Sr_xCoO_3$.

Table 10.3 Reported characteristics of stacks developed by several companies in 2003–2007 [240, 241, 234, 243]

Institution/ Company	Power achieved (kW)	Power density (W/cm^2)	Number of stacks/cells	Temperature	Operation time (h)
CFCL	1.00	0.105	/140	900	1900
Toho gas	1.03	0.152	1/68	800	
TOTO	3.20	0.077		1000	
MMC/KEPCo	1.25	0.220	1/?	770	
Acumentrics/ SOFCo	5				
Chubu EPCo	15	0.240	30/10	1000	7500
Global	2.30	0.320	1/80	750	1100
Kyocera	1.50	0.300	4/50	780	210
Delphi/PNNL (Pacific Northwest National Lab.)	1.60		2/30	750	
GE	1.4	0.420	1/30	800	x
Sulzer Hexis	1.00	0.180	1/50	950	4500
TMI	1.20				
FZJ	13.1	0.210	1/60	800	3000
FCT (Fuel Cell Technology)/ SWPC	4.36	0.120	88 tubes with 85 cm	950	4550
MHI (Mitsubishi Heavy Industries)	21.00	0.180	No complete system	900	
SWPC (Siemens Westinghouse Power Corporation)	110	0.130	1/1152	950	20,400
SWPC	191	0.120	2/1152	950	1100

5.4. High-temperature steam electrolysis

Hydrogen production through alkaline water electrolysis is a well-established technology, which however, suffers from high hydrogen production cost due to the high electricity consumption and expensive membranes (energy conversion efficiency of ~30%). High-temperature electrolysis (HTE) would enhance energy conversion efficiency up to 45–50% by using substantial external heat from an inexpensive source. The practicability and advantages of HTE compared to alkaline one have been demonstrated since 1980 (HOT ELLY experiments) [251]. The HTE uses a

Table 10.4 Capital costs of power generation systems and electrical efficiencies [250]

Technology	Current typical installed capital cost ($/kW)	Electrical efficiency (%)	Plant power (estimated kW)
Gas turbine	700–900	20–40	1000–100,000
Microturbine	700–1300	25–30	1–100
Steam turbine	800–1000	35–50	10,000–1,000,000
Wind turbine	800–1250	–	–
Natural gas engine	800–1500	30–40	50–5000
Solar photovoltaic	>5000	–	–
Fuel cell:	>2300	50–55	80–10,000
(SECA DOE program 2006)	767$/kW for Delphi (4.24 kW stack) 800$/kW for GE (6 kW stack)		

cell device very similar to an SOFC, so that the manufacturing process may take advantages of recent advances achieved in the latter field (Table 10.5).

Emerging concepts that sustain increasing interest integrate HTE and nuclear technology, and particularly, the high-temperature helium-cooled reactor (HTR). For HTRs, the expected efficiency of electricity production is 50%, and the cost of thermal energy for HTR is estimated to be 50% less. In 2004, INEEL and Ceramatec proposed an integrated HTE-HTGR (High Temperature Gas-cooled reactor) concept [252], using an electrolyzer with an oxygen ion-conducting electrolyte for operation at 800–1000 °C [253]. Differently from INEEL, Areva investigates HTE operation at 400–600 °C by use of a ceramic proton conducting electrolyte [253, 254]. Compared to the oxygen ion electrolyte-based electrolyzers, this

Table 10.5 Energy performance of HTE and alkaline electrolysis [252]

	HOT ELLY results		
	Alkaline electrolysis	HT autothermal electrolysis	HT allothermal electrolysis
Electrical energy supplied for producing 1 Nm^3 d'H_2	4.3 kWh	3.2 kWh	2.6 kWh
Heat supplied to convert water (liter) to water (g)	0	0.6 kWh	0.6 kWh
Heat supplied to reach 1000°C	–	0	0.5 kWh
Conversion yield	81%	117%	137%
Energy yield	32%	39%	45%

technology produces only a hydrogen stream whereas the former produces hydrogen and steam. The technology faces several specific challenges related to HTE materials and design concept including reduction of manufacturing cells costs, development of suitable interconnects, and increase in units lifetime.

5.5. Other applications

Considering the early stage of development of proton conductors compared to oxygen ion conductors, less focus has been given to their application in energy-related technologies. However, further development of these materials is expected to address technological issues in a broad range of applications, as for instance, those listed in table 10.6.

6. Concluding Remarks

The enormous prospects of high-temperature membrane separation technology will continue to be a strong driving force for further research and development. During the last three decades industry has been strongly involved in the development of dense ceramic membranes and SOFCs. Historically, this started with the industrial efforts to develop SOFC technology. Tremendous progress has been made, but still some fundamental challenges related to stability and efficiency remain. Furthermore, cost of production must come down before broad deployment of the technology will take place.

Table 10.6 Applications of ceramic proton conductors

	Applications of proton conductors	References
Signal	Hydrogen gas sensor	[255]
	Steam sensor	[244]
	Hydrocarbons sensor	[256]
	Hydrogen sensor for molten salts	[257]
	H^+ D^+ isotope sensors	[257]
Separation	Hydrogen extractor	[257]
	Hydrogen pump, pressurization of H_2 gas	[258]
	Regulator of hydrogen gas pressure	[254]
	Steam pump	[253]
	Isotope concentrator	[255]
Electrolysis	H_2S electrolyzer for desulfurization	[232]
	HCl electrolyzer for Cl_2 recovery	[244, 257, 259]
	Electrolyzer for NO_x removal	[244, 257, 259]
	Tritium purifier (nuclear-related applications)	[259]

To some extent, a similar development as seen for SOFC technology is taking place in the area of dense ceramic gas separation membranes. It is often argued that this technology is "easier" to establish due to the less complicated structure. Even if this may be true, the companies and research organizations involved are facing considerable challenges. The delicate nature of membrane technology, consisting of multicomponent membrane materials as thin layers on porous supports, which is integrated in modules with high-temperature sealing technology, is just from a production perspective not an easy task to achieve. In addition to this fact, harsh working conditions apply, which actually force the membrane material to react by setting up a high ionic diffusion rate. Material wise, the desire to have high stability contradicts to some extent the whish to obtain high flux. This inherent challenge of membrane technology is addressed in the design of materials composition, their structure, and shape in modules. Future work should, however, to a larger degree than previously address the chemical and mechanical stability in design of membrane materials and modules.

Besides focusing on performance in an isolated process, future work to develop membrane and SOFC technology should address process integration. Since industry is interested in the total cost of the process, integration could allow a relatively high cost for the membrane technology if it leads to a total cost reduction. We believe that process design including the membrane technology still is in its infancy and that further work should be pursued to develop novel solutions. The increasing focus on energy efficiency and CO_2 capture opens up for membrane technology in many large-scale processes. Technology based on high-temperature oxygen separation and hydrogen separation membranes, and SOFCs will play an important future role in many of these.

REFERENCES

[1] Bredesen, R., and Norby, T. (2000). On phase relations, transport properties and defect structure in mixed conducting $SrFe_{1.5-x}Co_xO_z$. *Solid State Ionics* **129,** 285–297.

[2] Thomas, D. G., and Lander, J. J. (1956). Hydrogen as a donor in zinc oxide. *J. Chem. Phys.* **25,** 1136–42.

[3] Kofstad, P. (1972). Non-Stoichiometry, Diffusion, and Electrical Conductivity in Binary Metal Oxides. Wiley, New York.

[4] Norby, T., and Haugsrud, R. (2006). Dense ceramic membranes for hydrogen separation. *In* "Nonporous Inorganic Membranes" (A. F. Sammells and M. V. Mundschau, eds.), pp. 1–48. Wiley-VCH, Weinheim ISBN 3-527-31342-7.

[5] Bonhoffer, K. F., and Farkas, F. (1932). On adsorption and reflection processes in the interaction of hydrogen and metals. *Trans. Faraday Soc.* **28,** 242–247.

[6] Couper, A., and Eley, D. D. (1950). The parahydrogen conversion on palladium-gold alloys. *Discuss. Faraday Soc.* **8,** 172–174.

[7] Ozaki, A., Nozaki, F., Maruya, K. I., and Ogasawara, S. (1967). Gas chromatographic determination of reversible adsorption of hydrogen: I. Reversible adsorption over a nickel catalyst. *J. Catal.* **7,** 234–239.

[8] Amenomiya, Y. (1971). Adsorption of hydrogen and H_2-D_2 exchange reaction on alumina. *J. Catal.* **22,** 109–122.

[9] Dowden, D. A., Mackenzie, N., and Trapnell, B. W. (1956). The catalysis of H_2 chembond {1,0} D_2 exchange by oxides. *Proc. R. Soc.* **237A,** 245–254.

[10] Ashmead, D. R., Eley, D. D., and Rudham, R. (1964). The catalytic activity of the rare earth oxides for parahydrogen conversion and hydrogen-deuterium equilibration. *J. Catal.* **3,** 280–288.
[11] Fergus, J. X. (2006). Electrolytes for solid oxide fuel cells. *J. Power Sources* **162,** 30–40.
[12] Kharton, V. V., Marques, F. M. B., and Atkinson, A. (2004). Transport properties of solid oxide electrolyte ceramics: A brief review. *Solid State Ionics* **174,** 135–149.
[13] Ralph, J. M., Rossignol, C., and Kumar, R. (2003). Oxygen reduction kinetics on mixed conducting SOFC model cathodes. *J. Electrochem. Soc.* **150,** 1518–1522.
[14] Ralph, J. M., Schoeler, A. C., and Krumpelt, M. (2001). Materials for lower temperature solid oxide fuel cells. *J. Mater. Sci.* **36,** 1161–1172.
[15] Ralph, J. M., Rossignol, C., and Kumar, R. (2003). Cathode materials for reduced-temperature SOFCs. *J. Electrochem. Soc.* **150,** 1518–1522.
[16] Nguyen, T. L., Kobayashi, K., Honda, T., Iimura, Y., Kato, K., Negishi, A., Nozaki, K., Tappero, F., Sasaki, K., Shirahama, H., Ota, K., and Dokiya, M. (2004). Preparation and evaluation of doped ceria interlayer on supported stabilized zirconia electrolyte SOFCs by wet ceramic processes. *Solid State Ionics* **174,** 163–174.
[17] Kawada, T., Suzuki, J., Sase, M., Kaimai, A., Yashiro, K., Nigara, Y., Mizusaki, J., Kawamura, K., and Yugami, H. (2002). Determination of oxygen vacancy concentration in a thin film of $La_{0.6}Sr_{0.4}coo_3$ by an electrochemical method. *J. Electrochem. Soc.* **149,** 252–259.
[18] Tuller, H. L. (1981). Mixed conduction in non-stoichiometric oxides. *In* "Nonstoichiometric Oxides, Materials Science and Technology" (O. T. Sorensen, ed.), pp. 271–275. Academic Press, New York.
[19] Schwarz, K. (2006). Materials design of solid electrolytes. *Proc. Natl. Acad. Sci. USA* **103,** 3497–3497.
[20] Andersson, D. A., Simak, S. I., Skorodumova, N. V., Abrikosov, I. A., and Johansson, B. (2007). Redox properties of CeO_2–MO_2 (M = Ti, Zr, Hf, or Th) solid solutions from first principles calculations. *Appl. Phys. Lett.* **90,** 6943–6951.
[21] Mogensen, M., Lybye, D., Bonanos, N., Hendriksen, P. V., and Poulsen, F. W. (2004). Factors controlling the oxide ion conductivity of fluorite and pervoskite structured oxides. *Solid State Ionics* **174,** 279–286.
[22] Capoen, E., Nowogrocki, G., Chater, R. J., Skinner, S. J., Kilner, J. A., Malys, M., Boivin, J. C., Mairesse, G., and Vannier, R. N. (2006). Oxygen permeation in bismuth-based materials. Part II: Characterisation of oxygen transfer in bismuth erbium oxide and bismuth calcium oxide ceramic. *Solid State Ionics* **177,** 489–492.
[23] Yu, T. H., and Tuller, H. L. (1998). Electrical conduction and disorder in the pyrochlore system $(Gd_{1-x}Ca_x)_2Sn_2O_7$. *J. Electroceram.* **2,** 49–55.
[24] Peña, M. A., and Fierro, J. L. G. (2001). Chemical structures and performance of perovskite oxides. *Chem. Rev.* **101,** 1981–2017.
[25] Kharton, V. V., Shaula, A. L., Vyshatko, N. P., and Marques, F. M. B. (2003). Electron-hole transport in LSGM electrolyte: Effects of ceramic microstructure. *Electrochim. Acta* **48,** 1817–1828.
[26] Steil, M. C., Ratajczak, F., Capoen, E., Pirovano, C., Vannier, R. N., and Mairesse, G. (2005). Thermal stability and preparation of dense membrane ceramics of BIMEVOX. *Solid State Ionics* **176,** 2305–2312.
[27] Capoen, E., Steil, M. C., Nowogrocki, G., Malys, M., Pirovano, C., Löfberg, A., Bordes-Richard, E., Boivin, J. C., Mairesse, G., and Vannier, R. N. (2006). Oxygen permeation in bismuth-based materials. Part I: Sintering and oxygen permeation fluxes. *Solid State Ionics* **177,** 483–488.
[28] Lacorre, P., Goutenoire, F., Bohnke, O., Retoux, R., and Laligant, Y. (2000). Designing fast oxide-ion conductors based on $La_2Mo_2O_9$. *Nature* **404,** 856–858.
[29] Cao, G. Z. (1994). Electrical conductivity and oxygen semipermeability of terbia and yttria stabilized zirconia. *J. Appl. Electrochem.* **24,** 1222–1227.
[30] Kharton, V. V., Kovalevsky, A. V., Viskup, A. P., Shaula, A. L., Figueiredo, F. M., Naumovich, E. N., and Marques, F. M. B. (2003). Oxygen transport in $Ce_{0.8}Gd_{0.2}O_{2\delta}$-based composite membranes. *Solid State Ionics* **160,** 247–258.

[31] Kharton, V. V., Kovalevsky, A. V., Viskup, A. P., Figueiredo, F. M., Yaremchenko, A. A., Naumovich, E. N., and Marques, F. M. B. (2000). Oxygen permeability of $Ce_{0.8}Gd_{0.2}O_{2-\delta}$-$La_{0.7}Sr_{0.3}MnO_{3-\delta}$ composite membranes. *J. Electrochem. Soc.* **147,** 2814–2821.

[32] Takamura, H., Okumura, K., Koshino, Y., Kamegawa, A., and Okada, M. (2004). Oxygen permeability of nanocrystalline $Ce_{0.8}Gd_{0.2}O_{1.9}$–$CoFe_2O_4$ mixed-conductive films. *J. Electroceram.* **13,** 613–618.

[33] Nigge, U., Wiemhöfer, H. D., Römer, E. W. J., Bouwmeester, H. J. M., and Schulte, T. R. (2002). Composites of $Ce_{0.8}Gd_{0.2}O_{1.9}$ and $Gd_{0.7}Ca_{0.3}CoO_{3\delta}$ as oxygen permeable membranes for exhaust gas sensors. *Solid State Ionics* **146,** 163–174.

[34] Shaula, A. L., Kharton, V. V., and Marques, F. M. B. (2004). Phase interaction and oxygen transport in $La_{0.8}Sr_{0.2}Fe_{0.8}Co_{0.2}O_3$–$(La_{0.9}Sr_{0.1})_{0.98}Ga_{0.8}Mg_{0.2}O_3$ composites. *J. Eur. Ceram. Soc.* **24,** 2631–2639.

[35] Vente, J. F., Haije, W. G., and Rak, Z. S. (2006). Performance of functional perovskite membranes for oxygen production. *J. Membr. Sci.* **276,** 178–184.

[36] Teraoka, Y., Zhang, H. M., Furukawa, S., and Yamazoe, N. (1985). Oxygen permeation through perovskite-type oxides. *Chem. Lett.* **11,** 1743–1746.

[37] Shao, Z., Yang, W., Cong, Y., Dong, H., Tong, J., and Xiong, G. (2000). Investigation of the permeation behavior and stability of a $Ba_{0.5}Sr_{0.5}Co_{0.8}Fe_{0.2}O_{3\delta}$ oxygen membrane. *J. Membr. Sci.* **172,** 177–188.

[38] Sirman, J. (2006). The evolution of materials and architecture for oxygen transport membranes. *In* "Nonporous Inorganic Membranes" (A. F. Sammells and M. V. Mundschau, eds.), pp. 165–184. Wiley-VCH, Weinheim ISBN 3-527-31342-7.

[39] Nagai, T., Ito, W., and Sakon, T. (2007). Relationship between cation substitution and stability of perovskite structure in $SrCoO_{3-\delta}$ based mixed conductors. *Solid State Ionics* **177,** 3433–3444.

[40] Carolan, M. (2006). Syngas membrane engineering design and scale up issues. Application of ceramic oxygen conducting membranes. *In* "Nonporous Inorganic Membranes" (A. F. Sammells and M. V. Mundschau, eds.), pp. 215–244. Wiley-VCH, Weinheim ISBN 3-527-31342-7.

[41] Balachandran, U., Dusek, J. T., Mieville, R. L., Poeppel, R. B., Kleefisch, M. S., and Pei, S. (1995). Dense ceramic membranes for partial oxidation of methane to syngas. *Appl. Catal. A Gen.* **133,** 19–29.

[42] Balachandran, U., Kleefisch, M. S., Kobylinski, T. P., Morissette, S. L., and Pei, S. (1996) US Patent 5, 580, 497.

[43] Balachandran, U., Dusek, J. T., Maiya, P. S., Ma, B., Mieville, R. L., Kleefisch, M. S., and Udovich, C. A. (1997). Ceramic membrane reactor for converting methane to syngas. *Catal. Today* **36,** 265–272.

[44] Balachandran, U., Ma, B., Maiya, P. S., Mieville, R. L., Dusek, J. T., Picciolo, J. J., Guan, J., Dorris, S. E., and Liu, M. (1998). Development of mixed-conducting oxides for gas separation. *Solid State Ionics* **108,** 363–370.

[45] Ikeguchi, M., Yoshino, Y., Kanie, K., Nomura, M., Kikuchi, E., and Matsukata, M. (2003). Effects of preparation method on oxygen permeation properties of $SrFeCo_{0.5}O_x$ membrane. *Sep. Purif. Technol.* **32,** 313–318.

[46] Fjellvåg, H., Hauback, B. C., and Bredesen, R. (1997). Crystal structure of the mixed conductor $Sr_4Fe_4Co_2O_{13}$. *J. Mater. Chem.* **7,** 2415–2419.

[47] Guggilla, S., and Manthiram, A. (1997). Crystal chemical characterization of the mixed conductor Sr (Fe, Co)$_{1.5}$O$_y$ exhibiting unusually high oxygen permeability. *J. Electrochem. Soc.* **144,** 120–122.

[48] de Arellano-López, A. R., Balachandran, U., Goretta, K. C., Ma, B., and Routbort, J. L. (2001). High-temperature deformation of $Sr(FeCo)_{1.5}O_x$ ceramics. *Acta Mater.* **49,** 3109–3116.

[49] Liu, W., Zhang, G. G., Xie, S., Chen, C. S., Meng, G. Y., and Peng, D. K. (2000). Electrical conduction and oxygen transport in $SrFeCo_{0.5}O_x$ oxide membranes. *Solid State Ionics* **135,** 727–730.

[50] Shaula, A. L., Yaremchenko, A. A., Kharton, V. V., Logvinovich, D. I., Naumovich, E. N., Kovalevsky, A. V., Frade, J. R., and Marques, F. M. B. (2003). Oxygen permeability of $LaGaO_3$-based ceramic membranes. *J. Memb. Sci.* **221,** 69–77.

[51] Yaremchenko, A. A., Kharton, V. V., Naumovich, E. N., Shestakov, D. I., Chukharev, V. F., Kovalevsky, A. V., Shaula, A. L., Patrakeev, M. V., Frade, J. R., and Marques, F. M. B. (2006).

Mixed conductivity, stability and thermomechanical properties of Ni-doped La(Ga,Mg)$O_{3\delta}$. *Solid State Ionics* **177,** 549–558.
[52] Yamaji, K., Horita, T., Sakai, N., and Yokokawa, H. (1999). Chemical stability of the $La_{0.9}Sr_{0.1}Ga_{0.8}Mg_{0.2}O_{2.85}$ electrolyte in a reducing atmosphere. *Solid State Ionics* **121,** 217–224.
[53] Figueiredo, F. M., Waerenborgh, J., Kharton, V. V., Näfe, H., and Frade, J. R. (2003). On the relationships between structure, oxygen stoichiometry and ionic conductivity of $CaTi_{1x}Fe_xO_{3-\delta}$ (x = 0.05, 0.20, 0.40, 0.60). *Solid State Ionics* **156,** 371–381.
[54] Chinarro, E., Jurado, J. R., Figueiredo, F. M., and Frade, J. R. (2003). Bulk and grain boundary conductivity of $Ca_{0.97}Ti_{1-x}Fe_xo_{3-\delta}$ materials. *Solid State Ionics* **160,** 161–168.
[55] Hamakawa, S., Sato, K., Inoue, T., Nishioka, M., Kobayashi, K., and Mizukami, F. (2006). Design of one-component ceramic membrane-reactor for natural gas conversion. *Catal. Today* **117,** 297–303.
[56] Figueiredo, F. M., Kharton, V. V., Viskup, A. P., and Frade, J. R. (2004). Surface enhanced oxygen permeation in $CaTi_{1-x}Fe_xO_{3-\delta}$ ceramic membranes. *J. Membr. Sci.* **236,** 73–80.
[57] Esaka, T., Fujii, T., Suwa, K., and Iwahara, H. (1990). Electrical conduction in $CaTi_{1-x}Fe_xO_{3-\delta}$ under low oxygen pressure and its application for hydrogen production. *Solid State Ionics* **40/41,** 544–547.
[58] Kharton, V. V., Figueiredo, F. M., Kovalevsky, A. V., Viskup, A. P., Naumovich, E. N., Jurado, J. R., and Frade, J. R. (2000). Oxygen diffusion in, and thermal expansion of $SrTiO_{3-\delta}$ and $CaTiO_{3-\delta}$-based materials. *Defect Diffusion Forum* **186/187,** 119–125.
[59] Dunyushkina, L. A., Demin, A. K., and Zhuralev, B. V. (1999). Electrical conductivity of iron-doped calcium titanate. *Solid State Ionics* **116,** 85–88.
[60] Smith, J. B., Fontaine, M. L., Larring, Y., and Bredesen, R. On the permeation of titanates oxides, in press.
[61] Zhu, X., Cong, Y., and Yang, W. (2006). Oxygen permeability and structural stability of $BaCe_{0.15}Fe_{0.85}O_{3-\delta}$ membranes. *J. Memb. Sci* **283,** 38–44.
[62] Zhu, X., Wang, H., and Yang, W. (2006). Structural stability and oxygen permeability of cerium lightly doped $BaFeO_{3-\delta}$ ceramic membranes. *Solid State Ionics* **177,** 2917–2921.
[63] Bannikov, D. O., and Cherepanov, V. A. (2006). Thermodynamic properties of complex oxides in the La–Ni–O system. *J. Solid State Chem.* **179,** 2721–2727.
[64] Greenblatt, M. (1997). Ruddlesden-Popper $Ln_{n+1}Ni_nO_{3n+1}$ nickelates: Structure and properties. *Curr. Opin. Solid State Mater. Sci.* **2,** 174–183.
[65] Bassat, J. M., Odier, P., Villesuzanne, A., Marin, C., and Pouchard, M. (2004). Anisotropic ionic transport properties in $La_2NiO_{4+\delta}$ single crystals. *Solid State Ionics* **167,** 341–347.
[66] Boehm, E., Bassat, J. M., Steil, M. C., Dordor, P., Mauvy, F., and Grenier, J. C. (2003). Oxygen transport properties of $La_2Ni_{1-x}Cu_xO_{4+\delta}$ mixed conducting oxides. *Solid State Sci.* **5,** 973–981.
[67] Boehm, E., Bassat, J. M., Dordor, P., Mauvy, F., Grenier, J. C., and Stevens, P. H. (2005). Oxygen diffusion and transport properties in non-stoichiometric $Ln_{2-x}NiO_{4+\delta}$ oxides. *Solid State Ionics* **176,** 2717–2725.
[68] Li, C., Hu, T., Zhang, H., Chen, Y., Jin, J., and Yang, N. (2003). Preparation and characterization of supported dense oxygen permeating membrane of mixed conductor $La_2NiO_{4+\delta}$. *J. Memb. Sci.* **226,** 1–7.
[69] Li, C., Yu, G., and Yang, N. (2003b). Supported dense oxygen permeating membrane of mixed conductor $La_2Ni_{0.8}Fe_{0.2}O_{4+\delta}$ prepared by sol–gel method. *Sep. Purif. Technol.* **32,** 335–339.
[70] Fontaine, M. L., Smith, J. B., Larring, Y., and Bredesen, R. Development of La_2NiO_4 supported membranes, in press.
[71] Zhu, D. C., Xu, X. Y., Feng, S. J., Liu, W., and Chen, C. S. (2003). La_2NiO_4 tubular membrane reactor for conversion of methane to syngas. *Catal. Today* **82,** 151–156.
[72] Rice, E., and Buttrey, D. J. (1993). An X-ray diffraction study of the oxygen content phase diagram of $La_2NiO_{4+\delta}$. *J. Solid State Chem.* **105,** 197–210.
[73] Ishigaki, T., Yamauchi, S., Kishio, K., Fueki, K., and Iwahara, H. (1986). Dissolution of deuterium into proton conductor $SrCe_{0.95}Yb_{0.05}O_3$. *Solid State Ionics* **21,** 239–241.
[74] Pope, J. M., and Simkovich, G. (1974). The use of $BaTiO_3$ as a solid electrolyte to determine water vapour effects upon electrical transport mechanisms. *Mater. Res. Bull.* **9,** 1111–1117.

[75] Uchida, H., Yoshikawa, H., and Iwahara, H. (1989). Dissolution of water vapor (or hydrogen) and proton conduction in $SrCeO_3$-based oxides at high temperature. *Solid State Ionics* **35,** 229–234.
[76] Yajima, T., Kazeoka, H., Yogo, T., and Iwahara, H. (1991). Proton conduction in sintered oxides based on $CaZrO_3$. *Solid State Ionics* **47,** 271–275.
[77] Yajima, T., and Iwahara, H. (1992). Studies on behavior and mobility of protons in doped perovskite-type oxides:(I) In situ measurements of hydrogen concentration in $SrCe_{0.95}Yb_{0.05}O_3$ at high temperature. *Solid State Ionics* **50,** 281–286.
[78] Yajima, T., and Iwahara, H. (1992). Studies on proton behavior in doped perovskite-type oxides: (II) Dependence of equilibrium hydrogen concentration and mobility on dopant content in Yb-doped $SrCeO_3$. *Solid State Ionics* **53–56,** 983–988.
[79] Yajima, T., Suzuki, H., Yogo, T., and Iwahara, H. (1992). Proton conduction in $SrZrO_3$ based oxides. *Solid State Ionics* **51,** 101–107.
[80] Iwahara, H. (1992). Oxide-ionic and protonic conductors based on perovskite-type oxides and their possible applications. *Solid State Ionics* **52,** 99–104.
[81] Taniguchi, N., Hatoh, K., Niikura, J., Gamo, T., and Iwahara, H. (1992). Proton conductive properties of gadolinium-doped barium cerates at high temperatures. *Solid State Ionics* **53–56,** 998–1003.
[82] Hibino, T., Mizutani, K., Yajima, T., and Iwahara, H. (1992). Evaluation of proton conductivity in $SrCeO_3$, $BaCeO_3$, $CaZrO_3$ and $SrZrO_3$ by temperature programmed desorption method. *Solid State Ionics* **57,** 209–303.
[83] Iwahara, H. (1992). High temperature protonic conductors based on perovskite-type oxides. *ISSI Lett.* **3,** 11–13.
[84] Iwahara, H., Yajima, T., Hibino, T., Ozaki, K., and Suzuki, H. (1993). Protonic conduction in calsium, strontium and barium zirconates. *Solid State Ionics* **61,** 65–69.
[85] Iwahara, H., Yajima, T., and Ushida, H. (1994). Effect of ionic radii of dopants on mixed ionic conduction ($H^+ + O^{2-}$) in $BaCeO_3$-based electrolytes. *Solid State Ionics* **70/71,** 267–271.
[86] Tsuji, T., Suzuki, T., and Iwahara, H. (1994). Electrical conduction in $BaThO_3$ doped with Nd_2O_3. *Solid State Ionics* **70/71,** 291–295.
[87] Iwahara, H. (1996). Proton conducting ceramics and their applications. *Solid State Ionics* **86–88,** 9–15.
[88] Shin, S., Huang, H. H., Ishigame, M., and Iwahara, H. (1990). Protonic conduction in the single crystals of $SrZrO_3$ and $SrCeO_3$ doped with Y_2O_3. *Solid State Ionics* **40/41,** 910–913.
[89] Yugami, H., Shibayama, Y., Matsuo, S., Ishigame, M., and Shin, S. (1996). Proton sites and defect-interactions in $SrZrO_3$ single crystals studied by infrared absorption spectroscopy. *Solid State Ionics* **85,** 319–322.
[90] Huang, H. H., Ishigame, M., and Shin, S. (1991). Protonic conduction in the single crystals of Y-doped $SrZrO_3$. *Solid State Ionics* **47,** 251–255.
[91] Scherban, T., Lee, W. K., and Nowick, A. S. (1988). Bulk protonic conductivity in Yb-doped $SrCeO_3$ and $BaCeO_3$. *Solid State Ionics* **28–30,** 585–588.
[92] Liu, J. F., and Nowick, A. S. (1992). The incorporation and migration of protons in Nd-doped $BaCeO_3$. *Solid State Ionics* **50,** 131–138.
[93] Scherban, T., Baikov, Yu. M., and Shalkova, E. K. (1993). H^+/D^+ isotope effect in Y-doped $BaCeO_3$ crystals. *Solid State Ionics* **66,** 159–164.
[94] Kreuer, K. D. (1996). Proton conductivity: Materials and Applications. *Chem. Mater.* **8,** 610–641.
[95] Kreuer, K. D., Schönherr, E., and Maier, J. (1994). Proton and oxygen diffusion in $BaCeO_3$ based compounds: A combined thermal gravimetric analysis and conductivity study. *Solid State Ionics* **70/71,** 278–284.
[96] Kreuer, K. D., Dippel, T., Baikov, Y. M., and Maier, J. (1996). Water solubility, proton and oxygen diffusion in acceptor doped $BaCeO_3$: A single crystal analysis. *Solid State Ionics* **86–88,** 613–620.
[97] Münch, W., Seifert, G., Kreuer, K. D., and Maier, J. (1996). A quantum molecular dynamics study of proton conduction phenomena in $BaCeO_3$. *Solid State Ionics* **86–88,** 647–652.

[98] Kruth, A., and Irvine, J. T. S. (2003). Water incorporation studies on doped barium cerate perovskites. *Solid State Ionics* **162/163,** 83–91.
[99] Krug, F., and Schober, T. (1997). The high-temperature proton conductor strontium zirconate: Thermogravimetry of water uptake. *J. Am. Ceram. Soc.* **80,** 794–796.
[100] Krug, F., Schober, T., Paul, R., and Springer, T. (1995). Investigation of the hydrogen uptake of doped $SrCeO_3$ by TEM, thermogravimetry and cold neutron prompt gamma activation analysis (CNPGAA). *Solid State Ionics* **77,** 185–188.
[101] Krug, F., Schober, T., and Springer, T. (1995). In situ measurements of the water uptake in Yb doped $SrCeO_3$. *Solid State Ionics* **81,** 111–118.
[102] Bonanos, N. (1992). Transport properties and conduction mechanism in high-temperature protonic conductors. *Solid State Ionics* **53–56,** 967–974.
[103] Matzke, T., and Cappadonia, M. (1996). Proton conductive perovskite solid solutions with enhanced mechanical stability. *Solid State Ionics* **86–88,** 659–663.
[104] Kosacki, I., Becht, J. G. M., van Landschoot, R., and Schoonman, J. (1993). Electrical properties of $SrCe_{0.95}Yb_{0.05}O_3$ in hydrogen containing atmospheres. *Solid State Ionics* **59,** 287–296.
[105] Kosacki, I., and Tuller, H. L. (1995). Mixed conductivity in $SrCe_{0.95}Yb_{0.05}O_3$ protonic conductors. *Solid State Ionics* **80,** 223–229.
[106] Stevenson, D. A., Jiang, N., Buchanan, R. M., and Henn, F. E. G. (1993). Characterization of Gd, Yb and Nd doped barium cerates as proton conductors. *Solid State Ionics* **62,** 279–285.
[107] Slade, R. C. T., Flint, S. D., and Singh, N. (1995). Investigation of protonic conduction in Yb- and Y-doped barium zirconates. *Solid State Ionics* **82,** 135–141.
[108] Flint, S. D., Hartmanová, M., Jones, J. S., and Slade, R. C. T. (1996). Microstructure of Ca-doped barium cerate electrolytes $BaCe_{1-x}Ca_xO_{3-\delta}$ (x = 0, 0.02, 0.05, 0.10 and 0.15). *Solid State Ionics* **86–88,** 679–683.
[109] Tanner, C. W., and Virkar, A. V. (1996). Instability of $BaCeO_3$ in H_2O-containing atmospheres. *J. Electrochem. Soc.* **143,** 1386–1389.
[110] Hempelmann, R., Eschenbaum, J., Altmayer, M., Groß, B., Grambole, D., Herrmann, F., Nagengast, D., Krauser, J., and Weidinger, A. (1997). Pressure/composition isotherms of proton conducting $SrYb_{0.05}Zr_{0.95}O_{2.975}/H_2O$ by means of nuclear resonance reaction analysis, Berichte der Bunsen-Gesellschaft—Physical Chemistry. *Chem. Phys.* **101,** 985–993.
[111] Hempelmann, R., Karmonik, C. h., Matzke, T. h., Cappadonia, M., Stimmung, U., Springer, T., and Adams, M. A. (1995). Quasielastic neutron scattering study of proton diffusion in $SrCe_{0.95}Yb_{0.05}H_{0.02}O_{2.985}$. *Solid State Ionics* **77,** 152–156.
[112] Lechner, R. E. (1993). Quasielastic neutron scattering and super-protonic conductors. *Solid State Ionics* **61,** 3–11.
[113] Jones, D. J., and Rozière, J. (1993). Complementarity of optical and incoherent inelastic neutron scattering spectroscopies in study of proton conducting materials. *Solid State Ionics* **61,** 13–22.
[114] Matzke, T., Stimmung, U., Karmonik, C., Soetratmo, M., Hempelmann, R., and Güthoff, F. (1996). Quasielastic thermal neutron scattering experiment on the proton conductor $SrCe_{0.95}Yb_{0.05}H_{0.02}O_{2.985}$. *Solid State Ionics* **86–88,** 621–628.
[115] Waser, R. (1987). Proton solubility in undoped and Fe-doped $SrTiO_3$: Temperature dependence and formation of defect associates. *Z. Naturforsch.* **42a,** 1357–1365.
[116] Furøy, K. A., Haugsrud, R., Hänsel, M., Magrasó, A., and Norby, T. (2007). Role of protons in the electrical conductivity of acceptor-doped $BaPrO_3$, $BaTbO_3$, and $BaThO_3$. *Solid State Ionics* **178,** 461–467.
[117] Waser, R. (1998). *In* "Science of Ceramics" (D. Taylor, ed.), British Ceramic Society, Stoke-on-Trent, England. **14,** 383-388.
[118] Steinsvik, S., Larring, Y., and Norby, T. (2001). Hydrogen ion conduction in iron-substituted strontium titanate, $SrTi_{1+x}Fe_xO_{3-x/2}$ (0 x 0.8). *Solid State Ionics* **143,** 103–116.
[119] Widerøe, M., Münch, W., Larring, Y., and Norby, T. (2002). Proton and apparent hydrid ion conduction in Al-substituded $SrTiO_3$. *Solid State Ionics* **154/155,** 669–677.
[120] Lee, W. K., Nowick, A. S., and Boatner, L. A. (1986). Protonic conduction in acceptor-doped $KTaO_3$ crystals. *Solid State Ionics* **18/19,** 989–993.

[121] Scherban, T., and Nowick, A. S. (1992). Protonic conduction in Fe-doped $KTaO_3$ crystals. *Solid State Ionics* **53–56,** 1004–1008.
[122] Liang, K. C., and Nowick, A. S. (1993). High-temperature protonic conduction in mixed perovskite ceramics. *Solid State Ionics* **61,** 77–81.
[123] Nowick, A. S., and Du, Y. (1995). High temperature protonic conductors with perovskite-related structures. *Solid State Ionics* **77,** 137–143.
[124] Nowick, A. S., and Vaysleyb, A. V. (1997). Isotope effect and proton hopping in high-temperature protonic conductors. *Solid State Ionics* **97,** 17–26.
[125] Du, Y., and Nowick, A. S. (1996). Galvanic cell measurements on a fast proton conducting complex perovskite electrolyte. *Solid State Ionics* **91,** 85–91.
[126] Krug, F., and Schober, T. (1996). The high-temperature proton conductor $Ba_3(Ca_{1.18}Nb_{1.82})O_{9-\delta}$: Thermogravimetry of the water uptake. *Solid State Ionics* **92,** 297–302.
[127] Mono, T., and Schober, T. (1996). Lattice parameter change in water vapor exposed $Ba_3Ca_{1.18}Nb_{1.82}O_{9-\delta}$. *Solid State Ionics* **91,** 155–159.
[128] Ruiz-Trejo, E., and De Souza, R. A. (2005). Dopant substitution and oxygen migration in the complex perovskite oxide $Ba_3CaNb_2O_9$: A computational study. *J. Solid State Chem.* **178,** 1959–1967.
[129] Ruiz-Trejo, E., Baikov, Y.u.M., and Kilner, J. A. (1998). A SIMS study of deuterium in $Ba_3(Ca_{1.18}Nb_{1.82})O_{9-\delta}$. *J. Mater. Chem.* **8,** 429–432.
[130] Larring, Y., and Norby, T. (1994). Protons in $LaErO_3$. *Solid State Ionics* **70/71,** 305–310.
[131] Glerup, M., Poulsen, F. W., and Berg, R. W. (2002). Vibrational spectroscopy on protons and deuterons in proton conducting perovskites. *Solid State Ionics* **148,** 83–92.
[132] Dubok, V. A., Donets, I. G., and Lashneva, V. V. (1983). Microstructure and electrical conductivity of ceramic with the compositions $LaErO_3$ and $LaYbO_3$, Izvestiya Akademii Nauk SSSR. *Neorg. Mater.* **19,** 845–846.
[133] Ruiz-Trejo, E., Islam, M. S., and Kilner, J. A. (1999). Atomistic simulation of defects and ion migration in $LaYO_3$. *Solid State Ionics* **123,** 121–129.
[134] Ruiz-Trejo, E., and Kilner, J. A. (1997). Oxygen diffusion and proton conduction in $La_{1-x}Sr_xYO_{3-\delta}$. *Solid State Ionics* **97,** 529–534.
[135] Ruiz-Trejo, E., and Kilner, J. A. (2000). Deuterium conductivity, diffusion and implantation in Sr-doped $LaYO_3$. *Solid State Ionics* **130,** 313–324.
[136] Yamamura, H., Yamazaki, K., Kakinuma, K., and Nomura, K. (2002). The relationship between crystal structure and electrical conductivity in the $LaY_{1-x}In_xO_3$ ($x = 0.0$–0.7) system. *Solid State Ionics* **150**(3–4), 255–261.
[137] Hatchwell, C., Bonanos, N., and Mogensen, M. (2003). Fabrication and electrical characterisation of strontium and titanium-doped lanthanum scandate. *Solid State Ionics* **162–163,** 93–98.
[138] Hatchwell, C., Bonanos, N., and Mogensen, M. (2004). The role of dopant concentration, A-site deficiency and processing on the electrical properties of strontium- and titanium-doped lanthanum scandate. *Solid State Ionics* **167,** 349–354.
[139] Lybye, D., and Bonanos, N. (1999). Proton and oxide ion conductivity of doped $LaScO_3$. *Solid State Ionics* **125,** 339–344.
[140] Nomura, K., Takeuchi, T., Kamo, S., Kageyama, H., and Miyazaki, Y. (2004). Proton conduction in doped lasco$_3$ perovskites. *Solid State Ionics* **175,** 553–555.
[141] Nomura, K., Takeuchi, T., Tanase, S., Kageyama, H., Tanimoto, K., and Miyazaki, Y. (2002). Proton conduction in $(La_{0.9}Sr_{0.1})M^{III}O_{3-\delta}$ (M^{III} = Sc, In, and Lu) perovskites. *Solid State Ionics* **154/155,** 647–652.
[142] Nomura, K., and Tanase, S. (1997). Electrical conduction behavior in $(La_{0.9}Sr_{0.1})M^{III}O_{3-\delta}$ (M^{III} = Al, Ga, Sc, In, and Lu) perovskites. *Solid State Ionics* **98,** 229–236.
[143] Gorelov, V. P., Balakireva, V. B., and Sharova, N. V. (1999). Proton-oxygen conductance in substituted perovskites $ATi_{0.95}M_{0.05}O_{3-\delta}$ (A = Ca, Sr, Ba; M = Sc, Mg) in the reducing hydrogen-containing atmospheres. *Russ. J. Electrochem.* **35,** 438–444.
[144] Norby, T., and Kofstad, P. (1984). Electrical conductivity and defect structure of Y_2O_3 as a function of water vapor pressure. *J. Am. Ceram. Soc.* **67,** 786–792.
[145] Norby, T., and Kofstad, P. (1985). Hydrogen defects in Y_2O_3 at high temperature. *In* "Reactivity of Solids" (P. Barret and L.-C. Dufour, eds.), pp. 439–443. Elsevier, Amsterdam.

[146] Norby, T., and Kofstad, P. (1996). Proton and native-ion conductivities of Y_2O_3 at high temperatures. *Solid State Ionics* **20,** 169–184.
[147] Norby, T., and Kofstad, P. (1986). Direct-current conductivity of Y_2O_3 as a function of water vapor pressure. *J. Am. Ceram. Soc.* **69,** 780–783.
[148] Norby, T., and Kofstad, P. (1986). Electrical conductivity of Y_2O_3 as a function of oxygen partial pressure in wet and dry atmospheres. *J. Am. Ceram. Soc.* **69,** 784–789.
[149] Larring, Y., Norby, T., and Kofstad, P. (1991). Protonic conductivity in Ca-doped yttria. *Solid State Ionics* **49,** 73–77.
[150] Norby, T., Dyrlie, O., and Kofstad, P. (1992). Protons in Ca-doped La_2O_3, Nd_2O_3 and $LaNdO_3$. *Solid State Ionics* **53–56,** 446–452.
[151] Norby, T., Dyrlie, O., and Kofstad, P. (1992). Protonic conduction in acceptor-doped cubic rare-earth sesquioxides. *J. Am. Ceram. Soc.* **75,** 1176–1181.
[152] Larring, Y., and Norby, T. (1995). Protons in rare earth oxides. *Solid State Ionics* **77,** 147–151.
[153] Norby, T., and Larring, Y. (1994). Protonic and native ionic conductivity in oxides. *In* "Proc. Electroceram. IV" (R. Waser, ed.), pp. 703–712. Verlag der Augustinus Buchhandlung, Aachen ISBN 3-86073-287-0.
[154] Norby, T., and Larring, Y. (1996). Concentration and transport of protons and oxygen defects in oxides. *In* "British Ceramic Proceedings 56" (B. C. H. Steele, ed.), pp. 83–94. The University Press Cambridge, UK ISBN 1-86125-003-7.
[155] Larring, Y., and Norby, T. (1997). The equilibrium between water vapour, protons, and oxygen vacancies in rare earth oxides. *Solid State Ionics* **97,** 523–528.
[156] Balakireva, V. B., Neuimin, A. D., Arzhannikov, V. A., and Palyguev, S. F. (1981). Electrical conductivity of solid solutions based on rare earth oxides with added calcium oxide. *Izvestiya Akademii Nauk SSSR, Neorg. Mater.* **17,** 1226–1231.
[157] Balakireva, V. B., Neuimin, A. D., and Palyguev, S. F. (1981). Periodicity of the electrophysical properties of rare-earth oxides and solid solutions based on them. *Dokl. Akad. Nauk SSSR* **261,** 121–124.
[158] Balakireva, V. B., and Gorelov, V. P. (1989). Proton and hole transport in CaO-doped ErBiA. *Solid State Ionics* **36,** 217–218.
[159] Gorelov, V. P., and Balakireva, V. B. (1990). Proton and hole transport in Yb_2O_3 with added CaO. *Akad. Nauk CCCP, Inorg. Mater.* **26,** 102–106.
[160] Balakireva, V. B., and Gorelov, V. P. (1991). Proton and hole conductivities of rare-earth-oxides doped with CaO in humid atmosphere. *Akad. Nauk CCCP, Inorg. Mater.* **27,** 31–35.
[161] Haugsrud, R., Larring, Y., and Norby, T. (2005). Proton conductivity of Ca-doped Tb_2O_3. *Solid State Ionics* **176,** 2957–2961.
[162] Haugsrud, R., and Norby, T. (2006). Proton conduction in rare-earth ortho-niobates and ortho-tantalates. *Nat. Mater.* **5,** 193–196.
[163] Haugsrud, R., and Norby, T. (2006). High-temperature proton conductivity in acceptor-doped $LaNbO_4$. *Solid State Ionics* **177,** 1129–1135.
[164] Haugsrud, R., Ballesteros, B., Lira-Cantú, M., and Norby, T. (2006). Ionic and electronic conductivity of 5% Ca-Doped $GdNbO_4$. *J. Electrochem. Soc.* **153,** 87–90.
[165] Omata, T., Takagi, M., and Otsuka-Yao-Matsuo, S. (2004). O–H stretching vibrations of proton conducting alkaline-earth zirconates. *Solid State Ionics* **168,** 99–109.
[166] Glerup, M., Poulsen, F. W., and Berg, R. W. (2002). Vibrational spectroscopy on protons and deuterons in proton conducting perovskites. *Solid State Ionics* **148,** 83–92.
[167] Kruth, A., Mather, G. C., Jurado, J. R., and Irvine, J. T. S. (2005). Anomalous variations of unit cell parameters with composition in proton conducting, $ACeO_3$-type perovskite solid solutions. *Solid State Ionics* **176,** 703–712.
[168] Ahmed, I., Knee, C. S., Karlsson, M., Eriksson, S. G., Henry, P. F., Matic, A., Engeberg, D., and Börjesson, L. (2006). J. Of alloys and compounds, Location of deuteron sites in the proton conducting perovskite $BaZr_{0.50}In_{0.50}O_{3y}$. *Solid State Ionics* **177,** 1395–1403.
[169] Ito, T., Nagasaki, T., Iwasaki, K., Yoshino, M., Matsui, T., Igawa, N., and Ishii, Y. (2007). Location of deuterium atoms in $BaSn_{0.5}In_{0.5}O_{2.75+\alpha}$ by neutron powder diffraction at 10 K. *Solid State Ionics* **178,** 13–17.

[170] Ahmed, I., Enriksson, S., Ahlberg, E., Knee, C. S., Berastegui, P., Johanson, L. G., Rundlöf, H., Karlsson, M., Matic, A., Börjesson, L., Engberg, D. (2006). Synthesis and structural characterization of perovskite type proton conducting $BaZr_{1-x}In_xO_{3-\delta}(0.0 \leq x \leq 0.75)$. *Solid State Ionics* **177,** 1395–1403.
[171] Tsidilkovski, V. I., Gorelov, V. P., and Balakireva, V. B. (2003). Thermoelectric power of proton conducting oxides. *Solid State Ionics* **162/163,** 55–61.
[172] Tsidilkovski, V. I. (2003). Thermodynamic isotope effect H/D/T in proton-conducting oxides. *Solid State Ionics* **162/163,** 47–53.
[173] Cheng, S., Gupta, V. K., and Lin, J. Y. S. (2005). Synthesis and hydrogen permeation properties of asymmetric proton-conducting ceramic membranes. *Solid State Ionics* **176,** 2653–2662.
[174] Hamakawa, S., Li, L., Li, A., and Iglesias, E. (2002). Synthesis and hydrogen permeation properties of membranes based on dense $SrCe_{0.95}Yb_{0.05}O_{3-\alpha}$ thin films. *Solid State Ionics* **148,** 71–81.
[175] Guan, J., Dorris, S. E., Balachandran, U., and Liu, M. (1998). Transport properties of $SrCe_{0.95}Y_{0.05}O_{3-\delta}$ and its application for hydrogen separation. *Solid State Ionics* **110,** 303–310.
[176] Song, S. J., Wachsman, E. D., Rhodes, J., Dorris, S. E., and Balachandran, U. (2003). Numerical modeling of hydrogen permeation in chemical potential gradients. *Solid State Ionics* **164,** 107–116.
[177] Song, S. J., Wachsman, E. D., Rhodes, J., Dorris, S. E., and Balachandran, U. (2004). Hydrogen permeability of $SrCe_{1-x}M_xO_{3-\delta}$ (x = 0.05, M = Eu, Sm). *Solid State Ionics* **167,** 99–105.
[178] Larring, Y., Haugsrud, R., Smith, J. B., Norby, T., and Bredesen, R. (2006). Hydrogen flux trough mixed ionic-electronic membranes. *In* "Procs. Ninth International Conference on Inorganic Membranes" (R. Bredesen and H. Ræder, eds.), pp. 492–495. June. Lillehammer, Norway (ISBN 13: 978-82-14-04026-5).
[179] Haugsrud, R., and Norby, T. (2006). Mixed ionic-electronic conductivity of Nd_6WO_{12}. *In* "Proceedings of the Ninth International Conference on Inorganic Membranes" (R. Bredesen and H. Ræder, eds.), pp. 566–569. June. Lillehammer, Norway (ISBN 13: 978-82-14-04026-5).
[180] Haugsrud, R. (2007). Defects and transport properties in Ln_6WO_{12} (Ln = La, Nd, Gd, Er). *Solid State Ionics* **178,** 555–560.
[181] Haugsrud, R., Fjeld, H., Haug, K. R., and Norby, T. (2007). Mixed ionic and electronic conductivity of undoped and acceptor-doped Er_6WO_{12}. *J. Electrochem. Soc.* **154,** 77–81.
[182] Balachandran, U., Lee, T. H., Wang, S., and Dorris, S. E. (2007). Use of mixed conducting membranes to produce hydrogen by water dissociation. *Int. J. Hydrogen Energy* **32,** 451–456.
[183] Zhang, G., Dorris, S. E., Balachandran, U., and Liu, M. (2003). Interfacial resistances of Ni–BCY mixed-conducting membranes for hydrogen separation. *Solid State Ionics* **159,** 121–134.
[184] Siriwardane, R. V., Poston, J. A., Fisher, E. P., Lee, T. H., Dorris, S. E., and Balachandran, U. (2000). Characterization of ceramic hydrogen separation membranes with varying nickel concentrations. *Appl. Surf. Sci.* **167,** 34–50.
[185] Siriwardane, R. V., Poston, J. A., Jr., Fisher, E. P., Lee, T. H., Dorris, S. E., and Balachandran, U. (2003). Characterization of ceramic-metal composite hydrogen separation membranes consisting of barium oxide, cerium oxide, yttrium oxide and palladium. *Appl. Surf. Sci.* **217,** 43–49.
[186] Stølen, S., and Grande, T. (2004). Chemical Thermodynamics of Materials—Macroscopic and Microscopic Aspects John Wiley & Sons, Ltd, Chichester, England.
[187] Bakken, E., Norby, T., and Stølen, S. (2002). Redox energetics of perovskite-related oxides. *J. Mater. Chem.* **12,** 317–323.
[188] Shirsat, A. N., Kaimal, K. N. G., Bharadwaj, S. R., and Das, D. (2005). Thermochemical studies on $RE_2O_2CO_3$ (RE = Gd, Nd) decomposition. *J. Phys. Chem. Solids* **66,** 1122–1127.
[189] Diakonov, I. I., Ragnarsdottir, K. V., and Tagirov, B. R. (1998). Standard thermodynamic properties and heat capacity equations of rare earth hydroxides: II. Ce(III)-, Pr-, Sm-, Eu(III)-, Gd-, Tb-, Dy-, Ho-, Er-, Tm-, Yb-, and Y-hydroxides. Comparison of thermochemical and solubility data. *Chem. Geol.* **151,** 327–347.
[190] Cheng, J. H., Navrotsky, A., Zhou, X. D., and Anderson, H. U. (2005). Enthalpies of Formation of $LaMO_3$ Perovskites. *J. Mater. Res.* **20,** 191–200.
[191] Kreuer, K. D. (1997). On the development of proton conducting materials for technological applications. *Solid State Ionics* **97,** 1–15.

[192] Lein, H. L., Wiik, K., and Grande, T. (2006). Kinetic demixing and decomposition of oxygen permeable membranes. *Solid State Ionics* **177,** 1587–1590.

[193] Smith, J. B., and Norby, T. (2006). Cation self-diffusion in $LaFeO_3$ measured by the solid state reaction method. *Solid State Ionics* **177,** 639–646.

[194] Palcut, M., Wiik, K., and Grande, T. (2007). Cation self-diffusion in $LaCoO_3$ and La_2CoO_4 studied by diffusion couple experiments. *J. Phys. Chem. B* **111,** 2299–2308.

[195] Palcut, M., Wiik, K., and Grande, T. (2007). Cation self-diffusion and nonstoichiometry of lanthanum manganite sudied by diffusion couple measurements. *J. Phys. Chem. C* **111,** 813–822.

[196] Atkinson, A., and Ramos, T. M. G. M. (2000). Chemically-induced stresses in ceramic oxygen ion-conducting membranes. *Solid State Ionics* **129,** 259–269.

[197] Chen, X. Y., Yu, J. S., and Adler, S. B. (2006). Thermal and chemical expansion of Sr-doped lanthanum cobalt oxide ($La_{1-x}Sr_xCoO_{3-}$). *Chem. Mater.* **17,** 4537–4546.

[198] Sammes, N. M., and Ratnaraj, R. (1995). High temperature mechanical properties of $La_{1-x}Sr_xCr_{1y}Co_yO_3$ for SOFC interconnect. *J. Mater. Sci.* **30,** 4523–4526.

[199] Montross, C. S., Yokokawa, H., Dokiya, M., and Bekessy, L. (1995). Mechanical properties of magnesia-doped lanthanum chromite versus temperature. *J. Am. Ceram. Soc.* **78,** 1869–1872.

[200] Paulik, S. W., Baskaran, S., and Armstrong, T. R. (1998). Mechanical properties of calcium- and strontium-substituted lanthanum chromite. *J. Mater. Sci.* **33,** 2397–2404.

[201] Orlovskaya, N., Kleveland, K., Grande, T., and Einarsrud, M. A. (2000). Mechanical properties of $LaCoO_3$ based ceramics. *J. Euro. Ceram. Soc.* **20,** 51–56.

[202] Kleveland, K., Orlovskaya, N., Grande, T., Moe, A. M. M., Einarsrud, M. A., Breder, K., and Gogotsi, G. (2001). Ferroelastic behavior of $LaCoO_3$-based ceramics. *J. Am. Ceram. Soc.* **84,** 2029–2033.

[203] Fossdal, A., Einarsrud, M. A., and Grande, T. (2005). Mechanical properties of $LaFeO_3$ ceramics. *J. Eur. Ceram. Soc.* **25,** 927–933.

[204] Lein, H. L., Andersen, Ø. S., Vullum, P. E., Lara-Curzio, E., Holmestad, R., Einarsrud, M. A., and Grande, T. (2006). Mechanical properties of mixed conducting $La_{0.5}Sr_{0.5}Fe_{1-x}co_xo_{3-\delta}$ (0*x*1) materials. *J. Solid State. Electrochem.* **10,** 635–642.

[205] Lea Lein, H., Lara-Curzio, E., Wiik, K., Einarsrud, M. A., and Grande, T. (2006). High-temperature creep behavior of mixed conducting $La_{0.5}Sr_{0.5}Fe_{1x}Co_xO_{3-}$ (0.5-*x*-1) materials. *J. Am. Ceram. Soc.* **89,** 2895–2898.

[206] Majkic, G., Wheeler, L., and Salama, K. (2000). Creep of polycrystalline $srco_{0.8}Fe_{0.2}O_{3-\delta}$. *Acta Mater.* **48,** 1907–1917.

[207] Majkic, G., Wheeler, L. T., and Salama, K. (2002). High-temperature deformation of $La_{0.2}Sr_{0.8}$-$Fe_{0.8}Cr_{0.2}O_{3-\delta}$-mixed ionic-electronic conductor. *Solid State Ionics* **146,** 393–404.

[208] Majkic, G., Wheeler, L. T., and Salama, K. (2003). Stress-induced diffusion and defect chemistry of $La_{0.2}Sr_{0.8}Fe_{0.8}Cr_{0.2}O_{3-\delta}$: Part 1—creep in controlled-oxygen atmosphere. *Solid State Ionics* **164,** 137–148.

[209] Dyer, P. N., Richards, R. E., Russek, S. L., and Taylor, D. M. (2000). Ion transport membrane technology for oxygen separation and syngas production. *Solid State Ionics* **134,** 21–33.

[210] Sander, F., Foeste, S., and Span, R. (2006). Model of an oxygen transport membrane for coal fired power cycles. *In* "Proceedings of the Ninth International Conference on Inorganic Membranes" (R. Bredesen and H. Ræder, eds.), pp. 224–227. June 25–29. Lillehammer, Norway.

[211] Stiegel, G. J. (1999). Mixed conducting ceramic membranes for gas separation and reaction. *Membrane Technology* **110,** 5–7.

[212] Armstrong, P. A., Bennet, D. L., Foster, E. P. T., and Stein, V. E. (2002). 'Ceramic membrane development for oxygen supply to gasification applications'. *In* "Presentation at the Gasification Technologies Conference 2002" San Francisco, USA.

[213] Winkler, D., Naas, T., Vigeland, B., Budd, M., Julsrud, S., and Sundkvist, S. G. (2006). 'Development and testing of AZEP reactor components'. *In* "Eighth International Conference on Greenhouse Gas Control Technologies (GHGT-8)" Trondheim, Norway.

[214] Sundkvist, S. G., Klang, Å., Sjödin, M., Wilhelmsen, K., Åsen, K., Tintinelli, A., McCahey, S., and Ye, H. (2004). 'Azep gas turbine combined cycle power plants. Thermal optimisation and

LCA analysis'. *In* "Seventh International Conference on Greenhouse Gas Control Technologies (GHGT-7)" (E. S. Rubin, E. W. Keith, and C. F. Gilboy, eds.), Elsevier Science Oxford, UK, Vancouver Canada. 263–271.

[215] Sundkvist, S. G., Julsrud, S., Vigeland, B., Naas, T., Budd, M., and Leistner, H. (2007). Dieter Winkler, Development and testing of AZEP reactor components. *Int. J Greenhouse Gas Control* **1,** 180–187.

[216] Fredriksson-Möller, B., Torisson, T., Assadi, M., Sundkvist, S. G., Sjödin, M., Klang, Å., and Wilhemsen, K. (2006). AZEP gas turbine combined cycle power plants-thermoeconomic analysis. *Int. J. Thermodyn.* **9,** 21–28.

[217] Bredesen, R., and Sogge, J. (1996). A technical and economic assessment of membrane reactors for hydrogen and syngas production. *In* "Paper Presentation at the United Nation Seminar on the Ecological Applications of Innovative Membrane Technology in the Chemical Industry, Working Party on the Chemical Industry," pp. 25. 13 March. Cetraro, Italy.

[218] Sammells, A. F., Schwarts, M., Mackay, R. A., Barton, T. F., and Peterson, D. R. (2000). Catalytic membrane reactors for spontaneous synthesis gas production. *Catal. Today* **56,** 325–328.

[219] Bennet, D. L., Carolan, M. F., Chen, C. M., Taylor, D. M., Wilson, M. A., and Ohrn, T. R. (2005). *In* "Proceedings of the AIChE Spring National Meeting," April 12, Atlanta, Georgia.

[220] Åsen, K., Vigeland, B., Norby, T., Larring, Y., and Mejdell, T. (2004). *In* "7th International Conference on Greenhouse Gas Control Technologies (GHGT-7)" (E. S. Rubin, E. W. Keith, and C. F. Gilboy, eds.), Elsevier Science, Oxford, UK Vancouver, Canada.

[221] Sammells, A. F., White, J. H., and Mackay, R. (2006). Membranes for promoting partial oxidation chemistry. *In* "Nonporous Inorganic Membranes" (A. F. Sammells and M. V. Mundschau, eds.), pp. 165–184. Wiley-VCH.

[222] Rebeilleau-Dassonneville, M., Rosini, S., van Veen, A. C., Farrusseng, D., and Mirodatos, C. (2005). Oxidative activation of ethane on catalytic modified dense ionic oxygen conducting membranes. *Catal. Today* **104,** 131–137.

[223] Akin, F. T., and Lin, Y. S. (2002). Selective oxidation of ethane to ethylene in a dense tubular membrane reactor. *J. Membr. Sci.* **209,** 457–467.

[224] Wang, H., Tablet, C., Schiestel, T., and Caro, J. (2006). Hollow fiber membrane reactors, for the oxidative activation of ethane. *Catal. Today* **118,** 98–103.

[225] Kokkofotos, C., Ouzounidou, M., Skodra, A., and Stoukides, M. (2007). High temperature proton conductors: applications in catalytic processes. *Solid State Ionics* **178,** 507–513.

[226] Weber, A., and Ivers-Tiffée, E. (2004). Materials and concepts for solid oxide fuel cells (SOFCs) in stationary and mobile applications. *J. Power Sources* **127,** 273–283.

[227] Tietz, F., Haanappel, V. A. C., Mai, A., Mertens, J., and Stöver, D. (2006). Performance of LSCF cathodes in cell tests. *J. Power Sources* **156,** 20–22.

[228] Fontaine, C., Laberty-Robert, C., Ansart, F., and Tailhades, P. (2006). Composition and porosity graded $La_{2-x}NiO_{4+\delta}$ (x 0) interlayers for SOFC: Control of the microstructure via a sol–gel process. *J. Power Sources* **156,** 33–38.

[229] Elangovan, S., and Hartvigsen, J. (2007). Intermediate temperature reversible fuel cells. *Int. J. Appl. Ceram. Technol.* **4,** 109–118.

[230] Minh, N. Q., and Takahashi, T. (1995). Science and Technology of Ceramic Fuel Cell. Elsevier, Amsterdam.

[231] Coors, W. G. (2003). Protonic ceramic fuel cells for high-efficiency operation with methane. *J. Power Sources* **118,** 150–156.

[232] Ranran, P., Yan, W., Lisia, Y., and Zongqiang, M. (2006). Electrochemical properties of intermediate-temperature SOFCs based on proton conducting Sm-doped $BaCeO_3$ electrolyte thin film. *Solid State Ionics* **177,** 389–393.

[233] Takahashi, T., and Iwahara, H. (1980). Solid-state ionics: Protonic conduction in perovskite type oxide solid solution. *Rev. Chim. Miner.* **17,** 243–253.

[234] Yamaguchi, S., Shishido, T., Yugami, H., Yamamoto, S., and Hara, S. (2003). Construction of fuel cells based on thin proton conducting oxide electrolyte and hydrogen-permeable metal membrane electrode. *Solid State Ionics* **162–163,** 291–296.

[235] Iwahara, H. (1995). Technological challenges in the application of proton conducting ceramics. *Solid State Ionics* **77,** 289–298.

[236] Schober, T. (2003). Applications of oxidic high-temperature proton conductors. *Solid State Ionics* **162/163,** 277–281.

[237] Ito, N., Iijima, M., Kimura, K., and Iguchi, S. (2005). New intermediate temperature fuel cell with ultra-thin proton conductor electrolyte. *J. Power Sources* **152,** 200–203.

[238] Mather, G. C., Figueiredo, F. M., Fagg, D. P., Norby, T., Jurado, J. R., and Frade, J. R. (2003). Synthesis and characterisation of Ni–$SrCe_{0.9}Yb_{0.1}O_{3-\delta}$ cermet anodes for protonic ceramic fuel cells. *Solid State Ionics* **158,** 333–342.

[239] McFarlan, A., and Pelletier, L. (2004). An intermediate-temperature ammonia fuel cell using Gd-doped barium cerate electrolyte. *J. Electrochem. Soc.* **151,** 930–932.

[240] Blum, L., Meulenberg, W. A., Nabielek, H., and Steinberger-Wilckens, R. (2005). Worldwide SOFC technology overview and benchmark. *Int. J. Appl. Ceram. Technol.* **2,** 482–492.

[241] Sammes, N. M., and Du, Y. (2007). Fabrication and characterization of tubular SOFC. *Int. J. Appl. Ceram. Technol.* **4,** 89–102.

[242] Voisard, C. (2004). Manufacturing SOFC for the Sulzer Hexis Stationary System. *Int. J. Appl. Ceram. Technol.* **1,** 31–39.

[243] Delphi SOFC (2006). Delphi SOFC satisfies performance and cost requirements. *Fuel Cells Bulletin* **11,** 7–8.

[244] Sidwell, R. W., and Grover Coors, W. (2005). Large limits of electrical efficiency in hydrocarbon fueled sofcs. *J. Power Sources* **143,** 166–172.

[245] Maurstad, O., Bredesen, R., Bolland, A., Kvamsdal, H. M., and Schell, M. (2004). *In* "7th International Conference on Greenhouse Gas Control Technologies (GHGT-7)" (E. S. Rubin, E. W. Keith, and C. F. Gilboy, eds.), Elsevier Science, Oxford, UKVancouver, Canada.

[246] Liu, Z., Li, X., and Liu, Z. (2006). Thermodynamic modeling and analysis of the ratio of heat to power based on a conceptual CHP system. *In* "HVAC Technologies for Energy Efficiency, IV-6-1" Shenzhen, China.

[247] Aguiar, P., Adjiman, C. S., and Brandon, N. P. (2004). Anode-supported intermediate temperature direct internal reforming solid oxide fuel cell. I: Model-based steady-state performance. *J. Power Sources* **138,** 120–136.

[248] Omosun, A. O., Bauen, A., Brandon, N. P., Adjiman, C. S., and Hart, D. (2004). Modelling system and costs of two biomass fuelled SOFC systems. *J. Power Sources* **131,** 96–106.

[249] Ghosha, S., and De, S. (2006). Energy analysis of a cogeneration plant using coal gasification and solid oxide fuel cell. *Energy* **31,** 345–363.

[250] Seitarides, Th., Athanasiou, C., and Zabaniotou, A. (2007). Modular biomass gasification-based solid oxide fuel cells (SOFC) for sustainable development. *Renewable and Sustainable Energy Reviews.* in press.

[251] Quandt, K. H., and Streicher, R. (1985). Concept and design of a 3.5MW pilot plant for high temperature electrolysis of water vapour. *Int. J. Hydrogen Energy* **11,** 309–315.

[252] O'Brien, J. E., Stoots, C. M., Herring, J. S., and Lessing, P. A. (2004). Performance measurements of solid-oxide electrolysis cells for hydrogen production from nuclear energy. *In* "The 12th ICONE Meeting" April 25–29. Arlington, VA, USA.

[253] Py, J. P., and Capitaine, A. (2006). Hydrogen production by high temperature electrolysis of water vapour and nuclear reactors. *In* "Proc. 16th World Hydrogen Energy Conference (WHEC-2006)" June 6. Lyon.

[254] Sala, B., Willemin, S., Lacroix, O., Colomban, P., Rahmouni, K., Julbe, A., Takenouti, H., and Py, J. P. (2006). Characterisation of proton incorporation in perovskite-type structures. *In* "Proc. 16th World Hydrogen Energy Conference (WHEC-2006)" Lyon.

[255] Taniguchi, N., Kuroha, T., Nishimura, C., and Iijima, K. (2005). Characteristics of novel $BaZr_{0.4}Ce_{0.4}In_{0.2}O_3$ proton conducting ceramics and their application to hydrogen sensors. *Solid State Ionics* **176,** 2979–2983.

[256] Schober, T., and Condon, J. B. (1996). Possible applications of high temperature proton conductors. *Int. J. Ionics* **2,** 132–136.

[257] Sundmacher, K., Rihko-Struckmann, L. K., and Galvita, V. (2005). Solid electrolyte membrane reactors: Status and trends. *Catal. Today* **104,** 185–199.

[258] Matsumoto, H., Hamajima, S., Yajima, T., and Iwahara, H. (2001). Electrochemical hydrogen pump using $SrCeO_3$-based proton conductor: Effect of water vapor at the cathode on the pumping capacity. *J. Electrochem. Soc.* **148,** 121–124.
[259] Kawamura, Y., Isobe, K., and Yamanishi, T. (2007). Mass transfer process of hydrogen via ceramic proton conductor membrane of electrochemical hydrogen pump. *Fusion Eng. Des.* **82,** 113–121.

Index

MIX
Papier | Fördert gute Waldnutzung
FSC® C083411

Kolibri 360 GmbH
Ferdinand-Jühlke-Straße 7
99095 Erfurt, Deutschland
produktsicherheit@kolibri360.de